Holt McDougal Mathematics

Course 3: Pre-Algebra
Kentucky Mathematics Worktext

HOLT McDOUGAL
a division of Houghton Mifflin Harcourt

Printed in the United States of America

ISBN: 978-0-55-400247-7
ISBN: 0-55-400247-7
7 8 9 0868 14 13 12

4500370220

About the Worktext and Table of Contents

The Grade 8 **Kentucky Mathematics Worktext** is designed to help all students master the mathematics standards found in the Grade 8 **Combined Curriculum Document,** including the **Program of Studies** and the **Core Content for Assessment Version 4.1.** Each standard is taught in one or more lessons, providing plenty of practice to help you build confidence in your math skills. A correlation showing which worktext lessons address each standard appears on p. xvii.

Each chapter opens with a hands-on Project and closes with Chapter Summary discussion questions and KCCT Prep items. All Worktext lessons contain Exploration activities, Review for Mastery worksheets, and Homework and Practice worksheets. The column at right states which pages in ***Holt McDougal Mathematics*** *Course 3* support Worktext lessons.

Chapter 1 • Principles of Algebra

Lessons denoted with (*) include Hands-On Labs. Lessons with (†) include Technology Labs.

Chapter 2 • Rational Numbers

Worktext		Page	Program of Studies	Core Content	HM Math Course 3
Project	Eat Up	50			
2A	Rational Numbers	52	MA-8-NPO-S-NS3	MA-08-1.1.2, MA-08-1.1.3	2-1
2B	Comparing and Ordering Rational Numbers	56	MA-8-NPO-S-NS2, MA-8-NPO-S-NS3	MA-08-1.1.2, MA-08-1.1.3	2-2
2C	Adding and Subtracting Rational Numbers*	59	MA-8-NPO-S-NO1, MA-8-NPO-S-NS3	MA-08-1.1.2, MA-08-1.3.1, MA-08-5.2.2	2-3
2D	Multiplying Rational Numbers	64	MA-8-NPO-S-NO1, MA-8-NPO-S-NS3	MA-08-1.1.2, MA-08-1.3.1	2-4
2E	Dividing Rational Numbers*	67	MA-8-NPO-S-NO1, MA-8-NPO-S-NS3	MA-08-1.1.2, MA-08-1.3.1	2-5
2F	Adding and Subtracting with Unlike Denominators	72	MA-8-NPO-S-NS3	MA-08-1.1.2, MA-08-1.3.1	2-6
2G	Solving Equations with Rational Numbers	76	MA-8-AT-S-VEO3, MA-8-NPO-S-NO1	MA-08-1.2.1, MA-08-5.1.2	2-7
2H	Solving Two-Step Equations*	79	MA-8-AT-S-EI1, MA-8-AT-S-EI4, MA-8-AT-S-VEO1, MA-8-AT-S-VEO3, MA-8-NPO-S-NO1	MA-08-1.2.1, MA-08-5.1.2, MA-08-5.2.1, MA-08-5.2.2, MA-08-5.3.1	Lab 2-8, 2-8
Review	Chapter Summary	84			
Practice	KCCT Prep	85			

Lessons denoted with (*) include Hands-On Labs. Lessons with (†) include Technology Labs.

Chapter 3 • Graphs, Functions, and Sequences

Lessons denoted with (*) include Hands-On Labs. Lessons with (†) include Technology Labs.

Chapter 4 • Exponents and Roots

Worktext		Page	Program of Studies	Core Content	HM Math Course 3
Project	The Tiniest Pieces	126			
4A	Exponents	128	MA-8-AT-S-VEO1, MA-8-NPO-S-NO1	MA-08-1.3.1, MA-08-5.2.1	4-1
4B	Properties of Exponents*	131	MA-8-NPO-S-NO1	---	4-3
4C	Scientific Notation†	136	MA-8-NPO-S-NS2	---	4-4, Lab 4-4
4D	Squares and Square Roots	141	MA-8-NPO-S-NO2, MA-8-NPO-S-NS1, MA-8-NPO-S-NS3, MA-8-NPO-S-E1, MA-8-NPO-S-E2	MA-08-1.1.2, MA-08-1.3.2	4-5
4E	Estimating Square Roots†	146	MA-8-NPO-S-E1, MA-8-NPO-S-E2, MA-8-NPO-S-NS1, MA-8-NPO-S-NS2, MA-8-NPO-S-NS3	MA-08-1.2.1	4-6, Lab 4-6
4F	The Real Numbers*	152	MA-8-NPO-S-E1, MA-8-NPO-S-NS1, MA-8-NPO-S-NS2, MA-8-NPO-S-NS3	MA-08-1.1.1, MA-08-1.1.2, MA-08-1.1.3	4-7
4G	The Pythagorean Theorem*†	159	MA-8-M-S-MPA6	MA-08-2.1.6	4-8, Lab 4-8
Review	Chapter Summary	168			
Practice	KCCT Prep	169			

Lessons denoted with (*) include Hands-On Labs. Lessons with (†) include Technology Labs.

Chapter 5 • Ratios, Proportions, and Similarity

Lessons denoted with (*) include Hands-On Labs. Lessons with (†) include Technology Labs.

Chapter 6 • Percents

Lessons denoted with (*) include Hands-On Labs. Lessons with (†) include Technology Labs.

Chapter 7 • Foundations of Geometry

Worktext		Page	Program of Studies	Core Content	HM Math Course 3
Project	The Ultimate Playground	270			
7A	Points, Lines, Planes, and Angle*	272	MA-8-G-S-SR1, MA-8-M-S-MPA1, MA-8-M-S-MPA2, MA-8-M-S-MPA8	MA-08-2.1.2, MA-08-2.1.3, MA-08-3.1.1	7-1, Lab 7-1
7B	Measuring and Constructing Angles*	280	MA-8-G-S-SR1, MA-8-M-S-MPA1, MA-8-M-S-MPA2, MA-8-M-S-MPA8	MA-08-2.1.2, MA-08-2.1.3, MA-08-3.1.1	7-3
7C	Pairs of Angles	287	MA-8-G-S-SR1	MA-08-3.1.1	7-4, Lab 7-4
7D	Angles in Triangles*	292	MA-8-G-S-SR2, MA-8-M-S-MPA2	MA-08-2.1.3, MA-08-3.1.2	7-5
7E	Classifying Polygons*	299	MA-8-G-S-SR2, MA-8-M-S-MPA1	MA-08-3.1.2	7-6
7F	Coordinate Geometry†	309	MA-8-G-S-CG2	MA-08-3.3.1	7-7
7G	Congruence*	318	MA-8-G-S-SR4	MA-08-3.1.2, MA-08-3.1.4	7-8
7H	Transformations	324	MA-8-G-S-TS2, MA-8-G-S-TS3, MA-8-G-S-TS4	MA-08-3.1.4, MA-08-3.2.2, MA-08-3.2.3	7-9
7I	Symmetry†	330	MA-8-G-S-TS2	MA-08-3.2.1	
Review	Chapter Summary	335			
Practice	KCCT Prep	336			

Lessons denoted with (*) include Hands-On Labs. Lessons with (†) include Technology Labs.

Chapter 8 • Perimeter, Area, and Volume

Worktext		Page	Program of Studies	Core Content	HM Math Course 3
Project	Solid Guessing	338			
8A	Perimeter and Area of Rectangles and Parallelograms*	340	MA-8-AT-S-EI2, MA-8-AT-S-VEO2, MA-8-G-S-SR2, MA-8-M-S-MPA3, MA-8-M-S-MPA9	MA-08-2.1.2, MA-08-3.1.2	-
8B	Perimeter and Area of Triangles and Trapezoids*	349	MA-8-AT-S-EI2, MA-8-AT-S-VEO2, MA-8-G-S-SR2, MA-8-M-S-MPA4, MA-8-M-S-MPA9	MA-08-2.1.1, MA-08-3.1.2	8-1, Lab 8-1
8C	Circles*	356	MA-8-AT-S-EI2, MA-8-AT-S-VEO2, MA-8-G-S-SR2, MA-8-M-S-MPA5, MA-8-M-S-MPA9, MA-8-NPO-S-NS1, MA-8-NPO-S-NS3	MA-08-1.1.1, MA-08-1.1.2, MA-08-2.1.1, MA-08-2.1.2, MA-08-3.1.2	8-2
8D	Volume of Prisms and Cylinders*	362	MA-8-AT-S-EI2, MA-8-AT-S-VEO2, MA-8-G-S-SR3, MA-8-M-S-MPA7	MA-08-2.1.4, MA-08-3.1.3	Lab 8-5, 8-5
8E	Volume of Pyramids and Cones*	374	MA-8-AT-S-EI2, MA-8-AT-S-VEO2, MA-8-M-S-MPA7	MA-08-3.1.3	Lab 8-6, 8-6
8F	Surface Area of Prisms and Cylinders*†	382	MA-8-AT-S-EI2, MA-8-AT-S-VEO2, MA-8-G-S-SR3, MA-8-M-S-MPA4, MA-8-M-S-MPA7	MA-08-2.1.5, MA-08-3.1.3	Lab 8-7, 8-7
8G	Surface Area of Pyramids and Cones*	395	MA-8-AT-S-EI2, MA-8-AT-S-VEO2, MA-8-G-S-SR3, MA-8-M-S-MPA4, MA-8-M-S-MPA7	MA-08-3.1.3	Lab 8-8, 8-8
8H	Spheres*†	403	MA-8-AT-S-EI2, MA-8-AT-S-VEO2, MA-8-G-S-SR3	MA-08-3.1.3	8-9
Review	Chapter Summary	410			
Practice	KCCT Prep	411			

Lessons denoted with (*) include Hands-On Labs. Lessons with (†) include Technology Labs.

Chapter 9 • Data and Statistics

Worktext		Page	Program of Studies	Core Content	HM Math Course 3
Project	Poor Pizza	413			
9A	Samples and Surveys	415	MA-8-DAP-S-CD5, MA-8-DAP-S-ES1	MA-08-4.3.1	9-1, Lab 9-1
9B	Organizing Data†	418	MA-8-DAP-S-CD4, MA-8-DAP-S-DR1	MA-08-4.1.1, MA-08-4.1.2, MA-08-4.1.4	9-2
9C	Measures of Central Tendency†	430	MA-8-DAP-S-CD3	MA-08-4.2.1	9-3
9D	Outliers, Gaps, and Clusters	435	MA-8-DAP-S-CD1, MA-8-DAP-S-DR1	MA-08-4.1.1, MA-08-4.2.1	9-4, Lab 9-4
9E	Variation	438	MA-8-DAP-S-CD1, MA-8-DAP-S-CD3, MA-8-DAP-S-CD4, MA-8-DAP-S-DR1	MA-08-4.1.1, MA-08-4.1.2, MA-08-4.1.4, MA-08-4.1.5, MA-08-4.2.1	Lab 9-5, 9-5, Lab 9-5
9F	Displaying Data*	443	MA-8-DAP-S-CD4, MA-8-DAP-S-DR1, MA-8-NPO-S-E2	MA-08-4.1.1, MA-08-4.1.2, MA-08-4.1.4	9-6
9G	Misleading Graphs and Statistics	455	MA-8-DAP-S-CD5, MA-8-DAP-S-DR1, MA-8-DAP-S-DR4	MA-08-4.1.1, MA-08-4.1.2, MA-08-4.1.4	9-7, Lab 9-7
9H	Scatter Plots*†	461	MA-8-DAP-S-DR1, MA-8-DAP-S-CD4, MA-8-DAP-S-P1	MA-08-4.1.1, MA-08-4.1.2, MA-08-4.1.4	9-8, Lab 9-8
9I	Choosing the Best Representation of Data*†	471	MA-8-DAP-S-DR1, MA-8-DAP-S-DR2, MA-8-DAP-S-DR3	MA-08-4.1.1, MA-08-4.1.2, MA-08-4.1.4	
Review	Chapter Summary	483			
Practice	KCCT Prep	484			

Lessons denoted with (*) include Hands-On Labs. Lessons with (†) include Technology Labs.

Chapter 10 • Probability

Worktext		Page	Program of Studies	Core Content	HM Math Course 3
Project	Breaking the Code	486			
10A	Probability	488	MA-8-DAP-S-CD2, MA-8-DAP-S-P2, MA-8-DAP-S-P7	---	10-1
10B	Experimental Probability	492	MA-8-DAP-S-CD2, MA-8-DAP-S-P1, MA-8-DAP-S-P2	MA-08-4.4.2, MA-08-4.4.3	10-2
10C	Use a Simulation*	495	MA-8-DAP-S-CD2	MA-08-4.4.2, MA-08-4.4.3	10-3
10D	Theoretical Probability*	500	MA-8-DAP-S-P7, MA-8-DAP-S-P8	MA-08-4.4.2, MA-08-4.4.3, MA-08-4.4.4	10-4
10E	Independent and Dependent Events*	506	MA-8-DAP-S-CD2, MA-8-DAP-S-P2, MA-8-DAP-S-P6	MA-08-4.4.2, MA-08-4.4.4	10-5
10F	Making Decisions and Predictions	512	MA-8-DAP-S-P1, MA-8-DAP-S-P2, MA-8-DAP-S-P5	MA-08-4.4.2	10-6
10G	Counting Principles†	515	MA-8-DAP-S-P3, MA-8-DAP-S-P4, MA-8-DAP-S-P5	MA-08-4.4.1, MA-08-4.4.2	10-8
10H	Permutations and Combinations*†	521	MA-8-DAP-S-P3	MA-08-4.4.1	10-9
Review	Chapter Summary	532			
Practice	KCCT Prep	533			

Lessons denoted with (*) include Hands-On Labs. Lessons with (†) include Technology Labs.

Chapter 11 • Multi-Step Equations and Inequalities

Worktext		Page	Program of Studies	Core Content	HM Math Course 3
Project	Riding the Rapids	535			
11A	Simplifying Algebraic Expressions*	537	MA-8-AT-S-EI1, MA-8-AT-S-VEO1, MA-8-AT-S-VEO3	MA-08-5.1.2, MA-08-5.2.1, MA-08-5.3.1	11-1
11B	Solving Multi-Step Equations*	544	MA-8-AT-S-EI1, MA-8-AT-S-EI4, MA-8-AT-S-VEO1, MA-8-AT-S-VEO3	MA-08-5.1.2, MA-08-5.2.1, MA-08-5.3.1	11-2
11C	Solving Equations with Variables on Both Sides†	552	MA-8-AT-S-EI1, MA-8-AT-S-EI4, MA-8-AT-S-VEO1, MA-8-AT-S-VEO3	MA-08-5.1.2, MA-08-5.2.1, MA-08-5.3.1	11-3, Lab 11-3
11D	Solving Inequalities by Multiplying or Dividing*	559	MA-8-AT-S-EI3	MA-08-5.1.2, MA-08-5.3.1	11-4
11E	Solving Two-Step Inequalities	567	MA-8-AT-S-EI3, MA-8-AT-S-EI4	MA-08-5.1.2, MA-08-5.3.1	11-5
Review	Chapter Summary	572			
Practice	KCCT Prep	573			

Lessons denoted with (*) include Hands-On Labs. Lessons with (†) include Technology Labs.

Chapter 12 • Graphing Lines

Worktext		Page	Program of Studies	Core Content	HM Math Course 3
Project	Life Lines	575			
12A	Graphing Linear Equations*†	577	MA-8-AT-S-PRF3, MA-8-AT-S-PRF4, MA-8-AT-S-PRF5, MA-8-G-S-CG1	MA-08-3.3.1, MA-08-5.1.5	12-1
12B	Slope of a Line†	585	MA-8-AT-S-PRF6, MA-8-G-S-CG1, MA-8-G-S-CG2	MA-08-1.4.1, MA-08-3.3.1, MA-08-5.1.5	12-2
12C	Using Slopes and Intercepts*	591	MA-8-AT-S-EI1, MA-8-AT-S-PRF5, MA-8-AT-S-VEO3, MA-8-G-S-CG2	MA-08-3.3.1	12-3, Lab 12-3
12D	Direct Variation	600	MA-8-G-S-CG2	---	12-5
12E	Lines of Best Fit*	603	MA-8-AT-S-EI1, MA-8-AT-S-PRF5, MA-8-AT-S-PRF3	---	12-7
Review	Chapter Summary	610			
Practice	KCCT Prep	611			

Lessons denoted with (*) include Hands-On Labs. Lessons with (†) include Technology Labs.

Chapter 13 • Sequences and Functions

Worktext		Page	Program of Studies	Core Content	HM Math Course 3
Project	Growing and Growing	613			
13A	Terms of Arithmetic Sequences	615	MA-8-AT-S-PRF1	MA-08-5.1.1	13-1
13B	Terms of Geometric Sequences*†	620	MA-8-AT-S-PRF1	MA-08-5.1.1	13-2, Lab 13-2
13C	Linear Functions	631	MA-8-AT-S-PRF1, MA-8-AT-S-PRF2, MA-8-AT-S-PRF3, MA-8-AT-S-PRF5	MA-08-5.1.1	13-4
13D	Quadratic Functions*†	635	MA-8-AT-S-PRF2	MA-08-5.1.2	13-6
Review	Chapter Summary	645			
Practice	KCCT Prep	646			

Lessons denoted with (*) include Hands-On Labs. Lessons with (†) include Technology Labs.

Kentucky Program of Studies and Core Content

The table below lists the standards in the Grade 8 Program of Studies and the Core Content for Assessment and shows which Worktext lessons address each standard.

Grade 8 Program of Studies		Worktext Lessons
MA-8-NPO-S-E1	Students will estimate to solve real-world and/or mathematical problems with rational numbers and common irrational numbers, checking for reasonable and appropriate computational results.	4D, 4E, 4F, 6B
MA-8-NPO-S-E2	Students will estimate with large and small quantities of objects.	4D, 4E, 6B, 9F
MA-8-NPO-S-NO1	Students will add, subtract, multiply, divide and apply order of operations (including positive whole number exponents) using rational numbers to solve real-world problems.	1A, 1B, 1D, 1E, 1F, 1H, 1I, 2C, 2D, 2E, 2G, 2H, 4A, 4B
MA-8-NPO-S-NO2	Students will determine and explain the inverse relationship between addition and subtraction, multiplication and division, or raising to an exponent and taking the root of a number.	1H, 1I, 4D
MA-8-NPO-S-NS1	Students will continue to develop number sense to include irrational numbers (e.g., square roots, cube roots, pi).	4D, 4E, 4F, 8C
MA-8-NPO-S-NS2	Students will provide examples of, describe and compare irrational and rational numbers (e.g., magnitude, order on a number line, scientific notation, very large and very small integers, numbers close to zero).	1C, 2B, 4C, 4E, 4F

Grade 8 Program of Studies		Worktext Lessons
MA-8-NPO-S-NS3	Students will describe and provide multiple representations of numbers (rational, square roots, cube roots and pi) in a variety of equivalent forms using models, diagrams and symbols based on real-world and/or mathematical situations.	2A, 2B, 2C, 2D, 2E, 2F, 4D, 4E, 4F, 8C
MA-8-NPO-S-PNO1	Students will identify and use the commutative properties, the associative properties, the identity properties and the inverse properties for addition and multiplication, the distributive property and inverse relationships to justify a given step in solving problems.	1G
MA-8-NPO-S-RP1	Students will use percentages and proportions in problem solving, including consumer applications (e.g., simple interest, percentages of increase and decrease, discounts, unit pricing, sale prices).	5A, 5B, 6B, 6C, 6D, 6E, 6F, 6G
MA-8-NPO-S-RP2	Students will derive and use formulas for various rates (e.g., distance/time, miles per hour).	5C
MA-8-M-S-MPA1	Students will read and use measurement tools (e.g., rulers, scales, protractors, angle rulers).	5C, 7A, 7B, 7E
MA-8-M-S-MPA2	Students will estimate and find angle measures and segment measures.	5E, 7A, 7B, 7D
MA-8-M-S-MPA3	Students will determine measures of the lengths of sides and the perimeter both regular and irregular shapes, including lengths to the nearest sixteenth of an inch or the nearest millimeter.	8A, 8B
MA-8-M-S-MPA4	Students will determine the area of triangles and quadrilaterals.	8B, 8F, 8G
MA-8-M-S-MPA5	Students will determine the area and circumference of circles.	8C
MA-8-M-S-MPA6	Students will develop and apply the Pythagorean theorem.	4G

Grade 8 Program of Studies		Worktext Lessons
MA-8-M-S-MPA7	Students will develop and apply formulas for volume and surface area of cubes, cylinders and right rectangular prisms; investigate relationships between and among them.	8D, 8E, 8F, 8G
MA-8-M-S-MPA8	Students will estimate measurements in standard units in real world and/or mathematical situations.	7A, 7B
MA-8-M-S-MPA9	Students will explain how measurements and measurement formulas are related or different (perimeter and area; rate, time and distance; circumference and area of a circle).	8A, 8B, 8C
MA-8-M-S-SM1	Students will provide examples of and apply money, time and U.S. Customary and metric units of measurement to solve real-world problems.	5C
MA-8-G-S-CG1	Students will identify and graph ordered pairs on a coordinate system, identifying the origin, axes and ordered pairs; apply graphing in the coordinate system to solve real-world problems.	12A, 12B, 3A
MA-8-G-S-CG2	Students will analyze the graph of a line to determine the slope, *y*-intercept and equation of the line.	12B, 12C, 12D, 7F
MA-8-G-S-SR1	Students will describe and provide examples of basic geometric elements that include points, segments, rays, lines, angles and planes; use these elements in real-world and/or mathematical situations.	7A, 7B, 7C
MA-8-G-S-SR2	Students will identify and compare properties of two-dimensional figures (circles; triangles: acute, right, obtuse, scalene, isosceles, equilateral; quadrilaterals: square, rectangle, rhombus, parallelogram, trapezoid; regular/irregular polygons); apply these properties and figures to solve real-world problems.	7D, 7E, 8A, 8B, 8C

Grade 8 Program of Studies		Worktext Lessons
MA-8-G-S-SR3	Students will compare properties of three-dimensional figures (spheres, cones, cylinders, prisms, pyramids); apply these properties and figures to solve real-world problems.	8D, 8G, 8H
MA-8-G-S-SR4	Students will provide examples of and apply congruent and similar two-dimensional figures to solve real-world problems.	5E, 5G, 5H, 7G
MA-8-G-S-SR5	Students will apply proportional reasoning to solve problems involving scale models and real objects and scale drawings and similar two-dimensional figures.	5E, 5F, 5G, 5H
MA-8-G-S-TS1	Students will investigate the congruence, proportionality and/or similarity of pre-images and images of dilations (e.g., enlargements, reductions) in a coordinate plane.	5F
MA-8-G-S-TS2	Students will describe, provide examples of and apply to real-world and/or mathematical situations rotational symmetry (45°, 90°, 180°, 270°, 360°).	7H, 7I
MA-8-G-S-TS3	Students will rotate (clockwise or counterclockwise) shapes in a coordinate plane about the origin.	7H
MA-8-G-S-TS4	Students will transform figures in a coordinate plane (translations, reflections and dilations [magnifications and contractions] with the center of dilation at the origin); determine the new coordinates of the image after the transformation.	5F, 7H
MA-8-DAP-S-CD1	Students will determine and interpret clusters, quartiles, gaps and outliers in data.	9D, 9E
MA-8-DAP-S-CD2	Students will make predictions, draw conclusions and verify results from probability experiments or simulations, making use of technology as appropriate.	10A, 10B, 10C, 10E

Grade 8 Program of Studies		Worktext Lessons
MA-8-DAP-S-CD3	Students will determine and interpret the mean, median, mode and range of a set of data.	9C, 9E
MA-8-DAP-S-CD4	Students will compare sets of data.	9B, 9E, 9F, 9H
MA-8-DAP-S-CD5	Students will explore how statistics can be interpreted in many ways.	9A, 9G
MA-8-DAP-S-DR1	Students will collect, organize, construct, analyze and make inferences from data in a variety of graphical methods (e.g., drawings, tables/charts, pictographs, bar graphs, circle graphs, line plots, Venn diagrams, line graphs, stem-and-leaf plots, scatter plots, histograms, box-and-whiskers plots).	9B, 9D, 9E, 9F, 9G, 9H, 9I
MA-8-DAP-S-DR2	Students will select an appropriate graph to represent data and justify its use.	9I
MA-8-DAP-S-DR3	Students will compare similar data from various types of graphs.	9I
MA-8-DAP-S-DR4	Students will relate different representations of data (e.g., tables, graphs, diagrams, plots) and explain how misleading representations affect interpretations and conclusions about data.	Chapter 3 Project, 9G
MA-8-DAP-S-ES1	Students will explain how data gathering, bias issues or faulty data analysis can affect the results of data collection, data representation and data interpretation.	9A
MA-8-DAP-S-P1	Students will make predictions, draw conclusions and verify results from statistical data and probability experiments, making use of technology as appropriate.	10B, 10F, 9H
MA-8-DAP-S-P2	Students will analyze situations, such as games of chance, board games or grading scales and make predictions using knowledge of probability.	10A, 10B, 10E, 10F

Grade 8 Program of Studies		Worktext Lessons
MA-8-DAP-S-P3	Students will identify and describe the number of possible arrangements of several objects, using a tree diagram or the basic counting principle; make a list, picture, chart or tree diagram to represent a sample space.	10G, 10H
MA-8-DAP-S-P4	Students will investigate counting techniques (e.g., networks).	10G
MA-8-DAP-S-P5	Students will investigate and explain the role of probability in everyday decision making.	10F, 10G
MA-8-DAP-S-P6	Students will explore concepts of randomness and independent events.	10E
MA-8-DAP-S-P7	Students will determine theoretical (mathematical) probabilities (e.g., express probability as a ratio, decimal, percent, area model as appropriate for a given situation).	10A, 10D
MA-8-DAP-S-P8	Students will compare theoretical and experimental results and explain reasons why there might be differences.	10D
MA-8-AT-S-EI1	Students will use multiple representations to model and solve one- and two-variable linear equations.	11A, 11B, 11C, 12C, 12E, 1H, 1I, 2H
MA-8-AT-S-EI2	Students will solve problems using formulas.	13D, 6G, 8A, 8B, 8C, 8D, 8E, 8F, 8G, 8H
MA-8-AT-S-EI3	Students will investigate linear inequalities using a variety of methods and representations.	11D, 11E, 1J
MA-8-AT-S-EI4	Students will model and solve real-world problems with one- or two-step equations or inequalities (e.g., $4x + 2 = 22$, $x - 4 < -60$).	11B, 11C, 11D, 11E, 1H, 1I, 1J, 2H
MA-8-AT-S-PRF1	Students will recognize, create and extend patterns (generalize the pattern by giving the rule for the *n*th term and explain the generalization).	13A, 13B, 13C, 3E

Grade 8 Program of Studies		Worktext Lessons
MA-8-AT-S-PRF2	Students will represent, interpret and describe linear and simple quadratic functional relationships (input/output) through tables, graphs and symbolic rules.	12A, 12B, 12C, 13C, 13D, 3C, 3D
MA-8-AT-S-PRF3	Students will organize input-output coordinate pairs into tables, plot points in all four quadrants of a coordinate (Cartesian) system/grid and interpret resulting patterns or trends using technology as appropriate.	12A, 12E, 13C, 3A, 3C, 3D
MA-8-AT-S-PRF4	Students will interpret and explain relationships between tables, graphs, verbal rules and equations, using technology as appropriate.	12A, 3B, 3C, 3D
MA-8-AT-S-PRF5	Students will graph linear functions in a four quadrant (Cartesian) system/grid and interpret the results, using technology as appropriate.	12A, 12C, 12E, 13C
MA-8-AT-S-PRF6	Students will explain how change in the input affects change in the output (e.g., in $d = rt$, increasing the time (t) increases the distance (d)).	12B, 13C, 3C, 3D
MA-8-AT-S-VEO1	Students will apply order of operations to evaluate and simplify algebraic expressions.	11A, 11B, 11C, 1A, 1B, 2H, 4A
MA-8-AT-S-VEO2	Students will given a formula, substitute appropriate elements from a real-world or mathematical situation.	6G, 8A, 8B, 8C, 8D, 8E, 8F, 8G, 8H
MA-8-AT-S-VEO3	Students will describe, define and provide examples of variables and expressions with a missing value based on real-world and/or mathematical situations.	11A, 11B, 11C, 11D, 11E, 12C, 1H, 1I, 2G, 2H

Grade 8 Core Content for Assessment Version 4.1		Worktext Lessons
MA-08-1.1.1	Students will provide examples of and identify rational numbers and irrational numbers (square roots and π only).	4F, 8C
MA-08-1.1.2	Students will describe and provide examples of representations of numbers (rational, square roots, and π) and operations in a variety of equivalent forms using models, diagrams and symbols (e.g., number lines, 10 by 10 grids, rectangular arrays, number sentences) based on real-world and mathematical problems.	2A, 2B, 2C, 2D, 2E, 2F, 4D, 4F, 8C
MA-08-1.1.3	Students will convert, compare and order multiple numerical representations (e.g., fractions, decimals, percentages) of rational numbers and irrational numbers (square roots and π only).	2A, 2B, 4F, 6A
MA-08-1.2.1	Students will estimate to solve real-world and mathematical problems with rational numbers, checking for reasonable and appropriate computational results.	2G, 2H, 4E, 6B
MA-08-1.3.1	Students will add, subtract, multiply and divide rational numbers to solve real-world problems and apply order of operations (including positive whole number exponents) to simplify numerical expressions.	1A, 1B, 1D, 1E, 1F, 2C, 2D, 2E, 2F, 2H, 4A
MA-08-1.3.2	Students will explain how operations (additions and subtraction; multiplication and division; squaring and taking the square root of a number) are inversely related.	1H, 1I, 4D
MA-08-1.4.1	Students will apply ratios and proportional reasoning to solve real-world problems (e.g., percents, constant rate of change, unit pricing, percent of increase or decrease).	12B, 5A, 5B, 5C, 5D, 6B, 6C, 6D, 6E, 6F, 6G
MA-08-1.5.2	Students will identify the use of properties (commutative properties of addition and multiplication, the associative properties of addition and multiplication, the identity properties for addition and multiplication, inverse properties and the distributive property of multiplication over addition and subtraction) to justify a given step in solving problems.	1G

Grade 8 Core Content for Assessment Version 4.1		Worktext Lessons
MA-08-2.1.1	Students will measure lengths (to the nearest sixteenth of an inch or the nearest millimeter) and will determine and use in real-world or mathematical problems: • area and perimeter of triangles and quadrilaterals; • area and circumference of circles; • area and perimeter of compound figures composed of triangles, quadrilaterals and circles; • area from circumference or perimeter and • circumference or perimeter from area.	5F, 8A, 8B, 8C
MA-08-2.1.2	Students will estimate measurements in standard units in real-world and mathematical problems.	4E, 5E, 6B, 7A, 7B, 8C
MA-08-2.1.3	Students will evaluate the measures of angles by estimation, measurement with a protractor or angle ruler and determine angle measures in mathematical and/or real-world situations (e.g., supplementary, external, vertical).	7A, 7B, 7D
MA-08-2.1.4	Students will apply formulas to determine the volume of right rectangular prisms in real-world problems.	8D
MA-08-2.1.5	Students will use formulas to find surface area of right rectangular prisms in real-world and mathematical problems.	8F
MA-08-2.1.6	Students will apply the Pythagorean theorem to determine the length of a hypotenuse.	4G
MA-08-2.2.1	Students will convert units within the same measurement system and use these units to solve real-world problems.	5C
MA-08-3.1.1	Students will describe and provide examples of basic geometric elements that include points, segments, rays, lines, angles, and planes and will use these elements in real-world and mathematical problems.	7A, 7B, 7C

Grade 8 Core Content for Assessment Version 4.1		Worktext Lessons
MA-08-3.1.2	Students will identify and compare properties of two-dimensional figures (circles, triangles acute, right, obtuse, scalene, isosceles, equilateral], quadrilaterals [square, rectangle, rhombus, parallelogram, trapezoid], regular/irregular polygons), and will apply these properties and figures to solve real-world and mathematical problems.	7D, 7E, 7G, 8A, 8B, 8C
MA-08-3.1.3	Students will compare properties of three-dimensional figures (spheres, cones, cylinders, prisms, pyramids), and will apply these properties and figures to solve real-world and mathematical problems.	8D, 8E, 8F, 8G, 8H
MA-08-3.1.4	Students will: • provide examples of congruent and similar figures; • apply congruent and similar figures to solve real-world and mathematical problems and • apply proportional reasoning to solve problems involving scale drawings and proportional figures.	5E, 5F, 5G, 5H, 7G, 7H
MA-08-3.2.1	Students will describe, provide examples of, and apply to real-world and mathematical problems rotational symmetry (90°, 180°, 360°).	7I
MA-08-3.2.2	Students will transform (translations, reflections, and dilations with the center of dilation at the origin) figures in a coordinate plane and determine the new coordinates of the image after the transformation.	5F, 7H
MA-08-3.2.3	Students will identify rotations (clockwise or counterclockwise) of figures about the origin in a coordinate plane.	7H
MA-08-3.3.1	Students will identify and graph ordered pairs on a coordinate system, correctly identifying the origin, axes and ordered pairs; and will apply graphing in the coordinate system to solve real-world and mathematical problems.	12A, 12B, 12C, 12D, 3A, 3C, 3D, 7F

Grade 8 Core Content for Assessment Version 4.1		Worktext Lessons
MA-08-4.1.1	Students will analyze and make inferences from data displays (drawings, tables/charts, pictographs, bar graphs, circle graphs, line plots, Venn diagrams, line graphs, stem-and-leaf plots, scatter plots, histograms, box-and-whiskers plots).	9B, 9D, 9E, 9F, 9G, 9H, 9I
MA-08-4.1.2	Students will explain how different representations of data (e.g., tables, graphs, diagrams, plots) are related.	9B, 9E, 9F, 9G, 9H, 9I
MA-08-4.1.4	Students will: • construct data displays (Venn diagrams, tables, line graphs, stem-and-leaf plots, circle graphs, scatter plots); • explain why the type of display is appropriate for the data and • explain how misleading representations affect interpretations and conclusions about data (e.g., changing the scale on a graph).	9B, 9E, 9F, 9G, 9H, 9I
MA-08-4.1.5	Students will construct box-and-whiskers plots.	9E
MA-08-4.2.1	Students will: • determine the mean, median, mode, and range of a set of data; • identify clusters, gaps, and outliers and • apply these concepts to compare sets of data.	9C, 9D, 9E
MA-08-4.3.1	Students will explain how data gathering, bias issues, and faulty data analysis can affect the results of data collection.	9A
MA-08-4.4.1	Students will apply counting techniques to determine the size of a sample space for a real-world or mathematical situation.	10G, 10H
MA-08-4.4.2	Students will: • determine theoretical probabilities of simple events; • determine probabilities based on the results of an experiment and • make inferences from probability data.	10B, 10C, 10D, 10E, 10F, 10G
MA-08-4.4.3	Students will tabulate experimental results from simulations and explain how theoretical and experimental probabilities are related.	10B, 10C, 10D

Grade 8 Core Content for Assessment Version 4.1		Worktext Lessons
MA-08-4.4.4	Students will determine theoretical probabilities and represent them using area models.	10D, 10E
MA-08-5.1.1	Students will use variables to describe numerical patterns based on arithmetic sequences in real-world and mathematical problems (e.g., $f(N) = 2N + 3$).	13A, 13B, 13C, 3E
MA-08-5.1.2	Students will represent, analyze and generalize simple first and second degree relationships using tables, graphs, words and algebraic notations, and will apply the relationships to solve real-world and mathematical problems.	11A, 11B, 11C, 11D, 11E, 12A, 13D, 1H, 1I, 1J, 2G, 2H
MA-08-5.1.5	Students will explain how the change in one variable affects the change in another variable (e.g., if rate remains constant, an increase in time results in an increase in distance).	12A, 12B, 12D, 3D
MA-08-5.2.1	Students will evaluate and simplify algebraic expressions applying the order of operations.	11A, 11B, 11C, 1A, 1B, 2H, 4A
MA-08-5.2.2	Students will describe, define and provide examples of variables and expressions with a missing value based on real-world and mathematical problems.	11A, 11C, 11D, 11E, 1A, 1B, 1H, 1I, 2C, 2H
MA-08-5.3.1	Students will model and solve single variable, first-degree real-world and mathematical problems (e.g., $5x + 2 = x + 22$, $x - 4 < -60$).	11A, 11B, 11C, 11D, 11E, 1H, 1I, 1J, 2H

CHAPTER 1

Project
Algebra: Fire Drills

Career: Firefighter

A firefighter approaching a fire should be aware of ventilation, space, what is burning, and what could be ignited. Oxygen, fuel, heat, and chemical reactions are at the core of a fire, but the amounts and materials differ.

Before you start, work through the Fire Drills worksheet on the next page.

In this chapter project . . .

You research the amount of pollution in the air.

What to do:

- Research reports published by local governments on the amount of pollutants in the air of your town, county, or state.
- Learn about different toxic gases that can be released in a fire and how much of each is considered dangerous.
- Research three recent building fires or wildfires and find the total amount of pollutants released (or area burned) in each and the total amount of water or other fire-fighting agents used to control each fire.

What to present:

- A circle graph showing the types and amounts of pollutants found in the air in your community or a city or town of your choosing.
- A bar graph comparing the amounts of pollutants released by the three fires and the amount of water or other fire-fighting agents used to control each fire.

Where's the math?

- Use rates to express the amounts of pollutants in parts per million or parts per billion.
- Use a circle graph to compare types and amounts of pollutants.
- Use a bar graph to compare fires.

go.hrw.com
Chapter Project Online
KEYWORD: MT7 Ch1

Name ______________________ Date ____________ Class ____________

CHAPTER 1

Project Recording Sheet

Principles of Algebra

Firefighting

What toxic gases are created in a fire? The table lists some of the gases that are formed when different materials are burned.

Fire-Caused Toxic Gases

Gas	**Dangerous Level** (ppm)	**Burning Source**
Carbon dioxide (CO_2)	40,000	Complete burning of wood, paper
Carbon monoxide (CO)	1200	Incomplete burning
Hydrogen chloride (HCl)	50	Plastics
Hydrogen cyanide (HCN)	50	Wool, nylon, polyurethane foam, rubber, paper
Nitrogen dioxide (NO_2)	20	Grains, some plastics
Phosgene ($COCl_2$)	2	Refrigerants

The unit ppm stands for "parts per million." This means the number of molecules of gas there are in every million molecules of air. For example, 1200 parts per million of carbon monoxide means that for every one million molecules of air, 1200 molecules are carbon dioxide and the remaining 998,000 are similar to the air in the atmosphere.

Use the table for Exercises 1–4.

1. Which gas released by a fire is the most dangerous? Why do you think so?

____

2. How many times more dangerous is the most dangerous gas compared to the least dangerous?

____

3. Which toxic gas is most likely to be found in a common house fire? Why do you think so?

____

____

4. Write an algebraic equation that compares the toxicity of all of the gases. (For example: 50 HCL = 1200 CO.)

____

____

EXPLORATION

Program of Studies and Core Content

MA-8-AT-S-VE01, MA-8-NPO-S-NO1, MA-08-5.2.1, MA-08-5.2.2, MA-08-1.3.1

1A Variables and Expressions

Catherine's dance team is planning a spring trip to the coast. Catherine is saving money in a bank account to pay for the trip. Her parents started her account with $100. She sells Christmas plants and adds $2.50 to her account for each plant she sells.

How much will be in her account if she sells 50 plants?

Initial amount

Price × number of plants

100 + 2.5*n*

Evaluate the expression 100 + 2.5*n* by substituting 50 for *n*.

$$100 + 2.5(50)$$
$$100 + 125$$
$$225$$

There will be $225.00 in Catherine's account if she sells 50 plants.

Evaluate the expression 100 + 2.5*n* for each value of *n*.

1. $n = 10$ $\quad$ $100 + 2.5(10)$

2. $n = 25$ $\quad$ $100 + 2.5(\quad)$

3. $n = 75$ $\quad$ $100 + 2.5(\quad)$

Think and Discuss

4. Explain what *n* represents.

5. Describe how you evaluated the expression for different values of *n*.

Name ____________________ Date __________ Class __________

LESSON **1A**

Review for Mastery

Variables and Expressions

An **algebraic expression** uses at least one letter, or **variable,** which represents a value that can change.

A number that multiplies a variable is its **coefficient.** A **constant** is a specific number, whose value does not change.

To **evaluate** an algebraic expression, **substitute** a given number for a variable, and find the value of the resulting numerical expression.

Algebraic Expression

$$4x + 7$$

Coefficient (4) · Variable (x) · Constant (7)

Follow the order of operations:

1. Parentheses
2. Multiply or Divide
3. Add or Subtract

Evaluate $5(m + 1) + 8n$ for $m = 10$ and $n = 2$.

$5(m + 1) + 8n$	
$5(10 + 1) + 8(2)$	Substitute 10 for *m* and 2 for *n*.
$5(11) + 8(2)$	Parentheses, simplify inside.
$55 + 16$	Multiply, from left to right.
71	Add.

Complete to evaluate each expression.

1. $9 + 7z$ for $z = 3$

9 + 7 • ______

9 + ______

______

2. $5(q - 8)$ for $q = 17$

5 • (______ − 8)

5 • (______)

______

3. $25 - 2x$ for $x = 8$

25 − 2 • (______)

25 − ______

______

4. $2(x + 6) + 4$ for $x = 9$

2(______ + 6) + 4

2(______) + 4

______ + 4

______

5. $42 - 3(x + 1)$ for $x = 3$

42 − 3 • (______ + 1)

42 − 3 • (______)

42 − ______

______

6. $22 + 5(2z)$ for $z = 4$

22 + 5 • (2 • ______)

22 + 5 • (______)

22 + ______

______

Name ________________________ Date __________ Class __________

LESSON 1A

Homework and Practice

Variables and Expressions

Evaluate each expression.

1. $7(6 - n) + 3$ for $n = 2$

2. $5x + 3(x - 4)$ for $x = 5$

3. $10p - 2(2p - p)$ for $p = 1$

4. $a(18 - 20) + 6$ for $a = 7$

5. $7(b + 2.9) + 3$ for $b = 4.1$

6. $8xy$ for $x = 2$, $y = -3$

Simplify each algebraic expression.

7. Bart earns $12 an hour plus 7% commission on his sales. If *x* equals his total sales and *y* equals the number of hours he works in a week, then the expression $0.07x + 12y$ can be used to find the amount he earns. How much does Bart earn on $1,500 of sales and 4.5 hours of work?

8. If ℓ is the length of a room and *w* is the width, then the expression $2\ell + 2w$ can be used to find the perimeter of the room. Find the perimeter of a room with a length of 14 feet and a width of 11 feet.

9. A study group takes a dinner break and orders some pizza. If a large pizza costs $15 and a small pizza costs $11, then the expression $15x + 11y + 10$ can be used to find the total cost of the pizzas including a $10 tip. If the group orders 2 large pizzas and 3 small pizzas, what will be the total bill?

10. A local cell phone company offers two plans. Plan A charges $20 a month plus $0.05 a minute. Plan B charges $10 a month plus $0.10 a minute. If your average monthly use is 20 hours a month, what is the cost of each plan? Which plan is more cost effective?

EXPLORATION

Program of Studies and Core Content

MA-8-AT-S-VE01, MA-8-NPO-S-NO1, MA-08-5.2.1, MA-08-5.2.2, MA-08-1.3.1

1B Algebraic Expressions

The table shows the prices for different activities at a park.

Activity	Price
Park ride	\$1.25 per ride
Jet ski rental	\$25.00 + \$5.00 per hour
Scooter rental	\$10.00 + \$2.50 per hour
Bike rental	\$5.00 + \$2.00 per hour
Bay cruise	\$25.00 per person

An algebraic expression for this cost is 1.25*r*.

Use a variable to write an expression for each cost.

	Activity	Cost	Variable	Expression
	Park ride	\$1.25 per ride	r (rides)	$1.25r$
1.	Jet ski rental	\$25.00 + \$5.00 per hour	h (hours)	
2.	Scooter rental	\$10.00 + \$2.50 per hour	h (hours)	
3.	Bike rental	\$5.00 + \$2.00 per hour	h (hours)	
4.	Bay cruise	\$25.00 per person	p (people)	

To find the cost of renting a jet ski for 5 hours, evaluate the expression $25 + 5h$ by substituting 5 for h.

$$25 + 5(5) = 25 + 25 = 50$$

5. Use your expressions from above to evaluate the cost of a 5-hour scooter rental and of a 5-hour bike rental.

Think and Discuss

6. Describe a real-world situation in which you might use the expression $50 + 2x$.

7. Explain what you need to write an algebraic expression.

Name ____________________ Date __________ Class __________

LESSON 1B

Review for Mastery

Algebraic Expressions

What words tell you to add, subtract, multiply, or divide?

Add $n + 6$	**Subtract** $n - 6$	**Multiply** $6n$	**Divide** $n \div 6$
n plus 6 the sum of *n* and 6 6 added to *n* *n* increased by 6	*n* minus 6 the difference between *n* and 6 6 subtracted from *n* 6 less than *n* *n* decreased by 6	*n* times 6 the product of *n* and 6 6 multiplied by *n*	*n* divided by 6 the quotient of *n* and 6 6 equal shares of *n*

Write the algebraic expression that represents the word expression.

1. the quotient of *n* and 3

____

2. 3 more than *n*

____

3. the product of 3 and *n*

____

4. *n* decreased by 3

____

5. the product of *n* and 3

____

6. one-third times *n*

____

Write a word phrase for each algebraic expression.

7. 4n

____

8. $2 + n$

____

9. $n - 5$

____

10. $\frac{n}{2}$

____

11. Heather runs for some number of minutes then walks an additional 5 minutes 3 days a week. Write an algebraic expression to determine how many total minutes Heather exercises in a week. If each run is 30 minutes long, how many total minutes is Heather exercising during the week?

____

Name ______________________ Date ____________ Class ____________

LESSON 1B

Homework and Practice

Writing Algebraic Expressions

Write an expression for each word phrase.

1. 25 more than the quotient of x and 7

____

2. the sum of 10 and a number n

____

3. a number b decreased by 14

____

4. the product of a number x and 8

____

5. 12 times the difference of a number r and 7

____

6. 11 minus the product of 5 and a number k

____

7. a. Grace wants to divide the job of writing x invitations to the party equally among 4 committee members. Write an expression to determine the number of invitations she will have each committee member write.

____

b. If Grace has 220 invitations for the committee members to write, how many will she have each member write?

____

Write an algebraic expression and evaluate it to solve each word problem.

8. Arturo purchased a package of 50 blank CDs. After he used some of the CDs, he had that number less than 50 CDs remaining. Write an expression to represent this situation. How many CDs did he have left after using 31 CDs?

____

9. At the grocery store, Henry bought several pounds of cashews for $4.99 per pound. Write an expression to represent this situation. If he buys 2 pounds, how much will he pay?

____

EXPLORATION

Program of Studies and Core Content
MA-8-NPO-S-NS2

1C Integers and Absolute Value

You can use a number line to find the distance of a number from 0. For example, to find the distance of −3 from 0, start at −3 and count the number of units until you get to 0.

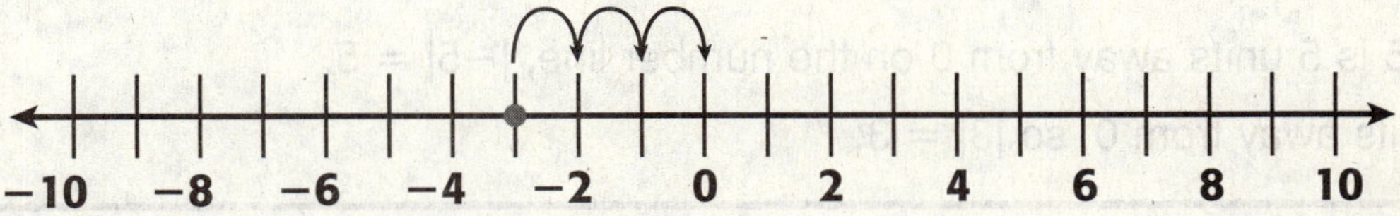

The distance of −3 from 0 is 3.

1. Use a number line to find the distance of each number from 0.

a. 2

b. 5

c. 8

d. 10

e. 0

2. Use a number line to find the distance of each number from 0.

a. −2

b. −5

c. −8

d. −10

Think and Discuss

3. Describe any patterns you notice in your results.

4. Explain whether it is possible for the distance of a number from 0 to be negative.

Name ______________________ Date __________ Class __________

LESSON **1C**

Review for Mastery

Integers and Absolute Value

Use a number line to find the absolute value of integers.

To simplify $|-5|$ and $|3|$, find the distance of each integer from 0 on a number line.

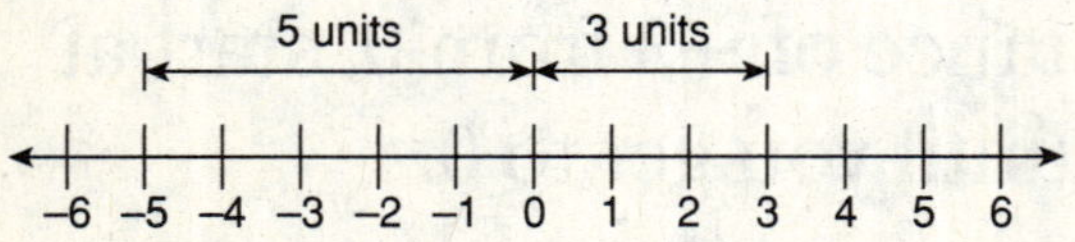

Since −5 is 5 units away from 0 on the number line, $|-5| = 5$.

3 is 3 units away from 0, so $|3| = 3$.

Simplify.

1. $|-2|$

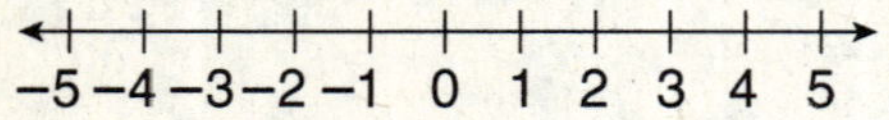

$|-2| =$ ☐

2. $|0|$

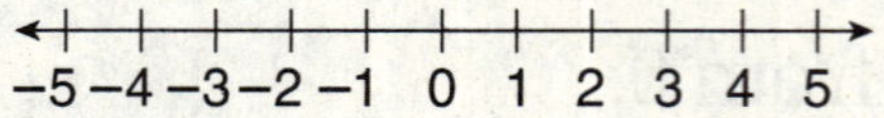

$|0| =$ ☐

To simplify the expression $|15 - 5|$, first simplify the expression inside the absolute value signs. Then find the absolute value of that integer.

Since $15 - 5 = 10$, and $|10|$ is 10 units away from 0 on the number line, $|15 - 5| = 10$.

To simplify the expression $|25| + |-7|$, first find each absolute value and then add.

Since $|25| = 25$, and $|-7| = 7$, $|25| + |-7| = 25 + 7 = 32$.

Evaluate each expression.

3. $|12 \cdot 10|$ ____________

4. $|66 - 18|$ ____________

5. $|-13| + |18|$ ____________

6. $|-2| + |3 \cdot 4|$ ____________

Name ______________________ Date ____________ Class ____________

LESSON **1C**

Review for Mastery

Integers and Absolute Value, continued

Many topics involve positive and negative numbers, such as temperature, sea level, banking, and objects moving in opposite directions. Applying absolute value can often help to solve problems involving such situations.

Sandra and Tom both leave the library at 3:00, and head in opposite directions. When Tom is 5 km north of the library, Sandra is 3.5 km south of the library. How far is Sandra from Tom?

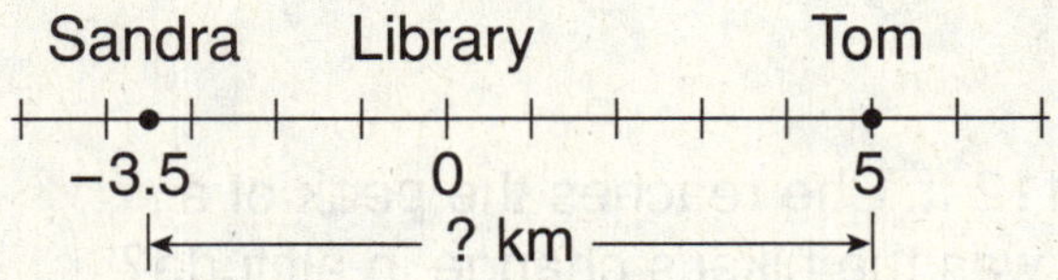

Draw a model. Place Tom at 5, and Sandra at −3.5.

Use absolute value to determine the distance: $|5| + |-3.5| = 8.5$

Sandra is 8.5 km from Tom.

Draw a model to solve each problem.

7. Walcott Middle School is playing West Island Prep at football. Walcott began on the 17 yard line, gained 6 yards, and then lost 9 yards. On what yard line is Walcott now?

____

8. The temperature at 11 P.M. was −4 °F. By 3 A.M., the temperature dropped 8 degrees, but then increased 3 degrees by 6 A.M. What was the temperature at 6 A.M.?

____

9. John had \$1,560 in his savings account. He deposited \$800, but then withdrew \$2,500 to buy a plasma TV. By how much did John overdraw his account? (This means John spent more than he had.)

____

Name ______________________ Date ____________ Class ____________

LESSON 1C

Homework and Practice

Integers and Absolute Value

Evaluate each expression.

1. $|-12| - |-3|$ ________

2. $|4| + |-12|$ ________

3. $|10 - 6|$ ________

4. $|12 + 24|$ ________

5. $|-5| \cdot |-6|$ ________

6. $|20| \div |-2|$ ________

Solve.

7. A hiker begins her hike at an altitude of –112 ft. She reaches the peak of a mountain with an altitude of 3821 ft. What was the hiker's change in altitude?

____

8. Trevor has \$12. He owes his friend Julie \$20, but he also knows that his brother owes him \$15. If Trevor settles his debts with Julie and his brother, how much money will he have?

____

9. On a "par 3" golf course, a golfer should require no more than 3 shots, or strokes, to complete each hole. Your score for a given hole is the number of strokes you actually need minus 3. On the first five holes, Sylvie scored 2, –1, 1, 0, and –1.

a. What does a score of 2 mean? A score of –1? Explain.

____

____

b. How many strokes did Sylvie take altogether? ________________

c. Which of the following expressions might best represent the total score? Explain your choice.

I. $|-1 + 1 + 0 + 2 + (-1)|$

II. $-1 + 1 + 0 + 2 + (-1)$

III. $|-1| + 1 + 0 + 2 + |-1|$

____

____

Program of Studies and Core Content

MA-8-NPO-S-N01, MA-08-1.3.1

1D Adding Integers

You can use a thermometer to model addition of integers.

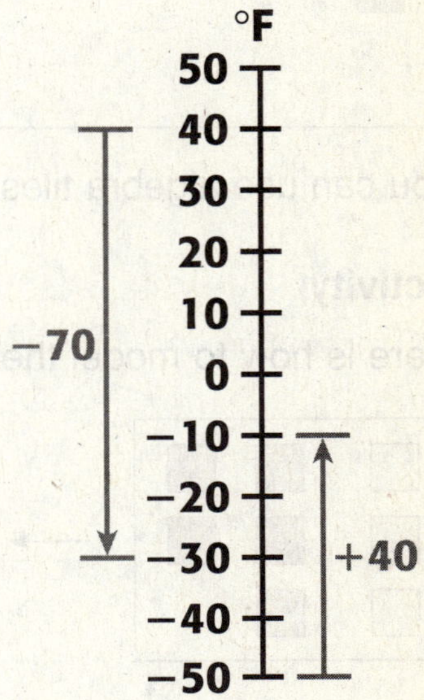

1. Suppose the temperature starts at −50°F and increases 40° during the day. Complete the addition statement to show the new temperature.

$$-50° + 40° = \_\_\_$$

2. Suppose the temperature starts at 40°F and drops 70° overnight. Complete the addition statement to show the new temperature.

$$40° + (-70°) = \_\_\_$$

Complete the addition statement modeled by each number line.

3. $-4 + \_\_\_ = 5$

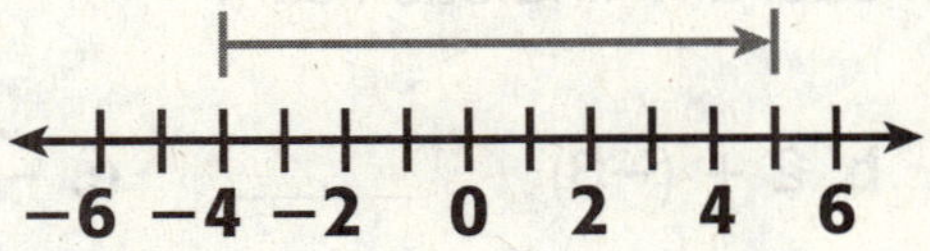

4. $-1 + (\_\_\_) = -8$

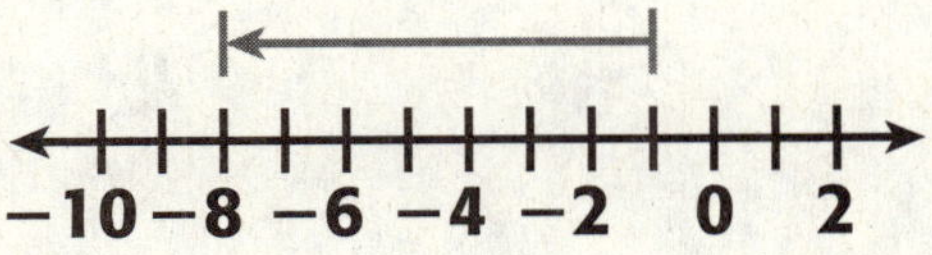

Think and Discuss

5. Explain how to add integers on a number line.

6. Explain whether 1 + (−3) has the same sum as −3 + 1. Is the same true for any two integers?

Name ______________________________ Date __________ Class __________

LESSON **1D**

Hands-On Lab

Model Adding Integers

KEY	REMEMBER:
 = 1  = −1	The sum of two or more numbers can be positive, negative, or zero. The sum of an integer and its opposite is zero.

You can use algebra tiles to help you add integers.

Activity

Here is how to model the solution to the problem 3 + (−5).

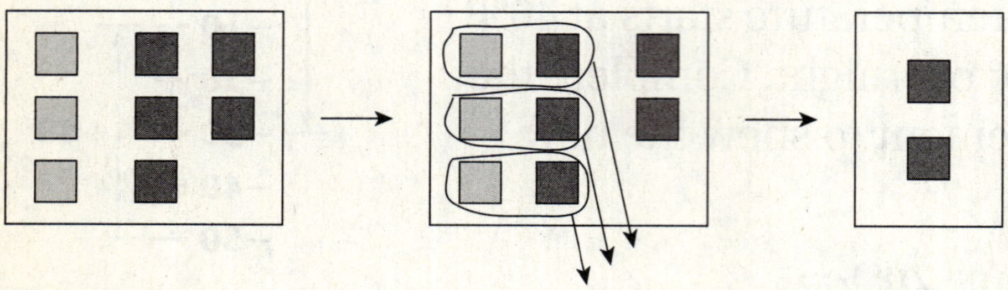

Remove pairs that equal zero

3 + −5

1. Use algebra tiles to model and find each sum

a. 4 + (−2) ______ **b.** 2 + (−3) ______ **c.** −3 + 3 ______

d. −2 + 5 ______ **e.** −3 + (−3) ______ **f.** −4 + (−1) ______

Name ______________________________ Date ____________ Class ____________

LESSON 1D

Hands-On Lab

Model Adding Integers, continued

The same technique works with problems involving three or more addends, such as −3 + 2 + (−4) + 1.

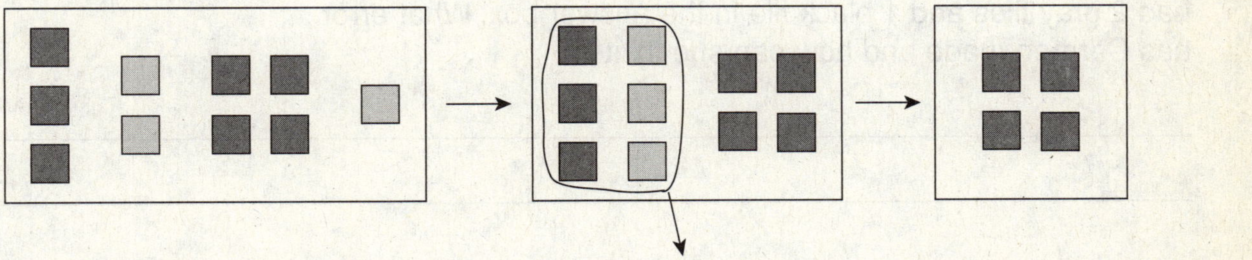

−3 + **2** + **−4** + **1** Regroup. Remove zero.

2. Use algebra tiles to model and find each sum

a. 3 + 2 + (−2) + (−2) ______

b. 2 + (−1) + (−4) + (−2) ______

c. −3 + (−1) + (−2) + (−3) ______

d. −2 + (−1) + 3 + (−3) + (−2) + 1 ______

Name ______________________ Date __________ Class __________

LESSON **1D**

Hands-On Lab

Model Adding Integers, continued

Think and Discuss

1. When Carmen completed a model of the sum of 3 + (−2), she had 2 gray tiles and 1 black tile in the answer box. What error has Carmen made and how can she fix it?

____

____

____

2. Without using a model, how do you know that the sum of 1 and −5 is a negative integer? How can you use this information to check the accuracy of your model?

____

____

____

3. What would you conclude about a sum if all of the algebra tiles were paired off and removed from a model?

____

4. Can three numbers sum to zero? If so, give an example. If not, explain why not.

____

Try This

Use algebra tiles to model and find each sum.

1. 5 + (−3) _____

2. 2 + (−4) + (−1) _____

3. −1 + 3 + (−2) + (−3) _____

4. −1 + (−1) + (−2) + 4 _____

5. −2 + (−4) + (−1) + (−1) _____

6. 2 + (−2) + 4 + (−1) + (−2) + 1 _____

Name ____________________ Date __________ Class __________

LESSON **1D**

Review for Mastery

Adding Integers

You can model integer addition using two-color counters. Use the yellow side for 1 and the red side for –1. Remember that one yellow counter and one red counter are opposites, so their sum is zero.

7 + 5 = ⓎⓎⓎⓎⓎⓎⓎ ⓎⓎⓎⓎⓎ = 12 yellow counters = 12

7 + (–5) = ⓎⓎⓎⓎⓎⓎⓎ ⓇⓇⓇⓇⓇ = 2 yellow counters = 2

–7 + (–5) = ⓇⓇⓇⓇⓇⓇⓇ ⓇⓇⓇⓇⓇ = 12 red counters = –12

–7 + 5 = ⓇⓇⓇⓇⓇⓇⓇ ⓎⓎⓎⓎⓎ = 2 red counters = –2

If the given integers were added, state whether the result would be positive or negative.

1. –4 + (–6) ____________

2. –3 + 8 ____________

3. –5 + 2 ____________

Notice if the counters are the same color, you add the absolute values of the integers. The answer is the sign of the integers. If the counters are both colors, you subtract the absolute values of the integers. Use the sign of the integer with the greater absolute value.

To add the given integers, state whether you need to add or subtract absolute values.

4. 8 + 3 ____________

5. –4 + (–1) ____________

6. 3 + (–6) ____________

Complete to find each sum. **7.** 5 + (–9) = ? **8.** –6 + (–4) = ?

Are the signs the same or different? ____________ ____________

Which sign will you use for the sum? ____________ ____________

Will you add or subtract absolute values? ____________ ____________

Write the sum. ____________ ____________

Name ______________________ Date ____________ Class ____________

LESSON 1D

Homework and Practice

Adding Integers

Use a number line to find the sum.

1. $2 + 3$

−7 −6 −5 −4 −3 −2 −1 0 1 2 3 4 5 6 7

2. $-5 + 3$

−7 −6 −5 −4 −3 −2 −1 0 1 2 3 4 5 6 7

Add.

3. $-6 + 15$

4. $-28 + (-7)$

5. $31 + (-19)$

6. $-34 + 21$

Evaluate each expression for the given value of the variables.

7. $-5 + r$ for $r = -7$

8. $-9 + w$ for $w = -12$

9. $m + 8$ for $m = -11$

10. $-6 + k$ for $k = -9$

11. $t + (-5)$ for $t = -8$

12. $b + (-7)$ for $b = -7$

13. $-11 + x$ for $x = 15$

14. $g + (-15)$ for $g = 15$

15. $-16 + j$ for $j = -18$

16. Two years ago the music club had 328 members. The following year there were 103 new members. This year 33 members left the club. How many current members are in the music club?

17. During one month, Nikki withdrew amounts of $25, 60, and $75. She deposited $100. If her starting balance was $430, find her new balance.

18. Rosetta is playing a board game with her friends. She rolls doubles of five on her first roll of the number cubes. If a player rolls doubles, they roll again after moving forward. However if they roll doubles again they must move backwards. Rosetta rolls doubles of six on her second roll. How many spaces is Rosetta from her starting place?

EXPLORATION

Program of Studies and Core Content

MA-8-NPO-S-N01, MA-08-1.3.1

1E Subtracting Integers

You can use a number line to model subtracting integers.

To subtract 10 from 50, begin at the number being subtracted, 10, and count the number of units to the number 50.

$50 - 10 = 40$

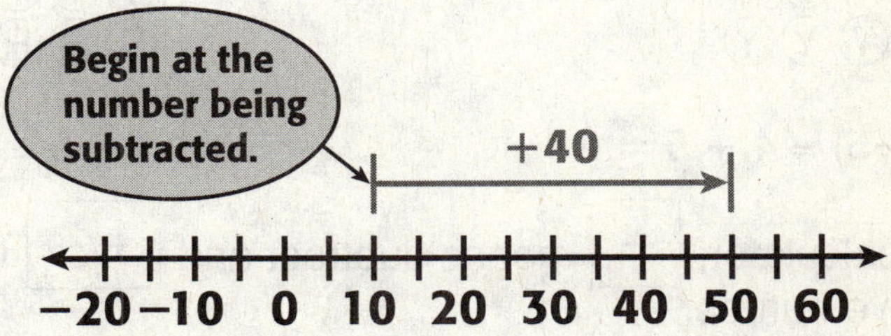

The direction is **right,** so the difference is **positive.**

To subtract −10 from −70, begin at the number being subtracted, −10, and count the number of units to the number −70.

$-70 - (-10) = -60$

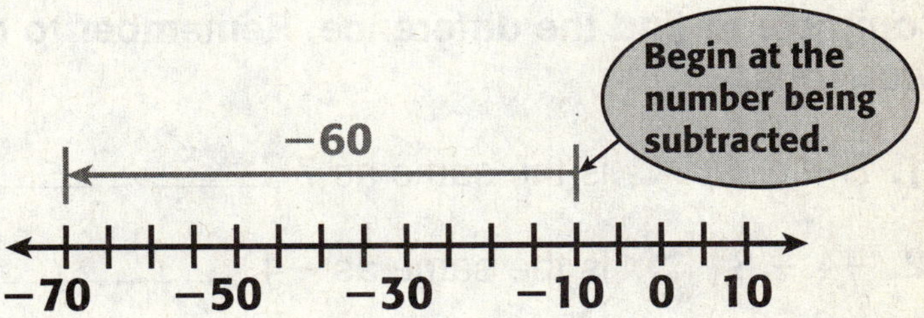

The direction is **left,** so the difference is **negative.**

Use the number line to complete each subtraction statement.

1. $5 - (-4) =$ ___

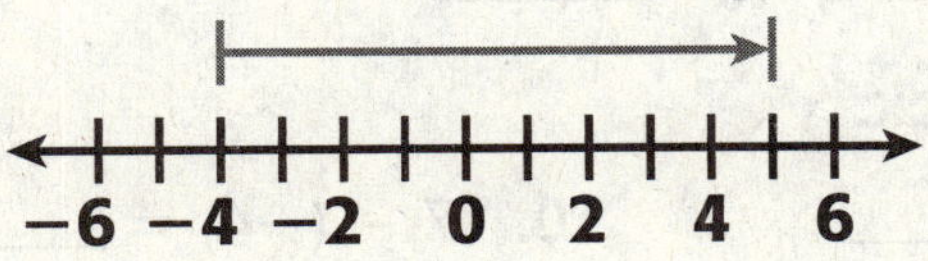

2. $-8 - (-1) =$ ___

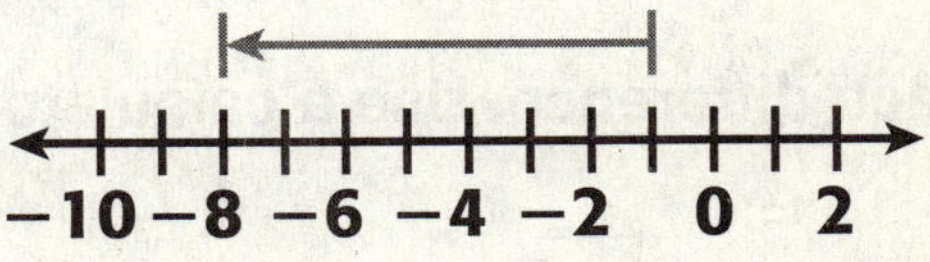

Think and Discuss

3. Discuss a different strategy for subtracting integers.

Name ________________________________ Date ____________ Class ____________

LESSON 1E

Review for Mastery

Subtracting Integers

To subtract one integer from another, rewrite the subtraction as the addition of an opposite. Then use the rules for adding integers.

4 − (−5)

Subtracting a Negative

Change subtraction to addition of a positive.

4 + (+5)

ⓎⓎⓎⓎ
ⓎⓎⓎⓎⓎ

4 − (−5) = 4 + 5 = 9

4 − 5
4 − (+5)
4 + (−5)

Subtracting a Positive

Change subtraction to addition of a negative.

ⓎⓎⓎⓎ
ⓇⓇⓇⓇⓇ

4 − 5 = 4 + (−5) = −1

On a calculator, [−] means subtract and [+/−] will enter a negative number.

To do 4 − (−5) on a calculator:

Input: 4 [−] 5 [+/−] [=]

Display: 9

To do 4 − 5 on a calculator:

Input: 4 [−] 5 [=]

Display: −1

Complete to find the difference. Remember to change two signs.

1. 7 − (−6) is the same as 7 _____ _____ 6 = _____
2. −4 − 3 is the same as −4 _____ _____ 3 = _____
3. −2 − (−9) is the same as −2 _____ _____ 9 = _____
4. 14 − 16 is the same as 14 _____ _____ 16 = _____
5. 7 − (−10) is the same as 7 _____ _____ 10 = _____
6. −8 − (−19) is the same as −8 _____ _____ 19 = _____
7. −5 − 12 is the same as −5 _____ _____ 12 = _____

Find each difference. Use a calculator to check.

8. 7 − 12 = _____
9. −3 − 8 = _____
10. 17 − (−4) = _____
11. −14 − (−3) = _____
12. 5 − 8 = _____
13. −6 − 4 = _____

Name ______________________ Date __________ Class __________

LESSON 1E

Homework and Practice

Subtracting Integers

Subtract.

1. 9 − 3 ________

2. 12 − 4 ________

3. 9 − 16 ________

4. 19 − 12 ________

5. 7 − 15 ________

6. 18 − 4 ________

7. −8 − 12 ________

8. −11 − 14 ________

9. 25 − 49 ________

10. 18 − 36 ________

11. −101 − 52 ________

12. −76 − (−12) ________

Evaluate each expression for the given value of the variables.

13. $a - 10$ for $a = 7$ ________

14. $x - 13$ for $x = 20$ ________

15. $17 - y$ for $y = 6$ ________

16. $15 + b$ for $b = -9$ ________

17. $14 - t$ for $t = -18$ ________

18. $d - 24$ for $d = -17$ ________

19. $-16 - w$ for $w = -15$ ________

20. $-10 - r$ for $r = -22$ ________

21. $-14 - g$ for $g = -19$ ________

22. $x - (-18)$ for $x = -25$ ________

23. $y - (-14)$ for $y = -14$ ________

24. $n - (-17)$ for $n = 21$ ________

25. The largest island in the world is Greenland with an area of 839,999 square miles. The second largest island is New Guinea with an area of 316,615 square miles. What is the difference in the areas of the two islands?

____

26. Justin weighed 223 pounds and lost 45 pounds. What is Justin's new weight?

____

27. At 6 A.M. the temperature was –7 °F. By noon, the temperature had risen to 1 °F. And by 6 P.M. the temperature had fallen to –3 °F.

a. By how many degrees did the temperature change between 6 A.M. and noon?

____

b. By how many degrees did the temperature change between noon and 6 P.M.?

____

EXPLORATION

Program of Studies and Core Content
MA-8-NPO-S-NO1,
MA-08-1.3.1

1F Multiplying and Dividing Integers

Imagine a person walking on a number line. If the person faced a **positive direction,** it would be **to the right.** If the person faced a **negative direction,** it would be **to the left.**

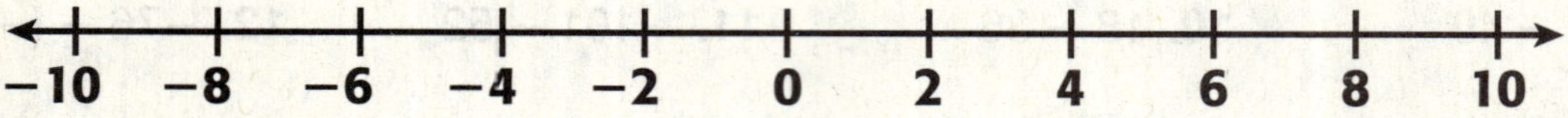

Suppose each step is 2 units long.

1. A person who is standing at 0 and facing a **positive direction** takes 3 steps **backward.** Complete the multiplication statement to find the person's location.

$$3 \cdot (-2) = \_\_\_$$

2. A person who is standing at 0 and facing a **negative direction** takes 4 steps **forward.** Complete the multiplication statement to find the person's location.

$$-4 \cdot 2 = \_\_\_$$

3. A person who is standing at 0 and facing a **negative direction** takes 5 steps **backward.** Complete the multiplication statement to find the person's location.

$$-5 \cdot (-2) = \_\_\_$$

Think and Discuss

4. Describe a situation for the multiplication statement $6 \cdot (-3) = -18$ by using a number line.

Name ______________________ Date __________ Class __________

LESSON **1F**

Technology Lab

Multiplying and Dividing Integers

As you divide integers by the number 9, you will notice a pattern developing in each quotient. Using a calculator will help you recognize these patterns. Eventually you will be able to predict the results of dividing by 9.

Activity

1. Divide the numbers 10, 11, and 12 by 9, first using mental math and then using a calculator.

Mental Math	Calculator
10 ÷ 9 = 1 R1	10 ÷ 9 = 1.11111. . .
11 ÷ 9 = 1 R2	11 ÷ 9 = 1.22222. . .
12 ÷ 9 = 1 R3	12 ÷ 9 = 1.33333. . .

In each case the remainder is shown by the calculator as a repeating decimal. The pattern continues to 18 ÷ 9, which equals 2.

2. Next use your calculator to find the quotient of 19 ÷ 9 and 28 ÷ 9.

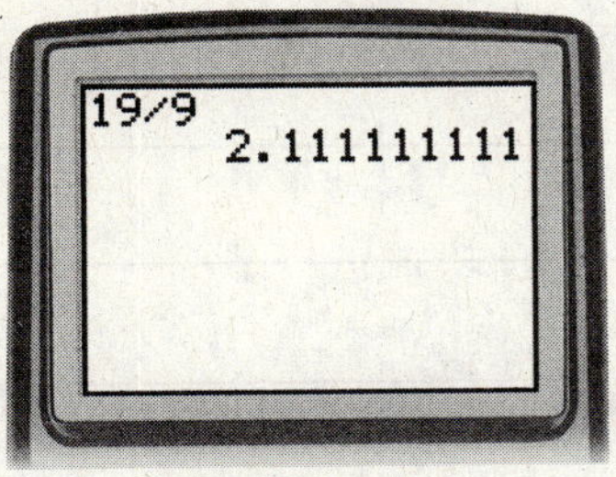

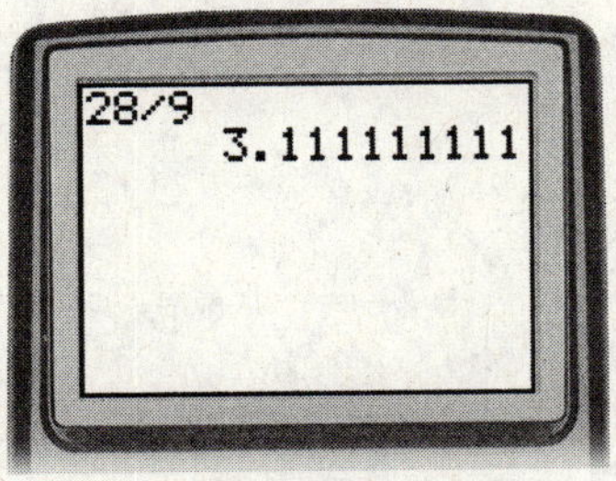

The quotient of any positive integer divided by 9 will always be a whole number (18 ÷ 9 = 2) or a number with a repeating decimal (28 ÷ 9 = 3.111111...). Calculators display the remainders of quotients as repeating decimals (3 R1 = 3.111111...).

Name ______________________________ Date ____________ Class ____________

LESSON 1F

Technology Lab

Multiplying and Dividing Integers, continued

Think and Discuss

1. Which of the following numbers is NOT a quotient obtained from dividing an integer by 9?
   934.11111...   23.88888...   8763.12345

____

____

Try This

1. Add together the digits of each number below, and then use long division to divide the total by 9. Then use your calculator to divide each number by 9. What pattern is displayed in each answer?
   82   169   211,458   1,006,937,092

____

____

____

____

____

____

Name ______________________ Date ____________ Class ____________

LESSON 1F

Review for Mastery

Multiplying and Dividing Integers

Since multiplication is a shortcut for addition, a pattern becomes apparent when you multiply or divide integers. Look at the two multiplication problems written as repeated addition.

$8 \times 3 = 8 + 8 + 8 = 24$		$-8 \times 3 = -8 + -8 + -8 = -24$	
$8 \times 3 = 24$		$-8 \times 3 = -24$	
$8 \times 2 = 16$	*Notice the product*	$-8 \times 2 = -16$	*Notice the product*
$8 \times 1 = 8$	*is decreasing by 8.*	$-8 \times 1 = -8$	*is increasing by 8.*
$8 \times 0 = 0$		$-8 \times 0 = 0$	
$8 \times -1 = -8$	*Keep decreasing*	$-8 \times -1 = 8$	*Keep increasing by*
$8 \times -2 = -16$	*by 8 even though*	$-8 \times -2 = 16$	*8 even though both*
$8 \times -3 = -24$	*one factor is negative.*	$-8 \times -3 = 24$	*factors are negative.*

Multiplication and division have the same rules for negative and positive integers.

Positive × Positive = Positive   Positive × Negative = Negative
Negative × Positive = Negative   Negative × Negative = Positive

Tell if the signs are the same or different and if the result will be positive or negative.

1. $9 \times (-5)$

signs are ____________

product is ____________

2. $-14 \div (-2)$

signs are ____________

quotient is ____________

3. $27 \div 3$

signs are ____________

quotient is ____________

4. $50 \div (-2)$

signs are ____________

quotient is ____________

Enter + or − for the product or quotient.

5. $(-14) \div (-2) =$ ______ 7

6. $8 \times 7 =$ ______ 56

7. $(-3) \times 5 =$ ______ 15

8. $16 \times (-3) =$ ______ 48

9. $25 \div (-5) =$ ______ 5

10. $(-100) \div 50 =$ ______ 2

Find the product or quotient.

11. $7 \times (-2) =$ ______

12. $-99 \div (-11) =$ ______

13. $8 \times (-6) =$ ______

14. $64 \div 16 =$ ______

15. $-12 \times (-3) =$ ______

16. $-48 \div 12 =$ ______

Name ______________________________ Date ____________ Class ____________

LESSON **1F**

Homework and Practice

Multiplying and Dividing Integers

Enter + or − for the product or quotient.

1. $9 \times (-9) =$ ______ 81
2. $-27 \div (-3) =$ ______ 9
3. $-10 \times 6 =$ ______ 60
4. $-15 \times (-2) =$ ______ 30
5. $26 \div (-13) =$ ______ 2
6. $-21 \div 7 =$ ______ 3

Find the product or quotient.

7. $20 \div (-5) =$ ______
8. $-14 \times 4 =$ ______
9. $28 \div (-7) =$ ______
10. $-54 \div (-2) =$ ______
11. $-22 \times (-5) =$ ______
12. $14 \times (-8) =$ ______
13. $-64 \div 4 =$ ______
14. $60 \times 6 =$ ______
15. $-56 \times (-2) =$ ______
16. $-72 \div (-9) =$ ______

17. Jack rides his bicycle for a total of 5 miles each afternoon for one week delivering newspapers. Write an expression which represents two-thirds of his total distance.

____

18. Jane is hiking in a state park. She descends 390 feet over a period of 30 minutes. If she descended at a constant rate, what is her change in depth per minute?

____

19. During one week in February, the daily high temperatures in degrees Fahrenheit were –3, 5, 3, –1, 2, 10, and 12. What was the average daily high temperature for the week?

____

20. Keith played 9 holes of golf and recorded his scores for each hole as –2, –1, 0, –1, 2, 0, –2, 1, –1. What was Keith's average score (rounded to the nearest integer)?

____

1G Properties

Program of Studies and Core Content
MA-8-NPO-S-PN01, MA-08-1.5.2

You can investigate some properties of whole numbers by answering the following questions.

1. Which of the following are true?
   a. $10 + 5 = 5 + 10$
   b. $10 - 5 = 5 - 10$
   c. $10 \cdot 5 = 5 \cdot 10$
   d. $10 \div 5 = 5 \div 10$
2. For which operations (addition, subtraction, multiplication, division) does the order of the numbers not matter?
3. Which of the following are true?
   a. $8 + (4 + 2) = (8 + 4) + 2$
   b. $8 - (4 - 2) = (8 - 4) - 2$
   c. $8 \cdot (4 \cdot 2) = (8 \cdot 4) \cdot 2$
   d. $8 \div (4 \div 2) = (8 \div 4) \div 2$
4. For which operations (addition, subtraction, multiplication, division) can you group the numbers together in any combination and get the same result?

Think and Discuss

5. **Describe** two different ways to group the numbers in the sum $6 + 7 + 14$. Do both groupings give the same result?
6. **Describe** two different ways to group the numbers in the product $5 \cdot 3 \cdot 2$. Do both groupings give the same result?

Name ______________________ Date ____________ Class ____________

LESSON 1G

Review for Mastery

Properties

You can use the following properties to simplify algebraic expressions.

	Addition	Multiplication
Commutative Property	$3 + 4 = 4 + 3$	$2 \bullet 5 = 5 \bullet 2$
	$a + b = b + a$	$a \bullet b = b \bullet a$
Associative Property	$(3 + 4) + 5 = 3 + (4 + 5)$	$(2 \bullet 4) \bullet 10 = 2 \bullet (4 \bullet 10)$
	$(a + b) + c = a + (b + c)$	$(a \bullet b) \bullet c = a \bullet (b \bullet c)$
Identity Property	$6 + 0 = 6$	$4 \times 1 = 4$
	$a + 0 = a$	$a \times 1 = a$
Inverse Property	$3 + (-3) = 0$	$5 \times \frac{1}{5} = 1$
	$a + (-a) = 0$	$a \times \frac{1}{a} = 1, a \neq 0$
Distributive Property	$2(5 + 9) = 2(5) + 2(9)$	
	$a(b + c) = a(b) + a(c)$	

Simplify 12 + *h* + 16.

12 + *h* + **16**	*Identify terms that can be combined.*
h + **12** + **16**	*Use the Commutative Property to rearrange the terms.*
h + **(12 + 16)**	*Use the Associative Property to group the numbers.*
h + 28	*Simplify.*

Simplify 4(*y* + 6) + 2*y*.

4(*y*) + 4(6) + 2*y*	*Use the Distributive Property to distribute the 4.*
4*y* + 24 + 2*y*	*Multiply.*
4*y* + **2*y*** + **24**	*Use the Commutative Property to rearrange the terms.*
6*y* + 24	*Simplify.*

Use the properties above to simplify each expression.

1. $\frac{1}{9} \times 9$ **2.** $(t - 3) + (t - 3)$ **3.** $13 - 2.4(6w + 3.5)$

____________ ____________ ____________

4. $\frac{1}{2}(2k - 4p) + 2p$ **5.** $6(m + \frac{2}{3}v) - m(4 + v)$ **6.** $1.2 + (-1.2)$

____________ ____________ ____________

Name ______________________ Date ____________ Class ____________

LESSON 1G

Homework and Practice

Properties

Identify the property shown.

1. $7x + 3y = 3y + 7x$

2. $\frac{2}{3}(g + 9) = \frac{2}{3}g + \left(\frac{2}{3}\right)(9)$

3. $4f \times \frac{1}{4f} = 1, (f \neq 0)$

____

4. $\frac{7}{8} \times \frac{8}{7} = 1$

5. $-15ab + 15ab = 0$

6. $\frac{2}{5} + \left(4 + \frac{3}{4}\right) = \left(\frac{2}{5} + 4\right) + \frac{3}{4}$

______________________ ______________________ ______________________

Simplify each algebraic expression. Justify each step.

7. $6(x + 3) - 2x$

$6(\square) + (\square)(3) - 2x$ ______________________ Property

$6x + \square - 2x$ Multiply.

$\square - \square + 18$ ______________________ Property

$\square + 18$ Simplify.

8. $(f - 3w) + (3w - 2f)$

$f + (\square + 3w) - 2f$ ______________________ Property

$(\square + 3w) + (f - 2f)$ ______________________ Property

$\square + \square$ Simplify.

$\square$ Add.

9. A rectangular desk has a length of $2(x + 3)$ and a width of $2x$. Another rectangular desk has a length of $3(x + 2)$ and a width of x. How do the perimeters of these two desks compare? Explain.

____

____

EXPLORATION

Program of Studies and Core Content

MA-8-AT-S-EI1, MA-8-AT-S-EI4, MA-AT-S-VEO3, MA-8-NPO-S-NO2, MA-8-NPO-S-NO1, MA-08-5.1.2, MA-08-5.2.2, MA-08-5.3.1, MA-08-1.3.2

1H Solving Equations by Adding or Subtracting

Evaluate the expressions for each given value of *x*.

1.

x	$x + 1$
0	
1	
2	
3	

2.

x	$x - 2$
3	
4	
5	
6	

Each expression below has been evaluated. Find the value of *x*.

3.

x	$x + 2$
	3
	4
	5
	6

4.

x	$x - 5$
	0
	1
	2
	3

Think and Discuss

5. Explain how you evaluated the expressions in Problems 1 and 2.

6. Explain how you found the values of *x* in Problems 3 and 4.

Name ______________________ Date ____________ Class ____________

LESSON **1H**

Hands-On Lab

Model Solving Equations by Subtracting

KEY	REMEMBER:
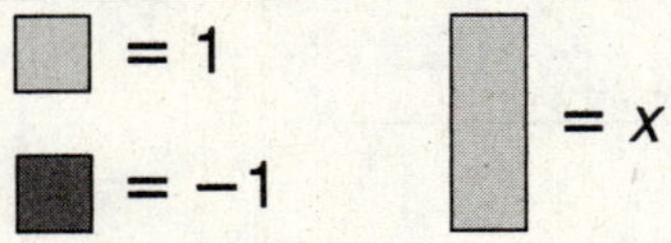	It will not change the value of an equation if you subtract the same number from each side.

You can use algebra tiles to help you solve equations.

Activity

To solve the equation $x + 3 = 5$, you need x alone on one side of the equal sign. You can achieve this goal by adding -3 to the left side of the equation. You can perform this operation if you add -3 to the right side as well.

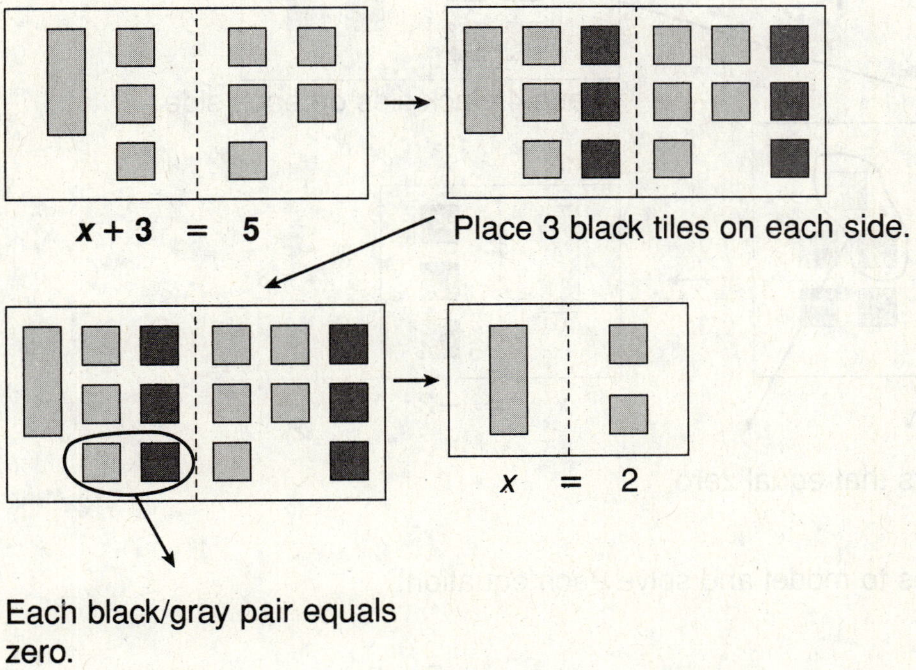

Name ______________________ Date ____________ Class ____________

LESSON **1H**

Hands-On Lab

Model Solving Equations by Subtracting, continued

1. Use algebra tiles to model and solve each equation.

a. $x + 2 = 4$ __________ **b.** $x + 3 = 5$ __________

c. $x + 1 = 5$ __________ **d.** $5 + x = 6$ __________

The same technique works to model the solution to the equation $x + 4 = 1$, where there are fewer tiles on the right side than the left.

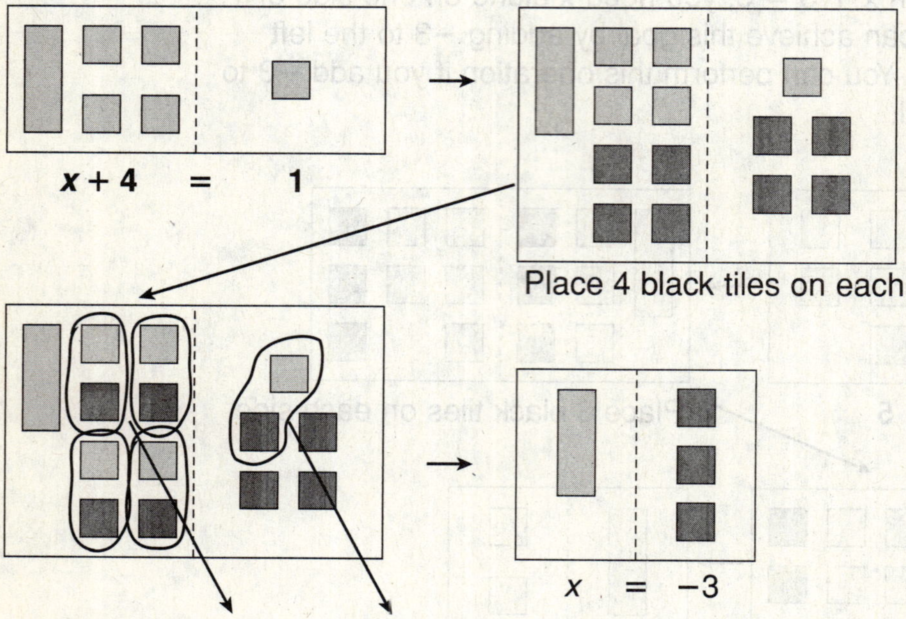

Remove all pairs that equal zero.

2. Use algebra tiles to model and solve each equation.

a. $x + 4 = 3$ __________ **b.** $x + 5 = 2$ __________

c. $x + 3 = 1$ __________ **d.** $x + 3 = 3$ __________

Name ______________________ Date __________ Class __________

LESSON 1H

Hands-On Lab

Model Solving Equations by Subtracting, continued

Think and Discuss

1. How do you know how many black tiles to place on each side?

____

____

2. Why is it important to add the same number of black tiles to both sides of the model?

____

____

3. Why are you allowed to remove pairs of black and gray tiles from the model?

____

4. How is solving the equation $x + 1 = 3$ different from solving $x + 3 = 1$? How is it the same?

____

____

____

____

Try This

Use algebra tiles to model and solve each equation.

1. $x + 1 = 2$ __________ **2.** $x + 3 = 5$ __________ **3.** $x + 2 = 7$ __________

4. $x + 3 = 2$ __________ **5.** $x + 4 = 2$ __________ **6.** $x + 2 = 1$ __________

7. $x + 1 = -1$ __________ **8.** $x + 1 = 0$ __________ **9.** $x + 2 = -1$ __________

Name ______________________ Date ____________ Class ____________

LESSON 1H

Review for Mastery

Solving Equations by Adding or Subtracting

Addition and subtraction are *inverse operations*, which means they "undo" each other.

To solve an addition equation, use subtraction.

Solve $x + 5 = 12$.

$$\begin{array}{rr} x + 5 = & 12 \\ \underline{-5} & \underline{-5} \\ x = & 7 \end{array}$$

Subtract the number added to the variable.

To solve a subtraction equation, use addition.

Solve $9 = w - 3$.

$$\begin{array}{rr} 9 = & w - 3 \\ \underline{+3} & \underline{+3} \\ 12 = & w \end{array}$$

Add the number subtracted from the variable.

Tell what you would add or subtract to solve the equation.

1. $a + 2 = 7$ ________
2. $x - 2 = 4$ ________
3. $6 + y = 9$ ________
4. $34 = b + 13$ ________
5. $21 = z - 9$ ________
6. $14 + r = 20$ ________
7. $18 = d - 11$ ________
8. $6 = 5 + p$ ________

Complete to solve the equation. In Exercises 9–11 check the solution.

9. $x - 2 = 15$
 $\underline{+2} \quad \underline{+2}$
 $x -$ ____ $\stackrel{?}{=}$ ____
 $x \stackrel{?}{=}$ ____
 Check: $x - 2 = 15$
 ____ $- 2 \stackrel{?}{=} 15$
 ____ $\stackrel{?}{=} 15$ ✓

10. $z + 6 = 14$
 ____ ____
 $z +$ ____ $\stackrel{?}{=}$ ____
 $z \stackrel{?}{=}$ ____
 Check: $z + 6 = 14$
 ____ $+ 6 \stackrel{?}{=} 14$
 ____ $\stackrel{?}{=} 14$ ✓

11. $7 = 4 + n$
 ____ ____
 $\stackrel{?}{=}$ ____ $+ n$
 ____ $\stackrel{?}{=} n$
 Check: $7 = 4 + n$
 $7 \stackrel{?}{=} 4 +$ ____
 $7 \stackrel{?}{=}$ ____ ✓

12. $t + 5 = 16$
 ____ ____
 $t +$ ____ $=$ ____
 $t =$ ____

13. $a - 7 = 13$
 ____ ____
 $a -$ ____ $=$ ____
 $a =$ ____

14. $28 = 9 + b$
 ____ ____
 ____ $=$ ____ $+ b$
 ____ $= b$

Name ______________________________ Date ____________ Class ____________

LESSON 1H

Review for Mastery

Solving Equations by Adding or Subtracting, continued

Sassafras Mountain in South Carolina is approximately 3,560 feet high. You can drive part of the way up the mountain and then take a foot trail to the summit. If the drive covers 1,660 vertical feet, how many vertical feet does the foot trail cover?

Step 1 **Understand the Problem**

The **answer** is the number of vertical feet the foot trail covers.

List the important information:

- The elevation of Sassafras Mountain is 3,560 ft.
- The drive covers a distance of 1,660 ft.

Show the relationship of the information:

Total height	=	Trail height	+	Drive height

Step 2 **Make a Plan**

Write an equation and solve it. Let x represent the foot trail height.

3,560	=	x	+	1,660

Step 3 **Solve**

$$\begin{aligned} 3{,}560 &= x + 1{,}660 \\ \underline{-\,1{,}660} & \quad \underline{-\,1{,}660} \end{aligned}$$ *Subtract 1,660 from both sides.*

$$1{,}900 = x$$

The foot trail covers 1,900 ft.

Step 4 **Look Back**

The problem states that the total elevation is 3,560 ft. This means that the height of the drive plus the height of the foot trail must equal 3,560 ft. Since 1,900 + 1,660 = 3,565, the answer is correct.

15. In 2006, the population of McCormick County was 10,226. The population of Jasper County in 2006 was 21,809. Write and solve an equation to find n, the difference in the populations of these two counties.

____

Name ______________________ Date ____________ Class ____________

LESSON **1H**

Homework and Practice

Solving Equations by Adding or Subtracting

Determine which value is a solution of the equation.

1. $y - 14 = 22$; $y = 8$, 12, or 36

2. $d + 17 = 52$; $d = 25$, 35, or 69

Solve.

3. $m - 9 = 14$

4. $7 + x = 16$

5. $k + 12 = 31$

6. $b + 39 = 57$

7. $p + 27 = 27$

8. $0 = w - 45$

9. $8.7 + s = 12.1$

10. $x - 3.7 = 0.8$

11. $n - 3.41 = 5.40$

12. The school library is combining books from two storage units into a newly designed area for the library. The first original unit held 186 books and the second unit had 307 books. Write and solve an equation to find the number of books in the new area of the library.

13. Amelia wanted to buy a new computer and printer which had a total cost of \$1099.95. She had saved \$389.85 for her purchase. Write and solve an equation to find how much more she need to save to buy the computer and printer she wants.

14. In 1990, the population of Greenville County, South Carolina, was 320,167. By 2000, the population was 379,616. Write and solve an equation to find n, the change in the population between 1990 and 2000.

15. Explain why addition and subtraction are inverse operations.

EXPLORATION

Program of Studies and Core Content

MA-8-AT-S-EI1, MA-8-AT-S-EI4, MA-8-AT-S-VE03, MA-8-NPO-S-NO2, MA-8-NPO-S-NO1, MA-08-5.1.2, MA-08-5.2.2, MA-08-5.3.1, MA-08-1.3.2

1I Solving Equations by Multiplying or Dividing

Evaluate the expressions for each given value of x.

1.

x	$2x$
0	
1	
2	
3	

2.

x	$\frac{x}{2}$
0	
2	
4	
6	

Each expression below has been evaluated. Find the value of x.

3.

x	$3x$
	3
	6
	9
	12

4.

x	$\frac{x}{4}$
	0
	1
	2
	3

Think and Discuss

5. Explain how you evaluated the expressions in Problems 1 and 2.

6. Explain how you found the values of x in Problems 3 and 4.

Name ______________________ Date __________ Class __________

LESSON 11

Hands-On Lab

Model Solving Equations by Dividing

KEY	REMEMBER:
 = 1  = −1  = *x*	It will not change the value of an equation if you divide both sides by the same number.

You can use algebra tiles to help you solve equations.

Activity

To solve the equation $2x = 4$, you need x alone on one side of the equal sign. You can achieve this goal by dividing the left side of the equation by 2. You can divide as long as you divide the same way on both sides.

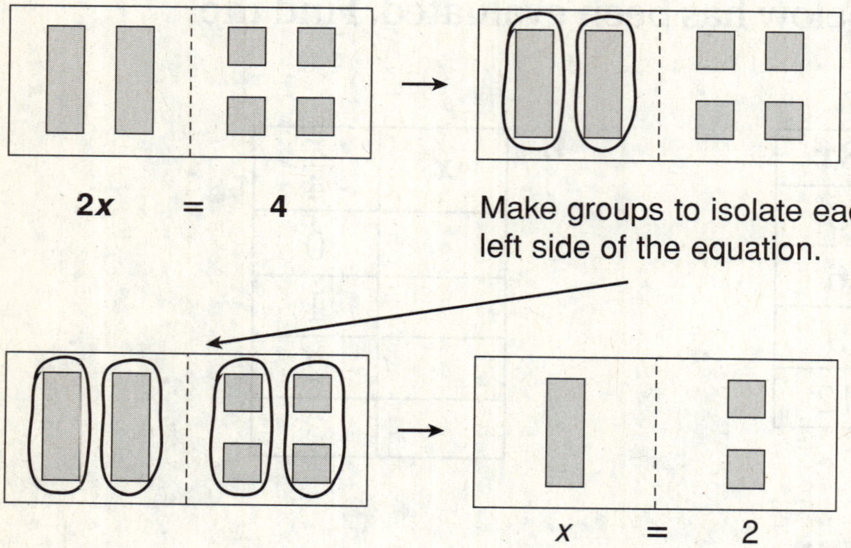

Make the same number of groups on the right side of the equation.

Name ______________________ Date ____________ Class ____________

LESSON 11

Hands-On Lab

Model Solving Equations by Dividing, continued

1. Use algebra tiles to model and solve each equation.

a. $2x = 8$ __________ **b.** $3x = 9$ __________

c. $4x = 8$ __________ **d.** $2x = 6$ __________

Solving the equation $2x = -2$ is not much different. You can use tiles to represent negative numbers just as you can use them for positive numbers. Once you set up the equation using the *x*-tiles and the negative tiles, the method for solving is the same.

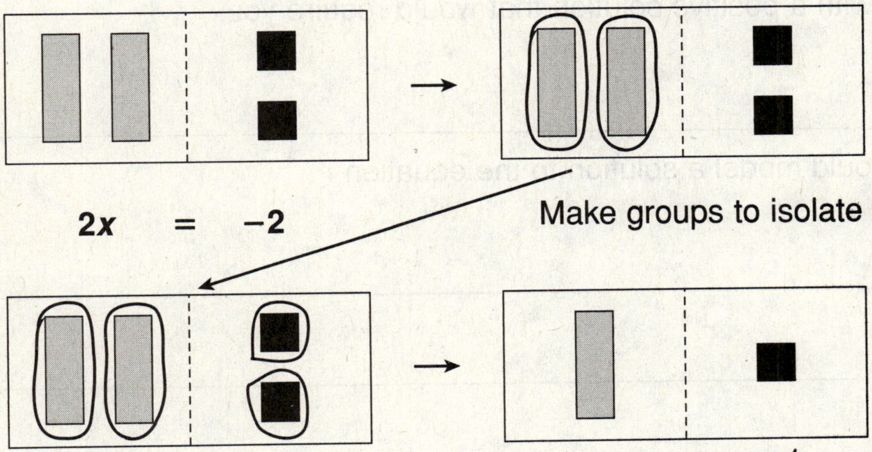

$x = -1$

Make equal groups on each side.

2. Use algebra tiles to model and solve each equation.

a. $3x = -6$ __________ **b.** $4x = -4$ __________

c. $2x = -4$ __________ **d.** $3x = -9$ __________

Name ______________________ Date __________ Class __________

LESSON 11

Hands-On Lab

Model Solving Equations by Dividing, continued

Think and Discuss

1. When you group tiles, how do you know how many groups to make?

____

____

2. How is solving the equation $2x = 4$ different from solving $2x = -4$? How is it the same?

____

____

3. Write an equation with a positive solution that would require you to make 3 groups.

____

4. Explain how you would model a solution to the equation $2x = -6$.

____

____

____

Try This

Use algebra tiles to model and solve each equation.

1. $2x = 6$ __________ 2. $3x = 6$ __________ 3. $4x = 12$ __________

4. $2x = -8$ __________ 5. $4x = -4$ __________ 6. $4x = -12$ __________

7. $3x = -6$ __________ 8. $2x = 8$ __________ 9. $4x = -8$ __________

Name ______________________ Date ____________ Class ____________

LESSON 11

Review for Mastery

Solving Equations by Multiplying or Dividing

To solve a multiplication equation, use division.

Solve $3x = 24$.

$$3x = 24$$
$$\frac{3x}{3} = \frac{24}{3}$$
$$x = 8$$

To solve a division equation, use multiplication.

Solve $\frac{x}{4} = 20$.

$$\frac{x}{4} = 20$$
$$4 \cdot \frac{x}{4} = 20 \cdot 4$$
$$x = 80$$

When an equation has two operations, undo addition or subtraction first. Then undo multiplication or division.

Solve $2x + 11 = 35$.

$$2x + 11 = 35$$
$$\underline{-11} \quad \underline{-11} \quad \text{Undo the addition.}$$
$$2x = 24$$
$$\frac{2x}{2} = \frac{24}{2} \quad \text{Undo the multiplication.}$$
$$x = 12$$

Tell what number you would multiply or divide by to solve the equation.

1. $5a = 60$ ____________

2. $\frac{x}{6} = 12$ ____________

3. $144 = 12f$ ____________

Solve.

4. $6x = 42$

$\frac{6x}{\_\_} = \frac{42}{\_\_}$

$x =$ ____

5. $\frac{a}{3} = 9$

____ $\cdot \frac{a}{3} = 9 \cdot$ ____

$a =$ ____

6. $25 = \frac{k}{5}$

____ $\cdot 25 = \frac{k}{5} \cdot$ ____

____ $= k$

7. $2x + 3 = 11$

____ ____

$\frac{2x}{\_\_} = \frac{\_\_}{\_\_}$

$x =$ ____

8. $\frac{b}{4} + 5 = 6$

____ ____

____ $\cdot \frac{b}{4} =$ ____ $\cdot$ ____

$b =$ ____

9. $5t - 9 = 36$

____

$\frac{5t}{\_\_} = \frac{\_\_}{\_\_}$

$t =$ ____

10. Explain the first step in solving the equation $\frac{y}{5} = 2$.

____

____

Name ______________________________ Date ____________ Class ____________

LESSON 11

Review for Mastery

Solving Equations by Multiplying or Dividing, continued

A student membership to the Culture & Heritage Museums is $20. A student membership is two-fifths the price of a family membership. What is the price of a family membership?

Step 1 **Understand the Problem**

The **answer** is the amount of a family membership.

List the important information:

- A student membership is $20.
- A student membership is $\frac{2}{5}$ of a family membership.

Show the relationship of the information:

Price of student membership	=	Fraction of the price of a family membership	•	Price of a family membership

Step 2 **Make a Plan**

Write an equation and solve it. Let *x* represent the cost of a family membership, and use the equation model.

20	=	$\frac{2}{5}$	•	x

Step 3 **Solve**

$$20 = \frac{2}{5}x$$

$$\left(\frac{5}{2}\right)20 = \left(\frac{5}{2}\right)\frac{2}{5}x \quad \textit{Multiply both sides by } \frac{5}{2}.$$

$$50 = x$$

Step 4 **Look Back**

The problem states that a student membership costs $20, which is two-fifths of a family membership. So, if you estimate that two-fifths is a little less than half, you can see that the $20 student membership is slightly less than the $50 family membership. So, the answer is reasonable.

11. Dylan estimates that he uses $\frac{1}{5}$ of the time he spends away from his house each day to commute back and forth to work. If Dylan spends 2 hours commuting, how much time does he spend away from his house each day?

____

Name ______________________ Date __________ Class __________

LESSON 1I

Homework and Practice

Solving Equations by Multiplying or Dividing

Solve.

1. $-x + 8 = 3$

2. $\frac{y}{3} - 4 = 9$

3. $\frac{n}{-2} = 8$

4. $\frac{a}{-6} = 7$

5. $-9 + 3k = -15$

6. $-9d = 36$

7. $\frac{c}{4} - 8 = -17$

8. $-8 + \frac{b}{5} = -11$

9. $-2x - 19 = 19$

10. $-5x = 65$

11. $\frac{r}{-6} = -4$

12. $\frac{h}{-5} - 17 = -12$

13. $\frac{x}{-6} = -6$

14. $3s - 15 = -15$

15. $-64 = -8v$

16. Peter went camping and rode his mountain bike 52 miles. This was 4 times as many miles as he rides in a normal week. How many miles does Peter ride in a normal week?

17. Gabrielle has saved $792. She started by saving $32 each week for 16 weeks. Then, she increased the amount she saved each week to $40. For how many weeks did Gabrielle save $40?

18. Twelve boys and 15 girls from a local middle school volunteer once a week at a senior center. These numbers represent 15% of the boys and 20% of the girls. What is the total number of students at the middle school?

19. Explain the inverse relationship between multiplication and division.

EXPLORATION

Program of Studies and Core Content

MA-8-AT-S-EI3, MA-8-AT-S-EI4, MA-08-5.1.2, MA-08-5.3.1

1J Introduction to Inequalities

To raise money for a school trip, Angela sells cookies for $2.50 per box. She has to raise at least $100 from the cookie sales in order to go on the trip.

1. Which of the numbers of boxes of cookies shown gives Angela the following amounts of money?

25 boxes
30 boxes
40 boxes
50 boxes

a. Less than her goal of $100

b. Exactly $100

c. More than her goal of $100

2. Name another number of boxes that Angela could sell that would give her less than her goal of $100.

3. Name another number of boxes that Angela could sell that would give her more than her goal of $100.

Think and Discuss

4. Describe the strategies you used to solve Problem 1.

5. Explain how you found numbers of boxes in order to solve Problems 2 and 3.

Name ______________________ Date __________ Class __________

LESSON 1J

Review for Mastery

Introduction to Inequalities

A **solution of an inequality** is a number that makes the inequality true. An inequality usually has more than one solution. All the solutions are contained in the **solution set.**

As with equations, solve a simple inequality by using inverse operations to isolate the variable.

Solve and graph $x + 4 > 9$.

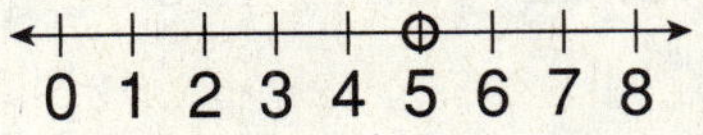

Draw an open circle at 5 to show that 5 is not included in the solution set.

$$\begin{aligned} x + 4 &> 9 \\ -4 \quad & \;\; -4 \quad \text{Subtract 4.} \\ x &> 5 \end{aligned}$$

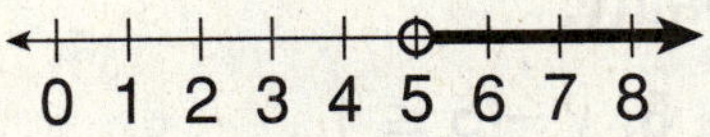

Draw an arrow to the right of 5 to show that all numbers greater than 5 are included in the solutions.

According to the graph, 6 should be a solution and 4 should not be a solution.

Check:

$x + 4 > 9$	$x + 4 > 9$
$6 + 4 \overset{?}{>} 9$	$4 + 4 \overset{?}{>} 9$
$10 > 9$	$8 \not> 9$

So, 6 *is* in the solution set and 4 *is not* in the solution set. Thus, the solution set for the inequality $x + 4 > 9$ is $x > 5$.

Write *true* or *false*.

1. $7 < 4$ ____________

2. $0 \leq 9$ ____________

3. $-3 > 4$ ____________

Using the variable *n*, write the inequality shown by each graph.

4.

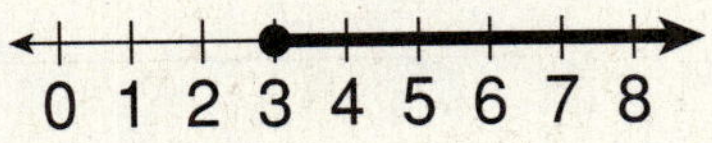

____

5.

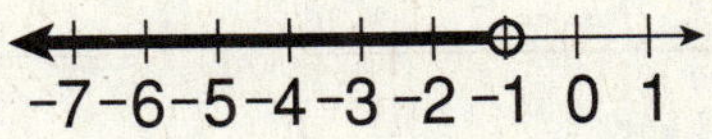

____

Complete. Is the given value in the solution set? Answer *is* or *is not.*

6. 3 ______ in the solution set of $x - 1 > 5$.

$x - 1 > 5$

$\_\_\_ - 1 \overset{?}{>} 5$

$\_\_\_\_\_ > 5$

7. 0 ______ in the solution set of $z + (-4) \geq -4$.

$z + (-4) \geq -4$

$\_\_\_ + (-4) \geq -4$

$\_\_\_ \geq -4$

8. 14 ______ in the solution set of $w + 10 \leq 25$.

$w + 10 \leq 25$

$\_\_\_ + 10 \leq 25$

$\_\_\_ \leq 25$

Name ______________________ Date ____________ Class ____________

LESSON **1J**

Homework and Practice

Introduction to Inequalities

Use <, >, or = to compare each inequality.

1. 8 + 13 ☐ 20

2. 23 ☐ 3(7)

3. 28 − 9 ☐ 18

4. 67 ☐ 9(8)

5. 52 − 37 ☐ 15

6. 78 ☐ 12(6)

Solve and graph each inequality.

7. $x + 3 > 7$

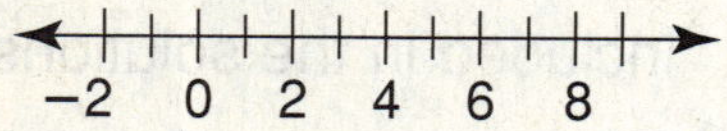

____

8. $y - 5 \leq 1$

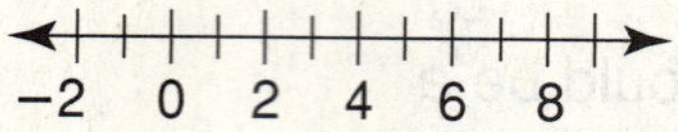

____

9. $4n \geq 20$

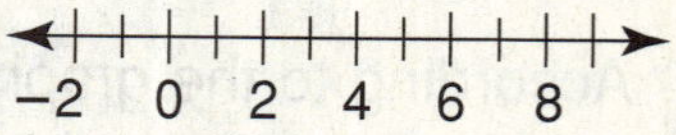

____

10. $h - 1 \geq 6$

____

11. $\frac{x}{5} < 3$

____

12. $\frac{t}{3} \geq 4$

____

13. $9d > 45$

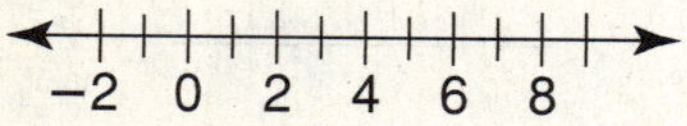

____

14. $17 + m < 23$

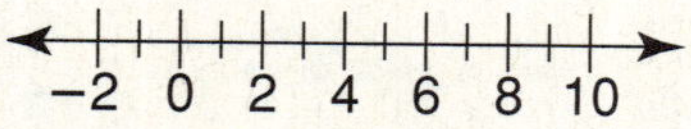

____

15. $128 \geq 16x$

____

16. Philippe wants to drink at least 72 ounces of water every day while at work. He works 6 hours a day. How many ounces of water must Philippe drink each hour? Write and solve an inequality to answer the question.

____

17. Glynn's car has a 18 gallon gas tank. He travels 450 miles. What is the least miles per gallon Glynn's car will get on this trip? Write and solve an inequality to answer the question.

____

CHAPTER 1

Chapter Summary

Answer these questions to summarize the important concepts from Chapter 1 in your own words.

1. Explain why $|7| = |-7|$.

2. Explain how to add integers when the signs are the same and when the signs are different.

3. Explain how to subtract integers.

4. Explain how to multiply and divide two integers.

5. Explain the difference between an open circle and a solid circle when graphing inequalities.

For more review of Chapter 1:

- Complete the Chapter 1 Study Guide and Review on pages 52–54 of your textbook.
- Complete the Ready to Go On quizzes on pages 30 and 48 of your textbook.

Name ______________________________ Date __________ Class __________

CHAPTER 1 **KCCT Prep**

Multiple Choice

1. What is the value of $4m - 4p + 7$ when $m = p$?

   (A) 4 (C) 11
   (B) 7 (D) 15

2. Which equation is an example of the Distributive Property?

   (A) $5(x + 6) = 5x + 5 \cdot 6$
   (B) $(7 + x) + 2 = 7 + (x + 2)$
   (C) $15 + m = m + 15$
   (D) $4(6z) = (4 \cdot 6)z$

3. In the morning, Pablo drives 120 miles. This is one-third of the total distance he needs to drive. Which equation can you solve to find the total distance Pablo must drive?

   (A) $3d = 120$ (C) $d = \frac{1}{3} \cdot 120$
   (B) $\frac{1}{3}d = 120$ (D) $d + \frac{1}{3} = 120$

4. This week, Andrea made \$62 babysitting. She wants the total amount she makes this week and next week to be at least \$100. Which inequality shows the amount y she must make next week?

   (A) $y > 38$ (C) $y < 38$
   (B) $y \geq 38$ (D) $y \leq 38$

5. Which of these has the greatest value when $a = -2$?

   (A) $a - 8$ (C) $a + 10$
   (B) $3a$ (D) $-6a$

6. The solution of the inequality $s + \square \leq 8$ is shown below. Which number is missing from the inequality?

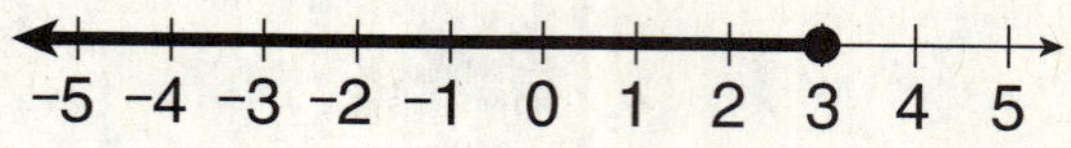

   (A) 3 (C) 8
   (B) 5 (D) 11

7. Which expression has the greatest value when $x = -3$ and $y = -4$?

   (A) $2x - y$ (C) $2y - x$
   (B) $2(x - y)$ (D) $2(y - x)$

8. The solution of the equation $c \square (-4) = 8$ is $c = -32$. Which operation is missing in the equation?

   (A) $\div$ (C) $+$
   (B) $\times$ (D) $-$

9. Suppose $a \neq b$. Which statement about the expression $|a - b|$ must be true?

   (A) Its value is greater than 1.
   (B) Its value is positive.
   (C) Its value is 0.
   (D) Its value is negative.

10. The temperature at 6 A.M. on a winter morning was −8 °F. The solution of the equation $12 = -8 + t$ gives the number of degrees by which the temperature increased from 6 A.M. to noon. What was the temperature at noon?

   (A) −4 °F (C) 12 °F
   (B) −2 °F (D) 20 °F

11. The equation $6 + 2n = 2n + 6$ is an example of which property?

Ⓐ Associative Property
Ⓑ Commutative Property
Ⓒ Distributive Property
Ⓓ Identity Property

12. After downloading 5 new songs to her mp3 player, Carrie's mp3 player had no less than 50 songs. Which inequality represents the original number of songs on her mp3 player?

Ⓐ $s > 45$
Ⓑ $s \geq 45$
Ⓒ $s < 45$
Ⓓ $s \leq 45$

13. What is the value of $pq - 3$ for $p = -4$ and $q = -1$?

Ⓐ -8
Ⓑ -7
Ⓒ -1
Ⓓ 1

14. Shauntay bought a skirt for $32 and a pair of sandals. The solution of the equation $61 = 32 + x$ gives the price of the sandals. What was the total amount that she paid for the skirt and the sandals?

Ⓐ $29
Ⓑ $61
Ⓒ $93
Ⓓ $125

15. Tickets to the school concert cost $8. David sold y tickets and he reached his goal of making at least $120 in ticket sales. Which inequality represents this situation?

Ⓐ $8y \geq 120$
Ⓑ $8y > 120$
Ⓒ $8y \leq 120$
Ⓓ $8y < 120$

Open Response

16. Anita is buying supplies for her office. A stapler costs $7, a pair of scissors costs $4, and a pack of sticky notes costs $3.

a. Anita buys p packs of sticky notes. The total cost is no more than $18. Write an inequality that represents this situation.

b. The total cost of two staplers, one pair of scissors, and one tape dispenser is $24. Write and solve an equation to find the price of the tape dispenser.

17. The highest temperature ever recorded in Kentucky was 114 °F. This was 151 °F greater than the lowest temperature ever recorded in Kentucky.

a. Write an equation to find the lowest temperature ever recorded in Kentucky.

b. Solve your equation from Part a.

c. The lowest temperature ever recorded in Alabama was $-27°$. The lowest temperature ever recorded in Connecticut is halfway between the record lows of Alabama and Kentucky. What was the lowest temperature ever recorded in Connecticut?

18. A rectangle has a length of 12 cm and a width of $(x - 10)$ cm.

a. Write an expression for the area of the rectangle.

b. Show how you can use the Distributive Property to write the expression a different way.

c. Serena claims that when x is any integer greater than 10, the area of the rectangle is an even number. Do you agree or disagree? Why?

CHAPTER 2

Project

Rational Numbers: Eat Up

Career: Nutritionist

Nutritionists use their knowledge of the nutrient content of food to help promote healthful eating. Together with food scientists, they develop guidelines for people who must follow medically necessary diets as well as for people who just want to improve their eating habits.

Before you start, work through the Food for Thought worksheet on the next page.

In this chapter project . . .

You calculate the amount of nutrients you would get from a junk food diet.

What to do:

- Research the nutrition requirements of protein, iron, and calcium for your age and gender.
- Make a list of healthy food items and use the nutrition labels to find out how many servings of each you would need to eat to satisfy your daily requirements.
- Make a list of junk food—such as chips, candy bars, and snack foods—and use the nutrition labels to find out how many servings of each you would need to satisfy your daily requirements.

What to present:

- A table displaying your daily requirements.
- A bar graph showing the number of servings you would need of healthy food and junk food to satisfy your requirements.
- A table showing the differences in calories, sugar, and fats from the two diets.

Where's the math?

- Use rational numbers to calculate the number of servings you need from healthy items and junk food items.
- Use a bar graph to compare the number of servings for items in each diet.

go.hrw.com
Chapter Project Online
KEYWORD: MT7 Ch2

Name ______________________ Date __________ Class __________

CHAPTER 2

Project Recording Sheet

Rational Numbers

Food For Thought

Nutritionists work in many situations in the development and use of foods. One of the nutrients that have been studied is sodium, which is part of common table salt. The recommendations for daily amounts of sodium are: Girls and Boys 9–13: 1900–3000 mg; Girls 14–18: 3000 mg; Boys 14–18: 2400–2700 mg

The chart lists the sodium content of some common food products.

Prepared Food	**Sodium Content** (mg)	**Prepared Food**	**Sodium Content** (mg)
Barbeque beans	570	Oatmeal cookies	150
Chicken noodle soup	890	Orange juice	0
Chili, canned	1220	Pasta sauce	500
Ice cream	35	Peas, canned	390
Cranberry juice	35	Peas, frozen	105
Donuts, powdered	280	Potato chips	180
French fries, frozen	270	Spaghetti, dry	10
Fruit cocktail	10	Sports drink	110
Hot dogs, beef	570	Turkey, sliced	300
Ketchup	190	Vegetable juice	630
Mustard	55	Waffle, frozen	420

Create some meals that meet the following requirements:

- 100% of the daily sodium requirement for your gender and age
- a low-sodium meal with $\frac{1}{2}$ of the normal requirement
- a meal with $\frac{3}{4}$ of the normal requirements
- a breakfast that contains about $\frac{1}{3}$ of your requirements
- a meal that contains at least twice as much sodium as the recommended daily requirement

Research: What other nutrients are important for good health? What are the daily minimum requirements? Create a chart that shows your results.

EXPLORATION

Program of Studies and Core Content
MA-8-NPO-S-NS3, MA-08-1.1.2, MA-08-1.1.3

2A Rational Numbers

Comparing numbers with $\frac{1}{2}$ is useful in many situations.

Phil surveyed 317 voters and found that 156 supported his reelection. He reasoned the following way to determine whether he had a majority.

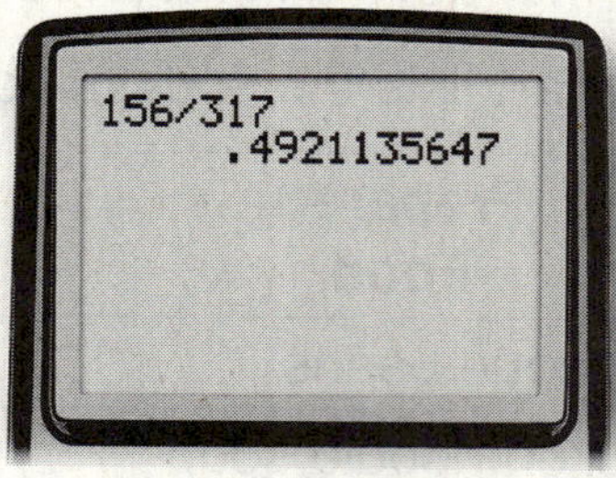

$$156 \times 2 = 150 \times 2 + 6 \times 2$$
$$= 300 + 12$$
$$= 312$$

Phil concluded that he did not have a majority. Look at how Phil checked his estimate with a calculator.

Double the numerator and compare it with the denominator of each fraction to determine whether the fraction is greater than or less than $\frac{1}{2}$. Check your work with a calculator.

	Fraction	$> \frac{1}{2}$?	$< \frac{1}{2}$?
1.	$\frac{51}{101}$		
2.	$\frac{221}{425}$		
3.	$\frac{260}{513}$		
4.	$\frac{578}{1152}$		

Think and Discuss

5. Explain how you checked your work with a calculator.

Name ______________________________ Date ____________ Class ____________

LESSON 2A

Review for Mastery

Rational Numbers

A **rational number** is a number that can be written as a fraction $\frac{x}{y}$, where x and y are integers and $y \neq 0$. Rational numbers include integers, fractions, decimals that repeat, and decimals that terminate.

To simplify a fraction, make the numerator and the denominator **relatively prime** (the only common factor between the numerator and denominator is 1).

Simplify $\frac{14}{21}$.

Method 1: Modeling

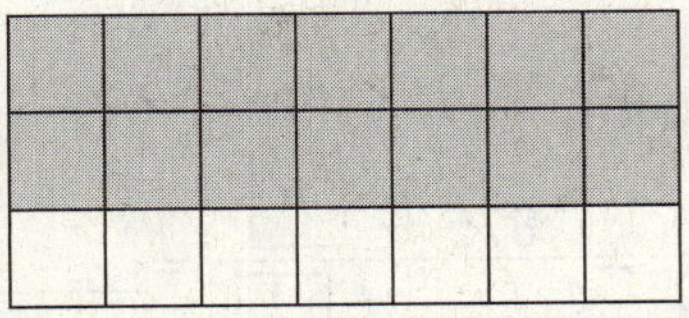

14 out of the 21 boxes are shaded.

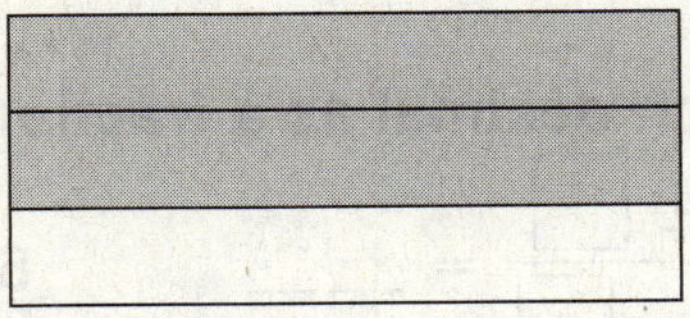

2 out of the 3 boxes are shaded.

Since the same area of both models is shaded, $\frac{14}{21} = \frac{2}{3}$.

Method 2: Greatest Common Factor

Rewrite the numerator and the denominator using prime factors. Then remove the common factors.

$$\frac{14}{21} = \frac{2 \cdot \cancel{7}}{3 \cdot \cancel{7}} = \frac{2}{3}$$

Simplify each fraction using models.

1. $\frac{6}{9}$

2. $\frac{10}{12}$

Simplify each fraction by finding the GCF.

3. $\frac{16}{24}$

4. $\frac{8}{48}$

Name ______________________ Date __________ Class __________

CHAPTER 2A

Review for Mastery

Rational Numbers, continued

To convert a decimal to a fraction, identify the place value of the digit farthest to the right. Then write a fraction using the digits to the right of the decimal point as the numerator and the place value as the denominator.

Convert 1.05 to a fraction.

$1.05 = 1\frac{5}{100}$ *5 is in the hundredths place.*

$= 1\frac{1}{20}$ *Divide by the GCF, 5.*

Complete to write each decimal as a fraction in simplest form.

5. $0.25 = \frac{25}{100} = \frac{25 \div \square}{100 \div \square} = \_\_\_\_$

6. $0.375 = \frac{375}{1000} = \frac{375 \div \square}{1000 \div \square} = \_\_\_\_$

Write each decimal as a fraction in simplest form.

7. 0.55 = __________

8. 0.32 = __________

To convert a fraction to a decimal, divide the numerator by the denominator.

Convert $\frac{3}{4}$ to a decimal. Convert $\frac{1}{3}$ to a decimal.

$$\frac{3}{4} = 4\overline{)3.00} = 0.75$$

```
    0.75
4)3.00
 -28↓
   20
  -20
    0
```

$$\frac{1}{3} = 3\overline{)1.00} = 0.\overline{3}$$

```
    0.3̄
3)1.00
  -9↓
   10
   -9
    1
```

Complete to write each fraction as a decimal.

9. $\frac{15}{4} = 4\overline{)15.00}$

10. $\frac{5}{6} = 6\overline{)5.00}$

11. $\frac{11}{3} = 3\overline{)11.00}$

Name ______________________ Date ____________ Class ____________

LESSON 2A

Homework and Practice

Rational Numbers

Simplify.

1. $\frac{6}{12}$ **2.** $\frac{6}{24}$ **3.** $\frac{12}{36}$ **4.** $-\frac{8}{32}$

5. $\frac{15}{27}$ **6.** $-\frac{16}{24}$ **7.** $-\frac{12}{96}$ **8.** $\frac{12}{54}$

Write each decimal as a fraction in simplest form.

9. 0.56 **10.** 3.2 **11.** 0.036 **12.** −2.05

13. 3.502 **14.** 0.064 **15.** −8.4 **16.** 0.004

17. −0.70 **18.** 5.25 **19.** 0.128 **20.** 0.0004

Write each fraction as a decimal.

21. $\frac{3}{8}$ **22.** $\frac{5}{9}$ **23.** $\frac{21}{8}$ **24.** $\frac{17}{18}$

25. $\frac{19}{3}$ **26.** $\frac{11}{20}$ **27.** $\frac{7}{5}$ **28.** $\frac{54}{36}$

29. Make up a fraction that cannot be simplified that has 25 as its denominator.

EXPLORATION

Program of Studies and Core Content

MA-8-NPO-NS2, MA-8-NPO-S-NS3, MA-08-1.1.2, MA-08-1.1.3

2B Comparing and Ordering Rational Numbers

You can use a number line to compare rational numbers. For example, the number line shows that $\frac{1}{8} < \frac{1}{4}$ because $\frac{1}{8}$ is to the left of $\frac{1}{4}$.

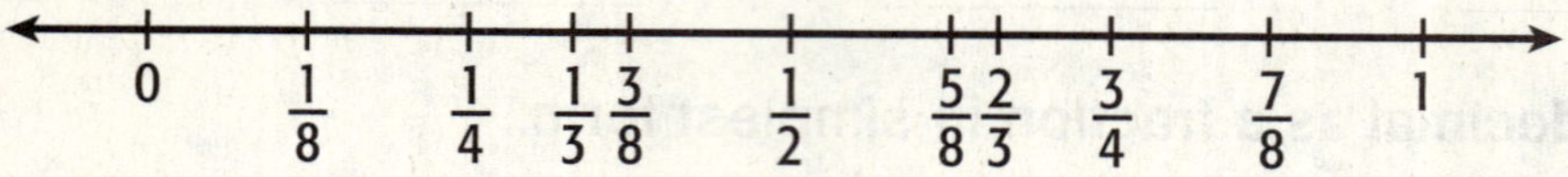

Use the number line to compare each pair of fractions by writing < or >. Then convert the fractions to decimals and use < or > to compare the decimals.

1. $\frac{5}{8}$ ☐ $\frac{2}{3}$

2. $\frac{3}{8}$ ☐ $\frac{1}{3}$

3. $\frac{7}{8}$ ☐ $\frac{3}{4}$

4. $\frac{3}{8}$ ☐ $\frac{2}{3}$

5. $\frac{3}{4}$ ☐ $\frac{2}{3}$

6. $\frac{1}{4}$ ☐ $\frac{1}{3}$

Think and Discuss

7. Explain how you know that $\frac{3}{8} < \frac{5}{8}$ without using the number line.

8. Describe how you could use the number line to compare $5\frac{3}{8}$ and $5\frac{1}{2}$.

Name ______________________ Date ____________ Class ____________

LESSON 2B

Review for Mastery

Comparing and Ordering Rational Numbers

Compare $-\frac{1}{2}$ and $-\frac{3}{4}$ using a number line. Write <, >, or =.

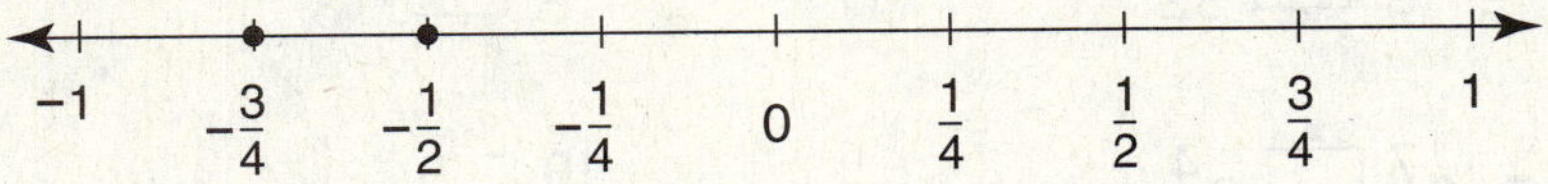

$-\frac{3}{4} < -\frac{1}{2}$

Compare using the number line above. Write <, >, or =.

1. $-\frac{1}{3}$ ☐ $-\frac{2}{3}$

2. $\frac{1}{2}$ ☐ $\frac{1}{4}$

Compare $\frac{3}{8}$ and $\frac{5}{6}$ by finding common denominators. Write <, >, or =.

Method 1: Multiply to find a common denominator

$\frac{3}{8} \bullet \frac{6}{6} = \frac{18}{48}$ *Multiply 6 and 8 to find a common denominator.*

$\frac{5}{6} \bullet \frac{8}{8} = \frac{40}{48}$ *Write the fractions with a common denominator.*

$\frac{18}{48} < \frac{40}{48}$, so $\frac{3}{8} < \frac{5}{6}$ *Compare the fractions.*

Method 2: Find the least common denominator

$\frac{3}{8} \bullet \frac{3}{3} = \frac{9}{24}$ *List multiples of 6 and 8. The LCM is 24.*

$\frac{5}{6} \bullet \frac{4}{4} = \frac{20}{24}$ *Write the fractions with a common denominator.*

$\frac{9}{24} < \frac{20}{24}$, so $\frac{3}{8} < \frac{5}{6}$ *Compare the fractions.*

Compare by finding a common denominator. Write <, >, or =.

3. $\frac{4}{9}$ ☐ $\frac{5}{12}$

4. $\frac{7}{15}$ ☐ $\frac{8}{21}$

Compare 0.7 and $\frac{7}{9}$ by using decimals. Write <, >, or =.

$\frac{7}{9} = 0.\overline{7}$ *Write $\frac{7}{9}$ as a decimal.*

$0.7 < 0.\overline{7}$, so $0.7 < \frac{7}{9}$ *Compare the decimals.*

Compare by using decimals. Write <, >, or =.

5. $2\frac{3}{5}$ ☐ $2\frac{4}{7}$

6. 4.08 ☐ $4\frac{4}{5}$

Name ______________________________ Date ____________ Class ____________

LESSON 2B

Homework and Practice

Comparing and Ordering Rational Numbers

Compare. Write <, >, or =.

1. $-\frac{2}{3}$ ☐ $-\frac{1}{4}$

2. 0.4 ☐ $\frac{15}{32}$

3. $1\frac{3}{5}$ ☐ $1\frac{2}{3}$

4. $2\frac{11}{7}$ ☐ $3\frac{4}{7}$

5. $6\frac{7}{8}$ ☐ $6\frac{4}{5}$

6. $\frac{10}{13}$ ☐ 0.09

7. $-5\frac{1}{5}$ ☐ $-5\frac{2}{7}$

8. $\frac{5}{12}$ ☐ $\frac{7}{16}$

9. 12.8 ☐ $12\frac{4}{5}$

10. −0.08 ☐ $-\frac{1}{125}$

11. $7\frac{3}{8}$ ☐ 7.375

12. $-\frac{8}{15}$ ☐ $-\frac{12}{19}$

13. Write a fraction that has a value between $-\frac{1}{4}$ and $-\frac{1}{2}$.

____

14. The approximate weights of various species of Wren, including the South Carolina state bird, the Carolina Wren, are shown in the table at the right.

a. List the birds in order from least to greatest weight.

____

b. The Rock Wren weighs about $\frac{3}{5}$ oz. Where would you place the Rock Wren in your list of weights from part **a**?

____

Bird	Weight (oz.)
Bewick's Wren	$\frac{3}{10}$
Canyon Wren	0.32
Carolina Wren	$\frac{16}{25}$
House Wren	$\frac{7}{20}$
Marsh Wren	0.49

EXPLORATION

Program of Studies and Core Content

MA-8-NPO-S-NO1, MA-8-NPO-S-NS3, MA-08-1.1.2, MA-08-1.3.1, MA-08-5.2.2

2C Adding and Subtracting Rational Numbers

Beckie has \$454.96 in a checking account. She needs to pay bills in the amounts \$25.95, \$313.00, \$45.76, and \$87.95.

Beckie estimates the following:

Account: \$454.96 ≈ \$455.00

Bills:
\$25.95 ≈ \$ 25.00
\$313.00 ≈ \$310.00
\$45.76 ≈ \$ 45.00
\$87.95 ≈ \$ 90.00
\$470.00

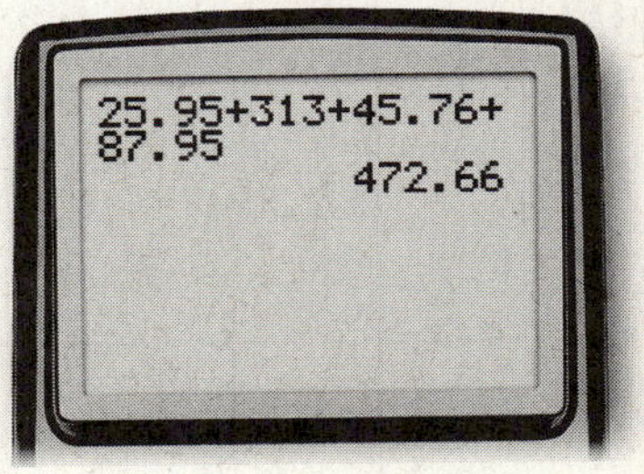

Beckie determines that she does not have enough money in her account to pay her bills. She then checks her estimate with a calculator.

Estimate the solution to each expression in the table. Then use a calculator to solve.

		Estimate	Actual
1.	120 − 9.8		
2.	45 − 17.8 + 15.9 + 16.1 − 1.07		
3.	88.10 + 109.85		
4.	34.12 − 18.30 + 65.25		

Think and Discuss

5. Describe the estimation strategies you used.

Name ________________________________ Date ______________ Class ______________

LESSON **2C**

Hands-On Lab

Adding and Subtracting Rational Numbers

Activity

Susi is a dolphin who likes to jump. She jumps $5\frac{1}{2}$ feet out of the water. Then she falls back into the water, covering a total vertical distance of $9\frac{1}{4}$ feet. How far is Susi now below the water's surface?

Draw Susi's jump.

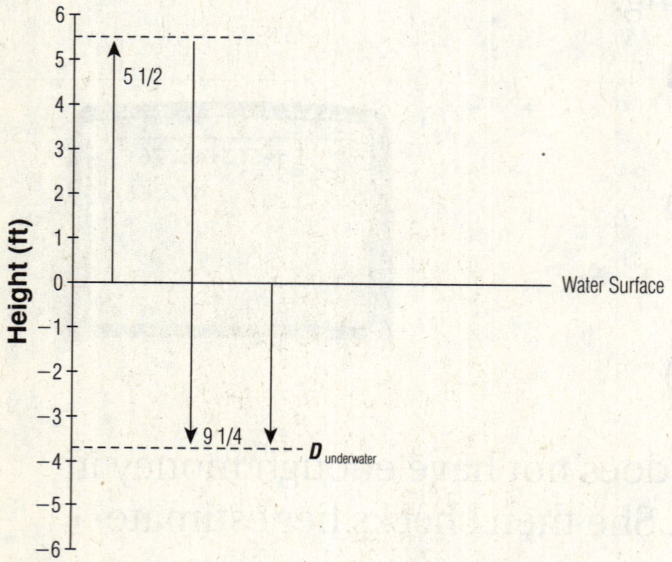

The graph shows that Susi is $3\frac{3}{4}$ feet under water.

Check your graph by writing and evaluating an expression for Susi's jump:

$$D_{\text{underwater}} = 5\frac{1}{2} + \left(-9\frac{1}{4}\right) = \frac{11}{2} - \frac{37}{4} = \frac{22}{4} - \frac{37}{4} = -\frac{15}{4} = -3\frac{3}{4}$$

Name ______________________ Date ____________ Class ____________

LESSON **2C**

Hands-On Lab

Adding and Subtracting Rational Numbers, continued

Think and Discuss

Susi continues to dive an additional 50 feet to the bottom of the ocean. How deep is the ocean? Explain.

____

____

____

Try This

A kingfisher dives from a height of 30 feet to catch a fish. The bird's dive covers a total vertical distance of $40\frac{1}{3}$ feet. Make a graph to describe the situation. How far is the fish under water? Check your graph by writing and evaluating an expression for the kingfisher's dive.

____

____

____

Name ______________________________ Date ___________ Class ___________

LESSON **2C**

Review for Mastery

Adding and Subtracting Rational Numbers

To add fractions that have the same denominator:

- Use the common denominator as the denominator for the sum.
- Add the numerators to get the numerator of the sum.
- Write the sum in simplest form.

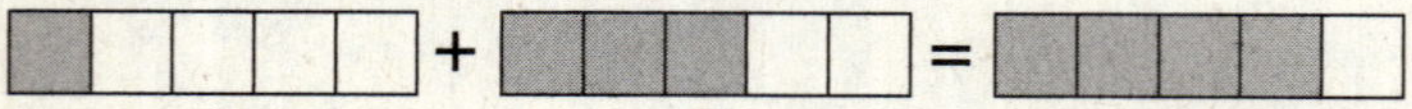

$\frac{1}{5}$ $\frac{3}{5}$ $\frac{1}{5} + \frac{3}{5} = \frac{4}{5}$

To subtract fractions that have the same denominator:

- Use the common denominator as the denominator for the difference.
- Subtract the numerators.
- Write the difference in simplest form.

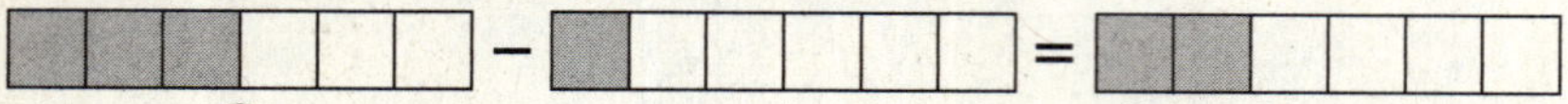

$\frac{3}{6}$ $\frac{1}{6}$ $\frac{3}{6} - \frac{1}{6} = \frac{2}{6}$ or $\frac{1}{3}$

Use fraction bars to add or subtract the fractions.

1. $\frac{3}{14} + \frac{4}{14} =$ _____ $=$ _____

2. $\frac{2}{10} + \left(-\frac{4}{10}\right) =$ _____ $=$ _____

3. $\frac{8}{9} - \frac{2}{9} =$ _____ $=$ _____

4. $\frac{9}{15} - \left(-\frac{3}{15}\right) =$ _____ $=$ _____

To add or subtract decimals, line up the decimal points and then add or subtract from right to left.

$$\begin{array}{r} 12.83 \\ +24.17 \\ \hline 37.00 \end{array} \qquad \begin{array}{r} 35.78 \\ -14.55 \\ \hline 21.23 \end{array}$$

Complete to add or subtract the decimals.

5. 44.02 + 8.07 = __________

6. 1.39 + 13.6 = __________

7. 33.47 − 0.6 = __________

8. 25.15 − 25.06 = __________

Name ______________________ Date __________ Class __________

LESSON **2C**

Homework and Practice

Adding and Subtracting Rational Numbers

Use a number line to find each sum.

1. $-0.3 + 0.8$

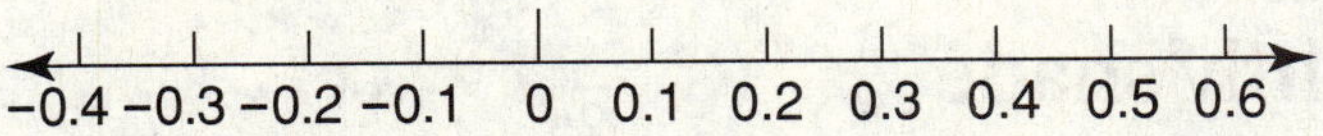

____

2. $-\frac{4}{5} + \frac{3}{5}$

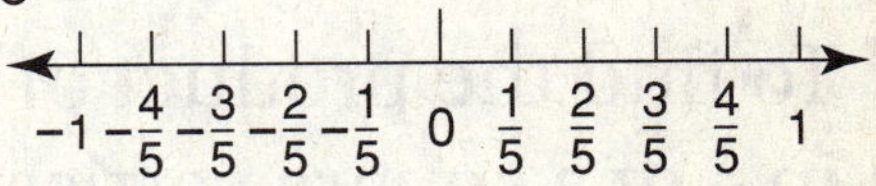

____

Use fraction bars to add or subtract the fractions.

3. $\frac{5}{7} + \left(-\frac{3}{7}\right)$ ______________

4. $\frac{3}{8} - \left(-\frac{4}{8}\right)$ ______________

Add or subtract. Write answers in simplest form.

5. $\frac{5}{7} + \frac{1}{7}$ ______________

6. $-\frac{1}{8} - \frac{5}{8}$ ______________

7. $\frac{5}{12} - \frac{1}{12}$ ______________

8. $\frac{3}{16} + \frac{5}{16}$ ______________

9. $\frac{2}{15} - \frac{8}{15}$ ______________

10. $\frac{7}{20} - \frac{3}{20}$ ______________

11. $-\frac{5}{18} - \frac{7}{18}$ ______________

12. $-\frac{3}{25} + \frac{11}{25}$ ______________

Evaluate each expression for the given value of the variable.

13. $92.7 + x$ for $x = 4.8$ ______________

14. $27.3 + x$ for $x = -9.5$ ______________

15. $\frac{5}{24} + x$ for $x = -\frac{7}{24}$ ______________

16. Benjamin received a cell phone bill that was $25.95 plus a late fee of $0.75. He wrote a check to cover the total expense. When he mailed it, Benjamin put a $0.37 stamp on the letter. What was the total cost for Benjamin to pay and mail his cell phone bill?

____

17. Dannika is preparing a recipe that calls for $2\frac{3}{4}$ cups of flour. If Dannika has only $2\frac{1}{4}$ cups left in the flour container, how much flour must she add from a new bag to complete the recipe?

____

EXPLORATION

Program of Studies and Core Content

MA-8-NPO-S-NO1, MA-8-NPO-S-NS3, MA-08-1.1.2, MA-08-1.3.1

2D Multiplying Rational Numbers

To find the product of $\frac{1}{2} \cdot \frac{1}{3}$, lightly shade in $\frac{1}{2}$ of a square vertically. Use a darker shade to fill in $\frac{1}{3}$ of the square horizontally. The product is represented by the area where the shading overlaps.

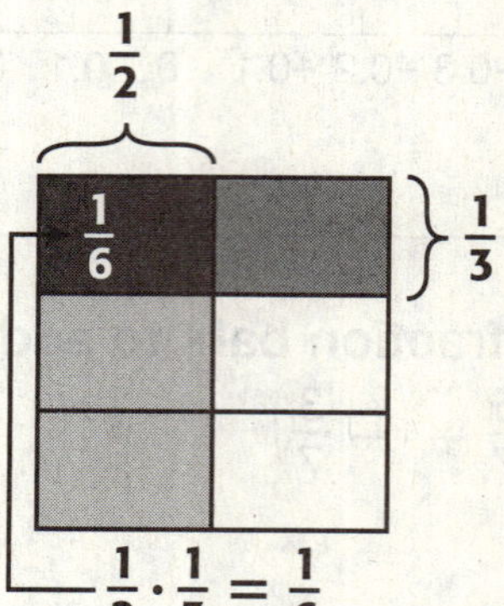

Use a similar model to find each product.

1. $\frac{1}{2} \cdot \frac{4}{6}$

2. $\frac{2}{3} \cdot \frac{3}{4}$

To multiply $2 \cdot 1\frac{1}{2}$, shade in 2 squares across and $1\frac{1}{2}$ down.

$$2 \cdot 1\frac{1}{2} = 3$$

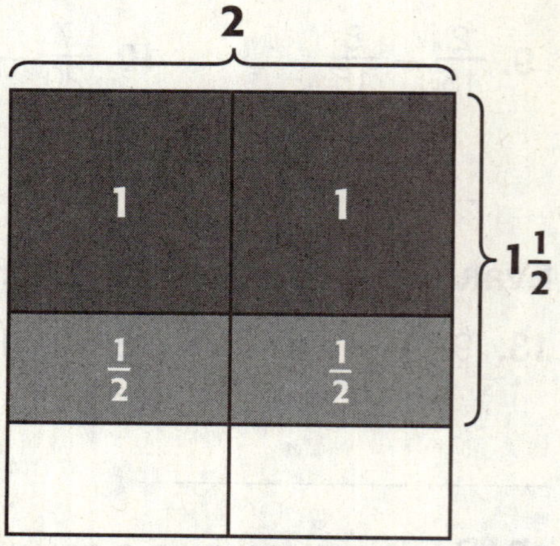

Use a similar model to find each product.

3. $3 \cdot 4\frac{1}{2}$

4. $8\frac{1}{4} \cdot 4$

Think and Discuss

5. Explain how the model shows that $2.5 \cdot 2.5 = 6.25$.

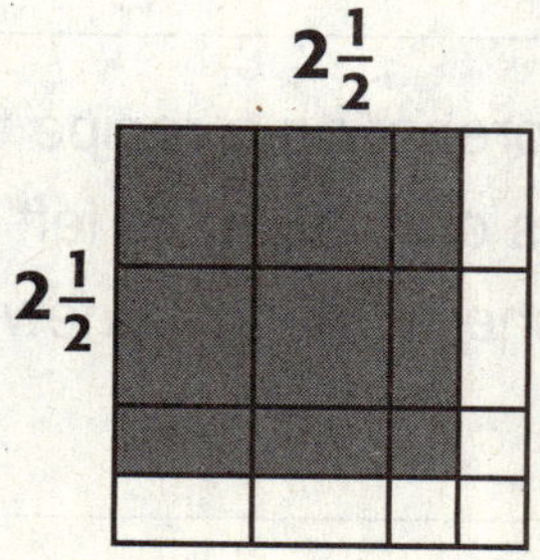

Name ______________________ Date __________ Class __________

LESSON 2D

Review for Mastery

Multiplying Rational Numbers

To model $\frac{1}{3} \times \frac{3}{4}$:

Divide a square into 4 equal parts. Lightly shade 3 of the 4.

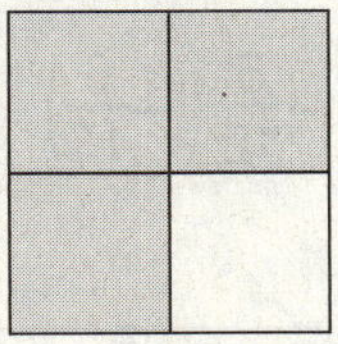

Darken 1 of the 3 shaded parts.

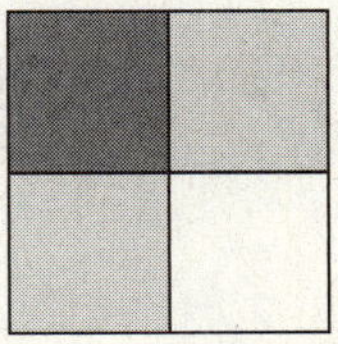

Compare the 1 darkened part to the original 4.

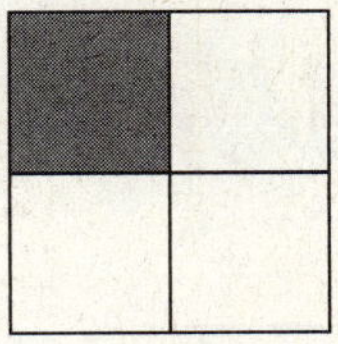

$\frac{1}{3} \times \frac{3}{4} = \frac{1}{4}$

Model each multiplication. Write the result.

1. $\frac{1}{2} \times \frac{2}{4} =$ ______

2. $\frac{3}{4} \times \frac{4}{6} =$ ______

3. $\frac{2}{3} \times \frac{3}{9} =$ ______

To multiply fractions:

- Cancel common factors, one in a numerator and the other in a denominator.
- Multiply the remaining factors in the numerator and in the denominator.
- If the signs of the factors are the same, the product is positive. If the signs of the factors are different, the product is negative.

$\frac{\cancel{3}^{1}}{\cancel{4}_{1}} \times \frac{\cancel{8}^{2}}{\cancel{9}_{3}} = \frac{1 \times 2}{1 \times 3} = \frac{2}{3}$

Multiply. Answer in simplest form.

4. $\frac{1}{2} \times \frac{4}{9} =$ ______

5. $\frac{2}{3} \times \frac{6}{7} =$ ______

6. $\frac{3}{5} \times \frac{15}{17} =$ ______

7. $\frac{2}{3} \times \left(-\frac{9}{10}\right) =$ ______

8. $\left(-\frac{2}{9}\right) \times \frac{27}{40} =$ ______

9. $\left(-\frac{4}{7}\right) \times \left(-\frac{21}{8}\right) =$ ______

Name ______________________ Date ____________ Class ____________

LESSON 2D

Homework and Practice

Multiplying Rational Numbers

Multiply. Write each answer in simplest form.

1. $6\left(\frac{2}{3}\right)$ ________
2. $-9\left(\frac{1}{3}\right)$ ________
3. $-8\left(-\frac{3}{16}\right)$ ________
4. $-14\left(\frac{3}{7}\right)$ ________
5. $\frac{5}{12}\left(-\frac{6}{25}\right)$ ________
6. $-\frac{1}{12}\left(\frac{18}{5}\right)$ ________
7. $\frac{1}{3}\left(\frac{18}{2}\right)$ ________
8. $-\frac{5}{12}\left(-\frac{6}{15}\right)$ ________

Complete the table.

	x	$\frac{2}{3} \cdot x$	Is the product less than, equal to, or greater than $\frac{2}{3}$?
9.	$-\frac{3}{2}$		
10.	$-\frac{2}{3}$		
11.	0		
12.	$\frac{2}{5}$		
13.	$\frac{1}{4}$		
14.	$\frac{3}{2}$		
15.	$\frac{7}{4}$		

Use your answers to Exercises 9–15 to help you answer Exercises 16–17.

16. What is the effect of multiplying a positive rational number by a number less than 0? By a number between 0 and 1? By a number greater than 1?

____

____

17. How would your answers to 16 change if the rational number being multiplied were negative?

____

EXPLORATION

Program of Studies and Core Content
MA-8-NPO-S-NO1, MA-8-NPO-S-NS3, MA-08-1.1.2, MA-08-1.3.1

2E Dividing Rational Numbers

You can use your calculator to investigate division of rational numbers.

1. Use your calculator to complete the table by dividing the rational numbers. The first one has been filled in as an example.

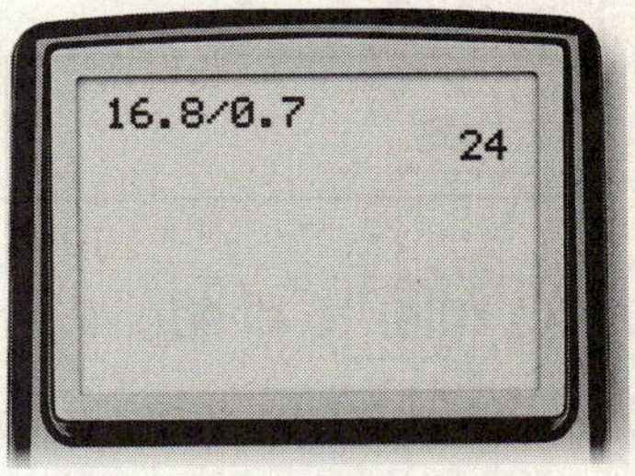

$\frac{16.8}{0.7} = 24$	$\frac{168}{7} =$	$\frac{1680}{70} =$
$\frac{4.1}{0.5} =$	$\frac{41}{5} =$	$\frac{410}{50} =$
$\frac{0.026}{0.04} =$	$\frac{0.26}{0.4} =$	$\frac{2.6}{4} =$
$\frac{75}{0.06} =$	$\frac{750}{0.6} =$	$\frac{7500}{6} =$
$\frac{16.32}{1.02} =$	$\frac{163.2}{10.2} =$	$\frac{1632}{102} =$

2. How are the division problems in each row related to each other?

Think and Discuss

3. Show how you can use the pattern in the first row of the table to write a new division problem whose quotient is equal to 24.

4. Explain how to write a division problem that is equivalent to $145 \div 16.5$ but that involves dividing by a whole number.

Name ______________________ Date __________ Class __________

LESSON **2E**

Hands-On Lab

Dividing Rational Numbers

You can use a ruler to divide rational numbers, both in fractional and in decimal form.

Activity

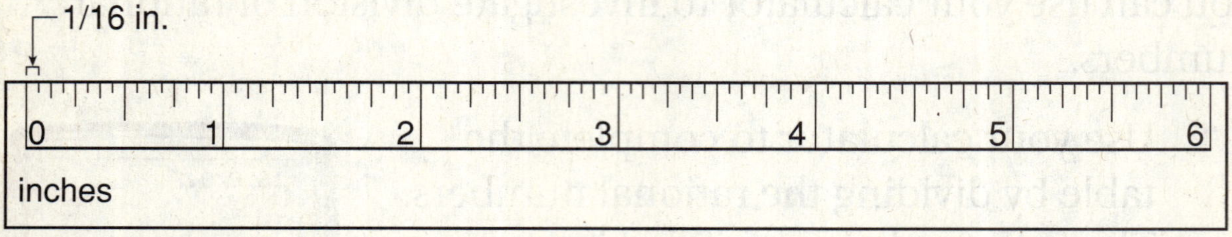

Use a ruler to divide $4\frac{1}{2}$ by $\frac{3}{4}$.

Use the ruler to draw a thin box $4\frac{1}{2}$ in. long and half an inch high. Measure $\frac{3}{4}$ in. from the left end of the box and place a vertical dividing line. Measure another $\frac{3}{4}$ in. from the dividing line and place a second dividing line. Continue measuring till you get to the end of the box. Did you come out even? How many segments did you divide the box into?

____

For decimal numbers, use a metric ruler to divide 13.3 by 1.9. Draw a box 13.3 cm long and 1 cm high. Use the metric ruler to divide it into segments 1.9 cm long. How many segments did you divide the box into?

____

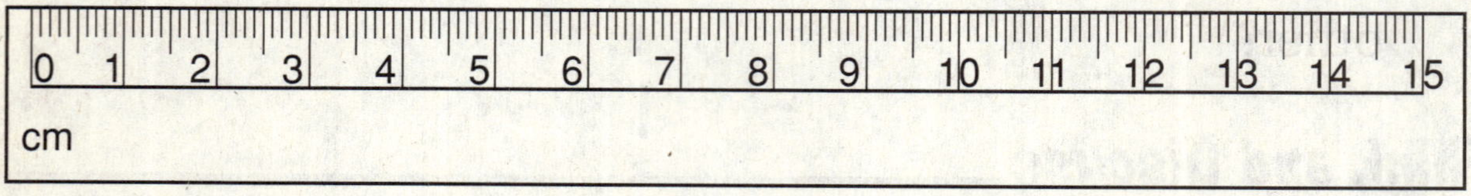

Name ______________________ Date __________ Class __________

LESSON 2E

Hands-On Lab

Dividing Rational Numbers, continued

Think and Discuss

1. When a positive number is divided by a positive rational number, does the number become smaller or larger? Under what circumstances will it become smaller or larger?

____

____

____

2. You should "divide by a fraction" by changing it to a multiplication problem. Multiply by the reciprocal of the fraction. Could you use that principle when dividing by 0.3?

____

____

Try This

Using the metric ruler draw a box 12 cm long. Make it 4 cm tall, with horizontal lines along the length dividing it into 4 sections, one on top of the other. Divide the first section into segments of 0.8 cm, the second into sections of 1.2 cm, the third into sections of 1.5 cm, and the fourth into sections of 2.0 cm. At the end of each row of segments write the division it represents: "$12 \div 0.8 = ?$", and so on.

Name ______________________ Date ____________ Class ____________

LESSON 2E

Review for Mastery

Dividing Rational Numbers

To write the **reciprocal** of a fraction, interchange the numerator and denominator.

The product of a number and its reciprocal is 1.

$\frac{2}{3} \leftrightarrow \frac{3}{2}$

Fraction Reciprocal

$\frac{2}{3} \times \frac{3}{2} = 1$

Write the reciprocal of each rational number.

1. The reciprocal of $\frac{3}{5}$ is: ____________

2. The reciprocal of 6 is: ____________

3. The reciprocal of $2\frac{1}{3}$ is: ____________

To divide by a fraction, multiply by its reciprocal.

$\frac{2}{3} \div 6$

$\frac{2}{3} \times \frac{1}{6}$

$\frac{2 \times 1}{3 \times 6} = \frac{2}{18} = \frac{1}{9}$

$\frac{3}{5} \div \frac{9}{10}$

$\frac{3}{5} \times \frac{10}{9}$

$\frac{\cancel{3}^{1} \times \cancel{10}^{2}}{\cancel{5}_{1} \times \cancel{9}_{3}} = \frac{2}{3}$

Complete to divide and simplify.

4. $\frac{3}{8} \div 12 = \frac{3}{8} \times$ ______ = ______

5. $\frac{4}{3} \div 16 =$ ______

6. $\frac{5}{7} \div \frac{20}{21} = \frac{5}{7} \times$ ______ = ______

7. $-\frac{3}{4} \div \left(\frac{9}{8}\right) = -\frac{3}{4} \times$ ______ = ______

Change a decimal divisor to a whole number. Using the number of places in the divisor, move the decimal point to the right in both the divisor and the dividend.

$0.7\overline{)4.34} \rightarrow 0.7.\overline{)4.3.4} \rightarrow \overset{6.2}{7\overline{)43.4}}$

Rewrite each division with a whole-number divisor. Then, do the division.

8. $0.6\overline{)1.14} \rightarrow$ ____________ = ______

9. $0.3\overline{)4.56} \rightarrow$ ____________ = ______

10. $0.02\overline{)7.12} \rightarrow$ ____________ = ______

11. $0.08\overline{)57.28} \rightarrow$ ____________ = ______

Name ______________________ Date __________ Class __________

LESSON 2E

Homework and Practice

Dividing Rational Numbers

Divide. Write each answer in simplest form.

1. $\frac{1}{4} \div \frac{5}{8}$

2. $-\frac{5}{6} \div \frac{5}{12}$

3. $-\frac{2}{5} \div \frac{7}{10}$

4. $-\frac{3}{14} \div \left(-\frac{6}{7}\right)$

5. $\frac{7}{9} \div \left(\frac{5}{6}\right)$

6. $\frac{9}{20} \div \left(-\frac{3}{4}\right)$

7. $-\frac{11}{12} \div \left(-\frac{44}{54}\right)$

8. $-\frac{13}{30} \div \left(\frac{39}{45}\right)$

Divide.

9. $32 \div 0.4$

10. $-6.58 \div 0.08$

11. $7.26 \div (-0.03)$

12. $3.333 \div 0.66$

13. $0.0096 \div (-1.2)$

14. $-25.28 \div 1.6$

15. $17.5 \div 0.07$

16. $279.4 \div 12.7$

Evaluate $\frac{18}{x}$ for each value of x.

17. $x = 4.5$

18. $x = 1.25$

19. $x = 0.5$

20. $x = 0.04$

21. What is the effect of dividing a positive rational number by a rational number greater than 1?

22. What is the effect of dividing a positive rational number by a rational number between 0 and 1?

23. Can a rational number be divided by 0? Explain.

24. Alison is installing a hard wood floor. If the room is $12\frac{1}{2}$ ft wide and each plank of wood is $3\frac{1}{4}$ in. wide, how many planks of wood does she need to cover the width of the room?

EXPLORATION

Program of Studies and Core Content

MA-8-NPO-S-NS3, MA-08-1.1.2, MA-08-1.3.1

2F Adding and Subtracting with Unlike Denominators

You can use models to show addition and subtraction of fractions with unlike denominators. Look at the models for $\frac{1}{2} + \frac{1}{3}$ and $\frac{1}{2} - \frac{1}{3}$.

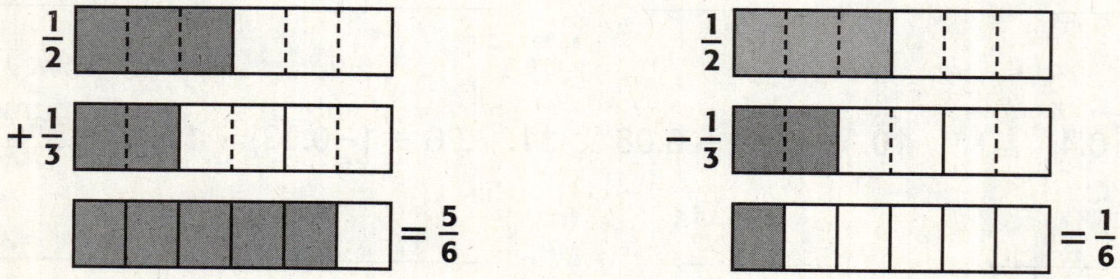

1. Draw a picture to show that $\frac{1}{2} + \frac{4}{5} = 1\frac{3}{10} = 1.3$.

2. Draw a picture to show that $\frac{1}{2} + \frac{2}{3} = \frac{7}{6} = 1\frac{1}{6}$.

3. Complete each addition problem. Simplify your answers.

a. $\frac{7}{10} + \frac{4}{5}$ **b.** $\frac{3}{5} + \frac{3}{4}$

Think and Discuss

4. Explain how to add and subtract fractions with unlike denominators.

5. Discuss why the fraction that results from adding or subtracting fractions with unlike denominators has a different denominator than the fractions.

Name ______________________ Date ____________ Class ____________

LESSON 2F

Review for Mastery

Adding and Subtracting with Unlike Denominators

To model $\frac{1}{2} + \frac{1}{3}$, use two rectangles of the same size and shape.

A. 1st rectangle: Shade $\frac{1}{2}$ vertically.

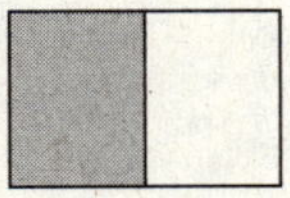

$\frac{1}{2}$

B. 2nd rectangle: Shade $\frac{1}{3}$ horizontally.

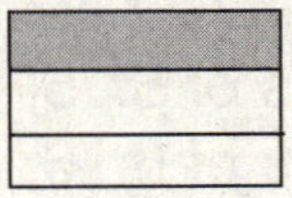

$\frac{1}{3}$

C. Separate the shaded portions into parts of equal size.

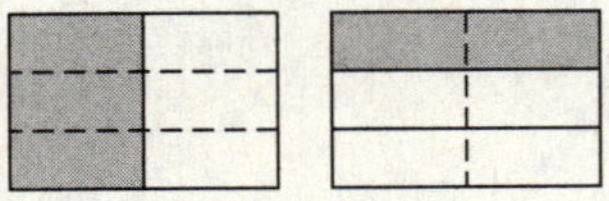

$\frac{1}{2} = \frac{3}{6}$ $\frac{1}{3} = \frac{2}{6}$

D. Use a new rectangle to show the sum, which is the total number of shaded portions of the two rectangles.

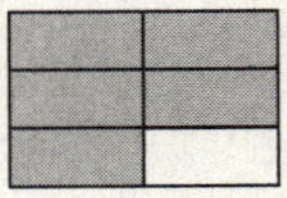

$\frac{1}{2} + \frac{1}{3} = \frac{3}{6} + \frac{2}{6} = \frac{5}{6}$

Model $\frac{1}{2} + \frac{2}{5}$. Write the result.

1.

Model $\frac{3}{5} - \frac{1}{3}$. Write the result.

2.

Name ______________________ Date __________ Class __________

LESSON 2F

Review for Mastery

Adding and Subtracting with Unlike Denominators, continued

To add fractions with different denominators, first rewrite the fractions using the lowest common denominator. To find the LCD, list the multiples of each denominator.

Find the LCD of 5 and 6.
Multiples of 5: 5, 10, 15, 20, 25, (30)
Multiples of 6: 6, 12, 18, 24, (30)
So, the LCD of 5 and 6 is 30.

Complete to find the LCD for each set of denominators.

3. The LCD of 6 and 4 is: ______________
Multiples of 6: ______________
Multiples of 4: ______________

4. The LCD of 3 and 7 is: ______________
Multiples of 3: ______________
Multiples of 7: ______________

To add fractions with different denominators:

Add: $\frac{1}{2} + \frac{1}{3}$

$\frac{1 \cdot 3}{2 \cdot 3} + \frac{1 \cdot 2}{3 \cdot 2}$

$\frac{3}{6} + \frac{2}{6} = \frac{5}{6}$

Complete to add fractions. Simplify.

5. $\frac{1}{4} = \frac{\ }{20}$

$+\frac{3}{5} = \frac{\ }{20}$

$=$ ______

6. $\frac{3}{4} = \frac{\ }{16}$

$+\frac{5}{16} = \frac{\ }{16}$

$=$ ______ $=$ ______

7. $5\frac{1}{3} = 5\frac{\ }{24}$

$+2\frac{5}{8} = 2\frac{\ }{24}$

$=$ ______

Add or subtract fractions. Simplify.

8. $\frac{1}{4} + \frac{7}{20} =$

9. $\frac{4}{9} - \frac{1}{5} =$

10. $\frac{8}{15} - \frac{1}{4} =$

Name ______________________ Date __________ Class __________

LESSON 2F

Homework and Practice

Adding and Subtracting with Unlike Denominators

Add or subtract. Write the answer in simplest form.

1. $\frac{1}{2} - \frac{3}{8}$
2. $\frac{3}{5} + \frac{1}{4}$
3. $\frac{7}{9} - \frac{5}{12}$
4. $-\frac{5}{9} + \frac{2}{3}$
5. $\frac{3}{10} - \left(-\frac{1}{2}\right)$
6. $\frac{7}{12} + \left(-\frac{5}{15}\right)$
7. $-\frac{4}{15} - \frac{2}{5}$
8. $\frac{5}{12} - \frac{2}{5}$
9. $-\frac{7}{8} + \frac{5}{12}$
10. $-2\frac{1}{8} - 3\frac{1}{4}$
11. $-4\frac{3}{5} + 1\frac{2}{3}$
12. $6\frac{1}{3} - 4\frac{5}{6}$
13. $7 - 5\frac{7}{8}$
14. $8\frac{1}{6} + \left(-\frac{7}{12}\right)$
15. $-16 + \left(-\frac{8}{9}\right)$
16. $4\frac{7}{10} - 11$

Evaluate each expression for the given value of the variable.

17. $-2\frac{5}{8} + x$ for $x = 3\frac{1}{4}$
18. $1\frac{1}{2} + x$ for $x = -2\frac{5}{6}$
19. $-7\frac{4}{9} - x$ for $x = -7$
20. $9\frac{5}{8} + x$ for $x = 6\frac{3}{4}$
21. $-15\frac{3}{8} + x$ for $x = 1\frac{3}{4}$
22. $-15\frac{2}{9} - x$ for $x = -12\frac{2}{3}$

23. Brendan practiced soccer for $1\frac{1}{2}$ hours on Monday, $1\frac{1}{4}$ hours on Tuesday, $1\frac{1}{6}$ hours on Wednesday and $\frac{3}{4}$ hours on Thursday in preparation for the game on Friday. How many total hours did Brendan practice soccer in this week?

24. Lucy ran $3\frac{1}{2}$ mi on Monday, $5\frac{3}{5}$ mi on Tuesday, $2\frac{3}{4}$ mi on Thursday, and $6\frac{2}{3}$ mi on Friday. Find the average distance Lucy ran each day.

EXPLORATION

Program of Studies and Core Content

MA-8-AT-S-VE03, MA-8-NPO-S-NO1, MA-08-5.1.2, MA-08-1.2.1

2G Solving Equations with Rational Numbers

A box of cookies has 6 servings. Each serving contains 40.5 Calories. How many Calories are in the whole box?

Estimate

40	40
40	40
40	40

$(40)(6) = 240$

Actual

$(40.5)(6) = C$

$243 = C$

Estimate the solution for each equation. Then use a calculator to solve.

	Equation	Estimate	Actual
1.	$124.75 - x = 50$		
2.	$x + 16.9 = 15.5$		
3.	$0.6x = 15$		
4.	$\frac{x}{1.25} = 8$		

Think and Discuss

5. Describe how you estimated the solutions for the equations in Problems **1–4.**

6. Discuss whether it is easier to estimate the solutions of some equations than the solutions of others.

Name ______________________ Date ____________ Class ____________

LESSON 2G

Review for Mastery

Solving Equations with Rational Numbers

Solving equations with rational numbers is basically the same as solving equations with integers or whole numbers:

Use inverse operations to isolate the variable.

$\frac{1}{4}z = -16$

$4 \cdot \frac{1}{4}z = -16 \cdot 4$

$z = -64$

Multiply each side by 4.

$y - \frac{3}{8} = \frac{7}{8}$

$+\frac{3}{8} \quad +\frac{3}{8}$

$y = \frac{10}{8} = 1\frac{2}{8} = 1\frac{1}{4}$

Add $\frac{3}{8}$ to each side.

$x + 3.5 = -17.42$

$-3.5 \quad -3.5$

$x = -20.92$

Subtract 3.5 from each side.

$-26t = 317.2$

$\frac{-26t}{-26} = \frac{317.2}{-26}$

$t = -12.2$

Divide each side by −26.

Tell what you would do to isolate the variable.

1. $x - 1.4 = 7.82$ ____________

2. $\frac{1}{4} + y = \frac{7}{4}$ ____________

3. $3z = 5$ ____________

Solve each equation.

4. $14x = -129.5$ ____________

5. $\frac{1}{3}y = 27$ ____________

6. $265.2 = \frac{z}{22.1}$ ____________

7. $x + 53.8 = -1.2$ ____________

8. $25 = \frac{1}{5}k$ ____________

9. $m - \frac{2}{3} = \frac{3}{5}$ ____________

10. Seven coworkers go out to lunch. Three people get the salad bar for $7.99 and 4 people get a sandwich for $8.99. All 7 people get drinks with their lunch for $1.50 each, and 5 people get dessert for $2.50 each. If they collect $100 to pay the bill, estimate the amount leftover for the tax and tip.

____

Name ______________________ Date __________ Class __________

LESSON 2G

Homework and Practice

Solving Equations with Rational Numbers

Solve.

1. $y - 3.2 = -4.7$ ________
2. $x + 5.61 = 3.89$ ________
3. $0.4d = 8.2$ ________
4. $b - 2 = 17.5$ ________
5. $a + 101 = -14.2$ ________
6. $3.3x = -108.9$ ________
7. $-19.48 + x = -28.03$ ________
8. $\frac{m}{10.1} = 3.5$ ________
9. $217.25 = 39.5w$ ________
10. $\frac{t}{4.7} = -14.7$ ________
11. $b - 29.15 = -17.73$ ________
12. $\frac{r}{-7.6} = 3.05$ ________

Solve. Write each in simplest form.

13. $\frac{5}{8} + y = 2\frac{3}{8}$ ________
14. $\frac{12}{25}m = -\frac{9}{50}$ ________
15. $\frac{6}{35}r = \frac{18}{25}$ ________
16. $x - \left(-\frac{7}{10}\right) = 3\frac{3}{5}$ ________
17. $y - \frac{11}{18} = 4\frac{2}{9}$ ________
18. $-\frac{9}{24}w = \frac{18}{32}$ ________

19. Jessica baked 10 dozen cookies. She left $\frac{1}{8}$ of the baked cookies home and took the rest to school for the bake sale. How many cookies did Jessica take to school?

____

20. Doug must take 1.25 milliliters of medicine every day. How many days will the medicine last if Doug was given a prescription of the medicine in a bottle containing 50 milliliters?

____

EXPLORATION

Program of Studies and Core Content

MA-8-AT-S-EI1, MA-8-AT-S-EI4, MA-8-AT-S-VEO3, MA-8-NPO-S-NO1, MA-8-AT-S-VEO1, MA-08-1.3.1, MA-08-5.1.2, MA-08-5.3.1, MA-08-1.2.1, MA-08-5.2.1, MA-08-5.2.2

2H Solving Two-Step Equations

Felipe has a 750-gallon tank full of water on his ranch for his livestock. The livestock use 15 gallons of water per day, and no rain is expected to refill the tank during the dry season.

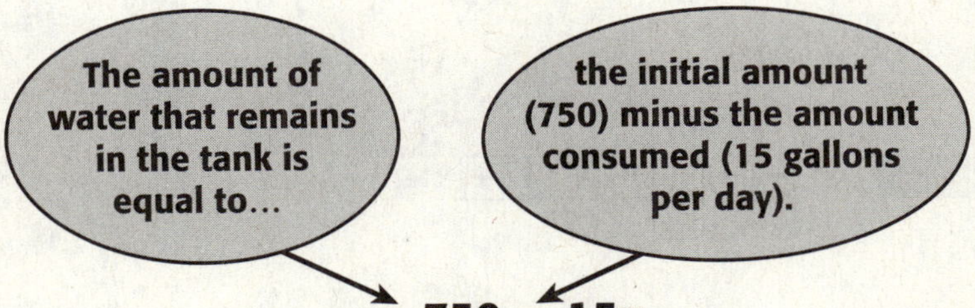

$$y = 750 - 15x$$

Use the equation $y = 750 - 15x$ to answer each question.

1. How much water is in the tank after the first day ($x = 1$)?

2. How much water is in the tank after 25 days ($x = 25$)?

3. What will be equal to zero in the equation $y = 750 - 15x$ when the tank is empty?

4. After how many days will the tank be empty?

Think and Discuss

5. **Discuss** what it means when 750 gallons (the initial amount) is exactly the same as $15x$ gallons (the amount consumed).

Name ______________________ Date __________ Class __________

LESSON **2H**

Hands-On Lab

Solving Two-Step Equations

Activity

Shelley shopped online in hopes of ordering cookies for a party. She found the cookies she wanted at 2 stores. Store X offered a discount of $5 if Shelley purchased 4 dozen cookies. Store Y will sell 3 dozen cookies at the same total price as the 4 dozen from Store X, including $4 for shipping. Using x as the cost of one dozen cookies, then $4x - 5$ is the total cost of the purchase from Store X, and $3x + 4$ is the total cost of the purchase from Store Y. Solve the equation below for x to find the price for a dozen cookies. Use algebra tiles to model the equation.

$4x - 5 = 3x + 4$

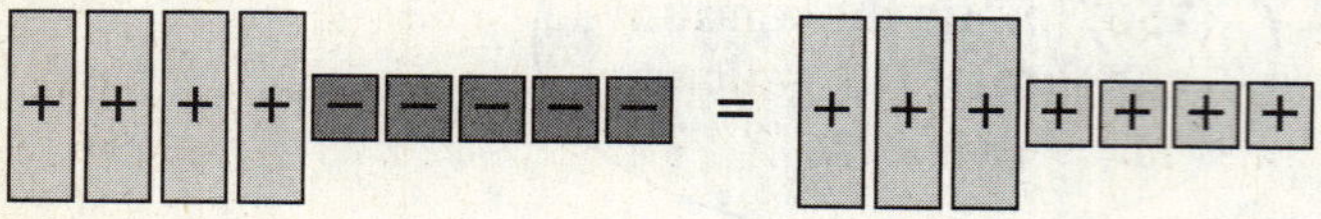

$4x - 5$ **=** **$3x + 4$**

Step 1	$4x - 5 = 3x + 4$	Given
Step 2	$4x - 5 - 3x = 3x + 4 - 3x$	Subtraction Property of Equality
Step 3	$x - 5 = 4$	Simplify

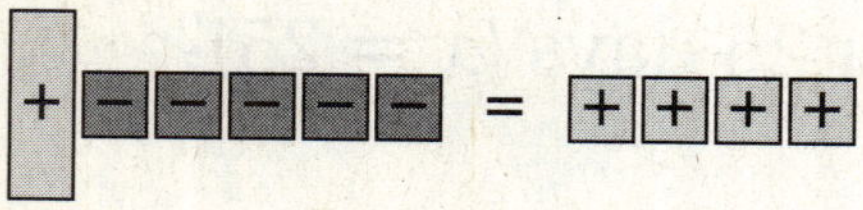

Step 4	$x - 5 + 5 = 4 + 5$	Addition Property of Equality

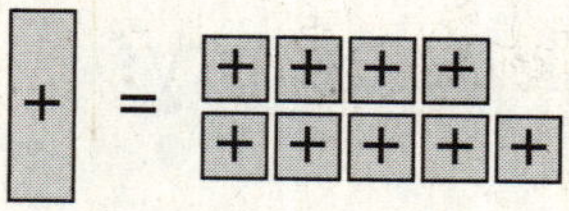

Step 5	$x = 9$	Addition Property of Equality

Name ______________________ Date __________ Class __________

LESSON 2H

Hands-On Lab

Solving Two-Step Equations, continued

Think and Discuss

What is meant by the Subtraction Property of Equality?

____

____

____

____

____

____

Try This

Solve and check the equation $7m - 3 = 2m + 27$, using algebra tiles.

____

Name ______________________ Date ____________ Class ____________

LESSON 2H

Review for Mastery

Solving Two-Step Equations

Maggie sold cupcakes at the music department's bake sale. She spent \$15 on supplies. She sold her cupcakes for 75¢ each and made a profit of \$22.50. Write and solve an equation to find how many cupcakes Maggie sold.

Understand the Problem

The answer is the number of cupcakes Maggie sold. List the important information: The supplies cost \$15, each cupcake sells for 75¢, and the profit is \$22.50.

Let x represent the number of cupcakes sold.

Profit = Cupcakes − Supplies

$22.50 = 0.75x - 15$

Make a Plan

Think: First the variable is multiplied by 0.75, and then 15 is subtracted from the result. Work backward to solve the equation. Undo the operations in reverse order: First add 15 to both sides of the equation, and then divide both sides of the new equation by 0.75.

Solve

$$22.5 = 0.75x - 15$$

$$22.5 + 15 = 0.75x - 15 + 15 \quad \textit{Add 15 to both sides.}$$

$$37.5 = 0.75x$$

$$\frac{37.5}{0.75} = \frac{0.75x}{0.75} \quad \textit{Divide both sides by 0.75.}$$

$$50 = x$$

Maggie sold 50 cupcakes.

Look Back

Estimate with compatible numbers: $25 = 1x - 15$; $40 = x$. The answer is reasonable.

Write and solve a two-step equation.

1. Colin's cell phone plan charges \$40 dollars a month, plus \$0.12 for each minute over the allowed minutes on his plan. Colin received a bill of \$50.80. How many minutes did Colin use his cell phone beyond the allowed minutes?

____

Name ______________________ Date __________ Class __________

LESSON 2H

Homework and Practice

Solving Two-Step Equations

Write and solve two-step equations to answer the following questions. Then use estimation to determine whether your answer is reasonable.

1. Elliot wants to buy a new printer that costs $189. He has $125 saved. He has a job that pays $8 an hour. How many hours must he work to earn enough to buy the printer?

   ____

2. Corbin's car payment is $289. This is $37 less than $\frac{1}{3}$ of his monthly income. What is Corbin's monthly income?

   ____

3. Mr. Cruz buys some wood and pays $245. Each bundle of wood costs $15 and there is a delivery charge of $35. How many bundles of wood did Mr. Cruz buy?

   ____

4. Madison is saving to take an SAT prep course that costs $400. She has saved $175 so far. She adds $30 a week to her savings. How many more weeks must she save to be able to afford the course?

   ____

5. Drew's sister is three years younger than twice his age. Drew's sister is 19. How old is Drew?

   ____

6. Jane belongs to a music club in which she pays an annual fee of $55. She then can download music for $0.75 a song. In one year, Jane spent $227.50. How many songs did Jane download?

   ____

7. The temperature outside Lauren's house at 6 A.M. is 29 °F less than twice the temperature at noon. The temperature at noon is 85 °F. What is the temperature outside Lauren's house at 6 A.M.?

   ____

CHAPTER 2

Chapter Summary

Answer these question to summarize the important concepts from Chapter 2 in your own words.

1. Explain how to compare two fractions with different denominators.

2. Explain how to multiply rational numbers in fraction form.

3. Explain how to divide rational numbers in fraction form.

4. Explain what it means to isolate a variable in an equation.

For more review of Chapter 2:

- Complete the Chapter 2 Study Guide and Review on pages 106–108 of your textbook.
- Complete the Ready to Go On quizzes on pages 90 and 102 of your textbook.

Name ______________________ Date ____________ Class ____________

CHAPTER 2

KCCT Prep

Multiple Choice

1. On a number line, which mixed number is closest to 8.4?

(A) $8\frac{1}{4}$ (C) $8\frac{4}{7}$

(B) $8\frac{3}{8}$ (D) $8\frac{3}{5}$

2. Which is the best estimate of the value of $|-3.9 - k|$ for $k = -5.02$?

(A) −1 (C) 2

(B) 1 (D) 8

3. A bag of dried fruit contains $1\frac{2}{3}$ lb of fruit. William pours dried fruit from $6\frac{1}{2}$ bags in a box. The box itself weighs $1\frac{1}{2}$ lb. What is the total weight of the fruit and the box?

(A) $9\frac{2}{3}$ lb (C) $12\frac{1}{3}$ lb

(B) $10\frac{5}{6}$ lb (D) $17\frac{1}{3}$ lb

4. Lauren drives 3.2 km farther than Cara, and Cara drives 4.9 km farther than Vance. If Lauren drives a total of 15.3 km, which equation can you use to find the distance d that Vance drives?

(A) $15.3 = d + 8.1$

(B) $15.3 = 3.2d$

(C) $3.2 = 4.9 + d + 15.3$

(D) $8.1 = 15.3 + d$

5. Peaches cost p dollars per pound. Zack buys $4\frac{1}{2}$ pounds of peaches and a jar of peanut butter that costs \$2. Altogether, he spends less than \$11. Which inequality represents the cost of the peaches?

(A) $4\frac{1}{2} + p < 11$ (C) $4\frac{1}{2}p < 11$

(B) $4\frac{1}{2} + p < 9$ (D) $4\frac{1}{2}p < 9$

6. A bucket can hold $2\frac{1}{4}$ gallons of liquid. Jake has 5 buckets and each is $\frac{1}{3}$ full. How much liquid is there altogether?

(A) $\frac{3}{4}$ gallon (C) $3\frac{3}{4}$ gallon

(B) $1\frac{2}{3}$ gallon (D) $11\frac{1}{4}$ gallon

7. The table shows the lengths of four insects at an insect zoo. Which insect is longest?

Insect Lengths

Type of Insect	Length
Beetle	$1\frac{3}{8}$ in.
Cricket	1.09 in.
Grasshopper	$1\frac{3}{4}$ in.
Waterbug	1.65 in.

(A) Beetle (C) Grasshopper

(B) Cricket (D) Waterbug

8. Miles wants to use the Associative Property to write an expression that is equivalent to $5 + (x + 9)$. Which of these expressions should he write?

(A) $5 + (9 + x)$ (C) $5x + 5 \cdot 9$

(B) $(5 + x) + 9$ (D) $5(x + 9)$

9. Vonda jogs the same distance each day. After 8 days, the total distance she has jogged is 33.6 kilometers. Paul jogs 2.7 km less than Vonda each day. How far does Paul jog each day?

Ⓐ 1.5 km Ⓒ 6.9 km
Ⓑ 2.7 km Ⓓ 30.9 km

10. Which expression has the greatest value when $x = -9$?

Ⓐ $2x + 9$ Ⓒ $-4x - 10$
Ⓑ $3x - 5$ Ⓓ $-3x$

11. Jocelyn's salary is $9.50 per hour. The total amount she earns this week is no less than $190. Which inequality represents the number of hours *h* that Jocelyn works this week?

Ⓐ $h \geq 20$ Ⓒ $h \leq 20$
Ⓑ $h \geq 180.5$ Ⓓ $h \leq 180.5$

12. Which is the best estimate of the value of $cd - 5.1$ when $c = -3.2$ and $d = -4.8$?

Ⓐ −20 Ⓒ 10
Ⓑ −11 Ⓓ 15

13. Albert orders 6 cases of dog food online. The shipping charge is $5.75 and he spends a total of $32.75 on the order. What does it cost to order 4 cans of dog food, assuming the same shipping charge?

Ⓐ $18 Ⓒ $23.75
Ⓑ $21.83 Ⓓ $27

14. How many integers are greater than $-2\frac{1}{2}$ and less than $3\frac{1}{2}$?

Ⓐ 2 Ⓒ 5
Ⓑ 4 Ⓓ 6

Open Response

15. The population density of Tennessee is 53.29 people per square kilometer. The population density of New Hampshire is $55\frac{1}{5}$ people per square kilometer.

a. Which state has the greater population density?

b. A region of New Hampshire has an area of 98 square kilometers. Explain how you can estimate the population of this region.

16. A tank contains 14 gallons of water. Raul turns on a faucet that adds 3.2 gallons of water to the tank every minute.

a. Write an expression that gives the total number of gallons of water in the tank after *m* minutes.

b. Use the expression to find the number of gallons of water in the tank after 20 minutes.

c. Raul turns off the faucet when there are 158 gallons of water in the tank. Write and solve an equation to find out how long the faucet was running.

17. A factory produces boxes of granola. Each box is supposed to weigh 24 oz. An employee weighs four boxes at random and records how each box varies from the ideal weight of 24 oz.

Box	A	B	C	D
Variation from Ideal Weight (oz)	$+\frac{1}{8}$	−0.5	+0.1	$-\frac{2}{3}$

a. List the boxes in order from lightest to heaviest.

b. A box of granola is considered acceptable if it varies from the ideal weight by less than ±0.2 oz. Which of the four boxes, if any, are acceptable?

CHAPTER 3

Project

Graphs, Functions, and Sequences: Healthy Dosage

Career: Pharmacist

In addition to dispensing medicine that has been prescribed by doctors, pharmacists advise patients about the uses and possible side effects of medications. Pharmacists may also make recommendations to help patients manage conditions such as diabetes and high blood pressure. Although many pharmacists work in drugstores, others work closely with doctors and nurses in hospitals.

Before you start, work through the Painkiller Dosage worksheet on the next page.

In this chapter project . . .

You calculate the amount of medication appropriate for children.

What to do:

- Research the recommended dosages by weight for a common childhood medication, like allergy or flu medication.
- Write an equation based on the dosages you found. Use the equation to make a graph on the coordinate plane.
- Find the average weights for children at different ages. Create a table that lists the medication dosages by weight and corresponding age.

What to present:

- An equation and graph for the medication you researched.
- A table listing the dosages by weight and age.
- An explanation for why finding a dosage for an age might not work as well as dosage given by weight.

Where's the math?

- Use an equation to describe the relationship between dosage and weight.
- Use a graph to describe the relationship between dosage and weight.

go.hrw.com
Chapter Project Online
KEYWORD: MT7 Ch3

Name ______________________ Date ____________ Class ____________

CHAPTER 3

Project Recording Sheet

Graphs, Functions, and Sequences

Painkiller Dosage

Paracetamol and ibuprofen are commonly used to relieve pain and fever. The table shows the recommended dose of each drug for children of various weights.

Recommended Doses for Children		
Weight of Child (lb)	**Dose of Paracetamol (mg)**	**Dose of Ibuprofen (mg)**
25	135	55
30	162	66
35	189	77

1. To find the number of milligrams of paracetamol that should be given to a child who weighs x pounds, use the equation $y = 5.4x$. Make a table of ordered pairs to show the recommended dose of paracetamol for children who weigh 25 lb, 30 lb, 35 lb, 40 lb, 45 lb, and 50 lb.

2. Use your table to graph the equation on a coordinate plane.

3. A pharmacist is advising a doctor on the correct dose of paracetamol for a child who weighs 37 lbs. Explain how you can use the graph to estimate the dose.

____

____

4. To find the number of milligrams of ibuprofen that should be given to a child who weighs x pounds, use the equation $y = 2.2x$. Make a table of ordered pairs, and then graph the equation.

5. Explain how the graph of ibuprofen doses is similar to the graph of paracetamol doses. How are the graphs different?

____

____

EXPLORATION

Program of Studies and Core Content

MA-8-G-S-CG1, MA-8-AT-S-PRF3
MA-08-3.3.1

3A Graphing on a Coordinate Plane

At the GasCo station, gasoline costs $3 per gallon. Let x represent the number of gallons of gasoline that you buy and let y represent the total cost of the gasoline.

1. Complete the table.

Number of Gallons, x	Total cost, y
1	$3
2	
3	
4	
5	
6	

2. Each row of the table contains an ***ordered pair.*** An ordered pair lists an x-value and its corresponding y-value. For example, the first row of the table contains the ordered pair (1, 3). Write the ordered pairs in the other rows of your table.

3. The equation $y = 3x$ gives the cost of x gallons of gasoline. Use the equation to find the cost of 12 gallons of gasoline.

4. Write the ordered pair from Problem 3.

Think and Discuss

5. Describe how you could find a new ordered pair for the above situation.

6. Explain whether you think the ordered pair (1, 3) is the same as (3, 1).

Name ______________________________ Date ____________ Class ____________

LESSON 3A

Technology Lab

Create a Table of Solutions

The *Table* feature on a graphing calculator can help you make a table of values quickly.

Activity

1 Make a table of solutions of the equation $y = 2x - 3$. Then find the value of y when $x = 29$.

To enter the equation, press the **Y=**. Then press 2 **X,T,θ,n** **−** 3.

Press **2nd** **WINDOW** (TBLSET) to go to the Table Setup menu. In this menu, **TblStart** shows the starting x-value, and **ΔTbl** shows how the x-values increase. If you need to change these values, use the arrow keys to highlight the number you want to change and then type a new number.

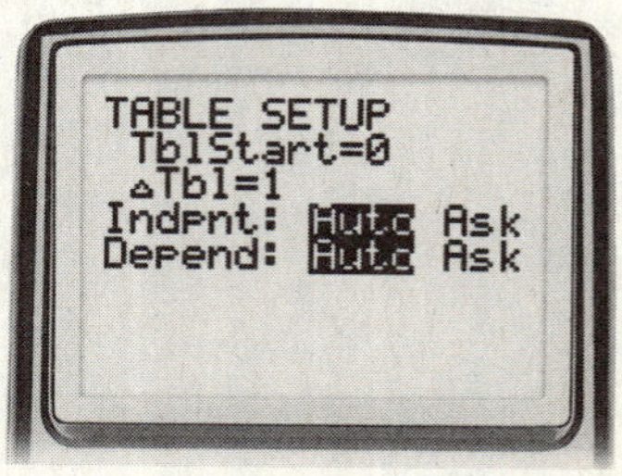

Press **2nd** **GRAPH** (TABLE) to see the table of values.

On this screen, you can see that $y = 7$ when $x = 5$.

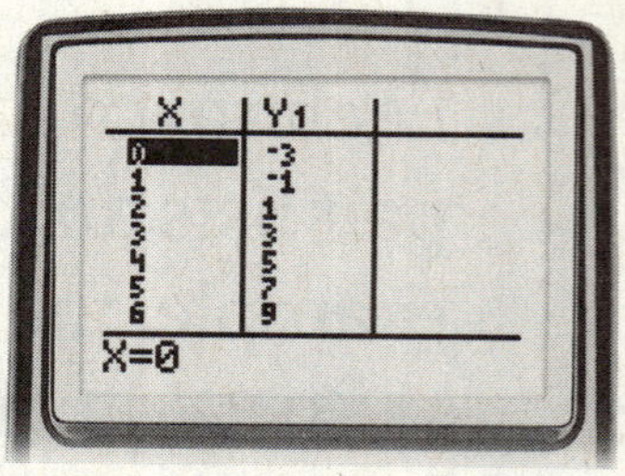

Use the arrow keys to scroll down the list. You can see that $y = 55$ when $x = 29$.

To check, substitute 29 into $y = 2x - 3$.

$y = 2x - 3$

$= 2(29) - 3 = 58 - 3 = 55$

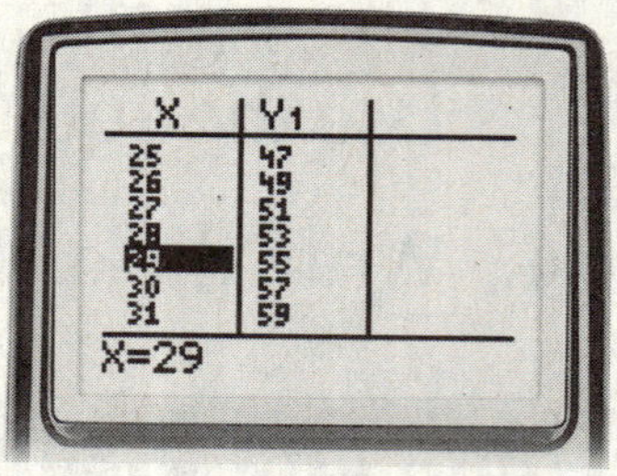

Name ______________________ Date ____________ Class ____________

LESSON 3A

Technology Lab

Create a Table of Solutions, continued

Think and Discuss

1. On an Internet site, pencils can be purchased for 17¢ each, but they only come in boxes of 12. You decide to make a table to compare *x*, the number of pencils, to *y*, the total cost of the pencils. What **TblStart** and **ΔTbl** values will you use? Explain.

____

____

____

Try This

For each equation, use a table to find the *y*-values for the given *x*-values. Give the **TblStart** and **ΔTbl** values you used.

1. $y = 3x + 6$ for $x = 1, 3,$ and 7

____

____

2. $y = \frac{x}{4}$ for $x = 5, 10, 15,$ and 20

____

____

____

Name ______________________ Date __________ Class __________

LESSON 3A

Review for Mastery

Graphing on a Coordinate Plane

An **ordered pair** can be used to write a solution for a two-variable equation. For the equation $y = x + 5$, a solution is (0, 5). When the *x*-value is 0, the *y*-value is 5.

(0, 5) — *x*-value, *y*-value

Which of the ordered pairs (5, 3) or (3, 5) is a solution of $y = 2x - 1$?

$y = 2x - 1$
$3 \stackrel{?}{=} 2(5) - 1$ Substitute 5 for *x* and 3 for *y*.
$3 \stackrel{?}{=} 10 - 1$
$3 \neq 9$

So, (5, 3) *is not* a solution of $y = 2x - 1$.

$y = 2x - 1$
$5 \stackrel{?}{=} 2(3) - 1$ Substitute 3 for *x* and 5 for *y*.
$5 \stackrel{?}{=} 6 - 1$
$5 = 5$

So, (3, 5) *is* a solution of $y = 2x - 1$.

Determine whether each ordered pair is a solution of the given equation. Write *is* or *is not*.

1. $y = 5x$; (3, 15) ____________

2. $y = 3x - 4$; (5, 3) ____________

3. $y = 6 - x$; (4, 2) ____________

A two-variable equation has infinitely many solutions. You can use a table to find and record solutions to a given equation.

Use $x = 1$, 2, and 3, for example, to make a table of values for $y = 5x - 1$. Substitute each given value of *x* in the expression for *x*. Evaluate the expression to find the value of *y* that completes the ordered pair.

x	$5x - 9 = y$	(*x*, *y*)
1	5(1) − 1 = 4	(1, 4)
2	5(2) − 1 = 9	(2, 9)
3	5(3) − 1 = 14	(3, 14)

Complete each table.

4. $y = 4x$

x	$4x = y$	(*x*, *y*)
0	4(0) = __	(0, __)
1	4(__) = __	(1, __)
2	4(__) = __	(2, __)

5. $y = 5x - 3$

x	$5x - 3 = y$	(*x*, *y*)
1	5(__) − 3 = __	(1, __)
2	5(__) − 3 = __	(2, __)
3	5(__) − 3 = __	(3, __)

Name ______________________ Date ____________ Class ____________

LESSON 3A

Review for Mastery

Graphing on a Coordinate Plane, continued

Point *A* is described by the ordered pair (3, −2). The first number, 3, is the ***x*-coordinate** and the second number, −2, is the ***y*-coordinate.**

Using a grid to represent a **coordinate plane,** graph point *A* by starting at the **origin,** where the ***x*-axis** and ***y*-axis** intersect. The *x*-coordinate, 3, tells you to go *right* 3. The *y*-coordinate, −2, tells you to go *down* 2.

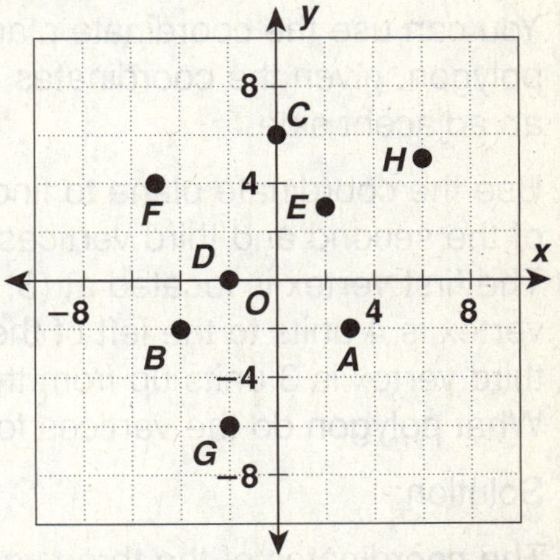

Write an ordered pair to describe the location of each point on the coordinate plane above.

6. point *C* ____________ **7.** point *F* ____________ **8.** point *G* ____________

Graph each ordered pair.

9. *J*(2, 1) **10.** *K*(−3, 0)

11. *L*(5, −4) **12.** *M*(−4, −4)

13. *N*(−2, 6) **14.** *P*(3, −2)

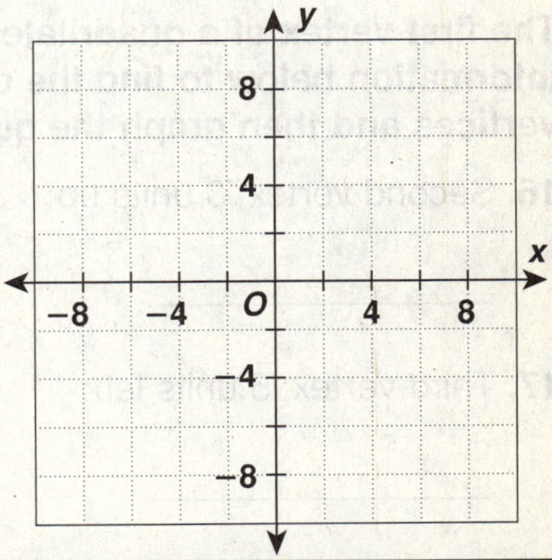

The graph of an equation is the set of all ordered pairs that are solutions of the equation.

Complete the table. Graph each ordered pair. Draw a line through the points.

15. $y = -x - 1$

x	***−x − 1***	***y***	**(*x*, *y*)**
−1	−(−1) − 1	____	(__, __)
0	−(__) − 1	−1	(__, __)
1	−(__) − 1	____	(__, __)

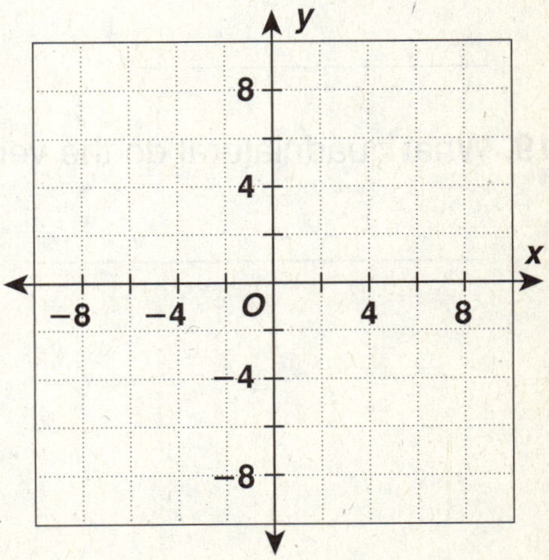

Name ______________________ Date ____________ Class ____________

LESSON 3A

Review for Mastery

Graphing on a Coordinate Plane, continued

You can use the coordinate plane to find the vertices of a polygon, given the coordinates of a one vertex and the length of an adjacent side.

Use the coordinate plane to find the coordinate of the second and third vertices of the polygon. The first vertex is located at (3, –1). The second vertex is 4 units to the left of the first vertex. The third vertex is 3 units up from the second vertex. What polygon do the vertices form?

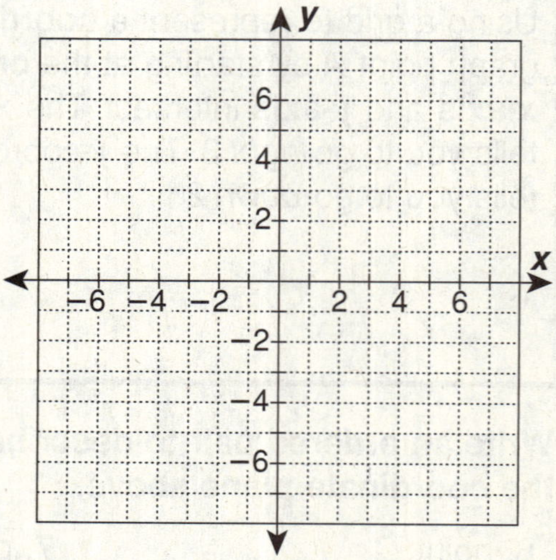

Solution:

The coordinates of the three vertices are (3, –1), (–1, –1), and (–1, 2). The vertices form a right triangle.

The first vertex of a quadrilateral is located at (6, 1). Use the information below to find the coordinates of the missing vertices and then graph the quadrilateral.

16. Second vertex; 3 units up:

____

17. Third vertex; 5 units left:

____

18. Fourth vertex; 3 units down:

____

19. What quadrilateral do the vertices form?

____

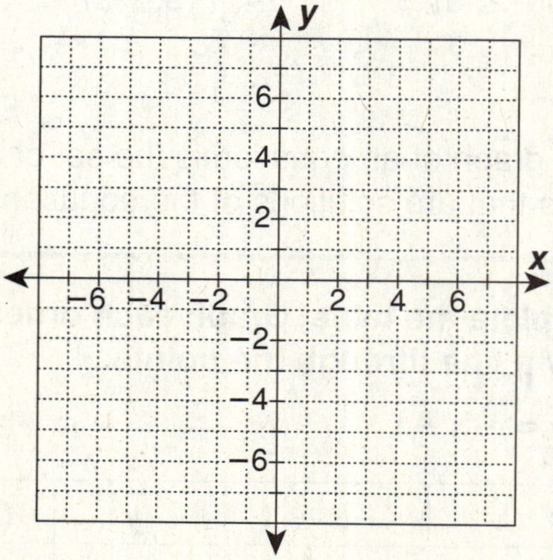

Name ______________________ Date ____________ Class ____________

LESSON 3A

Homework and Practice

Graphing on a Coordinate Plane

Determine whether each ordered pair is a solution of $y = 5 + 3x$.

1. (1, 8) ________

2. (3, 7) ________

3. (2, 10) ________

4. (0, 5) ________

Determine whether each ordered pair is a solution of $y = 4x - 1$.

5. (0, 1) ________

6. (1, 3) ________

7. (3, 11) ________

8. (5, 19) ________

Use the given values to complete the table of solutions.

9. $y = \frac{x}{3} - 2$ for $x = 0, 3, 6, 9, 12$

x	$\frac{x}{3} - 2$	y	(x, y)
0			
3			
6			
9			
12			

10. $y = \frac{1}{2}x + 2$ for $x = -4, -2, 0, 2, 4$

x	$\frac{1}{2}x + 2$	y	(x, y)
−4			
−2			
0			
2			
4			

11. $y = 4x - 1$ for $x = 1, 2, 4, 5, 8$

x	$4x - 1$	y	(x, y)
1			
2			
4			
5			
8			

12. $y = 5x + 2$ for $x = 0, 2, 4, 6, 8$

x	$5x + 2$	y	(x, y)
0			
2			
4			
6			
8			

13. Mrs. Frank had 150 customers when she began her delivery route. Each month she adds 5 new customers. The equation that gives the total number of customers, t, in her route is $t = 150 + 5m$, where m is the number of months since she began the route. How many customers will Mrs. Frank have after 12 months?

Name ______________________ Date __________ Class __________

LESSON **3A**

Homework and Practice

Graphing on a Coordinate Plane, continued

Give the coordinates of each point.

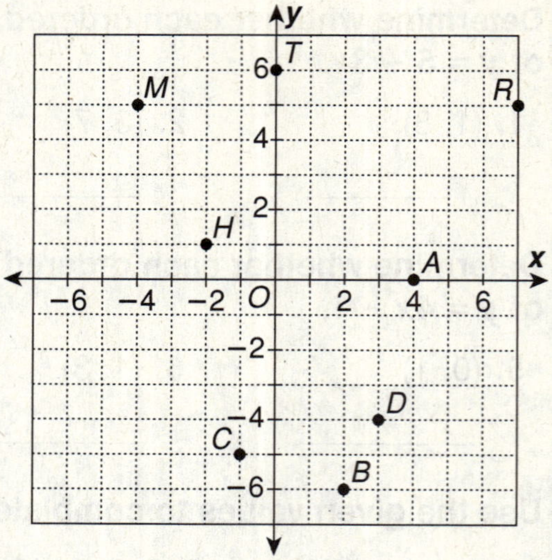

14. *A* __________

15. *D* __________

16. *H* __________

17. *C* __________

18. *B* __________

19. *T* __________

20. *M* __________

21. *R* __________

Graph each point on a coordinate plane. Label points *A–F*.

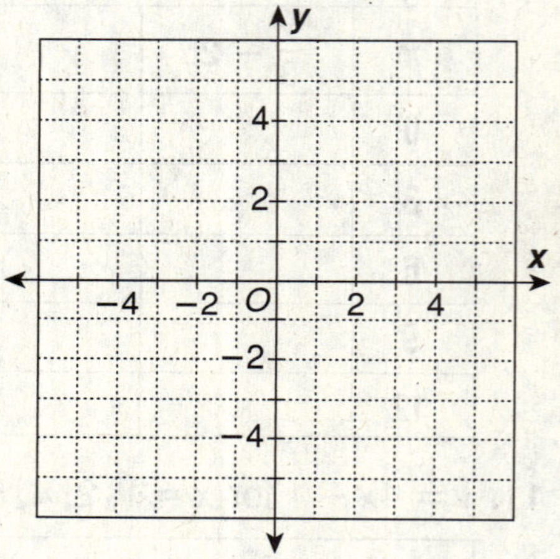

22. *A*(3, 5)

23. *B*(–6, 3)

24. *C*(0, –4)

25. *D*(–4, –6)

26. *E*(5, –2)

27. *F*(2, 0)

Complete the table of ordered pairs. Graph the equation on a coordinate plane.

28. $y = 2x$

x	*x*	*y*	(*x*, *y*)
1			
2			
3			

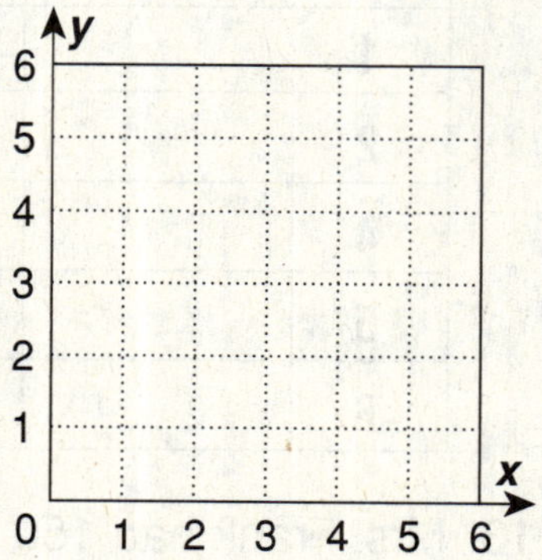

29. A doctor sees 4 patients each hour. Use the equation $y = 4x$, where the *y*-value represents the many patients that the doctor may see in *x* hours. Draw a graph. How many patients will the doctor see in a 5-hour day?

____

Name ______________________ Date __________ Class __________

LESSON **3A**

Homework and Practice

Graphing on a Coordinate Plane, continued

Find and plot the coordinates of all the vertices of each polygon given the coordinates of the first vertex. Then classify the polygon.

First vertex: (–4, –3)

30. Second vertex; 6 units right:

____

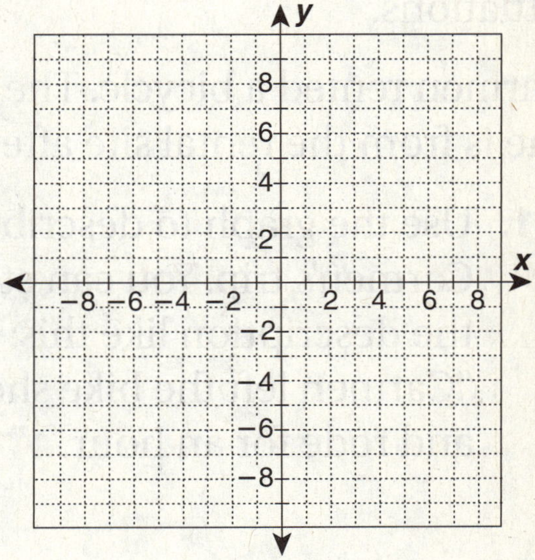

31. Third vertex; 4 units down, 2 units left:

____

32. Fourth vertex; 2 units left:

____

33. What polygon do the vertices form?

____

First vertex: (6, 2)

34. Second vertex; 4 units left:

____

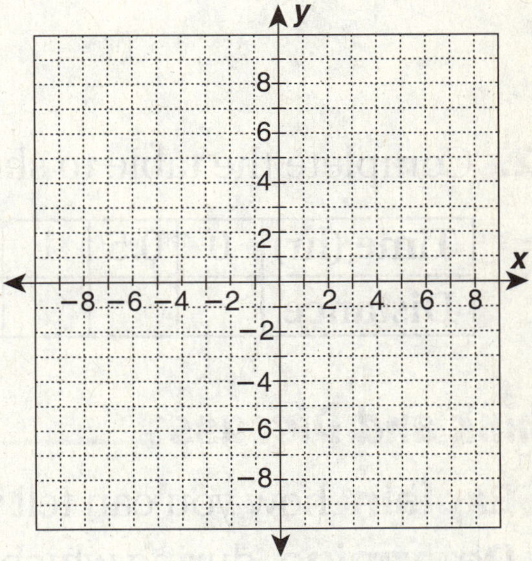

35. Third vertex; 1 unit left, 4 units down:

____

36. Fourth vertex; 4 units right:

____

37. What polygon do the vertices form?

____

EXPLORATION

Program of Studies and Core Content

MA-8-AT-S-PRF4

3B Interpreting Graphs and Tables

Graphs and tables can be used to represent many different situations.

Carmen rented a bicycle. The graph shows how far away she is from the rental site after each half hour of riding.

1. Use the graph to describe Carmen's trip. You can start the description like this: "Carmen left the bike shop and rode for an hour..."

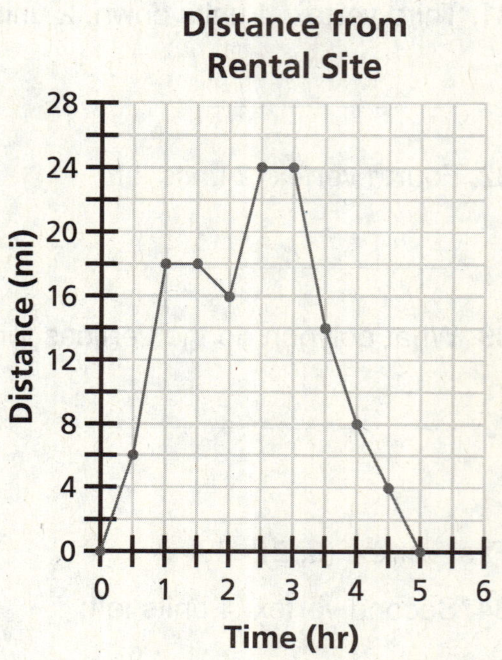

2. Complete the table to show her distance at each half hour.

Time (hr)	0	0.5	1	1.5	2	2.5	3	3.5	4	4.5	5
Distance											

Think and Discuss

3. Explain how you can tell when Carmen took a break.

4. Determine during which half hour Carmen covered the greatest distance.

Name ______________________ Date ____________ Class ____________

LESSON 3B

Technology Lab

Display Data in Different Kinds of Graphs

Each type of graph has its own advantages for displaying data. Line graphs, for example, show change over time while circle graphs show how a part relates to the whole. In some cases, data can be displayed effectively in more than one way. The spreadsheet program Excel creates different kinds of graphs and makes it easy to change from one style to another.

Activity

Create a worksheet in Excel using the data. This data table shows the unemployment rate over twelve months in two imaginary countries.

Month	Country AA	Country ZZ
January	3.2	3.5
February	3.7	3.7
March	2.9	3.7
April	3.4	3.8
May	3.3	4.0
June	3.4	3.9
July	3.6	3.9
August	3.8	3.7
September	3.9	3.6
October	4.2	3.7
November	4.2	3.5
December	4.1	3.6

Select all of the data in your table, then click on the Chart Wizard button.

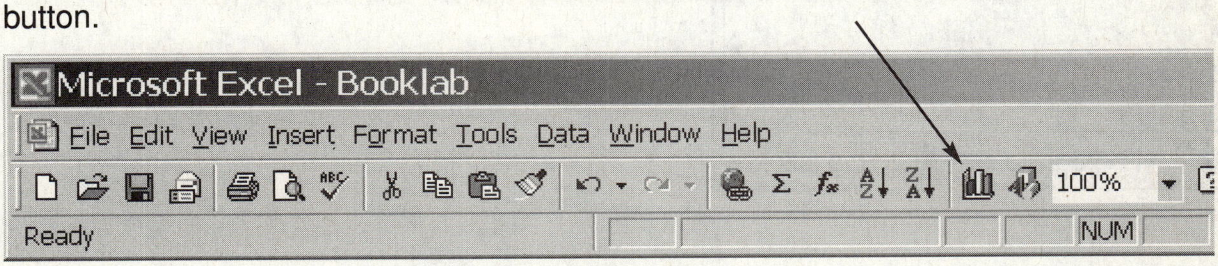

Name ______________________ Date ____________ Class ____________

LESSON 3B

Technology Lab

Display Data in Different Kinds of Graphs, continued

In Step 1, select "clustered column" as the graph type. Then click "Next."

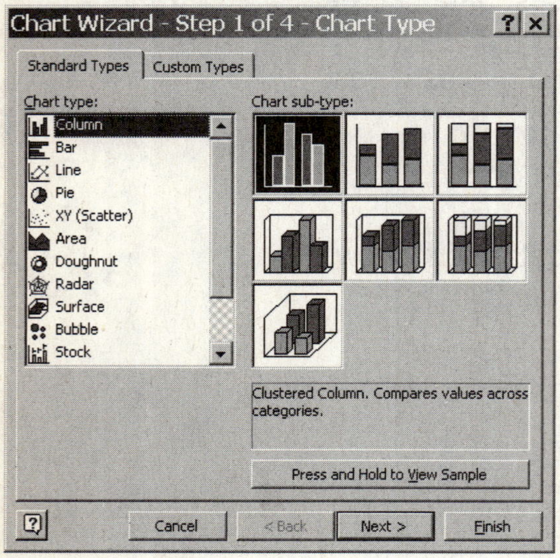

You do not need to do anything in Step 2, since you have already selected the necessary rows and columns. Click "Next."

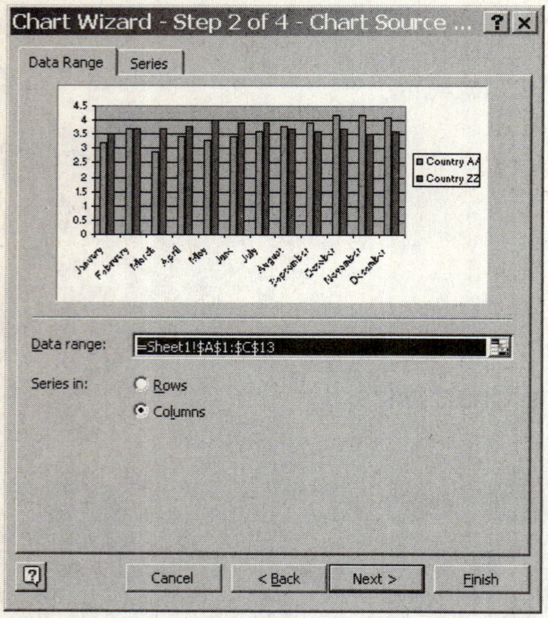

Name ______________________ Date ____________ Class ____________

LESSON 3B

Technology Lab

Display Data in Different Kinds of Graphs, continued

In Step 3, give the graph a title such as "Unemployment Rates" and then click "Finish." Your bar graph will appear on the screen.

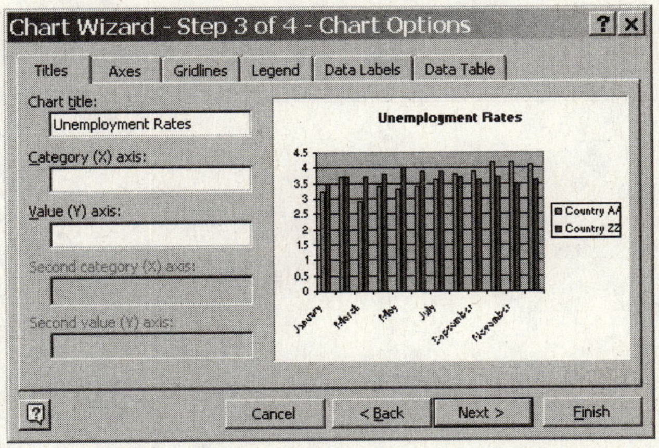

You can change the column graph to a line graph by clicking on the chart wizard button again and selecting line graph as your graph choice. Follow the steps as before.

Think and Discuss

1. Compare and contrast the line graph and column graph you created in the activity.

____

____

____

____

____

Name ______________________________ Date ____________ Class ____________

LESSON 3B

Technology Lab

Display Data in Different Kinds of Graphs, continued

Try This

1. Use the unemployment data to create a column graph, then a bar graph. Compare and contrast these two types of graph.

____

2. Create an Excel worksheet that shows the high temperature for a week in an imaginary town. Make a line graph. Then show the data as a column graph.

____

Name ______________________ Date ____________ Class ____________

LESSON 3B

Technology Lab

Display Data in Different Kinds of Graphs, continued

3. Survey the class by asking each student to pick red, yellow, blue or green as a favorite color. Summarize the data in a table in Excel, then create a circle (pie) graph and a column graph. Compare the two displays.

Name ______________________________ Date ____________ Class ____________

LESSON **3B**

Review for Mastery

Interpreting Graphs and Tables

Tell which table best corresponds to the given situation.
Situation: Mr. Savoy begins math lessons by going over yesterday's homework assignment. The class discussion about the new topic is lengthy. The remaining class time is spent in group work.

Table 1

Activity	Minutes
Review	12
New topic	24
Group work	24

Table 2

Activity	Minutes
Review	12
New topic	36
Group work	12

Table 3

Activity	Minutes
Review	12
New topic	15
Group work	33

Focus on review. Since all three tables indicate the same time for review, no tables can be eliminated. More time is spent on the new topic than on group work. Eliminate Table 1 since equal time is spent on each. Eliminate Table 3 since more time is spent on group work than on the new topic. So, Table 2 best corresponds to the given situation.

Which table best corresponds to the given situation?

At home today, Tim spent two hours on his schoolwork.
He had no science homework today.
His French assignment was to go over the errors made on yesterday's test.
Tim had only one minor error on that test.
Tomorrow is a math quiz and Tim is confident about the current topic.
Tim needed to do some research for his civics report, due tomorrow.

Table 1

Subject	Minutes
Civics	30
French	60
Math	30
Science	0

Table 2

Subject	Minutes
Civics	20
French	5
Math	100
Science	0

Table 3

Subject	Minutes
Civics	90
French	5
Math	30
Science	0

1. No table can be eliminated based on science time, since ____________________.

2. The French review should not have taken long, so eliminate Table ____________.

3. Given the circumstance, Tim would have spent more time on ____________________ than on ____________________.

4. So, the table that best corresponds is Table ____________________.

Name ______________________ Date __________ Class __________

LESSON 3B

Review for Mastery

Interpreting Graphs and Tables, continued

Tell which graph best corresponds to the given situation.
Situation: James is going to take a bath. After letting only hot water into the tub, he turns off the faucet while he is shaving. Then he adds warm water until the tub is three-quarters full. He takes a bath and empties the tub.

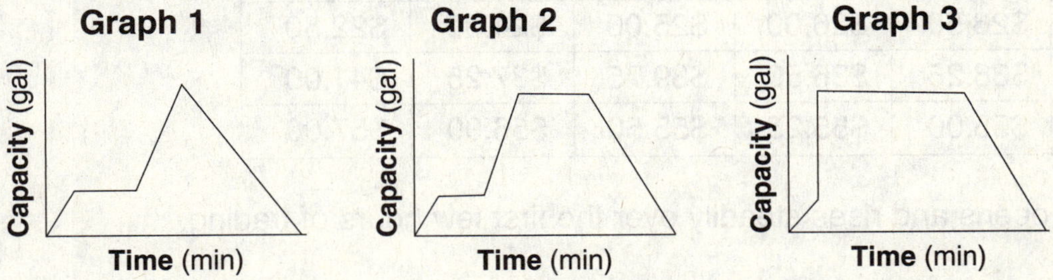

The graph should show two intervals when the amount of water in the tub remains constant: when James was shaving and when he was bathing. Eliminate Graphs 1 and 3 since each has only one interval when the amount of water in the tub remains constant. So, Graph 2 best corresponds to the given situation.

Which graph best corresponds to the given situation?

A car on a dino-coaster ride for children climbs to the top of a hill. At the top of the hill, it slows down and comes to a stop. It goes down the hill, and then slows down as it starts up another hill. It goes down the second hill and then comes to a stop at the end of the ride.

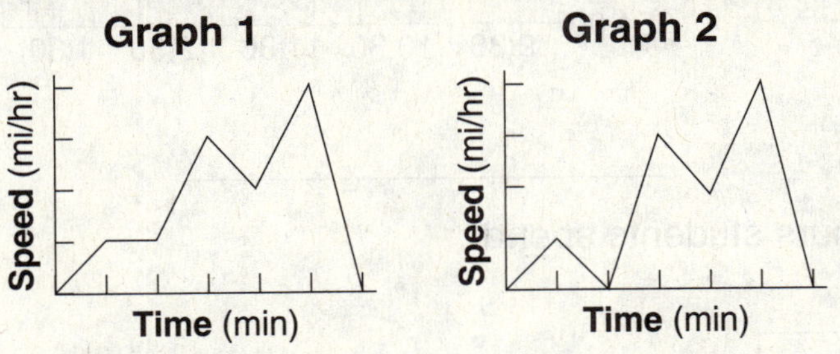

5. In the situation, the car comes to a stop ________ time(s) before the end of the ride.

6. When the car is stopped, the speed is ________ miles per hour.

7. Since it does not reach the stopping speed before the end of the ride, eliminate Graph ________.

8. Since the car slows down before it goes up the second hill, eliminate Graph ________.

9. So, the graph that best corresponds is Graph ________.

Name ______________________ Date ____________ Class ____________

LESSON 3B

Homework and Practice

Interpreting Graphs and Tables

The table gives the prices of three different stocks over the first few hours of trading. Tell which stock corresponds to each situation described below.

Time	9:30	10:30	11:30	12:30	1:30
Stock 1	$26.50	$26.00	$25.00	$24.00	$22.50
Stock 2	$38.25	$36.50	$39.75	$37.25	$41.00
Stock 3	$55.00	$55.25	$55.50	$56.00	$57.00

1. A stock opens and rises steadily over the first few hours of trading.

____

2. A stock opens and declines steadily over the first few hours of trading.

____

Tell which situation in Exercises 1–2 corresponds to each graph.

3. 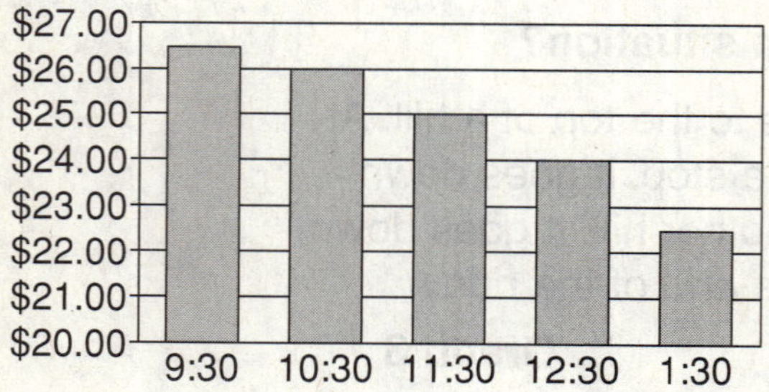

4. 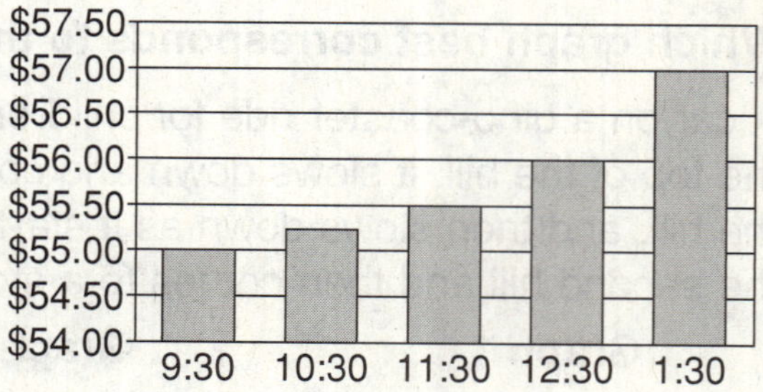

____________________ ____________________

5. Create a graph that illustrates the hours students spend studying for exams.

Student	Number of Hours Studying for Exams
Samantha	3 hours
Seth	3.5 hours
Simon	4 hours
Susan	2.5 hours

EXPLORATION

Program of Studies and Core Content

MA-8-AT-S-PRF2, MA-8-AT-S-PRF3, MA-8-AT-S-PRF4, MA-8-AT-S-PRF6, MA-08-3.3.1

3C Functions

A ***function*** is a rule that assigns one ***output*** value for each ***input*** value. An input/output table is a convenient way of representing functions.

1. Complete the table by applying the rule to each input value.

Input x	Rule: $2000 - 25x$	Output y
0	$2000 - 25(0) = 2000$	2000
1		
2		
3		

2. Determine the rule that produces the following output values from the given input values.

Input x	Rule	Output y
2		0
4		2
6		4
8		6

Think and Discuss

3. Explain the relationship between the input values, a rule, and the output values.

4. Discuss whether the output values have to be different from the input values.

Name ______________________ Date __________ Class __________

LESSON **3C**

Review for Mastery

Functions

From a given rule for a relation, you can write a table of values.

Choose convenient x-values (domain or input).
Get corresponding y-values (range or output).

$y = 2x - 1$

x	$2x - 1$	y
−2	$2(-2) - 1 = -5$	−5
−1	$2(-1) - 1 = -3$	−3
0	$2(0) - 1 = -1$	−1
1	$2(1) - 1 = 1$	1
2	$2(2) - 1 = 3$	3

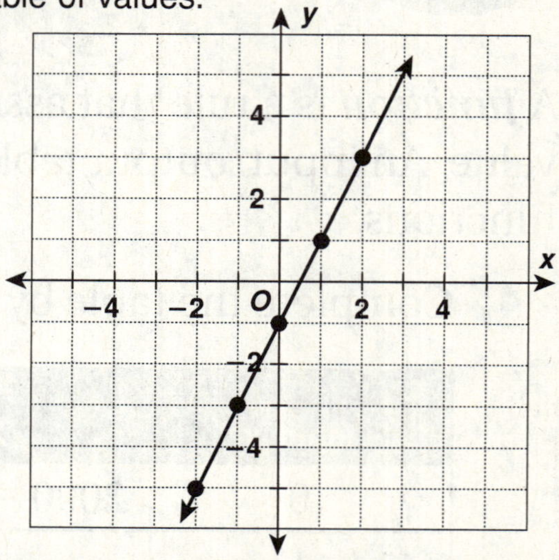

Write a table of values for each function and graph.

1. $y = x - 1$

x	$x - 1$	y
−2	$-2 - 1 = -3$	−3
−1	$-1 - 1 =$	
0	$- 1 =$	
1	$- 1 =$	
2	$- 1 =$	

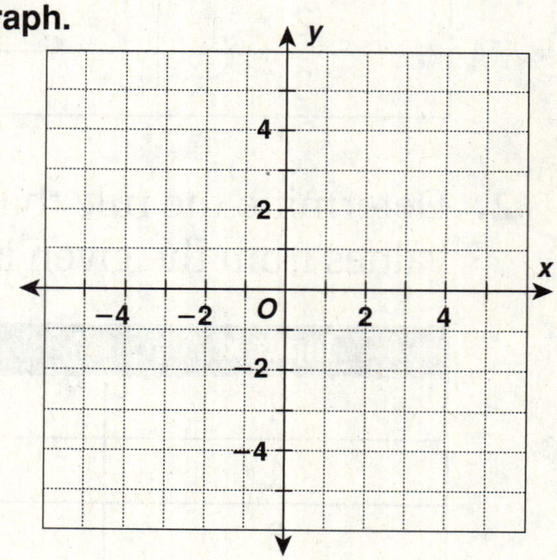

2. $y = 2x + 1$

x	$2x + 1$	y
−2	$2(\quad) + 1 =$	
−1		
0		
1		
2		

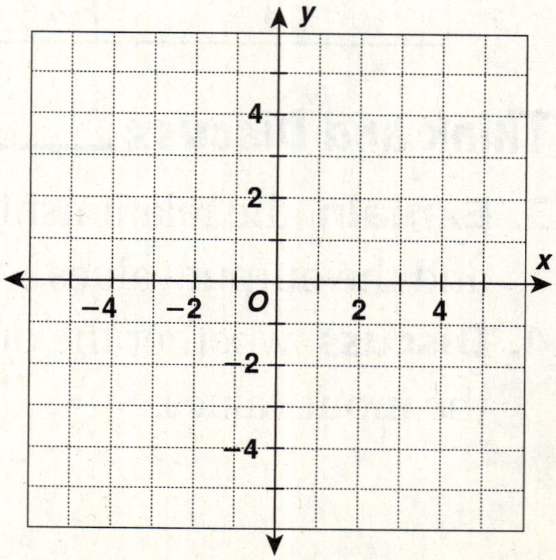

Name ______________________ Date __________ Class __________

LESSON 3C

Review for Mastery

Functions, continued

A function has exactly one output for each input. You can check to see if a relation is a function by making sure each value in the domain is associated with only one value in the range. You can use the vertical line test on a graph.

Determine if the relation is a function. Explain.

x	*y*
1	10
2	20
3	30
4	40
5	50

The relation is a function.
Each *x* value has only one *y* value.

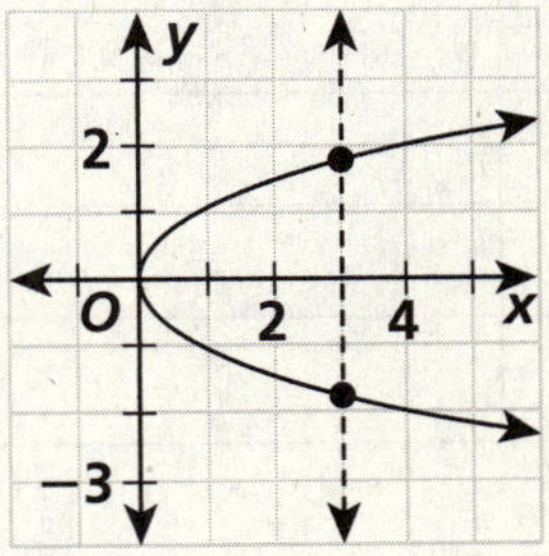

The relation is not a function.
A vertical line intersects the graph at two points.

Determine if the relation is a function. Explain.

3.

x	*y*
−2	5
−1	7
0	9
−1	6
−2	4

____

____

4.

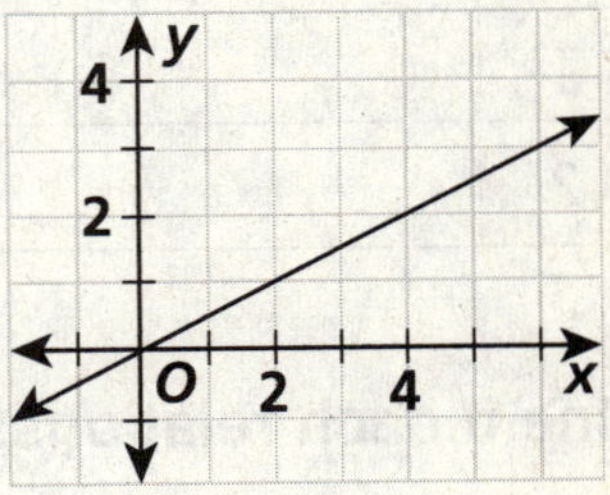

____

____

5.

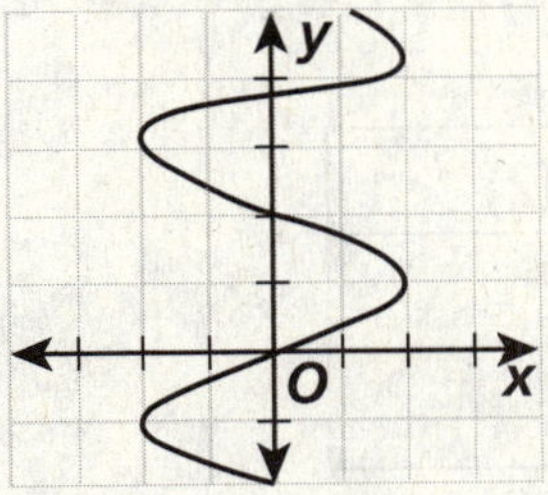

____

____

6.

x	*y*
1	1
2	1
3	1
4	1
5	1

____

____

Name ______________________ Date ____________ Class ____________

LESSON 3C

Homework and Practice

Functions

Complete the table and graph each function.

1. $y = 3x + 4$

x	$3x + 4$	y
−3		
−2		
−1		
0		
1		

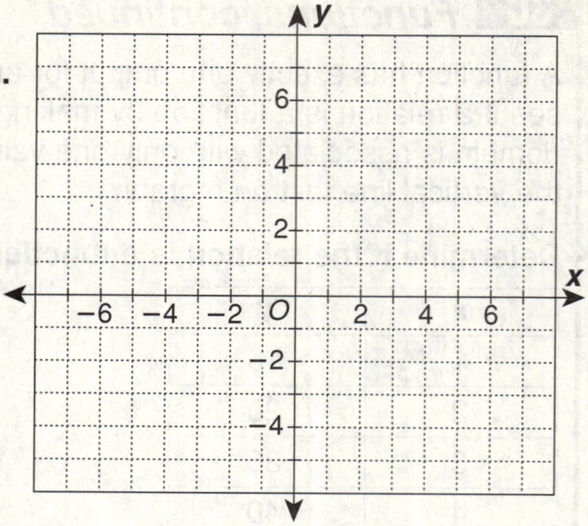

2. $y = 2x^2 - 1$

x	$2x^2 - 1$	y
−2		
−1		
0		
1		
2		

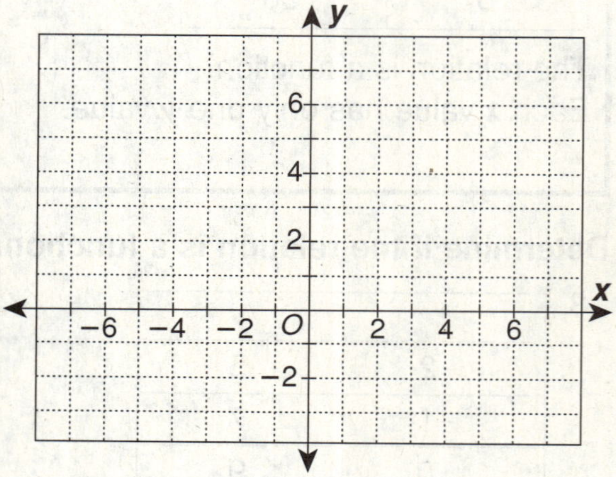

Determine if each relationship represents a function.

3. $y^2 = 4x - 3$

____

4.

x	2	4	6	8
y	−1	−2	−3	−4

____

5. For each function, find $f(0)$, $f(-1)$, and $f(2)$.

Function	$f(0)$	$f(-1)$	$f(2)$
$y = -5x + 2$			
$y = 3x^2 - 8$			

6. Given the function $f(x) = -4x^2 - 5$ and the domain $\{-3, -2, -1, 0, 1, 2, 3\}$. Find the range of the function.

____

EXPLORATION

Program of Studies and Core Content

MA-8-AT-S-PRF2, MA-8-AT-S-PRF3, MA-8-AT-S-PRF4, MA-8-AT-S-PRF6, MA-08-3.3.1, MA-08-5.1.5

3D Equations, Tables, and Graphs

A small plane flies at a constant speed of 120 mi/h. You can use an equation, a table, and a graph to represent the distance d that the plane travels in t hours.

1. Complete the table.

Time (hr), t	distance (mi), d
0.5	
1	120
1.5	
2	
2.5	
3	

2. Plot the points in your table and then connect the points to make a graph that shows the plane's distance as a function of time.

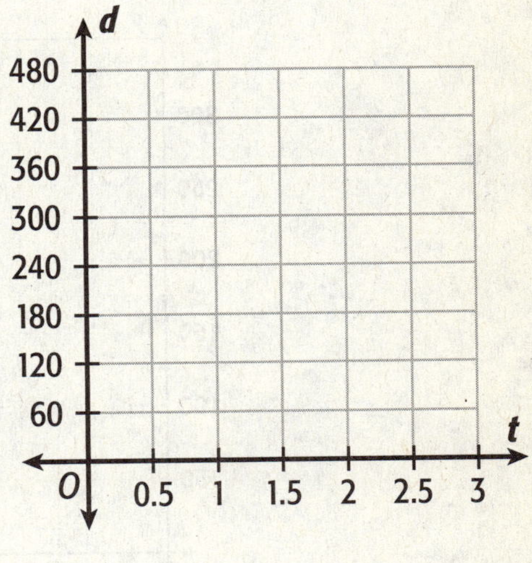

Think and Discuss

3. Describe the graph that you made.

4. Explain how distance is related to time. How can you use this to write an equation relating d and t?

Name ______________________________ Date __________ Class __________

LESSON 3D

Hands-On Lab

Equations, Tables, and Graphs

Activity

Relationships can be represented as equations, tables, or graphs. Each representation shows the same data, but in a different way.

A cheetah can run 50 miles an hour. You can show that relationship in an equation, a table, and a graph. First, fill in the table to show the relationship. Then use the table to make a graph and write an equation.

Hour (*h*)	50*h*	Distance (*d*)
0	0	0
1	50(1)	50
2		
3		
4		
5		

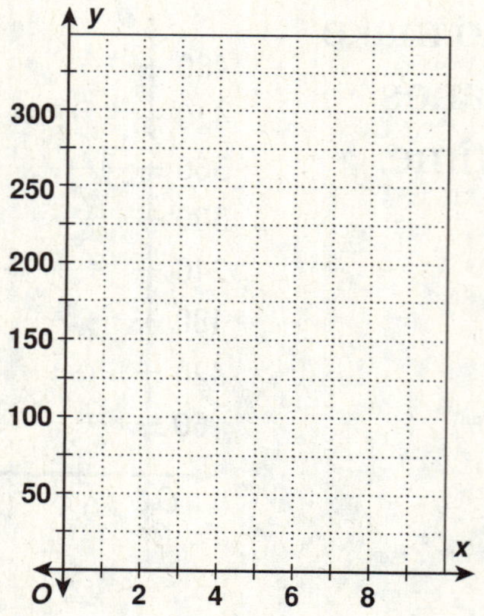

To make the equation, look for a pattern in the values.

Remember: The hour, *h*, is the input value. The distance, *d*, is the output value.

Name ______________________ Date ____________ Class ____________

LESSON 3D

Hands-On Lab

Equations, Tables, and Graphs, continued

Think and Discuss

1. What pattern did you see in the table?

____

____

2. If the cheetah ran 60 miles an hour instead of 50 miles per hour, how would that affect the graph, the table, and the equation?

____

____

____

Try This

A factory can make 25 basketballs a minute. How many basketballs can the factory make in 6 minutes? Make a table, a graph, and an equation to show that relationship.

Minutes	Basketballs made

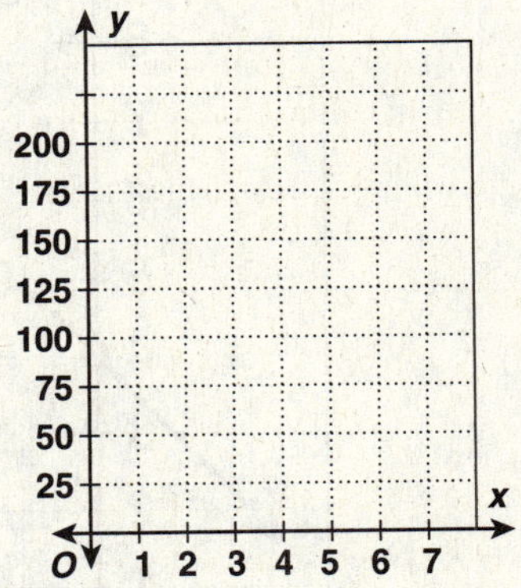

1. Equation:

____

2. How many basketballs can the factory make in 6 minutes?

____

Name ______________________________ Date ____________ Class ____________

LESSON 3D

Review for Mastery

Equations, Tables, and Graphs

Follow these steps to write an equation from its graph.

Step 1 Select points on the graph to make a table of values. Use the *x*-coordinate for the *x*-value and the *y*-coordinate for the *y*-value.

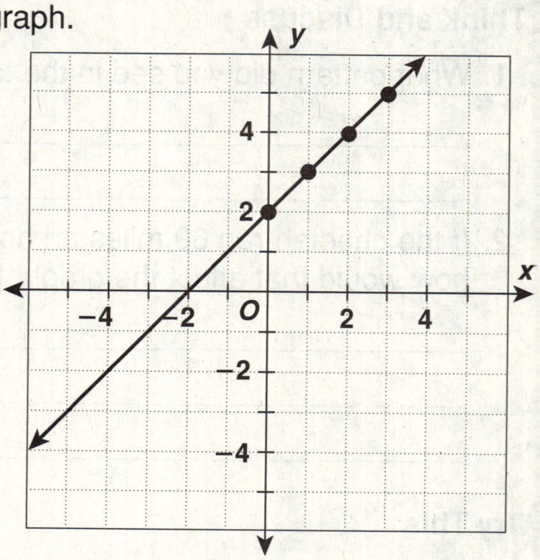

x	*y*
0	2
1	3
2	4
3	5

Step 2 Look for a pattern in the values that relates *y* to *x*.

Each value of *y* is 2 more than the value of *x*.

x *y*

0 + **2** = 2

1 + **2** = 3

2 + **2** = 4

3 + **2** = 5

Step 3 Use the pattern to write an equation.

$y = x + 2$

Complete the table and find a pattern to write an equation of the graph.

1.

x	*y*
0	4
1	
2	
3	

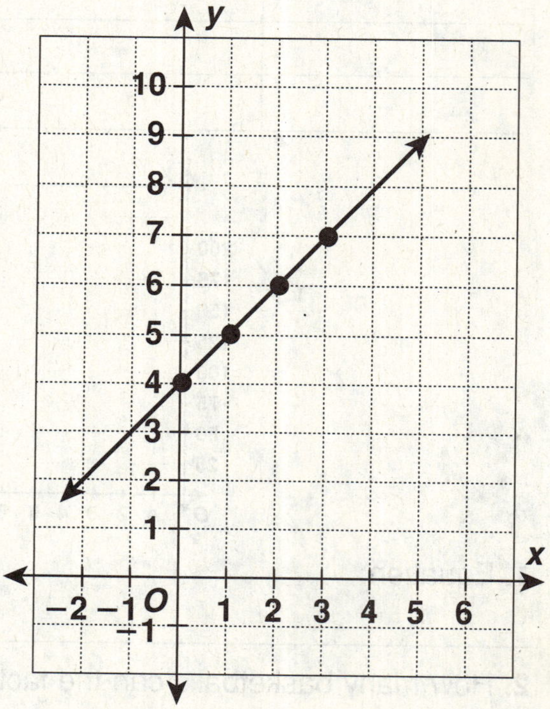

2. *x* *y*

0 + ________ = 4

1 + ________ = ________

____

____

3. Write the equation.

____

Name ______________________________ Date ____________ Class ____________

LESSON 3D

Review for Mastery

Equations, Tables, and Graphs, continued

You can use patterns to translate between tabular and algebraic representations of linear functions.

Grace makes her own bracelets. Each bracelet has 12 beads. Determine a relationship between the *x*- and *y*-values. Then write an equation representing the number of beads Grace needs if she makes *x* bracelets.

Number of bracelets *x*	1	2	3	4	5
Number of beads *y*	12	24	36	48	60

Step 1 List possible relationships between the first *x*- and *y*-values.

$1 + 11 = 12$ (x = 1, y = 12) or $1 \bullet 12 = 12$ (x = 1, y = 12)

Step 2 Test the possible relationships to see if one of them works for the remaining pairs of values.

x	*x* + 11	*y*
2	13	24
3	14	36
4	15	48
5	16	60

x	*x* • 12	*y*
2	24	24
3	36	36
4	48	48
5	60	60

In the first relationship, the outputs do not equal the *y*-values. In the second relationship, the outputs do equal the *y*-values.

Step 3 Write an equation.

$y = x \bullet 12$, or $y = 12x$

Determine a relationship between the *x*- and *y*-values. Write an equation.

4.

x	–2	–1	0	1	2
y	4	2	0	–2	–4

y = ____________

5. Sam is renting a bicycle. The rate is $11 each hour. Write an equation representing the cost of renting the bicycle. How much would it cost Sam to rent the bicycle for 8 hours?

x	0	1	2	3
y				

y = ____________

____

Name ______________________ Date __________ Class __________

LESSON 3D

Homework and Practice

Equations, Tables, and Graphs

Write an equation.

1.

x	–6	–3	0	3	6
y	–2	–1	0	1	2

y = ______________

2.

x	–2	–1	0	1	2
y	5	6	7	8	9

y = ______________

3.

x	–2	–1	0	1	2
y	15	10	5	0	–5

y = ______________

4. Morgan charges $20 plus $10 an hour to cut lawns on the weekend. How much will Morgan earn if he works 6 hours?

a. Complete the table.

Number of hours *x*	1	2			
Amount earned *y*					

b. Make a graph.

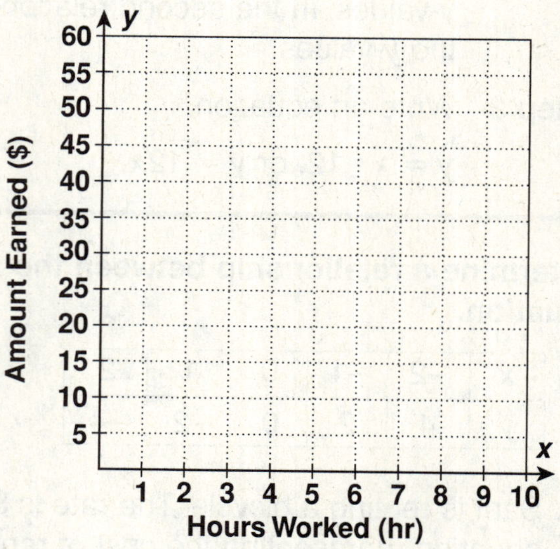

c. Write an equation representing Morgan's earnings.

____

Name ______________________ Date ____________ Class ____________

LESSON 3D

Homework and Practice

Equations, Tables, and Graphs, continued

Use the graph to complete each table. Write an equation representing each graph.

5.

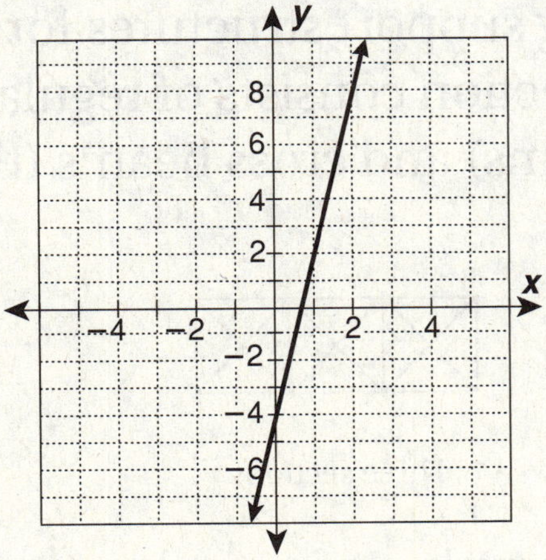

x	0	1	
y	−4		8

$y =$ ______________

6.

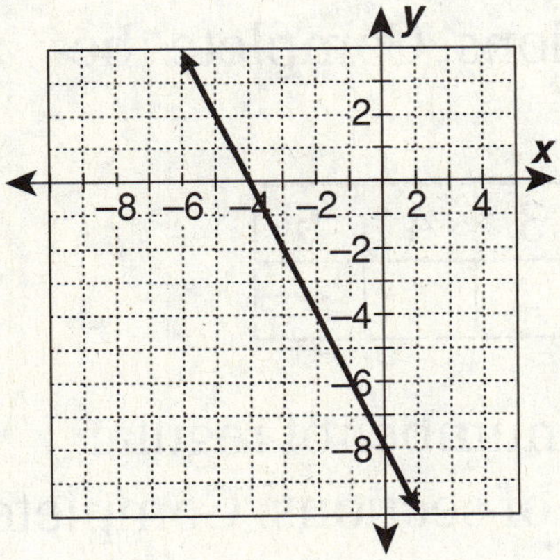

x	−4		−2
y		−8	

$y =$ ______________

7. Chester is participating in a charity road race. Before he runs he collects donations. Each person donates $15 plus $2 for each mile Chester completes.

a. Complete the table.

Number of miles x	1	2		
Amount donated y				

b. Make a graph.

c. Write an equation representing the situation.

____

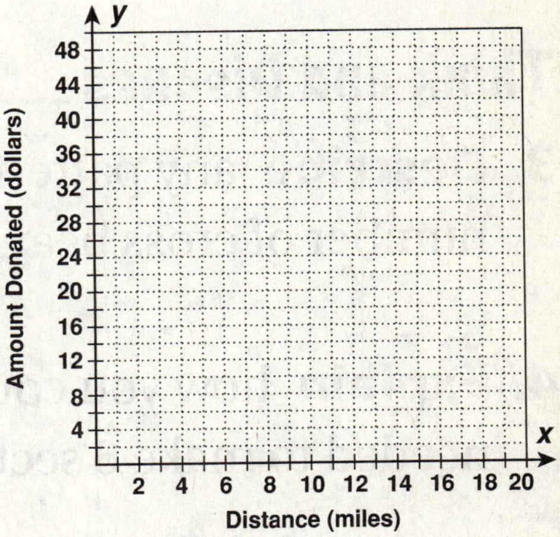

EXPLORATION

Program of Studies and Core Content

MA-8-AT-S-PRF1, MA-08-5.1.1

3E Arithmetic Sequences

Maya is a structural engineer designing support structures for a bridge. As shown in the figures, each section consists of regular beams (the horizontal and vertical beams) and cross beams (the diagonal beams).

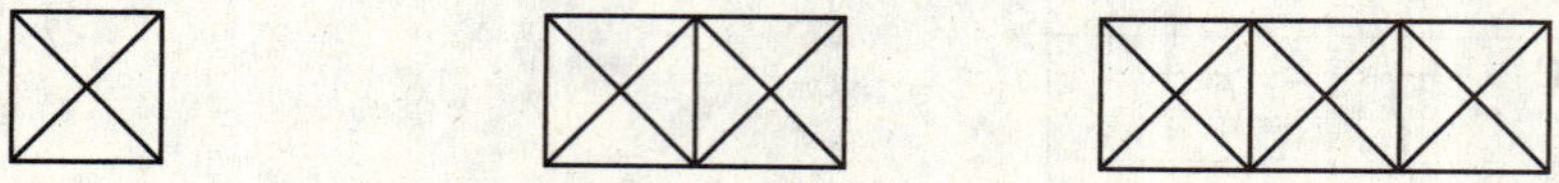

One section Two sections Three sections

1. Maya makes a table showing the number of cross beams needed for various numbers of sections. Complete the table.

Number of Sections	1	2	3	4	5
Number of Cross Beams	2	4			

2. She also makes a table showing the number of regular beams needed for various numbers of sections. Complete the table.

Number of Sections	1	2	3	4	5
Number of Regular Beams	4	7			

Think and Discuss

3. Describe any patterns you notice in the table for the number of cross beams.

4. Explain how you could find the number of regular beams needed to make 6 sections based on the patterns in your table.

Name ______________________ Date ____________ Class ____________

LESSON 3E

Hands-On Lab

Arithmetic Sequences

Activity

A **sequence** is an ordered list of numbers or objects, called **terms**. In an **arithmetic sequence**, the difference between one term and the next is always the same. This difference is called the **common difference**, and it is added to each term to get the next term.

A group of 32 students wants to order pizza for dinner, one large pizza for every 4 students. Find an ad for a large pizza in the paper, or call a local pizza company to find how much a large pizza costs. Use this information to fill in the chart and use an arithmetic sequence to determine the total cost for the pizzas.

Number of students	4	8	16	20	24	28	32
Total cost of pizza							

Use the grid below to model an arithmetic sequence. This time 3 students will share a pizza. Each square in the grid represents one student. Lightly color in groups of 3 to represent each pizza, until you have colored in enough groups of 3 to represent the total number of pizzas the students will need. Then write the sequence of numbers in the group of 3 squares that represent each pizza.

____

____

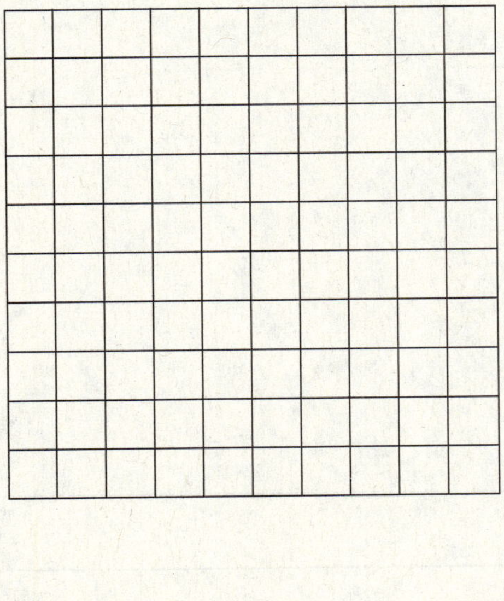

Name ______________________ Date ____________ Class ____________

LESSON 3E

Hands-On Lab

Arithmetic Sequences, continued

Think and Discuss

1. How does an arithmetic sequence work?

____

____

____

2. How can you find the common difference in an arithmetic sequence?

____

____

____

Try This

1. Find the missing terms in this arithmetic sequence.

17		23	26		32		

What is the common difference in this sequence?

____

2. Find a function in the arithmetic sequence below. Use y to identify each term in the sequence and *n* to identify each term's position.

Sequence:

4, 8, 12, 16, 20

n	*y = f (n)*
1	4
2	8
3	12
4	16
5	20

What is the function?

____

____

Name ______________________ Date ____________ Class ____________

LESSON 3E

Review for Mastery

Arithmetic Sequences

In an **arithmetic sequence**, the difference between one term and the next term is always the same. That difference is called the **common difference**.

If the terms increase, the common difference is positive.

2, 6, 10, 14, …

Look at differences between pairs of terms:

$6 - 2 = 4$
$10 - 6 = 4$
$14 - 10 = 4$

The common difference is 4.

To find next term in the sequence, add 4 to the last term.

$14 + 4 = 18$

So, the next term is 18.

If the terms decrease, the common difference is negative.

50, 47, 44, 41, …

Look at differences between pairs of terms:

$47 - 50 = -3$
$44 - 47 = -3$
$41 - 44 = -3$

The common difference is –3.

To find next term in the sequence, add –3 to the last term.

$41 + (-3) = 38$

So, the next term is 38.

Complete to find the common difference of each arithmetic sequence. Then find the next term in the sequence.

1. 3, 9, 15, 21, …

9 – 3 = ________

15 – 9 = ________

21 – 15 = ________

Common difference:

____

Next term:

21 + ________ = ________

2. 44, 42, 40, 38, …

42 – 44 = ________

____

____

Common difference:

____

Next term:

____

3. 4, 9, 14, 19, …

____

____

____

Common difference:

____

Next term:

____

Name ______________________ Date ____________ Class ____________

LESSON 3E

Homework and Practice

Arithmetic Sequences

Find the common difference in each arithmetic sequence.

1. 6, 19, 32, 45, ...

____

2. 12, 19, 26, 33, ...

____

3. 2, 3.4, 4.8, 6.2, ...

____

4. 0, −8, −16, −24, ...

____

5. 25, 16, 7, −2, ...

____

6. −33, −20, −7, 6, ...

____

Find the next three terms in the arithmetic sequence.

7. 14, 21, 28, 35, ...

____

8. $\frac{1}{4}, \frac{3}{4}, 1\frac{1}{4}, 1\frac{3}{4}, \ldots$

____

9. 31, 42, 53, 64, ...

____

10. 8, 4.3, 0.6, −3.1, ...

____

11. $-5\frac{2}{3}, -5\frac{1}{3}, -5, -4\frac{2}{3}$

____

12. 25, 42, 59, 76, ...

____

Find the missing term in the arithmetic sequence.

13. 21, 37, 53, 69, ___, ...

____

14. −30, −21, ___, −3, 6, ...

____

15. ___, 2.9, −1, −4.9, −8.8, ...

____

16. 0.7, ___, 3.5, 4.9, 6.3, ...

____

17. $\frac{1}{8}, \frac{3}{8}, \frac{5}{8},$ ___, $1\frac{1}{8}, \ldots$

____

18. $\frac{5}{6}, 1\frac{1}{6},$ ___, $1\frac{5}{6}, 2\frac{1}{6}$

____

CHAPTER 3

Chapter Summary

Answer these questions to summarize the important concepts from Chapter 3 in your own words.

1. Explain why (4, 37) is a solution to the function $y = 3x + 25$.

2. Explain how to graph the ordered pair $(-5, 2.5)$. Give the quadrant of the point.

3. Explain how to find the solution of two linear equations from a graph of both equations.

4. Explain how to write an equation from data in a table.

For more review of Chapter 3:

- Complete the Chapter 3 Study Guide and Review on pages 150–152 of your textbook.
- Complete the Ready to Go On quizzes on pages 132 and 146 of your textbook.

Name ______________________ Date __________ Class __________

CHAPTER 3

KCCT Prep

Multiple Choice

1. Which equation matches the ordered pairs in the table?

x	0	2	4	6
y	10	8	6	4

(A) $y = x + 10$ (C) $y = 10 - x$
(B) $y = x - 10$ (D) $y = 10x$

2. Given that a is a positive number, which point lies on the x-axis?

(A) $(0, -a)$ (C) $(0, a)$
(B) $(-a, 0)$ (D) $(a, -a)$

3. Which expression is equal to $8(m + 9)$ by the Distributive Property?

(A) $(8 + m) + 9$ (C) $8m + 9$
(B) $8(9 + m)$ (D) $8m + 8 \cdot 9$

4. Jason has $5000 in his savings account. Each month he withdraws $15. Which equation shows the amount y in his account after x years?

(A) $y = 5000 - 180x$
(B) $y = 5000 + 180x$
(C) $y = 500x - 15$
(D) $y = 15 - 500x$

5. Lynda makes a table of ordered pairs for the equation $y = 4x - 5$. If she makes the table correctly, which of these ordered pairs will NOT be in the table?

(A) $(0, -5)$ (C) $(3, 5)$
(B) $(1, -1)$ (D) $(5, 15)$

6. The equation $\square - 2x = -4$ has the solution $x = 7$. What number is missing from the equation?

(A) -24 (C) 10
(B) -10 (D) 24

7. Which equation matches the graph of the line?

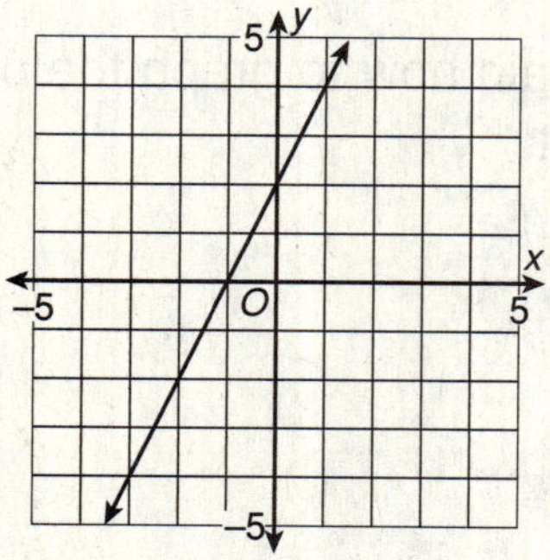

(A) $y = x + 2$ (C) $y = 2x - 2$
(B) $y = 2x$ (D) $y = 2x + 2$

8. Kate joins a DVD club. It costs $11.75 to join the club. Then she can buy as many DVDs as she likes for $6.50 each. Shipping costs $5.50, no matter how many DVD she buys. Which equation shows the total cost y to buy x DVDs?

(A) $y = 17.25x + 6.5$
(B) $y = 17.25 + 6.5x$
(C) $y = 6.5(11.75 + x)$
(D) $y = 5.5 + 11.75 + x$

9. A cup of cooked zucchini contains $2\frac{3}{5}$ grams of fiber. Which is the best estimate of the number of grams of fiber in 3 cups of cooked zucchini?

(A) 5 g (C) 7 g
(B) 6 g (D) 8 g

10. Which equation matches the ordered pairs in the table?

x	−1	0	2	3
y	−8	−3	7	12

Ⓐ $y = x - 7$　Ⓒ $y = 5x - 3$

Ⓑ $y = 8x$　Ⓓ $y = 3x + 1$

11. Which rational number lies closest to −3.8 on a number line?

Ⓐ $-3\frac{3}{4}$　Ⓒ $-3\frac{3}{8}$

Ⓑ −3.95　Ⓓ −3.7

12. The solution of the equation $(m + 4)$ □ $3 = 12$ is $m = 32$. What is the missing operation in the equation?

Ⓐ ÷　Ⓒ +

Ⓑ ×　Ⓓ −

13. Which of the following does NOT show the same relationship as the other three?

Ⓐ The value of y is 2 times the value of x plus 1.

Ⓑ $y = 2x + 1$

Ⓒ

x	0	1	3	5
y	0	2	6	10

Ⓓ

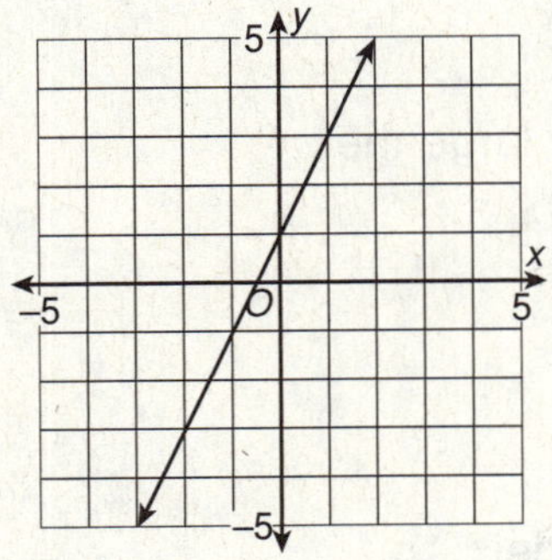

14. Given that b is a negative integer, which of these is not equal to the other three?

Ⓐ $|b|$　Ⓒ $\frac{b}{-1}$

Ⓑ $-b$　Ⓓ b

Open Response

15. Consider the table shown below.

x	0	1	2	3	4
y	−3	−2	−1	0	1

a. Write an equation for the relationship shown in the table.

b. Graph the equation.

16. The graph shows the cost y of renting rollerblades for x hours.

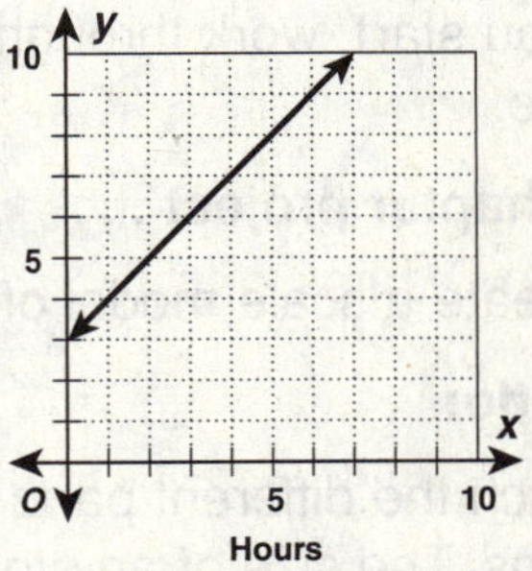

a. Make a table for the relationship.

b. Write an equation for the relationship.

c. Alicia rents rollerblades for 9 hours. Show how to use your equation to find the cost of the rental.

17. Toshio is driving across the country. He maintains an average speed of 58.5 miles per hour.

a. Make a table that shows how far Toshio drives in 1 hour, 2 hours, 3 hours, and 4 hours.

b. Write an equation that gives the total distance y that Toshio drives in x hours.

c. Toshio plans to drive a maximum of 8 hours per day. The distance from Boston to Pittsburgh is about 570 miles. Will Toshio be able to complete this part of his trip in one day? Explain why or why not.

CHAPTER 4

Project

Exponents and Roots: The Tiniest Pieces

Career: Nuclear Physicist

The atom was defined by the ancient Greeks as the smallest particle of matter. We now know that atoms are made up of many smaller particles. Nuclear physicists study these particles using large machines—such as linear accelerators, synchrotrons, and cyclotrons—that can smash atoms to uncover their component parts. Nuclear physicists use mathematics along with the data they discover to create models of the atom and the structure of matter.

Before you start, work through The Lives of Particles worksheet on the next page.

In this chapter project . . .

You create a scale model of an atom.

What to do:

- Research the different parts of an atom, including the nucleus and electrons. The size of an atom is difficult to measure because it is difficult to pinpoint the location of the electrons. For this activity, assume that an atom is a sphere with a radius of 5×10^{-9} m, and 5×10^{-14} m is the radius of the nucleus.
- Calculate how many times larger the atom is than the nucleus. If the nucleus has the same radius as a pencil lead (about 0.25 mm), how large would the atom be?
- To get an idea of how much empty space there is inside an atom, make a human model of the atom by choosing a spot outside for the nucleus, and positioning students using the distance you calculated in the previous step.

What to present:

- A table showing the size of an atom and a nucleus, and how large the atom would be if the nucleus was the size of a pencil lead.
- If possible, a photograph of your human atomic model.

Where's the math?

- Use scientific notation to describe the size of an atom.
- Use ratios to calculate the size of the atom in your scale model.

go.hrw.com
Chapter Project Online
KEYWORD: MT7 Ch4

Name ______________________ Date __________ Class __________

CHAPTER 4

Project Recording Sheet

Exponents and Roots

The Lives of Particles

Complete the table to show the life spans of the subatomic particles listed without using exponents.

Life Spans of Some Subatomic Particles

Subatomic Particle	Independent Life Span (s)	Life Span in Standard Notation (s)
Electron	Stable	Stable
Proton	Stable	Stable
Neutron	920	920
Mu meson	2.2×10^{-6}	
Pi meson	2.6×10^{-3}	
Sigma baryon	1.5×10^{-10}	
Tau lepton	3.4×10^{-13}	
Upsilon meson	1×10^{-20}	
Quark	Unknown	Unknown

1. List the atomic particles from the table in order from shortest life span to longest.

____

____

2. Which atomic particles have independent life spans longer than one second?

____

3. About how many life spans of a mu meson would be as long as a pi meson?

____

EXPLORATION

Program of Studies and Core Content

MA-8-AT-S-VE01, MA-8-NPO-S-N01, MA-08-5.2.1, MA-08-1.3.1

4A Exponents

You can multiply $(-5) \cdot (-5) \cdot (-5) \cdot (-5) \cdot (-5) \cdot (-5)$ using exponents and a calculator.

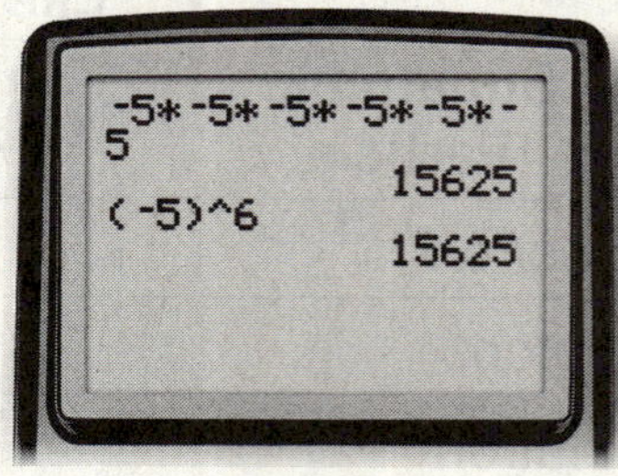

The number -5 is a factor 6 times, so you can write it as $(-5)^6$.

The expressions are equivalent because they have the same value.

$(-5) \cdot (-5) \cdot (-5) \cdot (-5) \cdot (-5) \cdot (-5) = 15{,}625$
and $(-5)^6 = 15{,}625$

Write each of the following using exponents. Then use a calculator to find the value of the expression.

1. $2 \cdot 2 \cdot 2 \cdot 2$

2. $8 \cdot 8 \cdot 8 \cdot 8 \cdot 8 \cdot 8$

3. $(-4) \cdot (-4) \cdot (-4)$

4. $(-5) \cdot (-5)$

5. $3 \cdot 3 \cdot 3 \cdot 3 \cdot 3 \cdot 3 \cdot 3 \cdot 3 \cdot 3$

6. $9 \cdot 9 \cdot 9$

Think and Discuss

7. Discuss whether 3^9 is the same as 9^3.

8. Explain why 4^6 is greater than 4^5.

Name ______________________________ Date ____________ Class ____________

LESSON 4A

Review for Mastery

Exponents

base → 3^5 ← exponent

$$3^5 = \underbrace{3 \cdot 3 \cdot 3 \cdot 3 \cdot 3}_{\text{3 used as a factor 5 times}}$$

3^5 can be read as "the fifth power of 3" or "3 to the fifth power."

Complete to write each expression using an exponent. State the power.

1. $5 \cdot 5 \cdot 5 \cdot 5 = 5^{\_\_}$

the ____________ power of 5

2. $(-7) \cdot (-7) \cdot (-7) = (-7)^{\_\_}$

the ____________ power of ____

Complete to simplify each expression.

3. $(-2)^3 = (-2)(-2)(-2) =$ ____________

4. $10^4 =$ ____ • ____ • ____ • ____ = ____________

5. $(-5)^4 =$ (____)(____)(____)(____) = ____________

When an expression is a product that includes a power, you simplify the power first.

$$3 \cdot 2^3 = 3 \cdot 2 \cdot 2 \cdot 2 = 3 \cdot 8 = 24$$

Complete to simplify each expression.

6. $4 \cdot (-2)^3 = 4$(____)(____)(____) = ____________

7. $5 \cdot 3^3 =$ ____ • ____ • ____ • ____ = ____________

8. $(3 \cdot 2)^3 = 6^3 =$ ____ • ____ • ____ = ____________

9. $(-4(-2))^3 =$ (____)^3 = (____)(____)(____) = ____________

10. $25 - 3(4 \cdot 3^2)$

= 25 − 3(4 • ______)

= 25 − 3(______)

= 25 − ______

= ______

11. $-100 - 2(3 \cdot 4)^2$

= −100 − 2(______)^2

= −100 − 2(______)

= −100 − ______

= ______

12. $15 - 4(3 + 3^2)$

= 15 − 4(3 + ______)

= 15 − 4(______)

= 15 − ______

= ______

Name ______________________ Date __________ Class __________

LESSON 4A

Homework and Practice

Exponents

Write using exponents.

1. $8 \times 8 \times 8 \times 8 \times 8 \times 8 \times 8$ ________

2. $(-5) \times (-5) \times (-5) \times (-5)$ ________

3. $9 \times 9 \times 9 \times 9 \times 9 \times 9$ ________

Evaluate.

4. 10^4 ________

5. $(-4)^3$ ________

6. $(-6)^2$ ________

7. 9^2 ________

8. 14^2 ________

9. $(-8)^2$ ________

10. $(-12)^2$ ________

11. $(-3)^4$ ________

Simplify.

12. $3^3 - 15$ ________

13. $(-4)^4 + 7$ ________

14. $7^2 - 2^5$ ________

15. $78 - 6^2$ ________

16. $11^2 - 3^4 + 3$ ________

17. $16 - 3^3 \times 4$ ________

18. $6^2 + 10 \times 2^4$ ________

19. $-3\,(4^3 + 9^2)$ ________

Evaluate for the given value of the variable.

20. x^3 for $x = -3$ ________

21. $6y^2$ for $y = 4$ ________

22. $w^4 - 17$ for $w = 3$ ________

23. Write an expression for eight times a number used as a factor four times.

____

24. If the length of a side of a regular cube is 12 cm, find its volume. (Hint: $V = l^3$).

____

EXPLORATION

Program of Studies and Core Content
MA-8-NPO-S-NO1

4B Properties of Exponents

You can use patterns to discover properties of exponents.

1. Complete the table.

Product of Powers	Write the Factors	Write As a Single Power
$3^2 \cdot 3^5$	$(3 \cdot 3) \cdot (3 \cdot 3 \cdot 3 \cdot 3 \cdot 3) = 3 \cdot 3 \cdot 3 \cdot 3 \cdot 3 \cdot 3 \cdot 3$	3^7
$4^3 \cdot 4^2$		
$7^4 \cdot 7^4$		
$5^2 \cdot 5^4$		

2. Look at the left-hand and right-hand columns of the table. What patterns do you notice?

3. Complete the table.

Quotient of Powers	Write the Factors	Write As a Single Power
$\frac{3^6}{3^2}$	$\frac{3 \cdot 3 \cdot 3 \cdot 3 \cdot 3 \cdot 3}{3 \cdot 3} = 3 \cdot 3 \cdot 3 \cdot 3$	3^4
$\frac{4^5}{4^3}$		
$\frac{5^4}{5^3}$		
$\frac{8^7}{8^3}$		

4. Look at the left-hand and right-hand columns of the table. What patterns do you notice?

Think and Discuss

5. Explain how you can use what you discovered to write $2^7 \cdot 2^{10}$ as a single power.

6. Explain how you can use what you discovered to write $\frac{2^{10}}{2^7}$ as a single power.

Name ______________________ Date ____________ Class ____________

LESSON 4B

Hands-On Lab

Use Tables to Explore the Properties of Exponents

You can use tables of numbers to see patterns in the value of exponents. These tables will help you see the properties of exponents and write a general form for them.

Activity

1. Complete the table.

$2^1 = 2$	$2^1 = 2$	$2^1 \times 2^1 = 4$	$2^2 = 4$
$2^1 =$ ___	$2^2 = 4$	$2^1 \times 2^2 =$ ___	$2^3 = 8$
$2^1 = 2$	$2^3 =$ ___	$2^1 \times 2^3 =$ ___	$2^4 = 16$
$2^1 =$ ___	$2^4 = 16$	$2^1 \times 2^4 = 32$	$2^5 =$ ___
$2^2 = 4$	$2^{\_\_} = 4$	$2^2 \times 2^2 = 16$	$2^4 =$ ___
$2^2 = 4$	$2^3 = 8$	$2^2 \times 2^{\_\_} = 32$	$\_\_^5 = 32$
$2^2 = 4$	$2^4 =$ ___	$2^{\_\_} \times 2^4 = 64$	$2^{\_\_} = 64$
$2^{\_\_} = 8$	$2^3 = 8$	$2^3 \times 2^3 = 64$	$2^6 =$ ___

Think and Discuss

1. What do all of the bases in the table have in common?

____

____

2. How are the exponents in the third column combined in the fourth column?

____

____

Name ______________________________ Date ____________ Class ____________

LESSON 4B

Hands-On Lab

Use Tables to Explore the Properties of Exponents, continued

3. What pattern do you notice about the products in columns three and four?

____

4. Use the observations from the table on the previous page to complete this equation. This equation summarizes the property for multiplying exponents with the same base: $x^a \times x^b = x^?$.

____

____

Try This

1. Complete the table below, which is similar to the one on the previous page but shows how exponents relate by division. Then complete the expression in the form $x^a \div x^b = x^?$ that summarizes the division property.

$2^4 = 16$	$2^1 = 2$	$2^4 \div 2^1 = 8$	$2^3 = 8$
$2^4 =$ ___	$2^2 = 4$	$2^4 \div 2^2 =$ ___	$2^2 = 4$
$2^4 = 16$	$2^3 =$ ___	$2^4 \div 2^3 =$ ___	$2^1 = 2$
$2^3 = 8$	$2^{\_\_} = 2$	$2^3 \div 2^1 = 4$	$2^2 =$ ___
$2^3 =$ ___	$2^2 = 4$	$2^3 \div 2^2 = 2$	$2^1 =$ ___
$2^2 = 4$	$2^1 = 2$	$2^2 \div 2^{\_\_} = 2$	$\_\_^1 = 2$

____

2. Determine which operation symbol makes this property true: $(x^a)^b = x^{a\,?\,b}$. Use a table if necessary.

____

Name ______________________ Date ____________ Class ____________

LESSON 4B

Review for Mastery

Properties of Exponents

To multiply powers with the same base, keep the base and add the exponents.	To divide powers with the same base, keep the base and subtract the exponents.	To raise a power to a power, keep the base and multiply the exponents.
$x^a \cdot x^b = x^{a+b}$	$x^a \div x^b = x^{a-b}$	$(x^a)^b = x^{ab}$
$4^5 \cdot 4^2 = 4^{5+2} = 4^7$	$4^5 \div 4^2 = 4^{5-2} = 4^3$	$(4^5)^2 = 4^{5(2)} = 4^{10}$
$8^3 \cdot 8 = 8^{3+1} = 8^4$	$8^3 \div 8 = 8^{3-1} = 8^2$	

Complete to see why the rules for exponents work.

1. $4^5 \cdot 4^2 = (\_\_)(\_\_)(\_\_)(\_\_)(\_\_) \cdot (\_\_)(\_\_) = 4^{\_\_}$

2. $8^3 \cdot 8 = (\_\_)(\_\_)(\_\_) \cdot (\_\_) = 8^{\_\_}$

3. $4^5 \div 4^2 = \frac{4^5}{4^2} = \frac{\cancel{4} \cdot \cancel{4} \cdot 4 \cdot 4 \cdot 4}{\cancel{4} \cdot \cancel{4}} = 4^{\_\_}$

4. $8^3 \div 8 = \frac{8^3}{8} = \frac{\cancel{8} \cdot 8 \cdot 8}{\cancel{8}} = 8^{\_\_}$

5. $(4^2)^3 = 4^2 \cdot 4^2 \cdot 4^2 = 4^{2+2+2} = 4^{2(3)} = 4^{\_\_\_}$

Complete to write each product or quotient as one power.

6. $12^3 \cdot 12^2 = 12^{3+2} = 12^{\_\_}$

7. $9^4 \cdot 9^3 = 9^{\_\_\_} = 9^{\_\_}$

8. $\frac{7^6}{7^2} = 7^{6-2} = 7^{\_\_}$

9. $\frac{12^6}{12^4} = 12^{\_\_\_} = 12^{\_\_}$

Write each product or quotient as one power.

10. $10^4 \cdot 10^6 =$ ________

11. $5^5 \cdot 5 =$ ________

12. $4^5 \cdot 4 \cdot 4^3 =$ ________

13. $\frac{15^6}{15^2} =$ ________

14. $\frac{9^5}{9} =$ ________

15. $\frac{2^{10}}{2^2} =$ ________

Simplify.

16. $(5^3)^4 = 5^{3(4)} =$ ________

17. $(6^2)^4 = 6^{2(4)} =$ ________

18. $(2^5)^2 =$ ________

Name ______________________ Date __________ Class __________

LESSON 4B

Homework and Practice

Properties of Exponents

Multiply. Write the product as one power.

1. $10^6 \times 10^9$ ________

2. $a^8 \times a^6$ ________

3. $15^6 \times 15^{12}$ ________

4. $11^{12} \times 11^7$ ________

5. $(-w)^8 \times (-w)^{12}$ ________

6. $(-12)^{18} \times (-12)^{13}$ ________

7. $13^{10} \times 13^{15}$ ________

8. $w^{14} \times w^{12}$ ________

Divide. Write the quotient as one power.

9. $\frac{a^{25}}{a^{18}}$ ________

10. $\frac{(-13)^{14}}{(-13)^9}$ ________

11. $\frac{14^{12}}{14^8}$ ________

12. $\frac{18^{11}}{18^5}$ ________

13. $\frac{19^{15}}{19^4}$ ________

14. $\frac{21^{22}}{21^{20}}$ ________

15. $\frac{(-x)^{17}}{(-x)^7}$ ________

16. $\frac{25^8}{25^3}$ ________

Write the product or quotient as one power.

17. $r^9 \times r^8$ ________

18. $\frac{16^{20}}{16^{10}}$ ________

19. $\frac{x^{15}}{x^9}$ ________

20. $(-17)^8 \times (-17)^7$ ________

21. $27^7 \times 27^6$ ________

22. $\frac{m^{16}}{m^{10}}$ ________

23. $(-b)^{21} \times (-b)^{14}$ ________

24. $\frac{26^{17}}{26^5}$ ________

25. $(-s)^{11} \times (-s)^4$ ________

26. Hampton has a baseball card collection of 5^6 cards. He organizes the cards into boxes that hold 5^4 each. How many boxes will Hampton need to hold the cards? Write the answer as one power.

____

27. Write the expression for a number used as a factor seventeen times being multiplied by a number used as a factor fourteen times. Then write the product as one power.

____

EXPLORATION

Program of Studies and Core Content
MA-8-NPO-S-NS2

4C Scientific Notation

1. Complete the table of values for the powers of ten.

Exponent	Power
−6	$10^{-6} =$
−5	$10^{-5} =$
−4	$10^{-4} =$
−3	$10^{-3} =$
−2	$10^{-2} = \frac{1}{10^2} = \frac{1}{10 \times 10} = 0.01$
−1	$10^{-1} = \frac{1}{10^1} = \frac{1}{10} = 0.1$
0	$10^0 = 1$
1	$10^1 = 10$
2	$10^2 = 10 \times 10 = 100$
3	$10^3 =$
4	$10^4 =$
5	$10^5 =$
6	$10^6 =$

Think and Discuss

2. Discuss the pattern you see in the table of values for powers of ten.

3. Explain how you know that $10^{-9} = 0.000000001$.

Name ______________________ Date __________ Class __________

LESSON 4C Technology Lab

Scientific Notation

Some numbers are so large that even calculators use scientific notation to display them. A scientific calculator displays numbers in standard form up to a certain number of digits. Beyond that limit, the calculator displays numbers in scientific notation. A scientific calculator makes it easy to multiply numbers in scientific notation.

Activity 1

Using a scientific or graphing calculator, you can determine the distance a ray of light travels in one day. The calculator will display the answer in scientific notation when it exceeds the display limit.

Light travels at a constant speed of 186,000 miles per second. To find out how far a ray of light travels in 1 minute, multiply 186,000 by 60 (1 minute = 60 seconds) to get 11,160,000.

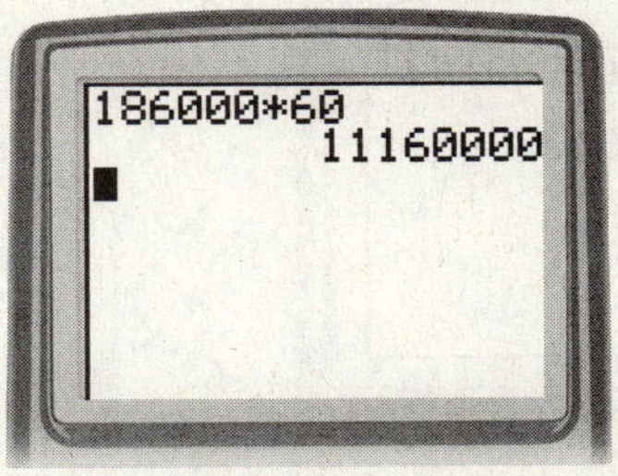

To find out how far the light travels in an hour, multiply 11,160,000 by 60 (1 hour = 60 minutes) to get 669,600,000 miles.

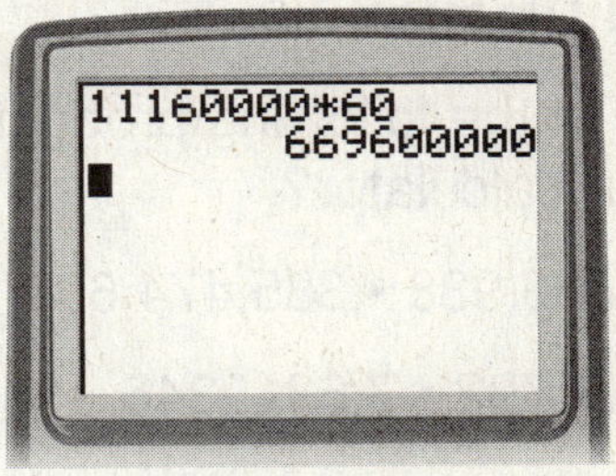

Next, multiply 669,600,000 by 24 (24 hours = 1 day).

The calculator automatically converts the final answer to scientific notation. This number can be used for further calculations.

Name ______________________ Date __________ Class __________

LESSON 4C

Technology Lab

Scientific Notation, continued

Activity 2

A scientific calculator allows you to easily multiply numbers written in scientific notation. Take the final answer from the previous activity (1.60704E10) and multiply that number by 8.96×10^3.

Enter the first number in scientific notation with the following keystrokes: 1.60704 [2nd] [EE ,] 10. Then press the multiplication button.

Enter the second number with the following keystrokes: 8.96 [2nd] [EE ,] 3. Then press "Enter" to obtain the product.

Think and Discuss

1. Which of the following products is displayed in scientific notation on your calculator?

   9,217,400,938 • 385,474,609 ______________

   771,247,792 • 7.6312345 ______________

   Were both of them displayed in scientific notation? Why or why not?

   ____

   ____

Try This

Perform each operation using a scientific calculator.

1. $(4.7 \cdot 10^9) \cdot (3.33 \cdot 10^6)$

   ____

2. $(9.92415 \cdot 10^3) \cdot (8.3 \cdot 10^{-7})$

   ____

3. $(4.3 \cdot 10^2) \div (6.3 \cdot 10^8)$

   ____

Name ______________________________ Date ____________ Class ____________

LESSON 4C

Review for Mastery

Scientific Notation

Standard Notation	Scientific Notation	
	(1st factor is between 1 and 10.)	(2nd factor is an integer power of 10.)
430,000	4.3×10^8	positive integer for large number
0.0000057	5.7×10^{-6}	negative integer for small number

To convert from scientific notation, look at the power of 10 to tell how many places and which way to move the decimal point.

Complete to write each in standard notation.

	1. 4.12×10^6	2. 3.4×10^{-5}
Is the exponent positive or negative?	________	________
Move the decimal point right or left? How many places?	________	________
Write the number in standard notation.	________	________

Write each number in standard notation.

3. 8×10^5 ________

4. 7.1×10^{-4} ________

5. 3.14×10^8 ________

To convert to scientific notation, determine the factor between 1 and 10. Then determine the power of 10 by counting from the decimal point in the first factor to the decimal point in the given number.

Complete to write each in scientific notation.

	6. 32,000,000	7. 0.0000000712
What is the first factor?	________	________
From its location in the first factor, which way must the decimal move to its location in the given number? How many places?	________	________
Write the number in scientific notation.	________	________

Write each number in scientific notation.

8. 41,000,000 ________

9. 0.0000000643 ________

10. 1,370,000,000 ________

Name ______________________ Date ____________ Class ____________

LESSON 4C

Homework and Practice

Scientific Notation

Write each number in standard notation.

1. 6.12×10^2 ____________

2. 7.9×10^{-3} ____________

3. 4.87×10^4 ____________

4. 9.3×10^{-2} ____________

5. 8.06×10^3 ____________

6. 5.7×10^{-4} ____________

7. 3.17×10^{-5} ____________

8. 9.00613×10^{-2} ____________

9. 9.85×10^{-5} ____________

10. 6.004×10^7 ____________

11. 8.23×10^4 ____________

12. 1.48×10^{-6} ____________

Write each number in scientific notation.

13. 108,000,000 ____________

14. 0.5943 ____________

15. 42 ____________

16. 0.0000673 ____________

17. 0.0056 ____________

18. 6004 ____________

19. 0.00852 ____________

20. 24,631,500 ____________

21. 89450 ____________

22. 0.005702 ____________

23. 8,005,000,000 ____________

24. 0.00012805 ____________

25. The mass of Earth is 5,980,000,000,000,000,000,000,000 kilograms. Write this number in scientific notation.

____

26. The mass of a dust particle is 7.53×10^{-10}. Write this number in standard notation.

____

EXPLORATION

Program of Studies and Core Content

MA-8-NPO-S-NO2, MA-8-NPO-S-NS1, MA-8-NPO-S-NS3, MA-8-NPO-S-E1, MA-8-NPO-S-E2, MA-08-1.1.2, MA-08-1.3.2

4D Squares and Square Roots

The sequence shows the square numbers 1, 4, 9, and 16.

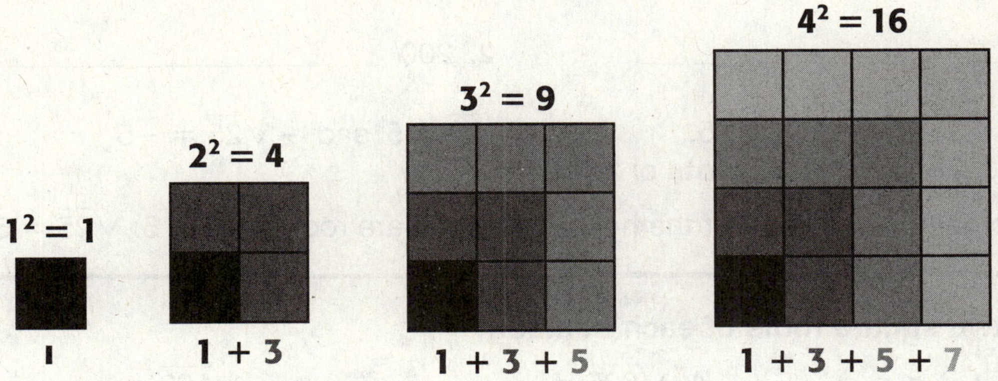

1. Draw a picture to show that $5^2 = 1 + 3 + 5 + 7 + 9$.

2. Add the odd numbers $1 + 3 + 5 + 7 + 9 + \cdots + 17 + 19$. What square number do you get?

3. The table starts with $11^2 = 1 + 3 + 5 + 7 + 9 + 11 + 13 + 15 + 17 + 19 + 21 = 121$. Complete the table by adding the next odd number to this sum.

11^2	12^2	13^2	14^2	15^2	16^2	17^2	18^2	19^2	20^2
121									

Think and Discuss

4. Explain how you can determine square numbers using sums of odd numbers.

5. Demonstrate that the value of 22^2 can be determined by adding odd numbers.

Name ______________________ Date __________ Class __________

LESSON 4D

Review for Mastery

Squares and Square Roots

A **perfect square** has two identical integer factors.

$25 = 5 \times 5 = 5^2$ **or** $25 = (-5) \times (-5) = (-5)^2$ 25 is a perfect square.

Tell if the number is a perfect square.
If yes, write its identical factors.

1. 121 ______________ 2. 200 ______________

Since $5^2 = 25$ and $(-5)^2 = 25$, both 5 and −5 are **square roots** of 25. $\sqrt{25} = 5$ and $-\sqrt{25} = -5$

The **principal square root,** or the nonnegative square root, of 25 is 5: $\sqrt{25} = 5$

Write the two square roots of each number.

3. $\sqrt{81} =$ ________ 4. $\sqrt{625} =$ ________ 5. $\sqrt{169} =$ ________

$-\sqrt{81} =$ ________ $-\sqrt{625} =$ ________ $-\sqrt{169} =$ ________

Write the principal square root of each number.

6. $\sqrt{144} =$ ________ 7. $\sqrt{6400} =$ ________ 8. $\sqrt{10{,}000} =$ ________

Use the principal square root when evaluating an expression. For the order of operations, do square root first, as you would an exponent.

$5\sqrt{100} - 3$
$5(10) - 3$
$50 - 3$
47

Complete to evaluate each expression.

9. $3\sqrt{144} - 20$

$3 \times$ ______ $- 20$

______ $- 20$

______

10. $\sqrt{25 + 144} + 13$

______ $+ 13$

______ $+ 13$

______

11. $\sqrt{\frac{100}{25}} + \frac{1}{2}$

$\sqrt{4} + \frac{1}{2}$

______ $+ \frac{1}{2}$

______

Name ______________________ Date ____________ Class ____________

LESSON **4D**

Review for Mastery

Squares and Square Roots, continued

Recall that the formula for the area of a square is $A = s^2$. A power in which the exponent is 2 is called a *square.*

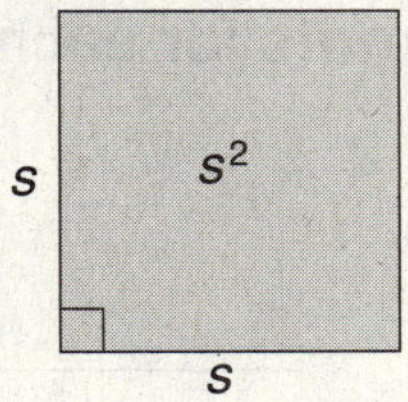

Use the model to help you simplify the square.

12. 5^2

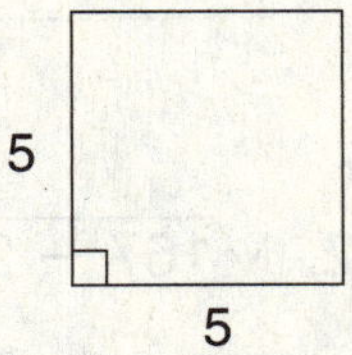

$A = s^2$

$A =$ ______^{2}

$A =$ ______ • ______

$A =$ ______

13. 9^2

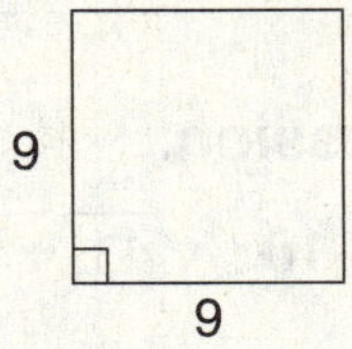

$A = s^2$

$A =$ (______)^2

$A =$ ______ • ______

$A =$ ______

Simplify.

14. 13^2 ______

15. 16^2 ______

16. 11^2 ______

17. 7^2 ______

The numbers 36 and 81 are *perfect squares.* **Perfect squares** are numbers that are the squares of integers.

The square root of 36 is 6. $\sqrt{36} = 6$ because $6 \cdot 6 = 36$.

A table of square roots can help you estimate the square root of a number that is not a perfect square.

Square Root	1	2	3	4	5	6	7	8	9	10
Perfect Squares	1	4	9	16	25	36	49	64	81	100

18. $\sqrt{44}$

Find the perfect square nearest 44.

44 is closest to ______.

Since $\sqrt{49} =$ ______,

$\sqrt{44}$ is closest to ______.

19. $\sqrt{87}$

Find the perfect square nearest 87.

87 is closest to ______.

Since $\sqrt{81} =$ ______,

$\sqrt{87}$ is closest to ______

Name ________________________ Date ______________ Class ______________

LESSON 4D

Homework and Practice

Squares and Square Roots

Find the two square roots for each number.

1. 16 ________

2. 9 ________

3. 64 ________

4. 121 ________

5. 36 ________

6. 100 ________

7. 225 ________

8. 400 ________

Evaluate each expression.

9. $\sqrt{27 + 37}$ ________

10. $\sqrt{41 + 59}$ ________

11. $\sqrt{122 - 41}$ ________

12. $\sqrt{167 - 23}$ ________

13. $3\sqrt{81} + 19$ ________

14. $25 - \sqrt{25}$ ________

15. $\sqrt{169} - \sqrt{36}$ ________

16. $\sqrt{196} + 25$ ________

17. $\frac{\sqrt{81}}{9}$ ________

18. $-4.9\sqrt{64}$ ________

19. $\frac{\sqrt{225}}{\sqrt{25}}$ ________

20. $\frac{\sqrt{100}}{-2.5}$ ________

21. Find the product of six and the sum of the square roots of 100 and 225.

____

22. Explain how you can verify that $\sqrt{289}$ is 17 without using a calculator.

____

23. If a replica of the ancient pyramids were built with a base area of 1,024 in.^{2}, what would be the length of each side?(Hint: $s = \sqrt{A}$)

____

24. The maximum displacement speed of a boat is found using the formula: Maximum Speed in km/h = $4.5\sqrt{\text{the waterline length of the boat in meters}}$. Find the maximum displacement speed of a boat that has a waterline length of 9 meters.

____

Name ______________________ Date __________ Class __________

LESSON 4D

Homework and Practice

Squares and Square Roots, continued

Model each power using a square. Then evaluate the power.

25. 7^2 ______ **26.** $(2.6)^2$ ______ **27.** $(5.8)^2$ ______ **28.** 26^2 ______

29. 21^2 ______ **30.** $(1.7)^2$ ______ **31.** $(9.9)^2$ ______ **32.** 13^2 ______

Estimate each square root to the nearest whole number. Use a calculator to check the reasonableness of your answers.

33. $\sqrt{33}$ ______ **34.** $\sqrt{15}$ ______ **35.** $\sqrt{37}$ ______ **36.** $\sqrt{24}$ ______

37. $\sqrt{51}$ ______ **38.** $\sqrt{8}$ ______ **39.** $\sqrt{148}$ ______ **40.** $\sqrt{102}$ ______

41. $\sqrt{46}$ ______ **42.** $\sqrt{171}$ ______ **43.** $\sqrt{90}$ ______ **44.** $\sqrt{19}$ ______

Simplify each expression.

45. $\sqrt{9^2}$ ______ **46.** $(\sqrt{9})^2$ ______ **47.** $(\sqrt{15})^2$ ______ **48.** $\sqrt{15^2}$ ______

49. Squaring and taking the square root are said to be inversely related. Explain what this means.

____

50. The area of a square tetherball court is 260 ft^2. What is the approximate length of each side of the court? Find your answer to the nearest foot.

____

51. Brian jogs one time around a square park with an area of 4 mi^2. How far does Brian jog?

____

52. Steve wants to make a curtain to cover a square window with an area of 9 ft^2. How long should each side of the curtain be?

____

EXPLORATION

Program of Studies and Core Content

MA-8-NPO-S-E1, MA-8-NPO-S-E2, MA-8-NPO-S-NS1, MA-8-NPO-S-NS2, MA-8-NPO-S-NS3, MA-08-1.2.1, MA-08-2.1.2

4E Estimating Square Roots and Cube Roots

Knowing the square numbers can help you estimate square roots.

1. Complete the table of squares.

1^2	2^2	3^2	4^2	5^2	6^2	7^2	8^2	9^2	10^2
1									

11^2	12^2	13^2	14^2	15^2	16^2	17^2	18^2	19^2	20^2
121									

Use the table of squares above to help you estimate each square root to the nearest tenth. Use a calculator to check your estimates. Round to two decimal places.

	Square Root	Estimate	Calculator
2.	$\sqrt{10}$		
3.	$\sqrt{20}$		
4.	$\sqrt{200}$		
5.	$\sqrt{300}$		
6.	$\sqrt{57}$		
7.	$\sqrt{130}$		

Think and Discuss

8. Discuss your strategy for estimating square roots.

Name ______________________________ Date __________ Class __________

LESSON 4E

Technology Lab

Finding Square Roots

If you know the area of a square or circle and need to find the length of the side or the radius, you can find your answer by using square roots.

Activity

The area of a square is 289 cm^2. What is the length of a side?

$s = \sqrt{289}$

On your graphing calculator use the following keystrokes to find the square root:

[2nd] [$\sqrt{\ }$ / x^2] 289 [ENTER]

The length of a side is 17 cm.

The area (A) of a circle is 520 cm^2. What is the radius of the circle?

$$A = \pi r^2 \qquad r = \sqrt{\frac{A}{\pi}} = \sqrt{\frac{520}{\pi}} \approx \sqrt{165.52} \approx 12.87$$

On your calculator use the following keystrokes:

[2nd] [$\sqrt{\ }$ / x^2] 520 [÷] [2nd] [π / ^] [ENTER]

The radius is about 12.87 cm.

Think and Discuss

1. Enter [π] on your calculator. Is π a rational number? Explain why or why not.

____

____

2. Use your calculator to find the square root of pi. Is the number you see in your display the actual square root or an approximate square root? How do you know?

____

____

Name ______________________________ Date ____________ Class ____________

LESSON 4E

Technology Lab

Finding Square Roots, continued

Try This

1. Find the length (s) of a side of a square with an area A of 12.25 cm^2. Show the buttons you pressed on your calculator to find the answer.

____

____

____

2. Find the radius of a circle with an area of 144 cm^2. Show the buttons you pressed on your calculator to find the answer.

____

____

____

Name ______________________________ Date ____________ Class ____________

LESSON 4E

Review for Mastery

Estimating Square Roots and Cube Roots

To locate a square root between two whole numbers, refer to the table.

Number	1	2	3	4	5	6	7	8	9	10
Square	1	4	9	16	25	36	49	64	81	100
Number	11	12	13	14	15	16	17	18	19	20
Square	121	144	169	196	225	256	289	324	361	400

Locate $\sqrt{260}$ between two whole numbers.

260 is between the perfect squares 256 and 289: $256 < 260 < 289$

So: $\sqrt{256} < \sqrt{260} < \sqrt{289}$

And: $16 < \sqrt{260} < 17$

Use the table to complete the statements.

1. ____ $< 39 <$ ____

____ $< \sqrt{39} <$ ____

____ $< \sqrt{39} <$ ____

2. ____ $< 130 <$ ____

____ $< \sqrt{130} <$ ____

____ $< \sqrt{130} <$ ____

After locating a square root between two whole numbers, you can determine which of the two whole numbers the square root is closer to.

27 is between the perfect squares 25 and 36: $25 < 27 < 36$

So: $\sqrt{25} < \sqrt{27} < \sqrt{36}$

And: $5 < \sqrt{27} < 6$

The difference between 27 and 25 is 2; the difference between 36 and 27 is 9. So, $\sqrt{27}$ is closer to 5.

$25 < 27 < 36$

2   9

Complete the statements.

3. $100 < 106 < 121$

____ $< \sqrt{106} <$ ____

____ $< \sqrt{106} <$ ____

$106 - 100 =$ ____

$121 - 106 =$ ____

$\sqrt{106}$ is closer to ____ than ____

4. ____ $< 250 <$ ____

____ $< \sqrt{250} <$ ____

____ $< \sqrt{250} <$ ____

$250 -$ ____ $=$ ____

____ $- 250 =$ ____

$\sqrt{250}$ is closer to ____ than ____

Name ______________________ Date ____________ Class ____________

LESSON 4E

Review for Mastery

Estimating Square Roots and Cube Roots, continued

Just as the square root of a number is one of two equal factors of the number, a *cube root* of a number is one of three equal factors of the number. The symbol for a cube root is $\sqrt[3]{\ }$.

Simplify the expression $\sqrt[3]{8}$.

$\sqrt[3]{8}$

$= \sqrt[3]{2 \cdot 2 \cdot 2}$ *Rewrite 8 using three equal factors: 2 • 2 • 2.*

$= 2$ *Pull the factors out from under the radical sign.*

Simplify each expression.

5. $\sqrt[3]{1{,}000}$ ________ **6.** $\sqrt[3]{216}$ ________ **7.** $\sqrt[3]{64}$ ________

You can locate the cube root of a number between two whole numbers using the table below.

Number	1	2	3	4	5	6	7	8	9	10
Cube	1	8	27	64	125	216	343	512	729	1000

The cube root, $\sqrt[3]{90}$, is between two whole numbers. Name them. Explain your answer.

The perfect cubes on either side of 90 are 64 and 125. Since $\sqrt[3]{64} = 4$ and $\sqrt[3]{125} = 5$ and 90 is between 64 and 125, $\sqrt[3]{90}$ is between 4 and 5.

Each cube root is between two whole numbers. Name the whole numbers. Explain your answer.

8. $\sqrt[3]{325}$

____

____

____

9. $\sqrt[3]{510}$

____

____

____

Name ______________________ Date __________ Class __________

LESSON 4E

Homework and Practice

Estimating Square Roots and Cube Roots

Each square root is between two whole numbers. Name the whole numbers. Determine which of the two whole numbers the square root is closer to.

1. $\sqrt{51}$

____

2. $\sqrt{39}$

____

3. $\sqrt{240}$

____

4. $\sqrt{155}$

____

Simplify each expression.

5. $\sqrt[3]{729}$ __________

6. $\sqrt[3]{343}$ __________

7. $\sqrt[3]{125}$ __________

8. $\sqrt[3]{512}$ __________

9. $\sqrt[3]{1}$ __________

10. $\sqrt[3]{27}$ __________

Each cube root is between two whole numbers. Name the whole numbers. Explain your answer.

11. $\sqrt[3]{600}$

____

____

12. $\sqrt[3]{289}$

____

____

13. $\sqrt[3]{354}$

____

____

14. $\sqrt[3]{132}$

____

____

15. $\sqrt[3]{644}$

____

____

16. $\sqrt[3]{859}$

____

____

EXPLORATION

Program of Studies and Core Content

MA-8-NPO-S-E1, MA-8-NPO-S-NS1, MA-8-NPO-S-NS2, MA-8-NPO-S-NS3, MA-08-1.1.1, MA-08-1.1.2, MA-08-1.1.3

4F The Real Numbers

The Set of Real Numbers

Rational numbers can be written as fractions or as decimals that terminate or repeat.	*Irrational numbers* are decimals that do not terminate or repeat.
Examples: $\frac{1}{4}$ $5 = \frac{5}{1}$ $\sqrt{1.69} = 1.3$ $\frac{5}{3} = 1.\overline{6}$	Examples: $\sqrt{7} = 2.645751311\ldots$ $\pi = 3.141592653\ldots$

Classify each number as rational or irrational.

		Rational	Irrational
1.	$2\frac{1}{2}$		
2.	$\sqrt{24}$		
3.	$\sqrt{16}$		
4.	$7.\overline{7}$		
5.	$\sqrt{\frac{4}{9}}$		

Think and Discuss

6. Explain how to classify numbers as rational.

7. Explain how to classify numbers as irrational.

Name ______________________ Date __________ Class __________

LESSON 4F

Hands-On Lab

Real Numbers

The square roots of numbers that are not perfect squares are impossible to express exactly as decimals, because they involve endless non-repeating decimals. They are definite quantities and can be used in calculations and in modeling figures.

Activity

Take a sheet of graph paper and turn it on its side. In the lower left corner, draw two squares side by side, each 5 units on a side. Label the first square ABDC and the second CDFE, as shown.

Take a compass and open it so that the point is on point B and the pencil is on point C, diagonally opposite. The width of the opening of the compass is equal to the length of the side of the square multiplied by the square root of two. With the compass at that opening, draw a square of that height and width. The new square should have an area twice that of the first square you drew.

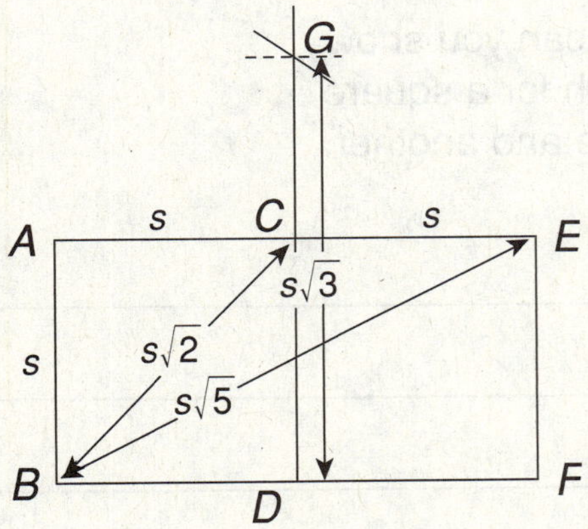

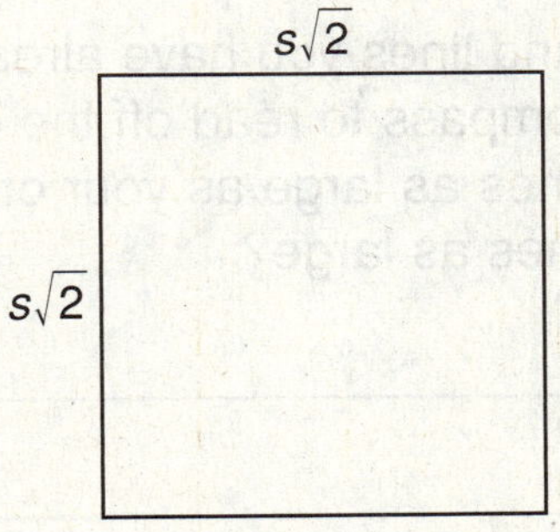

Between the first two squares you drew, extend a vertical line upwards from CD. Now put the point of the compass on point B and the pencil on point F. Keeping the point of the compass where it is, swing the pencil up to intersect the vertical line you just drew. Label the intersection point as point G. Now open the compass with its point on point D and the pencil on point G. The opening of the compass is equal to the side of the original square multiplied by the square root of three. Use that compass opening to draw a square of that side length. Its area should be three times that of the first square you drew.

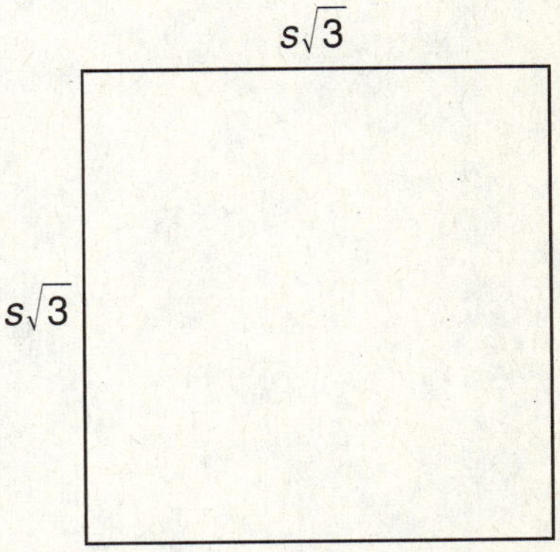

Name ______________________ Date ____________ Class ____________

LESSON 4F

Hands-On Lab

Real Numbers, continued

Think and Discuss

1. How do you know that the squares you constructed have two and three times the area, respectively, of the first square you drew?

____

____

____

2. Could you perform these operations just as well using a ruler and not a compass?

____

____

Try This

Using the points and lines you have already labeled, can you show how to use the compass to read off the correct length for a square that will be four times as large as your original square and another that will be five times as large?

____

____

____

Name ______________________________ Date ___________ Class ___________

LESSON 4F

Technology Lab

The Real Numbers

Activity

You can use a calculator to verify the density property of real numbers. To find a number between two given numbers, add the two numbers together and divide by two.

Find a real number between

$1\frac{3}{4}$ and $3\frac{1}{4}$

On your calculator, enter the equivalent of $1\frac{3}{4}$ by pressing:

[3] [÷] [4] [+] [1]

And the equivalent of $3\frac{1}{4}$ by pressing:

[1] [÷] [4] [+] [3]

Add each fraction equivalent together and divide by 2.

[(] [3] [÷] [4] [+] [1] [+] [1] [÷] [4] [+] [3] [)] [÷]

[2] [ENTER]

Name ______________________ Date ___________ Class ___________

LESSON **4F**

Technology Lab

The Real Numbers, continued

Think and Discuss

1. Why do each of the fractions need to be put into the calculator as a division problem first?

____

____

Try This

Use the calculator to find a real number between the given numbers.

1. $1\frac{6}{8}$ and $2\frac{2}{8}$ **2.** 1 and $2\frac{1}{3}$ **3.** 1.4 and $3\frac{6}{10}$ **4.** 2.45 and 3.55

__________ __________ __________ __________

Name ______________________ Date ____________ Class ____________

LESSON 4F

Review for Mastery

The Real Numbers

The set of **rational numbers** contains all integers, all fractions, and decimals that end or repeat.

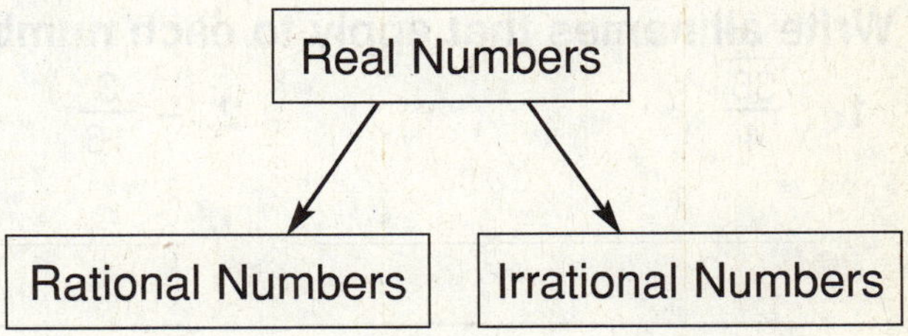

Irrational numbers can only be written as decimals that do not end or repeat.

Together, the rational numbers and the irrational numbers form the set of **real numbers.**

Square roots of numbers that are perfect squares are rational.

$\sqrt{25} = 5$

Square roots of numbers that are not perfect squares are irrational.

$\sqrt{3} = 1.732050807\ldots$

Tell if each number is rational or irrational.

1. $\sqrt{7}$ ____________
2. $\sqrt{81}$ ____________
3. $\sqrt{169}$ ____________
4. $\sqrt{101}$ ____________

The square of a nonzero number is positive. $3^2 = 9$ and $(-3)^2 = 9$
So, the square root of a negative number is not a real number.
$\sqrt{-9}$ is not a real number.

Tell if each number is real or not real.

5. -8 ____________
6. $-\sqrt{8}$ ____________
7. $\sqrt{-8}$ ____________
8. $\sqrt{-25}$ ____________

Between any two real numbers, there is always another real number. One way to find a number between two numbers is to find the number halfway between the two.

To find a real number between $7\frac{1}{5}$ and $7\frac{2}{5}$,
divide their sum by 2: $7\frac{1}{5} + 7\frac{2}{5} = \left(14\frac{3}{5}\right) \div 2 = 7\frac{3}{10}$

Find a real number between each pair.

9. $8\frac{3}{7}$ and $8\frac{4}{7}$ ____________
10. -1.6 and -1.7 ____________
11. $-3\frac{7}{9}$ and $-3\frac{2}{9}$ ____________
12. $6\frac{1}{2}$ and $6\frac{3}{4}$ ____________

Name ______________________ Date ____________ Class ____________

LESSON 4F

Homework and Practice

The Real Numbers

Write all names that apply to each number.

1. $\sqrt{\dfrac{36}{4}}$

2. $-\dfrac{3}{16}$

3. $\sqrt{0.81}$

4. -81

5. $-7.23\overline{3}$

6. $\sqrt{95}$

State if the number is rational, irrational, or not a real number.

7. $\sqrt{49}$

8. $-\sqrt{144}$

9. $\dfrac{9}{\sqrt{3}}$

10. $\dfrac{\sqrt{81}}{\sqrt{9}}$

11. $\dfrac{21}{0}$

12. $\dfrac{20}{8}$

13. $\sqrt{-100}$

14. 8.67

Find a real number between each pair of numbers.

15. $4\frac{2}{5}$ and $4\frac{3}{5}$

16. 7.25 and $\dfrac{15}{2}$

17. $\dfrac{5}{8}$ and $\dfrac{3}{4}$

18. Give an example of a rational number between $-\sqrt{36}$ and $\sqrt{36}$

19. Give an example of an irrational number less than 0.

20. Give an example of a number that is not real.

EXPLORATION

Program of Studies and Core Content

MA-8-M-S-MPA6, MA-08-2.1.6

4G The Pythagorean Theorem

The model shows a visual example of the Pythagorean Theorem.

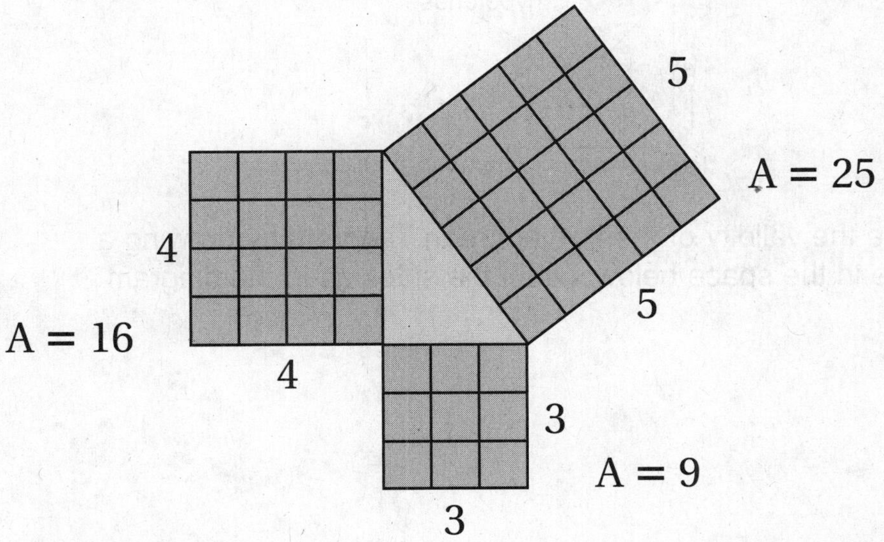

1. Using graph paper, cut out three squares with the same dimensions as the squares in the figure above.

2. Next, try to fit the two smaller squares completely inside the larger square without any overlap between the two smaller squares. (You may have to cut the small squares into sections.)

3. How does the area of the largest square compare to the combined area of the two smaller squares?

Think and Discuss

4. Explain how each side length of the triangle above is related to the area of the square that lies along that side of the triangle.

Name ______________________ Date __________ Class __________

LESSON 4G

Hands-On Lab

The Pythagorean Theorem

Activity

The Pythagorean Theorem states that in any right triangle, the sum of the squares of the lengths of the two legs is equal to the square of the length of the hypotenuse. In other words, $a^2 + b^2 = c^2$

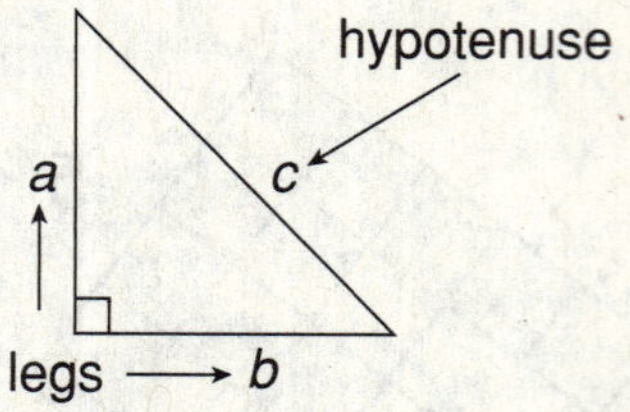

Demonstrate the validity of the Pythagorean Theorem by drawing a right triangle in the space below. Label the sides as in the diagram above.

Now, measure the 2 legs and fill in the measurements to find the length of the hypotenuse. Use a centimeter ruler if possible. Round to the nearest hundredth.

$a^2 + b^2 = c^2$

_____ + ______ $= c^2$

Find c. Write your answer as a whole number or as a radical expression.

____

Now measure the length of the hypotenuse: ______

____

Name ______________________________ Date ____________ Class ____________

LESSON 4G Hands-On Lab

The Pythagorean Theorem, continued

Think and Discuss

1. How could you use the Pythagorean Theorem to find a side in a right triangle if you knew the measurements of the other 2 sides?

____

____

____

2. Which side will always be the longest side in a right triangle?

____

____

____

Try This

Draw a right triangle. Measure the hypotenuse and one side. Use the Pythagorean Theorem to find the measurement of the third side. Round to the nearest hundredth.

$a^2 + b^2 = c^2$

$b^2 = c^2 - a^2$

$b^2 =$ ________ $-$ ________

$b =$

____

To check your work, measure the length of the third side: ________

____

Name ________________________________ Date ___________ Class ___________

LESSON **4G**

Technology Lab

Pythagorean Triples

Three positive integers *a*, *b*, and *c* that satisfy the equation $a^2 + b^2 = c^2$ are called **Pythagorean Triples**. You know that $3^2 + 4^2 = 5^2$. So, 3, 4, and 5 are Pythagorean triples.

Activity

You can generate Pythagorean triples *a*, *b*, and *c* by starting with two different whole numbers *m* and *n*, where *m* is the larger number. The Pythagorean triples will be as follows:

$a = m^2 - n^2$

$b = 2mn$

$c = m^2 + n^2$

Example: Using $m = 2$ and $n = 1$

$a = 2^2 - 1^2 = 4 - 1 = 3$

$b = 2(2)(1) = 4$

$c = 2^2 + 1^2 = 5$

Using a spreadsheet, enter the letters *m*, *n*, *a*, *b*, and *c*, respectively, in cells A1 to E1.

Then enter the formula for *a* as = **A2^2–B2^2** in cell C2.

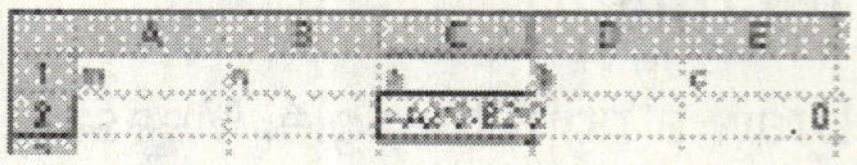

For *b*, enter = **2*A2*B2** in cell D2, and for *c*, enter = **A2^2 + B2^2** in cell E2.

Highlight cells C2, D2, and E2, and click the **Copy** button on the toolbar. Then select cells C2, D2, and E2, and drag down to highlight 7 rows. Click the **Paste** button on the toolbar.

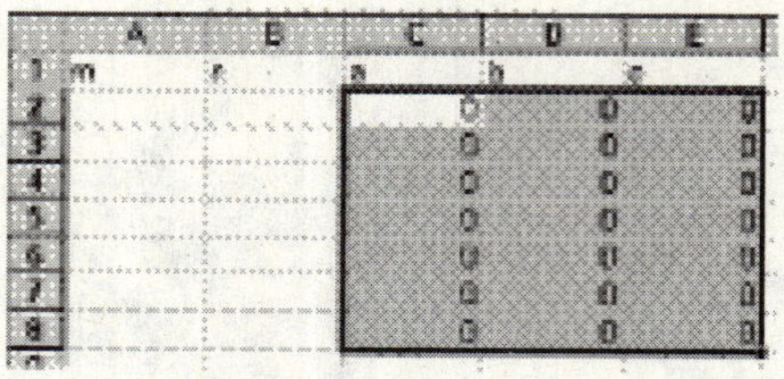

Next, enter 5 and 4 in cells A2 and B2, 5 and 3 in cells A3 and B3, and so forth until you complete the seventh row for patterns given.

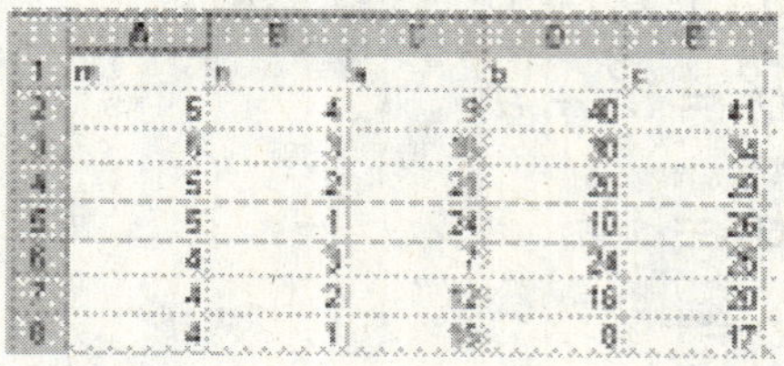

	A	B	C	D	E
1	m	n	a	b	c
2	5	4	9	40	41
3	5	3	16	30	34
4	5	2	21	20	29
5	5	1	24	10	26
6	4	3	7	24	25
7	4	2	12	16	20
8	4	1	15	8	17

All integers *a*, *b*, and *c* shown in the last three columns are Pythagorean triples.

Name ______________________________ Date ____________ Class ____________

LESSON 4G Technology Lab
Pythagorean Triples, continued

Think and Discuss

1. Is order important in a Pythagorean triple? Why?

____

____

Try This

1. Using a spreadsheet, generate 30 Pythagorean triples.

____

Name ______________________ Date __________ Class __________

LESSON **4G**

Review for Mastery

The Pythagorean Theorem

In a right triangle, *the sum of the areas of the squares on the legs* is equal to *the area of the square on the hypotenuse.*

$3^2 + 4^2 = 5^2$

$9 + 16 = 25$

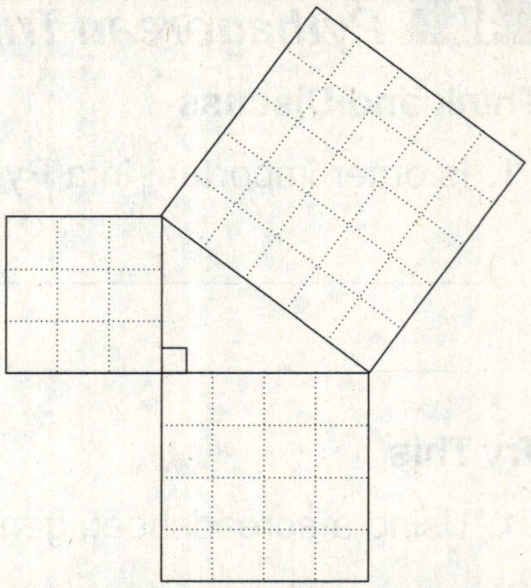

Given the squares that are on the legs of a right triangle, draw the square for the hypotenuse.

1. leg leg hypotenuse

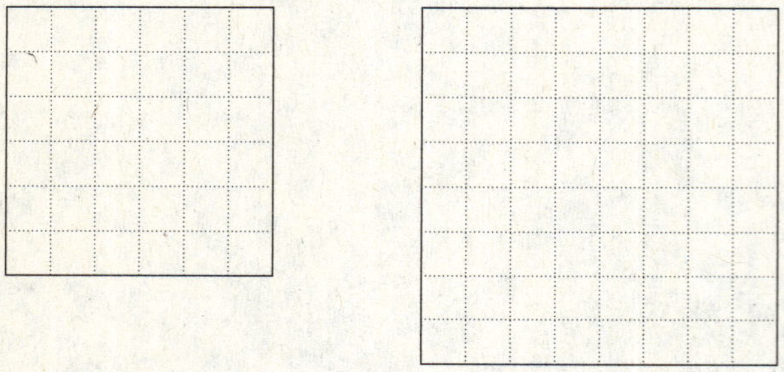

Without drawing the squares, you can find the length of a side.

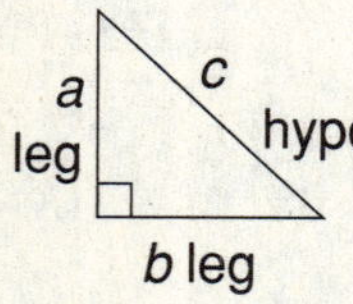

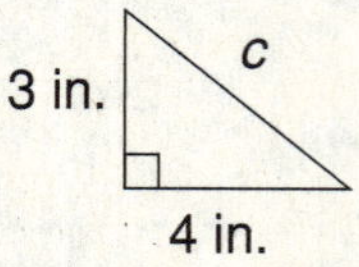

$a^2 + b^2 = c^2$

$3^2 + 4^2 = c^2$

$9 + 16 = c^2$

$25 = c^2$

$c = 5$ in.

Complete to find the length of each hypotenuse.

2.

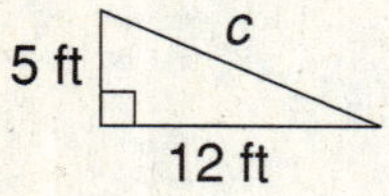

$a^2 + b^2 = c^2$

_____ + _____ $= c^2$

_____ + _____ $= c^2$

_____ $= c^2$

$c =$ _____ ft

3.

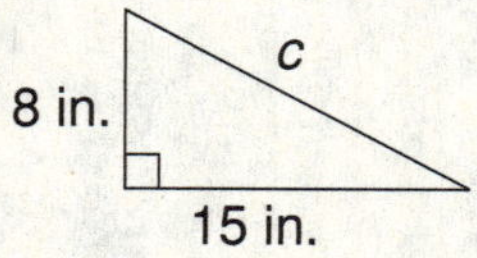

$a^2 + b^2 = c^2$

_____ + _____ $= c^2$

_____ + _____ $= c^2$

_____ $= c^2$

$c =$ _____ in.

Name ______________________ Date ___________ Class ___________

Review for Mastery

LESSON 4G

The Pythagorean Theorem, continued

You can use the Pythagorean Theorem to find the length of a leg if you know the length of the other leg and the hypotenuse.

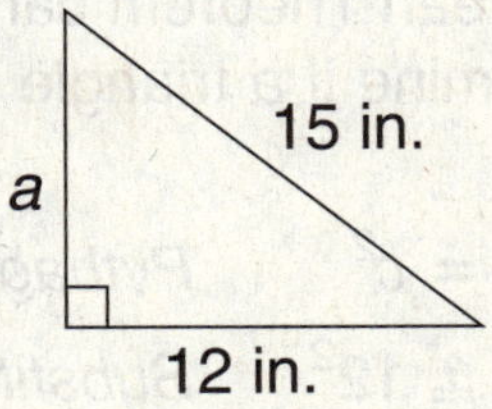

$a^2 + b^2 = c^2$

$a^2 +$ _____ = _____

$a^2 +$ _____ = _____

− _____ − _____

$a^2 =$ _____

$a =$ _____ in.

Complete to find the length of each leg.

4. $a^2 + b^2 = c^2$

_____ $+ b^2 =$ _____

_____ $+ b^2 =$ _____

− _____ − _____

$b^2 =$ _____

$b =$ _____ ft.

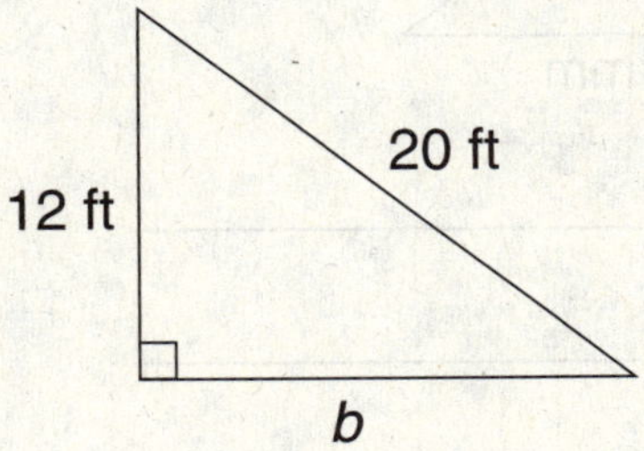

5. $a^2 +$ _____ = _____

$a^2 +$ _____ = _____

− _____ − _____

$a^2 =$ _____

$a =$ _____ cm

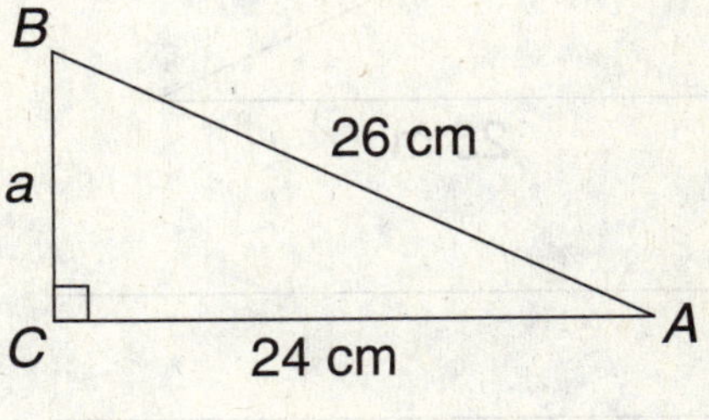

Name ____________________ Date __________ Class __________

LESSON 4G

Review for Mastery

The Pythagorean Theorem, continued

The Pythagorean Theorem can also be used to determine if a triangle is a right triangle.

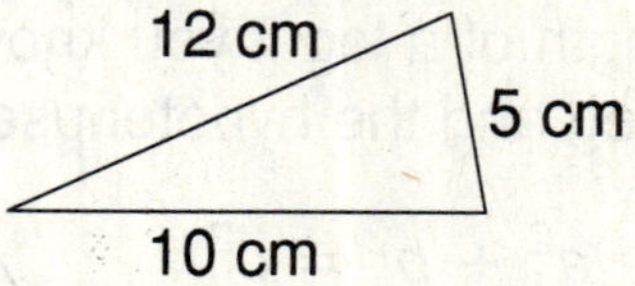

$a^2 + b^2 = c^2$ *Pythagorean Theorem*

$5^2 + 10^2 \overset{?}{=} 12^2$ *Substitute for a, b, and c.*

$25 + 100 \overset{?}{=} 144$ *Simplify powers.*

$125 \neq 144$

Since 125 does not equal 144, this triangle is not a right triangle.

Determine whether each triangle is a right triangle.

6.

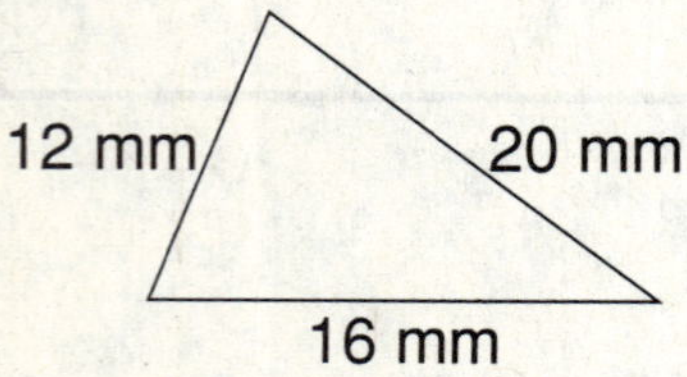

____

____

____

____

7.

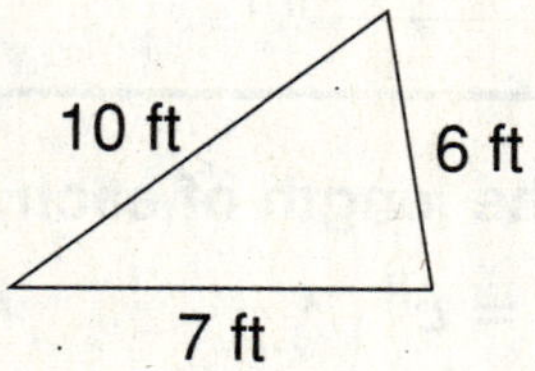

____

____

____

____

8.

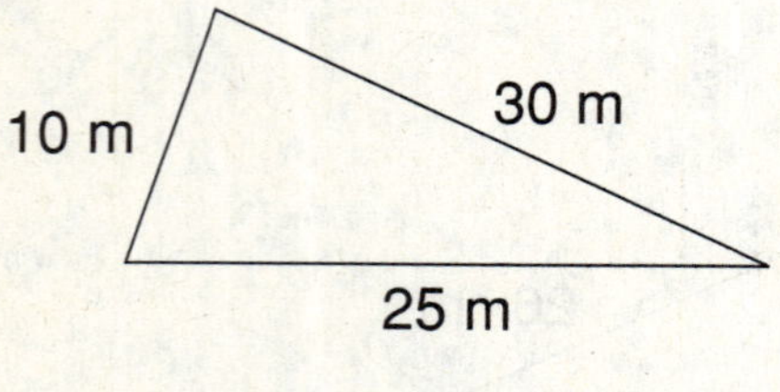

____

____

____

____

9.

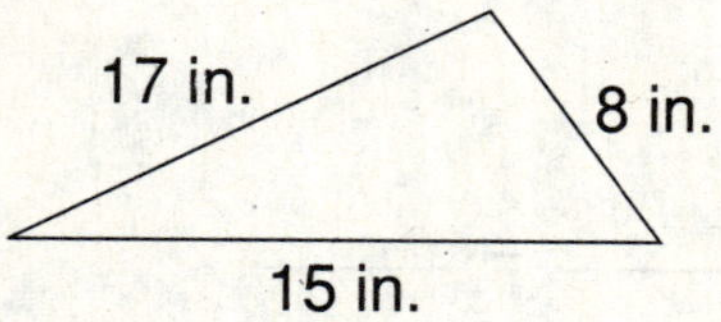

____

____

____

____

Name ______________________ Date ____________ Class ____________

LESSON 4G

Homework and Practice

The Pythagorean Theorem

Find the length of the hypotenuse in each triangle.

1.
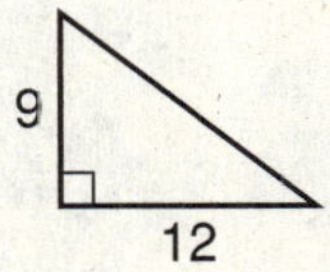

____

2.
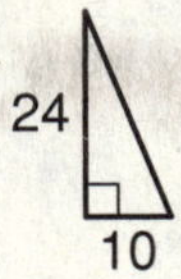

____

3.
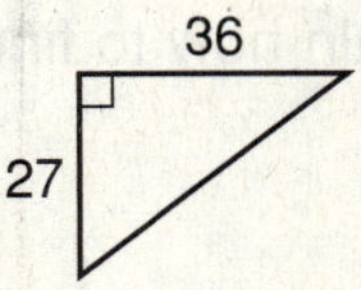

____

4.
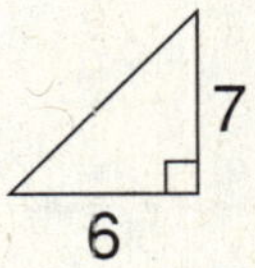

____

5.
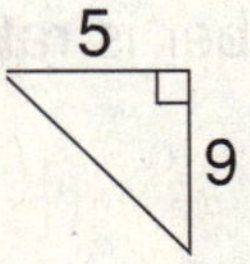

____

6.
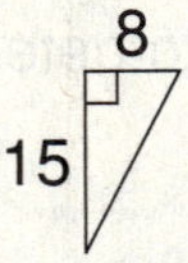

____

Solve for the unknown side in each right triangle.

7.
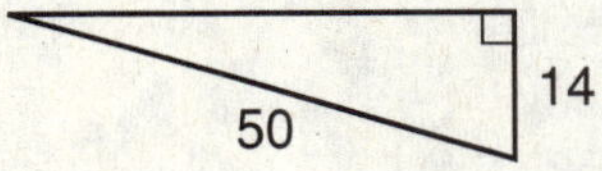

____

8.
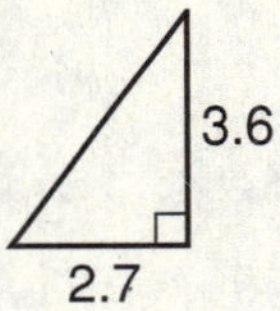

____

9.
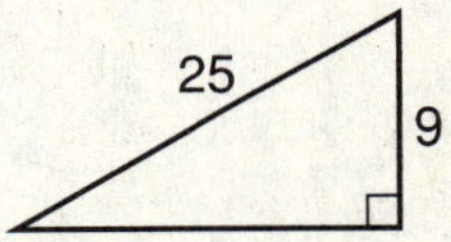

____

10.
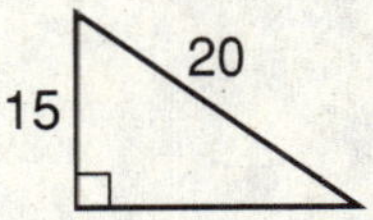

____

11.
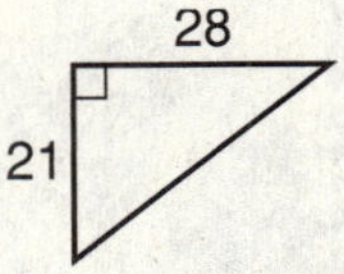

____

12.
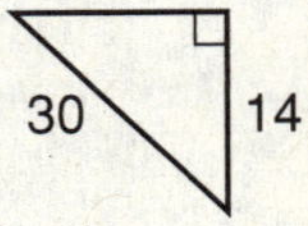

____

Determine whether each triangle is a right triangle.

13.
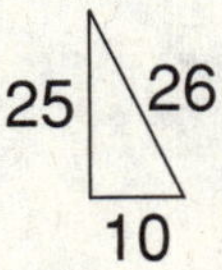

____

14.
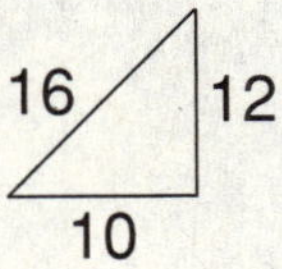

____

15.
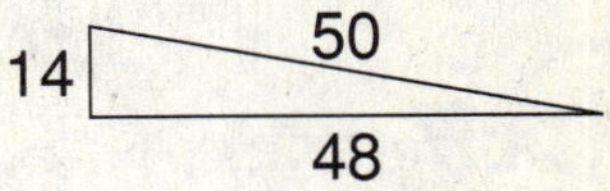

____

CHAPTER 4

Chapter Summary

Answer these questions to summarize the important concepts from Chapter 4 in your own words.

1. Explain how to find $\sqrt[3]{64}$.

2. Explain how to determine whether a number is rational or irrational.

3. Explain whether an equilateral triangle can be a right triangle.

4. Explain how to estimate $\sqrt{60}$.

5. Explain why $0.\overline{3}$ is a rational number.

For more review of Chapter 4:

- Complete the Chapter 4 Study Guide and Review on pages 204–206 of your textbook.
- Complete the Ready to Go On quizzes on pages 180 and 200 of your textbook.

Name ______________________ Date ____________ Class ____________

CHAPTER 4

KCCT Prep

Multiple Choice

1. Which of these is the best estimate of $\sqrt{45}$?

(A) 4.5 (C) 6.5
(B) 5.5 (D) 7.5

2. Kathy wants to plot $\sqrt{2}$ on a number line. Which point is closest to the location at which she should plot this number?

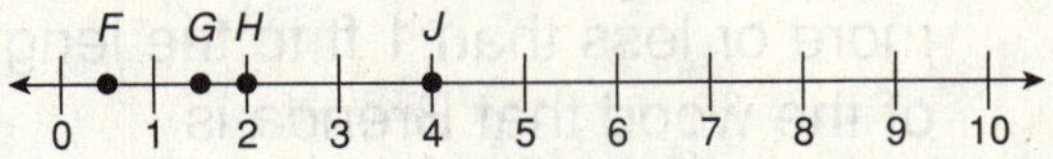

(A) Point *F* (C) Point *H*
(B) Point *G* (D) Point *J*

3. Which is the best estimate of the value of *c*?

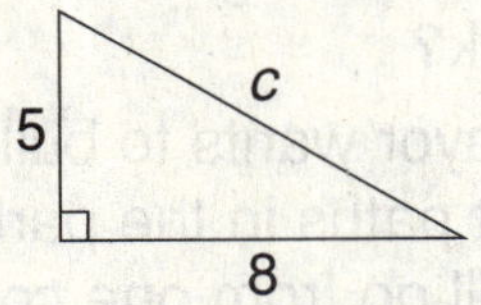

(A) 3.5 (C) 6.1
(B) 5.1 (D) 9.5

4. Suppose $a = 4$ and $b = 3$. Which of these is greater than 2 and less than 3?

(A) $\sqrt{a}$ (C) $\sqrt{a+b}$
(B) $\sqrt{b}$ (D) $\sqrt{ab}$

5. Which is the best estimate for 48.71×12.33?

(A) 450 (C) 550
(B) 500 (D) 600

6. A right triangle has legs that are both 3 cm long. Which is closest to the length of the triangle's hypotenuse?

(A) 2.4 cm (C) 9 cm
(B) 4.2 cm (D) 18 cm

7. Which equation matches the ordered pairs in the table?

x	1	3	5	7
y	1	5	9	13

(A) $y = 2x - 1$ (C) $y = x + 2$
(B) $y = 2x + 1$ (D) $y = 2x + 2$

8. Ricardo plots the points (0, 0), (6, 0), and (0, 8). Then he connects the points to form a right triangle. What is the length of the triangle's hypotenuse?

(A) 5 (C) 14
(B) 10 (D) 100

9. Which of these is closest to $\sqrt{27}$ on a number line?

(A) 3 (C) 5
(B) 4 (D) 6

10. The ordered pair (3, 14) appears in a table of values for the equation $y = \square x - 1$. What is the missing number in the equation?

(A) 2 (C) 11
(B) 5 (D) 13

11. The gas tank of Tim's car holds $12\frac{1}{2}$ gallons when it is full. Starting with a full tank, Tim uses $\frac{2}{5}$ of a tank. Then he adds $1\frac{1}{2}$ gallons of gas. How much gas is in the tank at this point?

Ⓐ 4 gallons Ⓒ $7\frac{1}{2}$ gallons
Ⓑ $6\frac{1}{2}$ gallons Ⓓ 9 gallons

12. What is the value of $1 + (m - n)^2 \div p$ when $m = n$ and $p \neq 0$?

Ⓐ –1 Ⓒ 1
Ⓑ 0 Ⓓ 2

13. Wei has $420 in her checking account. Each month, she writes a check for $19 to pay for her gym membership. Which equation shows the amount y in her account after x months?

Ⓐ $y = 19x$
Ⓑ $y = 19x - 420$
Ⓒ $y = 420x - 19$
Ⓓ $y = 420 - 19x$

14. Which of the following is NOT an irrational number?

Ⓐ $-\sqrt{2}$ Ⓒ $\sqrt{\frac{1}{9}}$
Ⓑ $\sqrt{3}$ Ⓓ $3\sqrt{10}$

15. Which of these points lies farthest from the y-axis?

Ⓐ (−5, 0) Ⓒ (4, 0)
Ⓑ (0, 6) Ⓓ (0, −9)

16. William drives 352.4 miles from San Jose to Los Angeles. The trip takes him 6.8 hours. Which is the best estimate of William's average speed during this trip?

Ⓐ 40 mi/hr Ⓒ 50 mi/hr
Ⓑ 45 mi/hr Ⓓ 60 mi/hr

Open Response

17. Brenda is building a skateboard ramp with the dimensions shown below.

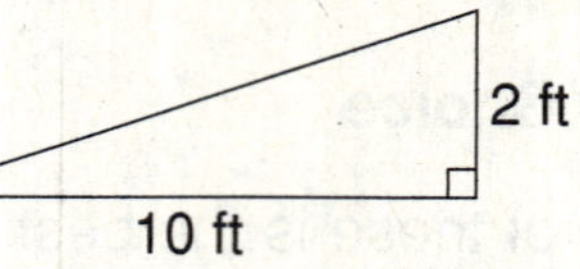

a. Brenda wants to order a piece of wood for the slanted part of the ramp. To the nearest tenth of a foot, what length should she give for this piece of wood?

b. Brenda's brother convinces her to add 1 ft to the height of the ramp so that it is 3 ft tall. Does this add more or less than 1 ft to the length of the wood that Brenda is ordering? Explain.

18. A town has a square park in front of city hall. The area of the park is 900 m^2.

a. What is the length of one side of the park?

b. The mayor wants to build two straight paths in the park. Each path will go from one corner of the park to the opposite corner. Explain how to estimate the total length of the two paths.

19. Consider the graph shown below.

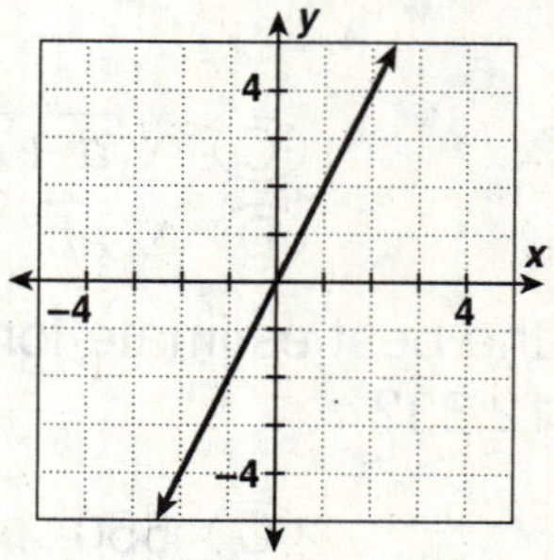

a. Make a table of values for the graph.

b. Write an equation for the graph.

c. Suppose the line were extended. Would the line pass through the point (10, 20) ? Why or why not?

CHAPTER 5

Project

Ratios, Proportions, and Similarity: Miniature Nature

Career: Horticulturist

Chances are that a horticulturist helped create many of the varieties of plants at your local nursery. Horticulturists work in vegetable development, fruit growing, flower growing, and landscape design. Horticulturists who are also scientists work to develop new types of plants or ways to control plant diseases. The art of *bonsai,* or making miniature plants, began in China and became popular in Japan. Now bonsai is practiced all over the world.

Before you start, work through the Growing Tiny Trees worksheet on the next page.

In this chapter project . . .

You use scale factors to draw a bonsai tree.

What to do:

- Choose a type of tree or large plant that you'd like to see as a bonsai. You can pick any type of tree, but a tree that grows locally would be best.
- Research your tree's dimensions, such as average height, trunk diameter, and leaf size. If the tree is local, gather some actual leaves for reference.
- The size of a bonsai tree can be anywhere from a few inches up to two feet high. Use a height of six inches so that the drawing of the tree will fit on a standard-size piece of paper.
- Find the scale factor to reduce your tree down to six inches high. Use the scale factor to find the reduced dimensions of the trunk and leaves. Do your best to draw the bonsai leaf so that it is similar to the normal-sized tree's leaf.

What to present:

- A drawing of your bonsai tree.
- The scale of your drawing, as well as the actual and scale measurements of several features of the tree.

Where's the math?

- Use proportions to convert the actual measurements to scale measurements.

go.hrw.com
Chapter Project Online
KEYWORD: MT7 Ch5

Name ______________________ Date ____________ Class ____________

CHAPTER 5

Project Recording Sheet

Ratios, Proportions, and Similarity

Growing Tiny Trees

Do you like to grow things? Bonsai is something that everyone can do if they have patience and the desire for perfection. The bonsai grower uses both science and art as he or she creates a small model of a large tree.

Complete the table.

$$\text{Scale} = \frac{\text{bonsai height in inches}}{\text{natural height in inches}}$$

Tree	Natural Height (ft)	Natural Height (in.)	Bonsai Height (in.)	Approximate Scale
Chinese Elm	60	720	10	$\frac{1}{72}$
Brush Cherry	50	600	8	$\frac{1}{75}$
Juniper	10		6	
Fig	30		29	
Japanese Maple	12		28	
Pitch Pine	200		14	
Red Cedar	80		20	
Eastern Hemlock	80		18	

Project: Find a picture of a tree that you like. Find out how tall the tree is in nature. Construct a scale model or drawing of the tree. Label the scale. Compare your model or drawing with those of your classmates. How does the scale help you understand the natural tree?

____

____

Extension: What is the weight comparison between a bonsai tree and the natural-sized tree?

____

Extension: Visit a nursery that specializes in bonsai and begin a project of your own.

EXPLORATION

Program of Studies and Core Content
MA-8-NPO-S-RP1, MA-08-1.4.1

5A Ratios and Proportions

Sam and Jill counted the number of foreign cars and the number of domestic cars that entered the parking lot of a movie theater between 5:30 P.M. and 5:45 P.M. for a week. Their results are listed in the table.

	Mon	Tue	Wed	Thu	Fri	Sat	Sun
Foreign	16	25	15	20	40	40	45
Domestic	8	25	10	25	20	50	45

They compared the number of foreign cars to the number of domestic cars by using ratios. For example, the ratio for Monday was $\frac{\text{foreign}}{\text{domestic}} = \frac{16}{8} = \frac{2}{1}$.

1. Write a ratio for each day.

2. Which ratios are greater than the Monday ratio?

3. Which ratios are less than the Monday ratio?

4. Which ratio is the same as the Monday ratio?

Think and Discuss

5. **Explain** what it means if two ratios are equal.

6. **Describe** how the ratios are different if you write them as $\frac{\text{domestic}}{\text{foreign}}$.

Name ______________________________ Date __________ Class __________

LESSON **5A**

Hands-On Lab

Ratios and Proportions

Activity

A ratio is a comparison of two quantities by division. Ratios that make the same comparison are equivalent ratios.

Missing terms are often computed by finding the cross-products of the terms:

$\frac{3}{4} = \frac{x}{8}$

$8 \cdot 3 = 4 \cdot x$

$4x = 24$

$x = 6$

You can find the same result graphically.

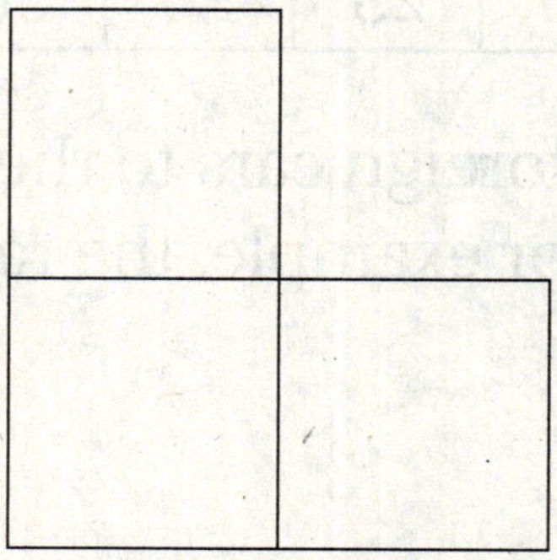

The above diagram represents three-fourths of a square. Each section represents $\frac{1}{4}$. Divide each section into 8. That is the equivalent of multiplying 8 times 3 in the calculation above. Now there are 24 sections. Then divide them up into 4 groups. This is the equivalent of dividing 24 by 4 in the calculation above. How many little sections are in each group?

____

Why are the first ratio, $\frac{3}{4}$, and this ratio, $\frac{6}{8}$, equivalent ratios?

____

____

____

Name ________________________ Date ____________ Class ____________

LESSON 5A

Hands-On Lab

Ratios and Proportions, continued

Think and Discuss

1. How could you change the model to make a different equivalent ratio?

____

____

____

2. What are two ways you could tell if ratios were not equivalent?

____

____

____

Try This

Create a figure on the first grid below and shade it in. Write the ratio of shaded to unshaded squares.

____

On the other grid, create a different figure with an equivalent ratio. Write the ratio of shaded to unshaded squares.

____

____

Name ______________________ Date ____________ Class ____________

LESSON 5A

Technology Lab

Ratios and Proportions

A classroom of 15 girls and 12 boys has the same ratio as a classroom of 10 girls and 8 boys. Even though they are different total amounts of students, each class has 5 girls for every 4 boys, so the ratio of girls to boys is 5 to 4, or 5:4.

Activity

A spreadsheet is used to record the numbers of green and red beads in 4 different bags.

	A	B	C	D	E
1		Bag 1	Bag 2	Bag 3	Bag 4
2	Green Beads	56	60	63	61
3	Red Beads	48	48	54	52

A final line computes the number of green beads for each red bead by dividing the value in row 2 by the value in row 3.

	A	B	C	D	E
1		Bag 1	Bag 2	Bag 3	Bag 4
2	Green Beads	56	60	63	61
3	Red Beads	48	48	54	52
4	Green/Red	1.166667	1.25	1.166667	1.173077

To express the ratio of green to red beads in Bag 1 as a simplified fraction, you can find what number can be evenly divided into both 56 and 48. Both of these numbers are evenly divisible by 2, 4, and 8. Take the greatest of these numbers and divide each color of bead by 8.

$$56 \div 8 = 7 \qquad 48 \div 8 = 6$$

The result is 7 green and 6 red beads. The number 8 is called the greatest common factor of 56 and 48. The resulting ratio is 7 to 6, or 7:6. If you entered 7 in cell B2 and 6 in cell B3, then the value of B4 would remain the same since $\frac{56}{48} = \frac{7}{6}$. This ratio is also expressed as 7:6. The rounded decimal equivalent of this ratio is expressed in cell B4, 1.166667.

Name ______________________________ Date ____________ Class ____________

LESSON 5A

Technology Lab

Ratios and Proportions, continued

Think and Discuss

Use the data in the table on the previous page to answer these questions:

1. Which bag has the same ratio of green to red beads as Bag 1?

____

2. Is the green/red ratio of Bag 2 greater or less than Bag 1?

____

3. What is the ratio of green to red of Bag 2?

____

Try This

1. Using a spreadsheet, compute the ratio of column A to column B for each of the following.

	A	B	C	D
1	Column A	Column B	Ratio	Decimal Equivalent
2	6	4		
3	10	8		
4	9	6		
5	12	9		
6	2	4		
7	12	8		
8	7	14		
9	15	10		
10	12	18		
11	14	21		
12	16	4		
13	10	30		

Name ______________________ Date ____________ Class ____________

LESSON 5A

Review for Mastery

Ratios and Proportions

A **ratio** compares two quantities by division. **Equivalent ratios** make the same comparison.

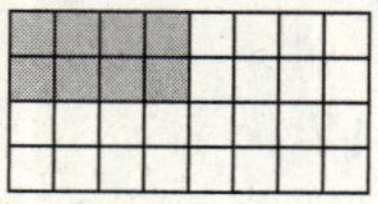
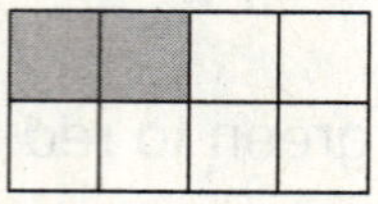
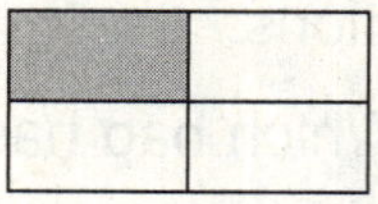

$\frac{8}{32} = \frac{2}{8} = \frac{1}{4}$

To find equivalent ratios:

Divide by a common factor.

$\frac{10 \div 5}{15 \div 5} = \frac{2}{3}$

Multiply by a common factor.

$\frac{10 \times 2}{15 \times 2} = \frac{20}{30}$

So, two ratios equivalent to $\frac{10}{15}$ are $\frac{2}{3}$ and $\frac{20}{30}$.

Complete to find two ratios equivalent to each given ratio.

1. $\frac{16 \div 4}{40 \div 4} =$ ______ $\frac{16 \times 3}{40 \times 3} =$ ______

2. $\frac{75 \div 25}{100 \div } =$ ______ $\frac{75 \times 2}{100 \times } =$ ______

Name two ratios equivalent to the given ratio.

3. $\frac{6}{9}$ ______

4. $\frac{1}{3}$ ______

5. $\frac{2}{5}$ ______

6. $\frac{16}{20}$ ______

When two ratios are equivalent, they form a **proportion.**

To tell whether two ratios form a proportion, write each in simplest form.

Does $\frac{16}{32} = \frac{50}{100}$?

$\frac{16 \div 16}{32 \div 16}$ | $\frac{50 \div 50}{100 \div 50}$

$\frac{1}{2}$ | $\frac{1}{2}$

The ratios form a proportion.

Does $\frac{8}{18} = \frac{10}{15}$?

$\frac{8 \div 2}{18 \div 2}$ | $\frac{10 \div 5}{15 \div 5}$

$\frac{4}{9}$ | $\frac{2}{3}$

The ratios do not form a proportion.

Tell whether the ratios form a proportion.

7. $\frac{14}{21} \stackrel{?}{=} \frac{50}{75}$ ______

8. $\frac{27}{48} \stackrel{?}{=} \frac{54}{72}$ ______

LESSON 5A

Homework and Practice

Ratios and Proportions

Find two ratios that are equivalent to each given ratio.

1. $\frac{7}{35}$ **2.** $\frac{4}{16}$ **3.** $\frac{16}{48}$

4. $\frac{7}{8}$ **5.** $\frac{24}{6}$ **6.** $\frac{1}{2}$

7. $\frac{14}{30}$ **8.** $\frac{16}{6}$ **9.** $\frac{25}{45}$

Simplify to tell whether the ratios form a proportion.

10. $\frac{10}{60}$ and $\frac{2}{12}$ **11.** $\frac{10}{24}$ and $\frac{30}{72}$ **12.** $\frac{8}{27}$ and $\frac{16}{56}$ **13.** $\frac{9}{36}$ and $\frac{13}{40}$

14. $\frac{14}{27}$ and $\frac{42}{81}$ **15.** $\frac{15}{25}$ and $\frac{20}{30}$ **16.** $\frac{16}{28}$ and $\frac{20}{35}$ **17.** $\frac{17}{68}$ and $\frac{14}{56}$

18. Mr. Blackwell had two ties for every five shirts he owned. If he owned 25 shirts, how many ties did he own?

19. The school had the classroom walls painted. Seven out of 10 people surveyed liked the new color of the walls. Of the 250 people surveyed, how many people liked the new color of the walls?

20. A medicine for cats indicates that the medicine should be given in the ratio of 1 tsp for every 8 lbs. If Laurie's cat, Felix, weighs 14 lbs, how many teaspoons of medicine should she give him?

21. Stella's mother recommends using a cleaning solution of $\frac{3}{4}$ cup bleach for each gallon of water. If Stella uses $3\frac{1}{4}$ cups of bleach, how much water does she need?

EXPLORATION

Program of Studies and Core Content

MA-8-NPO-S-RP1, MA-08-1.4.1

5B Ratios, Rates, and Unit Rates

The bar graph shows the number of acres of wilderness burned each year from 1991 to 2000. Each bar represents a ***unit rate,*** because it shows the number of acres burned in *one* year.

Acres of Wilderness Burned: 1991–2000

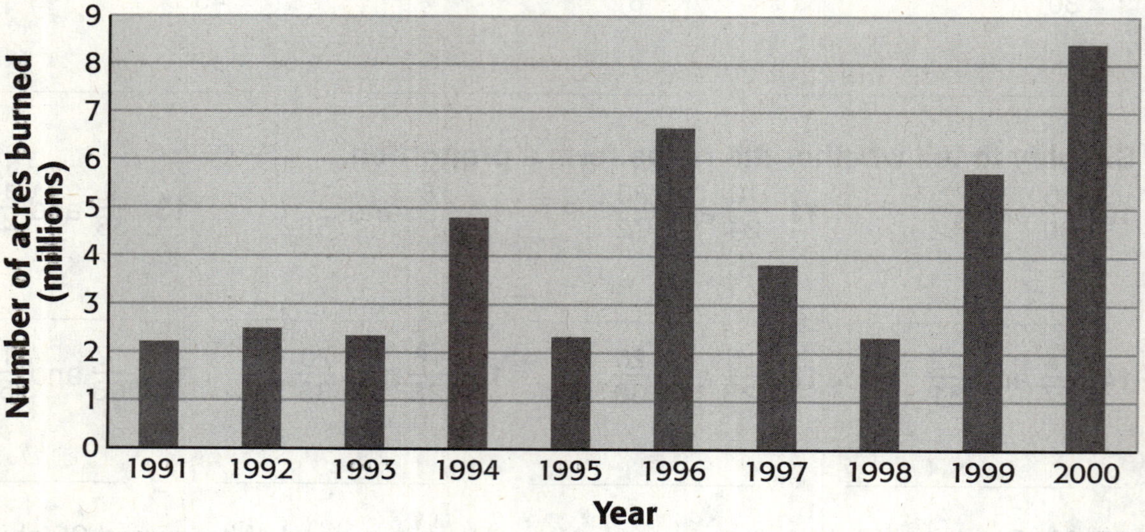

1. The National Interagency Fire Center reported that an average of 3,647,883 acres were burned per year from 1991 to 2000.

a. Is the average also a unit rate?

b. In what years were the number of acres burned above the average?

c. In what years were the number of acres burned below the average?

Think and Discuss

2. Explain why a unit rate such as 40 miles per hour may be more useful than an equivalent rate such as 60 miles per 1.5 hours.

Name ______________________________ Date ____________ Class ____________

Technology Lab

LESSON **5B** ***Ratios, Rates, and Unit Rates***

The concept of ratios applies to the idea of rates and unit rates. A ratio typically changes between objects, but a unit rate remains the same and becomes a standard of measurement. Rates are most commonly used as a conversion factor to change from one type of unit into another type of unit. If the understanding based on ratios is applied to rates it becomes much easier to perform the conversions.

Activity

Use a spreadsheet program to construct a table to convert feet to inches. The unit rate of feet to inches is 1:12, or 1 foot for every 12 inches. Setup the table to perform the calculations by labeling two columns as "feet" and "inches".

	A	B
1	Feet	Inches
2		

Now fill cell B2 with the formula "= A2*12". Since there are 12 inches in every foot, cell B2 should multiply the value in A2 by 12 to arrive at the total number of inches for the number of feet shown in cell A2. Next, put the value "1" into cell A2, press *enter,* and watch the value in B2 change to the appropriate value, 12. Try some other random values and B2 will change to the correct value each time.

	A	B
1	Feet	Inches
2	1	12

	A	B
1	Feet	Inches
2	4	48

Even if a very large value were entered for feet, such as the number of feet in a mile (5280) the spreadsheet still calculates the correct number of inches (63,360). Since the ratio of feet to inches remains the same, no matter how many feet are measured the formula always yields the correct conversion.

Name ______________________ Date __________ Class __________

LESSON 5B

Technology Lab

Ratios, Rates, and Unit Rates, continued

Think and Discuss

Answer the questions below based on the previous activity.

1. What would the *inches* value be if you filled in "0" for the *feet* value?

____

2. Would the formula still give the correct result if a fraction or decimal equivalent were entered for the *feet* value?

____

3. What formula could be used in A2 to convert inches to feet?

____

Try This

Answer the following questions by creating a spreadsheet similar to the one created above. Show the values in each cell of your spreadsheet.

1. There are 60 minutes in every hour. How many minutes are there in 24 hours?

Show your work.

____

2. There are 60 seconds in every minute. How many minutes are there in 1800 seconds?

Show your work.

____

Name ______________________ Date ___________ Class ___________

LESSON 5B

Review for Mastery

Ratios, Rates, and Unit Rates

A **rate** is a ratio that compares two *different kinds* of quantities.

2 aides for 18 students — $\frac{\text{2 aides}}{\text{18 students}}$

135 words in 3 minutes — $\frac{\text{135 words}}{\text{3 minutes}}$

7 ads per 4 pages of copy — $\frac{\text{7 ads}}{\text{4 pages of copy}}$

Express each comparison as a rate in ratio form.

1. 275 students per 11 teachers

____

2. 3 books in 2 months

____

3. 15 strike-outs in 6 innings

____

In a **unit** rate, the second quantity is 1.

300 miles in 6 hours

$$\frac{\text{300 miles}}{\text{6 hours}} = \frac{300 \div 6}{6 \div 6} = \frac{\text{50 miles}}{\text{1 hour}}$$

81 entries in 4 minutes

$$\frac{\text{81 entries}}{\text{4 minutes}} = \frac{81 \div 4}{4 \div 4} = \frac{\text{20.25 entries}}{\text{1 minute}}$$

Express each comparison as a unit rate.

4. 28 patients for 2 nurses

$$\frac{\text{28 patients}}{\text{2 nurses}} = \frac{28 \div 2}{2 \div 2} = \underline{\qquad\qquad}$$

5. 16 children in 7 families

$$\frac{\text{16 children}}{\text{7 families}} = \frac{16 \div \quad}{7 \div \quad} = \underline{\qquad\qquad}$$

A **unit price** tells the price per 1 unit.

$2.49 for 3 muffins

$$\frac{\$2.49}{\text{3 muffins}} = \frac{2.49 \div 3}{3 \div 3} = \frac{\$0.83}{\text{1 muffin}}$$

$0.83 for 1 muffin

$1.67 for 10 pencils

$$\frac{\$1.67}{\text{10 pencils}} = \frac{1.67 \div 10}{10 \div 10} = \frac{\$0.167}{\text{1 pencil}}$$

$0.17 for 1 pencil

Find each unit price.

6. $10.74 for 3 reams of paper

$$\frac{\$10.74}{\text{3 reams}} = \frac{10.74 \div 3}{3 \div 3}$$

____

7. $9.99 for 6 blank jewel cased CDs

$$\frac{\$9.99}{\text{6 CDs}} = \frac{9.99 \div \quad}{6 \div \quad}$$

____

8. $8.99 for a 12-pack of gel pens

$$\frac{\$8.99}{\text{12 pens}} = \frac{8.99 \div \quad}{12 \div \quad}$$

____

Name ______________________ Date ____________ Class ____________

LESSON 5B

Homework and Practice

Ratios, Rates, and Unit Rates

For evercises 1–3, use the bar graph to find each rate.

Yearly Gross Salary Based on 52 Weeks, 5 Days a Week, 8 Hours a Day

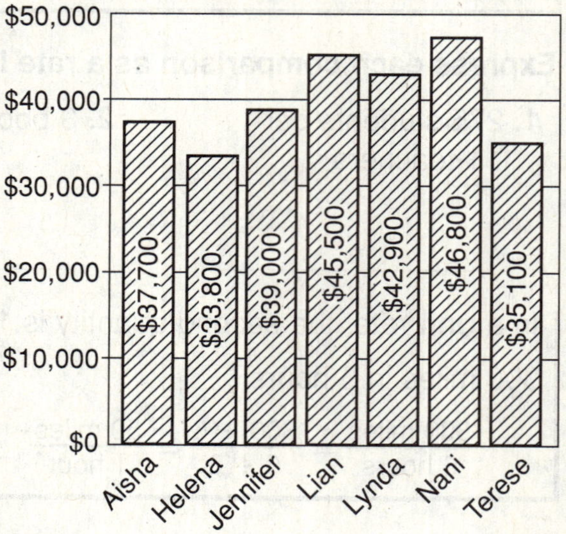

1. Helena's salary per week.

____

2. Aisha's salary per day.

____

3. Nani's salary per hour.

____

4. Of the 405 students who took the test, 162 of them passed. Write a ratio in simplest form for the number of students who passed the test to the number of students who took the test.

____

5. A lawn care company charges a homeowner $1,625 a year to maintain a lawn that is 65 ft by 50 ft. What is the maintenance fee per square foot?

____

6. A 12 oz container of soda has 180 calories, how many calories are in one ounce of the soda?

____

Determine the better buy.

7. a 500-count ream of paper for $8.25 or a 1250-count ream of paper for $20

____

8. a 1.5 lb bag of carrots for $2.85 or a 2 lb bag of carrots for $3.84

____

9. Morgan paid $16.74 for 12.4 gallons of gasoline. What was the cost per gallon?

____

Program of Studies and Core Content

MA-8-NPO-S-RP2, MA-8-NPO-S-SM1, MA-8-M-S-MPA1, MA-08-1.4.1, MA-08-2.2.1

5C Dimensional Analysis

The radius of a planet is called the ***equatorial radius*** and is the distance between the center of the planet and a point on its equator.

The table shows the radius of each planet in our solar system measured in two different units. In the second column, each radius is measured using the radius of Earth. In the third column, each radius is measured in kilometers.

Calculate the equatorial radius of each planet.

	Planet	Equatorial Radius (number of Earth radii)	Equatorial Radius (km)
1.	Mercury	0.38	
2.	Venus	0.95	
3.	Earth	1	6378.14
4.	Mars	0.53	
5.	Jupiter	11	
6.	Saturn	9	
7.	Uranus	4	
8.	Neptune	4	

Think and Discuss

9. Explain how you found the equatorial radius of each planet.

10. Discuss whether using Earth's radius or the kilometer makes comparing the radii of the planets easier.

Name ______________________________ Date ____________ Class ____________

LESSON 5C

Hands-On Lab

Dimensional Analysis

Activity

When measuring length in the customary system the common units used are inches, feet, yards, and miles. Units of length measure height, width, length, depth, and distance. Measure the sides of the square below in inches.

The square measures approximately 6 inches by 6 inches.

Now, turn your ruler over and measure the square in centimeters.

The square measures approximately 15.2 centimeters by 15.2 centimeters.

Name ______________________ Date ____________ Class ____________

LESSON 5C

Hands-On Lab

Dimensional Analysis, continued

You can convert inches to centimeters by multiplying the number of inches by 2.54. This gives you the following: 6 inches × 2.54 cm/in. = 15.24 cm.

The measurements you got with your ruler are close to the conversions, but they are not identical. What if you went the other way and converted the metric units to customary measurements?

15.2 centimeters ÷ 2.54 cm/in. = 5.984 in.

Think and Discuss

Does the activity demonstrate that the metric system is more precise than the customary system? Explain.

____

____

____

____

____

Try This

Using a metric ruler, measure your shoe. Walk the length of the classroom. How many centimeters long is your classroom? How many inches long is the classroom? What unit would be best for measuring the length of your classroom?

____

____

____

____

____

____

Name ____________________________ Date __________ Class __________

LESSON 5C

Review for Mastery

Dimensional Analysis

A **conversion relation** gives equivalent units.

1 foot = 12 inches   1 week = 7 days   1 pound = 16 ounces

Complete each conversion relation.

1. 1 gallon = ______ quarts

2. 1 meter = ______ centimeters

3. 1 yard = ______ feet

To change from one unit to another, multiply by a **conversion factor**.

Convert 7 ft to in.

Use 1 foot = 12 inches to write a conversion factor.

Put the unit to be changed in the denominator. $\frac{12 \text{ in.}}{1 \text{ ft.}}$

$7 \cancel{\text{ft}} \times \frac{12 \text{ in.}}{1 \cancel{\text{ft}}} = (7 \times 12) \text{ in.} = 84 \text{ in.}$

When converting from a larger unit to a smaller, the result is a larger number.

7 ft = 84 in.

When converting from a smaller unit to a larger, the result is a smaller number.

84 in. = 7 ft

Complete to convert each unit.

4. 72 inches to feet

1 ft = __________

denominator of conversion factor:

____

72 in. × ―――― = ______ ft

5. 40 ounces to pounds

1 lb = __________

denominator of conversion factor:

____

40 oz × ―――― = ______ lb

6. 735 meters to centimeters

1 m = __________

denominator of conversion factor:

____

735 m × ―――― = ________ cm

7. 150 pounds to kilograms

1 kg = 2.2 lb

denominator of conversion factor:

____

150 lb × ―――― ≈ ________ kg

Name ______________________ Date __________ Class __________

LESSON 5C

Review for Mastery

Dimensional Analysis, continued

When rates are involved, two conversions may be needed.

Convert 55 miles per hour to feet per second. $\frac{55\text{ mi}}{1\text{ hr}}$

First, change miles to feet.

Use 1 mile = 5280 feet to write a conversion factor.

Miles, the unit to be changed, is in the numerator of the given rate.

So, put *miles* in the denominator of the conversion factor. $\frac{5280\text{ ft}}{1\text{ mi}}$

$$\frac{55\text{ }\cancel{\text{mi}}}{1\text{ hr}} \times \frac{5280\text{ ft}}{1\text{ }\cancel{\text{mi}}} = \frac{(55 \times 5280)\text{ ft}}{1\text{ hr}} = \frac{290{,}400\text{ ft}}{1\text{ hr}}$$

Now, change hours to seconds.

Use 1 hour = 3600 seconds to write a conversion factor.

Hours, the unit to be changed, is in the denominator of the given rate.

So, put *hours* in the numerator of the conversion factor. $\frac{1\text{ hr}}{3600\text{ s}}$

$$\frac{290{,}400\text{ ft}}{1\text{ }\cancel{\text{hr}}} \times \frac{1\text{ }\cancel{\text{hr}}}{3600\text{ s}} = \frac{\overset{242}{\cancel{290{,}400}}\text{ ft}}{\underset{3}{\cancel{3600}}\text{ s}} \approx 80.7\ \frac{\text{ft}}{\text{s}}$$

Complete to convert the units of the given rate.

8. Convert 60 miles per hour to yards per minute.

In ratio form the rate is: __________

The conversion relation between miles and yards is: 1 mi = __________ yd

The unit *miles* is in the ______________ of the given rate.

What place will *miles* take in the conversion factor? ______________

To convert from miles to yards: $\frac{60\text{ mi}}{1\text{ hr}} \times$ __________ $= \frac{\_\_\_\_\text{ yd}}{1\text{ hr}}$

The conversion relation between hours and minutes is: 1 hr = __________ min

The unit *hours* is in the ______________ of the given rate.

What place will *hours* take in the conversion factor? __________

To convert from hours to minutes:

$\frac{\_\_\_\_\text{ yd}}{1\text{ hr}} \times$ __________ = __________ $\frac{\text{yd}}{\text{min}}$

Name ______________________ Date ____________ Class ____________

LESSON 5C

Homework and Practice

Dimensional Analysis

Find the appropriate factor for each conversion.

1. millimeters to meters

2. inches to yards

3. days to minutes

4. Oliver constructs a fence around a garden that is 24 feet by 36 feet. How many yards of fencing does Oliver need?

5. Marci drinks four 48 ounce glasses of water a day. How many pints of water does she drink every day?

6. Christopher wrestles for his high school in the light heavyweight division. Christopher weighs 175 pounds. How many ounces does he weigh?

7. Mari bought $9\frac{1}{2}$ yards of ribbon. How many inches of ribbon did she buy?

8. A bag of frozen vegetables weighs 64 ounces. How many pounds does the package of vegetables weigh?

9. The 18th hole on the local golf course is 543 yards long. How many feet is the distance of this par 5 hole?

A Boeing 777-300 commercial jetliner has a maximum take-off weight of 660,000 pounds, wing span of 199 feet 11 inches, overall length of 242 feet 4 inches, and an interior cabin width of 19 feet 3 inches. Find the following dimensions of the plane.

10. What is the maximum take-off weight in tons? ______________

11. What is the measure of the wing span in inches? ______________

12. What is the overall length of the plane in yards? ______________

EXPLORATION

Program of Studies and Core Content
MA-08-1.4.1

5D Solving Proportions

The grid shows three rectangles aligned by a diagonal line.

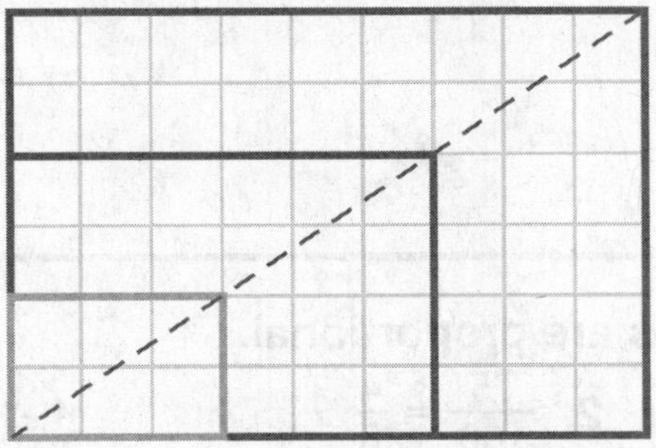

Complete the table. Use the grid to measure the length and width of each rectangle. Divide the length by the width and write this answer as a fraction in simplest form.

		Length	Width	$\frac{\text{Length}}{\text{Width}}$
1.	small rectangle			
2.	middle rectangle			
3.	large rectangle			

Think and Discuss

4. Compare the fractions you found in Problems 1–3.

5. Explain how you wrote each fraction in simplest form.

Name ______________________ Date __________ Class __________

LESSON 5D

Review for Mastery

Solving Proportions

In a proportion, the **cross products** are equal.

These ratios are proportional, since their cross products are equal.

$$\frac{7}{14} \stackrel{?}{=} \frac{8}{16}$$

$$7 \times 16 \stackrel{?}{=} 14 \times 8$$

$$112 = 112$$

These ratios are not proportional, since their cross products are not equal.

$$\frac{4}{9} \stackrel{?}{=} \frac{2}{3}$$

$$4 \times 3 \stackrel{?}{=} 9 \times 2$$

$$12 \neq 18$$

Complete to tell if the ratios are proportional.

1. $\frac{15}{45} \stackrel{?}{=} \frac{6}{18}$

$15 \times$ ___ $\stackrel{?}{=} 45 \times$ ___

____

The ratios ________ in proportion.

2. $\frac{75}{100} \stackrel{?}{=} \frac{4}{8}$

___ $\times$ ___ $\stackrel{?}{=}$ ___ $\times$ ___

____

The ratios ________ in proportion.

3. $\frac{7}{2} \stackrel{?}{=} \frac{21}{6}$

___ $\times$ ___ $\stackrel{?}{=}$ ___ $\times$ ___

____

The ratios ________ in proportion.

To solve for one member of a proportion, set the cross products equal.

$$\frac{n}{32} = \frac{9}{16}$$

$$16n = 32 \times 9$$

$$\frac{16n}{16} = \frac{288}{16}$$

$$n = 18$$

To check the result, substitute and see if the ratios are equivalent.

$$\frac{18}{32} \stackrel{?}{=} \frac{9}{16}$$

$$\frac{18 \div 2}{32 \div 2} \stackrel{?}{=} \frac{9}{16}$$

$$\frac{9}{16} = \frac{9}{16} \checkmark$$

Solve and check.

4. $\frac{8}{24} = \frac{2}{n}$

$8n =$ ________

$\frac{8n}{} =$ ________

$n =$ ________

Check: $\frac{8}{24} = \frac{2}{n}$

$\frac{8}{24} \stackrel{?}{=} \frac{2}{}$

$\stackrel{?}{=} \quad 2 \div$

________ $=$ ________

LESSON 5D

Homework and Practice

Solving Proportions

Tell whether each pair of ratios is proportional.

1. $\frac{5}{6} \stackrel{?}{=} \frac{30}{36}$ ______

2. $\frac{8}{20} \stackrel{?}{=} \frac{6}{14}$ ______

3. $\frac{12}{27} \stackrel{?}{=} \frac{8}{18}$ ______

4. $\frac{13}{15} \stackrel{?}{=} \frac{39}{45}$ ______

5. $\frac{8}{16} \stackrel{?}{=} \frac{50}{100}$ ______

6. $\frac{21}{49} \stackrel{?}{=} \frac{14}{36}$ ______

7. $\frac{18}{24} \stackrel{?}{=} \frac{9}{15}$ ______

8. $\frac{4}{18} \stackrel{?}{=} \frac{18}{81}$ ______

9. $\frac{16}{10} \stackrel{?}{=} \frac{26}{20}$ ______

10. $\frac{24}{14} \stackrel{?}{=} \frac{36}{21}$ ______

11. $\frac{25}{16} \stackrel{?}{=} \frac{5}{4}$ ______

12. $\frac{17}{11} \stackrel{?}{=} \frac{51}{33}$ ______

Solve each proportion.

13. $\frac{4}{15} = \frac{a}{45}$ ______

14. $\frac{16}{18} = \frac{8}{w}$ ______

15. $\frac{m}{11} = \frac{32}{44}$ ______

16. $\frac{12}{t} = \frac{4}{7}$ ______

17. $\frac{9}{b} = \frac{27}{72}$ ______

18. $\frac{x}{36} = \frac{21}{27}$ ______

19. $\frac{6}{y} = \frac{18}{51}$ ______

20. $\frac{10}{32} = \frac{15}{x}$ ______

21. $\frac{45}{81} = \frac{n}{27}$ ______

22. $\frac{28}{21} = \frac{44}{t}$ ______

23. $\frac{d}{42} = \frac{49}{147}$ ______

24. $\frac{17}{r} = \frac{102}{78}$ ______

25. The Orlemann family planted 3 rows of corn this year and harvested 270 ears. How many rows should they plant if they want to harvest 360 ears of corn?

______

26. Find the number of times a heart beats in 2 minutes if it beats 576 times in 8 minutes.

______

27. David is making chocolate-chip cookies. Chocolate chips cost \$1.79 for 12 ounces. David buys 60 ounces of chocolate chips for the cookies. How much did the chocolate chips cost?

______

EXPLORATION

Program of Studies and Core Content

MA-8-M-S-MPA2, MA-8-G-S-SR4, MA-8-G-S-SR5, MA-08-3.1.4, MA-08-2.1.2

5E Similar Figures

Similar figures have the same shape, but not necessarily the same size. Compare the two triangles shown below.

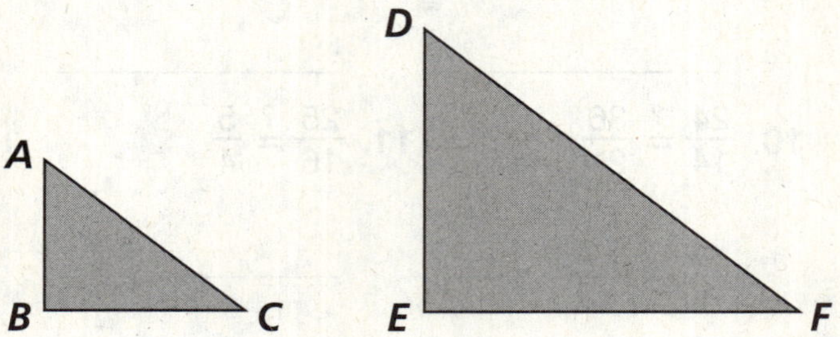

1. What appears to be true about each pair of angles below?

$\angle A$ and $\angle D$ $\angle B$ and $\angle E$ $\angle C$ and $\angle F$

2. Use a centimeter ruler to measure each side to the nearest tenth of a cm.

$AB =$ ____________ $BC =$ ____________ $AC =$ ____________

$DE =$ ____________ $EF =$ ____________ $DF =$ ____________

3. Calculate each of the ratios shown below. What do you notice?

$\frac{\text{length of } \overline{AB}}{\text{length of } \overline{DE}}$ $\frac{\text{length of } \overline{BC}}{\text{length of } \overline{EF}}$ $\frac{\text{length of } \overline{AC}}{\text{length of } \overline{DF}}$

4. Similar figures have corresponding angles that are congruent. They also have corresponding sides that form equivalent ratios. Are the triangles above similar?

Think and Discuss

5. Compare the ratio of the perimeters of the two triangles. What do you notice?

Name ______________________________ Date ____________ Class ____________

LESSON **5E**

Hands-On Lab 1

Similar Figures

Activity

For polygons to be similar, corresponding angles must be congruent and corresponding sides must have lengths that form equivalent ratios. Similar figures have the same shape, but not always the same size.

Measure the angles and the sides of this object and record the measurements.

Angles: ______________________________

Sides: ______________________________

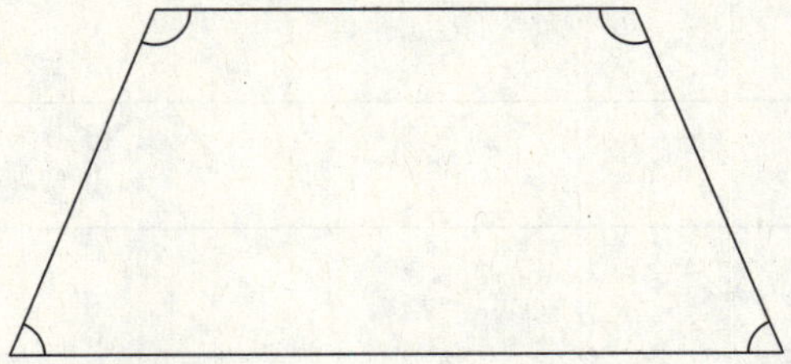

Follow these instructions to draw a larger and similar shape in the space below. In your drawing, label the lengths of the sides and the measurements of the angles.

To draw a similar figure, you need to find the scale factor, which is the ratio formed by the corresponding sides. To find the scale factor:

1. Draw a line in the box that will correspond with a line in the original figure. Measure the line.
2. Divide that measurement by the measurement of the corresponding line in the figure above to get the scale factor.
3. Multiply the lines in the original object by the scale factor to find how long the lengths should be in the similar figure.

   Use a protractor to draw lines that form the same angles.

Name ______________________ Date ____________ Class ____________

LESSON 5E

Hands-On Lab 1

Similar Figures, continued

Think and Discuss

1. How can you tell if two figures are similar?

____

____

____

2. A poster that is 24 in. high and 16 in. wide is being reduced so it is 9 in. high and will fit on a letter-sized page. How could you use a scale factor to find how wide the reduced poster will be?

____

____

____

____

Try This

Use what you know about scale factors to find the missing dimension.

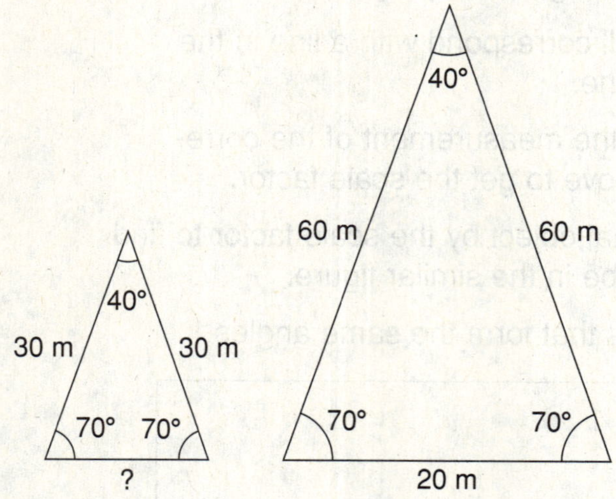

What is the unknown side length?

____

Name ________________________ Date ____________ Class ____________

LESSON **5E**

Hands-On Lab 2

Similar Figures

Activity

Two figures are similar if they have the same shape, but not necessarily the same size. For polygons to be similar, corresponding angles must be congruent and corresponding sides must have lengths that form equivalent ratios. In similar figures, the ratio formed by the corresponding sides is the scale factor.

Find an object in the classroom that has a surface that forms a polygon. Measure the angles and sides and record them here.

Angles: ____

Sides: ____

Now, draw a similar shape in the space below using the scale factor. In your drawing, label the lengths of the sides and the measurements of the angles.

To find the scale factor:

1. Draw a line in the box that will correspond with a line in the original figure. Measure the line.
2. Divide that measurement by the measurement of the corresponding line in the object you measured. That will give you the scale factor.
3. Multiply the lengths of the lines in the original object by the scale factor to find how long the remaining lines should be in the similar figure you are drawing.

Name ______________________ Date ____________ Class ____________

LESSON 5E

Hands-On Lab 2

Similar Figures, continued

Think and Discuss

1. A photographer takes a picture and prints it so it is 6 inches by 8 inches. The photographer later lays out the picture to appear in a magazine. The picture in the magazine is 5 inches by 5 inches. Are the two pictures similar?

   Explain why or why not.

   ____

   ____

   ____

Try This

Recreate the drawing you made for the previous page in this space, making it half the size and turning it 180 degrees to the right. Explain if the figure is similar or not.

____

____

____

Name ______________________ Date ____________ Class ____________

LESSON 5E

Review for Mastery

Similar Figures

Similar Polygons	Congruent Polygons
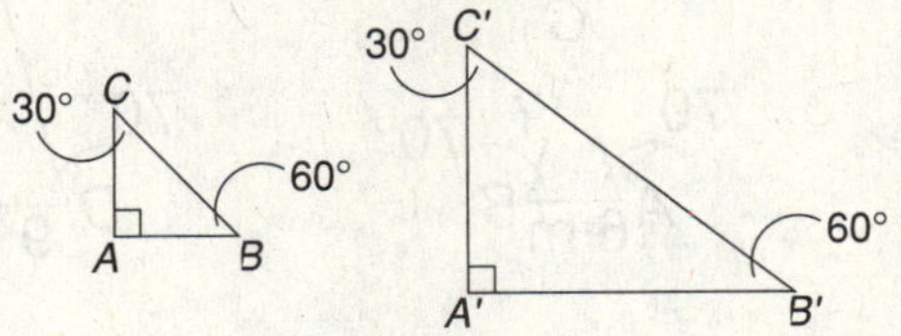	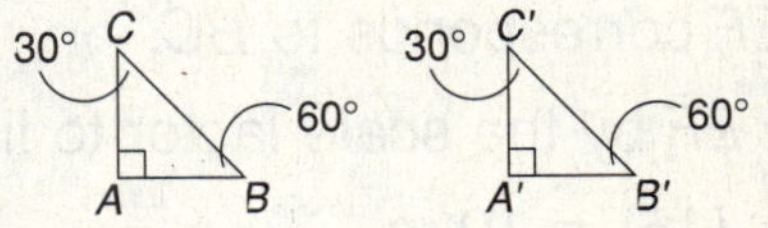
same shape corresponding angles are congruent $\angle A \cong \angle A'$, $\angle B \cong \angle B'$, $\angle C \cong \angle C'$	same shape corresponding angles are congruent $\angle A \cong \angle A'$, $\angle B \cong \angle B'$, $\angle C \cong \angle C'$
usually different size corresponding sides are in proportion $\frac{AB}{A'B'} = \frac{BC}{B'C'} = \frac{AC}{A'C'}$	same size corresponding sides are congruent $\overline{AB} \cong \overline{A'B'}$, $\overline{BC} \cong \overline{B'C'}$, $\overline{AC} \cong \overline{A'C'}$

Complete to tell if the polygons are similar.

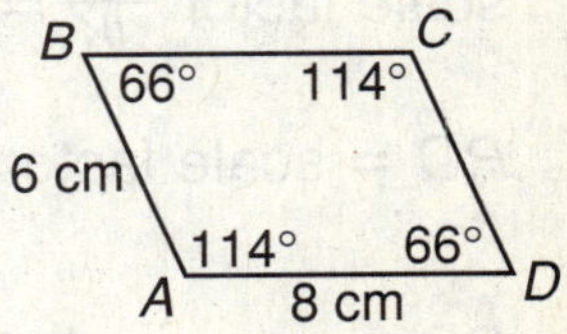

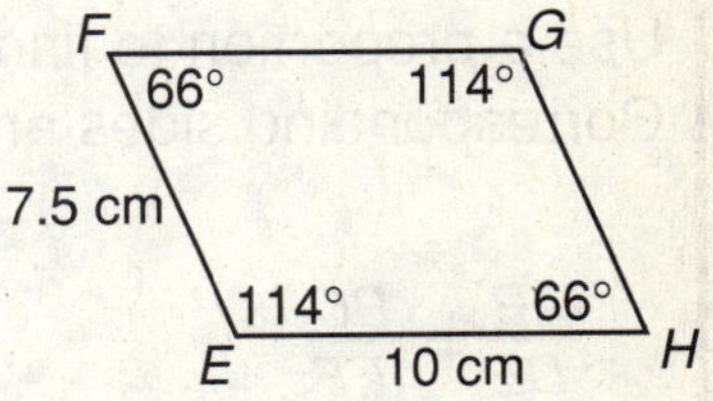

1. a. corresponding angles: $\angle A$ and ______, $\angle B$ and ______, $\angle C$ and ______, $\angle D$ and ______

b. all corresponding angles congruent? ______

c. corresponding sides: $\overline{AB}$ and ______, $\overline{BC}$ and ______, $\overline{CD}$ and ______, $\overline{AD}$ and ______

d. The ratio of each pair of corresponding sides is:

$\frac{AB}{}$ = ______, $\frac{BC}{}$ = ______, $\frac{CD}{}$ = ______, $\frac{AD}{}$ = ______

e. Are the corresponding sides proportional? Explain.

____

f. The **scale factor** is the simplest form of the ratio of the corresponding sides. What is the scale factor for these parallelograms? ____________

g. Are the parallelograms similar? Explain.

____

Name ______________________ Date ____________ Class ____________

LESSON 5E

Review for Mastery

Similar Figures, continued

Use the scale factor to find a missing dimension.

scale factor = $\frac{6}{9}$ or $\frac{2}{3}$

Since $\overline{EF}$ corresponds to $\overline{BC}$,

multiply EF by the scale factor to find BC.

$BC = \frac{2}{3}(15) = 10$ m

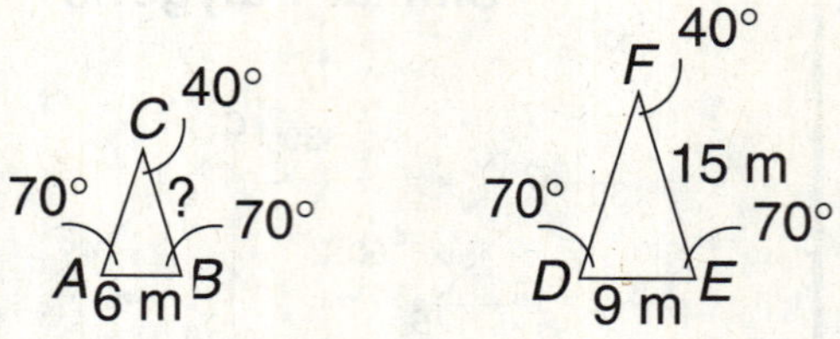

Complete to find each missing dimension.

2.

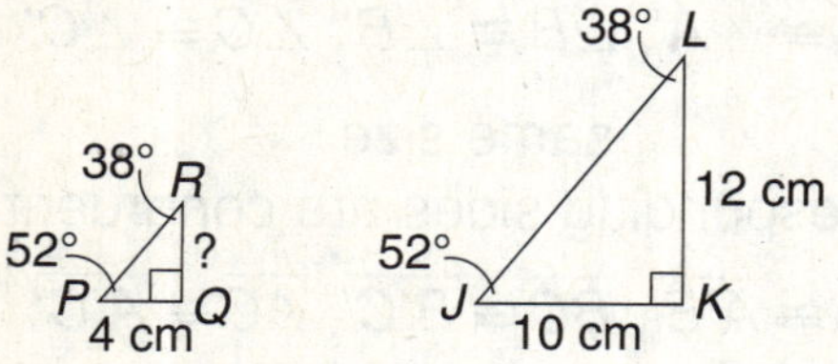

scale factor $\frac{PQ}{JK}$ = ______ or ______

RQ = scale factor • (LK)

RQ = ______ × ______ or ______cm

3.

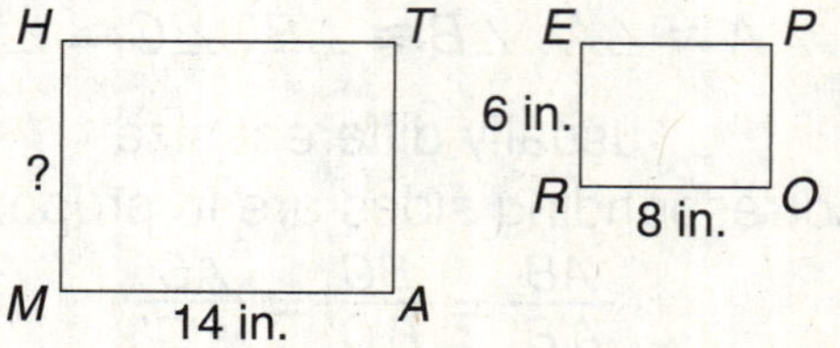

scale factor = $\frac{MA}{RO}$ = ______ or ______

MH = scale factor • (RE)

MH = ______ × ______ or ______in.

Use a proportion to find a missing dimension.
Corresponding sides are in proportion.

$\frac{AB}{DE} = \frac{BC}{EF}$

$\frac{6}{9} = \frac{BC}{15}$

$9(BC) = 90$

$BC = 10$ m

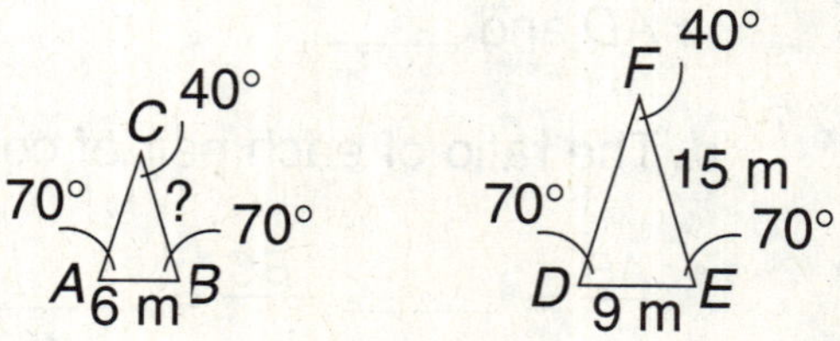

Complete to find the missing dimension.

4. $\frac{MH}{RE}$ = ______

$\frac{MH}{6}$ = ______

______ • (MH) = ______

MH = ______ in.

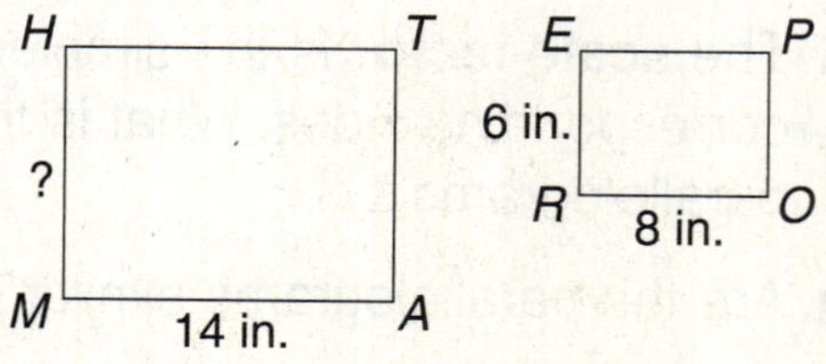

LESSON 5E

Homework and Practice

Similar Figures

1. Which figures are similar to *ABCD*?

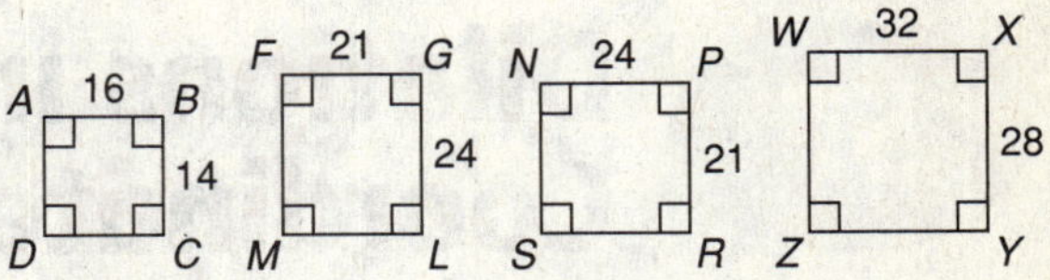

____

____

2. a. Measure the angles of the triangles. Then determine the corresponding sides.

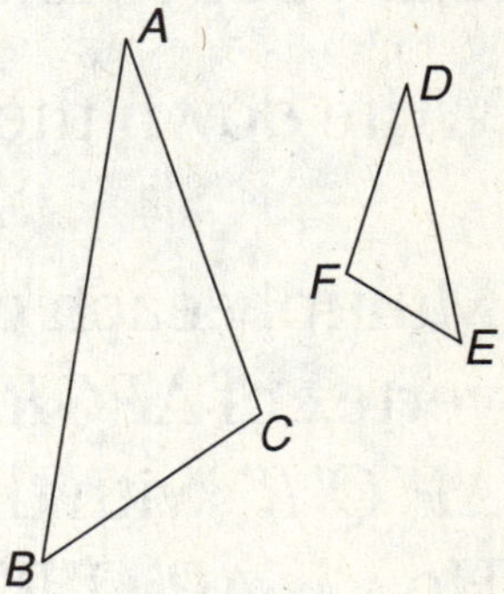

____

____

b. Find the ratios of the corresponding sides.

____

____

c. Based on your answers, are the triangles similar? Explain.

____

____

Find the length of the missing side in the similar figures.

3.

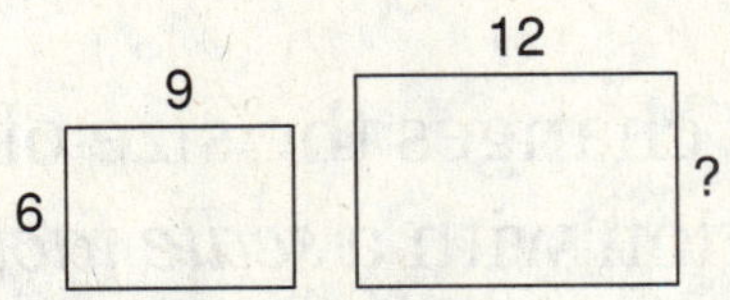

____

4.

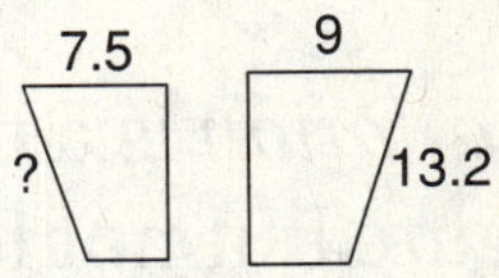

____

5. Lily is enlarging a photo that is 4 in. wid and 6 in. long to photo that is 24 in. wide. What would the perimeter of the new photo after the enlargement?

____

6. A 4 ft tall child is standing by The Leaning Tower of Pisa in Italy. The child casta 15 ft shadow at the same time the tower casts a 675 ft shadow. Approximate the height of the Leaning Tower of Pisa.

____

EXPLORATION

Program of Studies and Core Content

MA-8-G-S-SR5, MA-8-G-S-TS1, MA-8-G-S-TS4, MA-08-2.1.1, MA-08-3.1.4, MA-08-3.2.2

5F Dilations in the Coordinate Plane

Use graph paper to explore similarity in the coordinate plane.

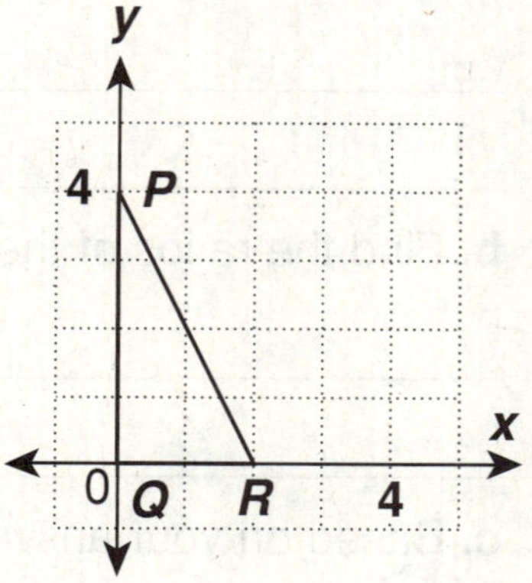

1. Write down the vertices of ΔPQR.

2. Multiply each coordinate of each vertex of ΔPQR by 3. Then graph $\Delta P'Q'R'$ with these new vertices. How is $\Delta P'Q'R'$ related to ΔPQR?

3. Now multiply each coordinate of each vertex of ΔPQR by $\frac{1}{2}$. Then graph $\Delta P''Q''R''$ with these new vertices. How is $\Delta P''Q''R''$ related to ΔPQR?

4. A *dilation* is a transformation that changes the size of a figure but not its shape. For a dilation with a *scale factor* of k, you can find the image of a point by multiplying each coordinate by k: $(a, b) \rightarrow (ka, kb)$. what is the scale factor of the dilation that mapped ΔPQR to $\Delta P'Q'R'$?

5. What is the scale factor of the dilation that mapped ΔPQR to $\Delta P''Q''R''$?

Think and Discuss

6. **Describe** how to find the image of the point (12, –15) under a dilation with scale factor $\frac{2}{3}$.

Name ____________________ Date __________ Class __________

LESSON 5F

Hands-On Lab 1

Dilations

Activity

A **dilation** is a transformation that changes the size, but not the shape, of a figure. The **scale factor** describes how much the figure is enlarged or reduced. The scale factor can be expressed in three ways: as a decimal, a fraction, or a percent.

Use a geoboard to create a small square, rectangle, or right triangle. Measure the polygon. Then dilate the figure by enlarging it by a scale factor of 2. To do that, multiply the length of each side by 2.

On the coordinate plane below, graph the polygon and its dilation.

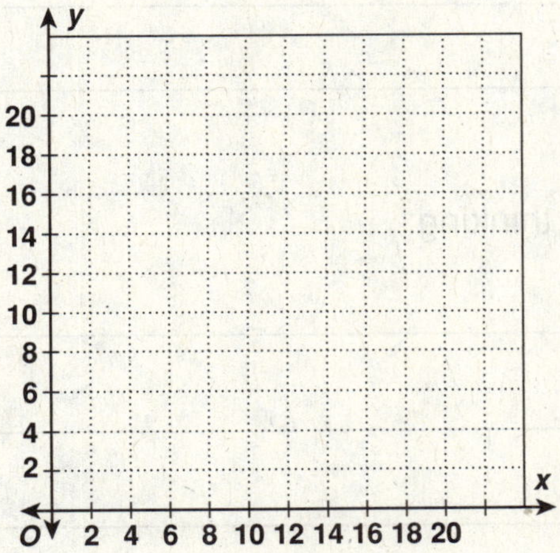

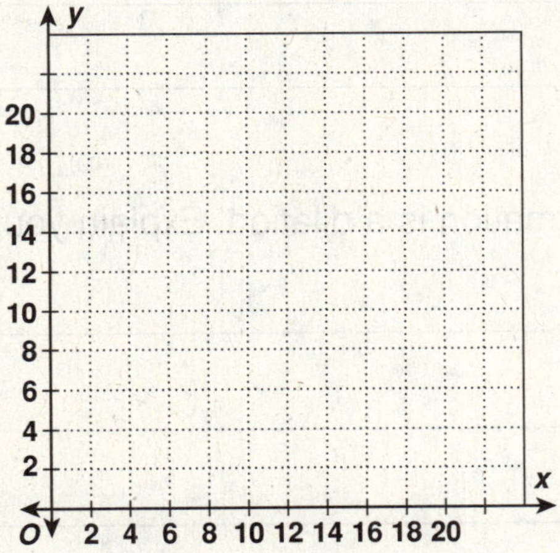

Label the points that map the figures. Then write the points that describe each figure below.

Small figure: ____________________

Large figure: ____________________

Name ______________________ Date ____________ Class ____________

LESSON 5F

Hands-On Lab 1

Dilations, continued

Think and Discuss

1. How can you tell if a transformation is a dilation?

2. What scale factor would show 2 figures that are the same size?

Try This

Decide if this transformation is a dilation. Explain your thinking.

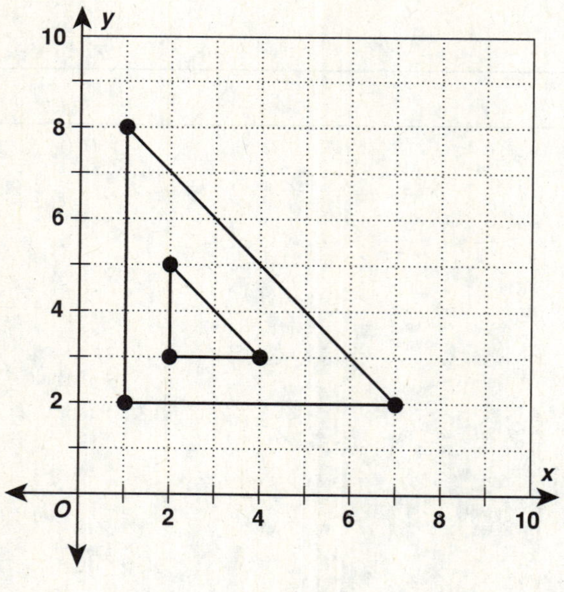

Name ______________________ Date ____________ Class ____________

LESSON **5F**

Hands-On Lab 2

Dilations

You can use graph paper or geometry software to explore dilations. If the scale factor is greater than 1, you are making a larger figure. If the scale factor is less than 1, you are making a smaller figure.

Activity

1. Using graph paper and a pencil, draw right triangle *ABC* with the following coordinates:

A(–2, –2), *B*(1, –1), *C*(0, 2)

2. Then use a colored pencil to graph a dilation *A′B′C′* using the origin as the center of dilation and a scale factor of 2.

Notice that the coordinates of the image are each equal to the original coordinates multiplied by 2.

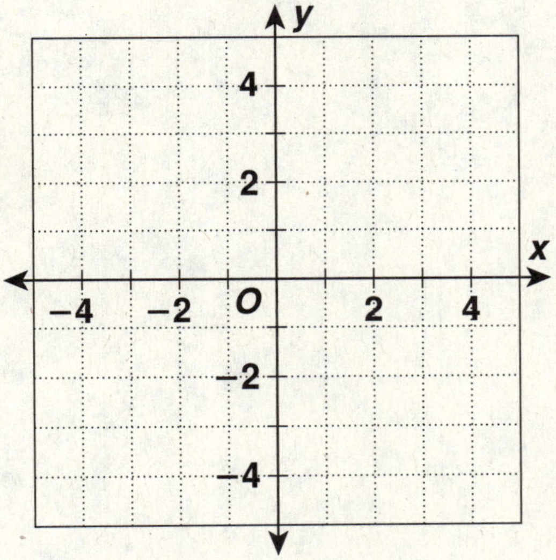

Name ______________________ Date ____________ Class ____________

LESSON 5F

Hands-On Lab 2

Dilations, continued

3. Next graph the rectangle *ABCD* with these coordinates.

$A(-6, -6)$, $B(-6, 6)$, $C(6, 6)$, $D(6, -6)$

4. Then graph a second rectangle with these coordinates.

$A'(-2, -2)$, $B'(-2, 2)$, $C'(2, 2)$, $D'(2, -2)$

You can see that the two figures share the origin as the center of dilation. You can calculate the scale factor by dividing one of the coordinates of the second rectangle by the corresponding coordinate of the original. For example, $\frac{B'}{B} = \frac{2}{6} = \frac{1}{3}$ for the *y*-coordinates.

So the scale factor is $\frac{1}{3}$.

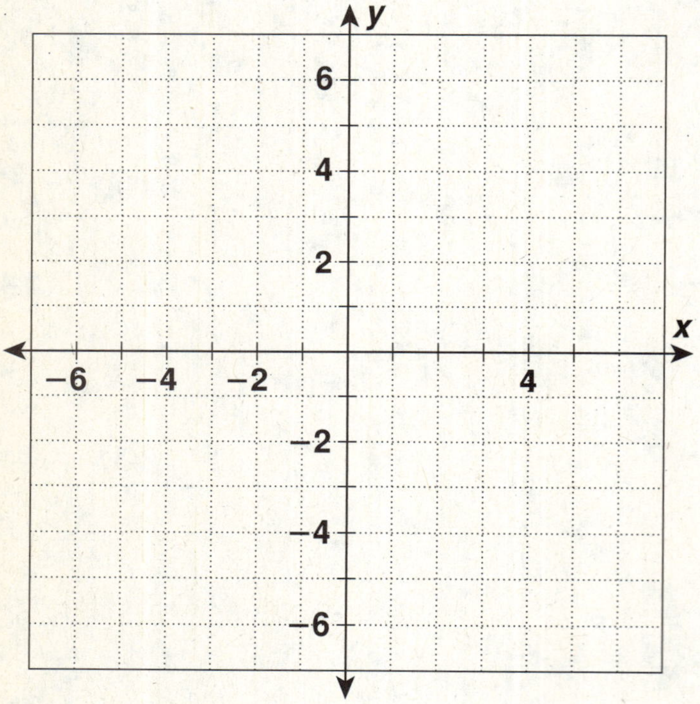

Think and Discuss

1. For what values of the scale factor will a dilation create a smaller image?

____

____

____

Name ______________________ Date __________ Class __________

LESSON **5F**

Hands-On Lab 2

Dilations, continued

Try This

1. Graph the rectangle *EFGH* with coordinates *E*(–2, –2), *F*(–2, 4), *G*(2, 4), *H*(2, –2). Then dilate the rectangle with a scale factor of $\frac{1}{2}$. Use point *E* as the center of dilation, which means that $E' = E$.

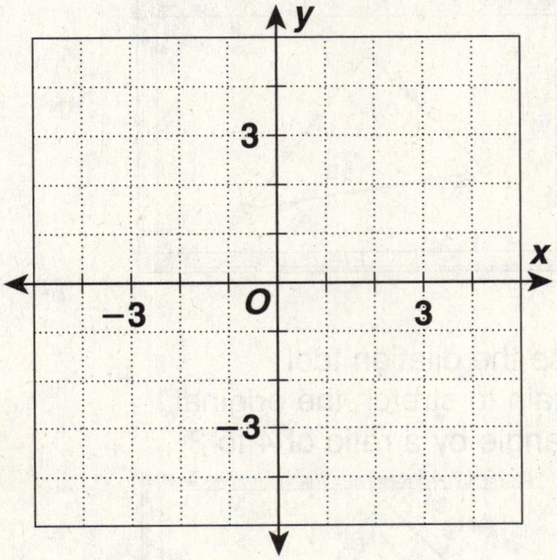

2. Can you calculate the scale factor for the figure in 1 by dividing a coordinate of the image by the corresponding coordinate of the original figure? Explain.

____

____

____

Name ______________________ Date __________ Class __________

LESSON 5F

Technology Lab
Dilations of Geometric Figures

A **dilation** is a geometric transformation that changes the size but not the shape of a figure.

Activity

1. Construct a triangle figure similar to the one shown below. Label the vertices *A*,*B*, and *C*.

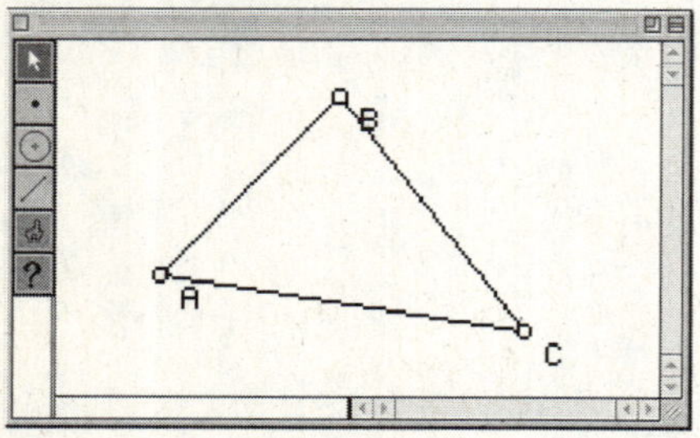

2. Next pick a center of dilation inside triangle *ABC* and label it point *D*.

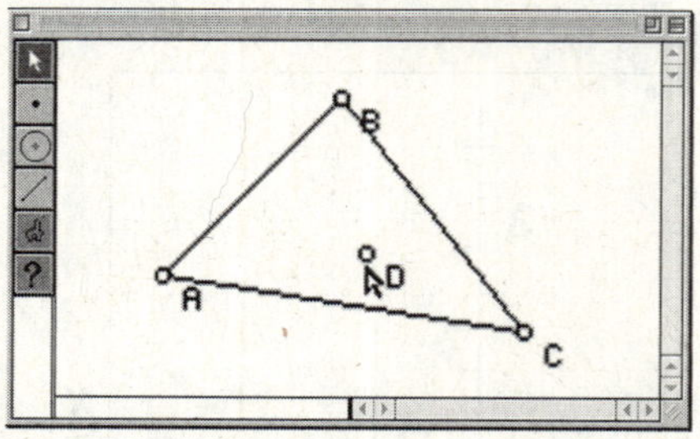

3. Use the dilation tool on your software to shrink the triangle by a ratio of 1 to 2.

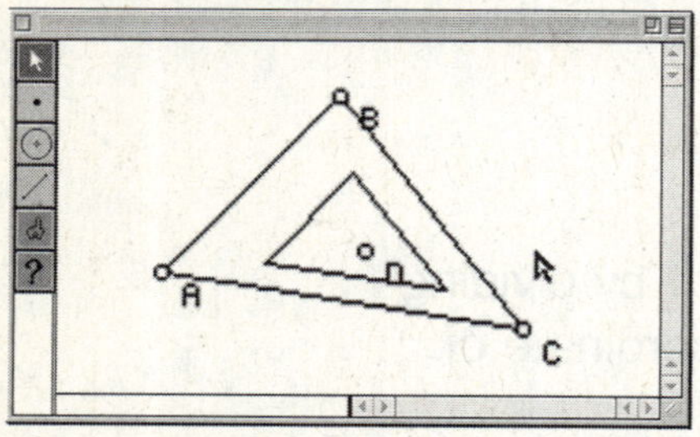

4. Use the dilation tool again to stretch the original triangle by a ratio of 4 to 3.

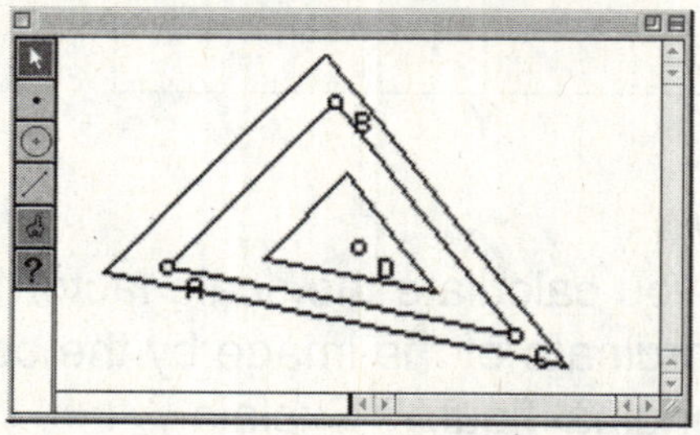

Notice that the dilations of triangle *ABC* are exactly the same *shape* as the original triangle, but they are different *sizes*.

Name ______________________ Date ____________ Class ____________

LESSON 5F

Technology Lab

Dilations of Geometric Figures, continued

Think and Discuss

1. Are all of the triangles shown in the last figure similar?

____

2. When the center of dilation is inside the triangle, and the dilated triangle is shrunk, is the smaller triangle always completely inside the original triangle?

____

Try This

1. Use geometry software to construct a rectangle *ABCD*.

   a. Choose a center of dilation inside *ABCD*. Shrink *ABCD* by a factor of 1 to 3.

   b. Choose a center of dilation outside *ABCD*. Stretch *ABCD* by a factor of 3 to 2.

Name ____________________ Date __________ Class __________

LESSON **5F**

Review for Mastery

Dilations

To dilate a figure in the coordinate plane, using the origin as the center of dilation, multiply each coordinate by the scale factor.

Using the origin O as the center of dilation, dilate $\triangle ABC$ by a scale factor of 2.5.

$A(2, 2) \longrightarrow A'(2.5 \cdot 2, 2.5 \cdot 2)$ or $A'(5, 5)$
$B(4, 0) \longrightarrow B'(2.5 \cdot 4, 2.5 \cdot 0)$ or $B'(10, 0)$
$C(4, 2) \longrightarrow C'(2.5 \cdot 4, 2.5 \cdot 2)$ or $C'(10, 5)$

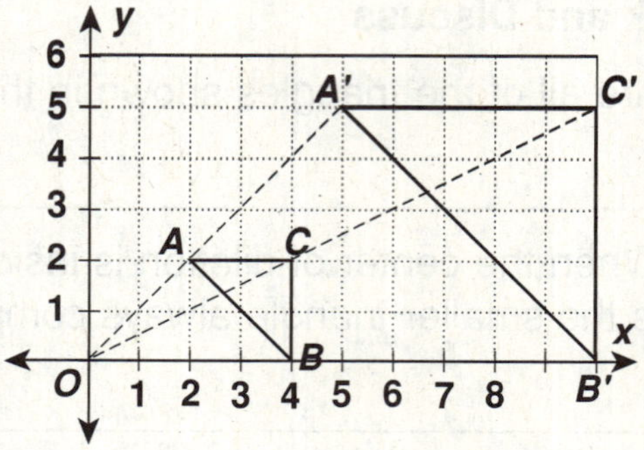

Using the origin as the center of dilation, complete to dilate $\triangle ABC$ by a scale factor of 2. Graph the dilation.

1. $A(1, 2) \longrightarrow A'(2 \cdot 1, 2 \cdot 2)$ or $A'($, $)$
$B(2, 0) \longrightarrow B'(\ \ \cdot 2,\ \ \cdot 0)$ or $B'($, $)$
$C(3, 3) \longrightarrow C'(\ \ \cdot 3,\ \ \cdot 3)$ or $C'($, $)$

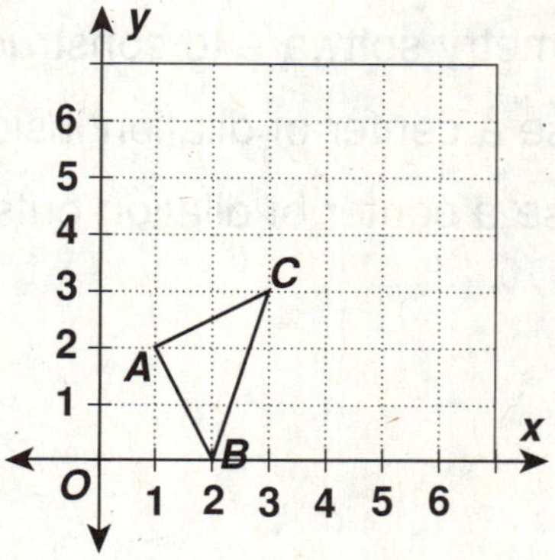

When the scale factor is a fraction between 0 and 1, the image is smaller than the original figure.

Using the origin O as the center of dilation, dilate $\triangle ABC$ by a scale factor of $\frac{1}{3}$.

$A(3, 3) \longrightarrow A'(\frac{1}{3} \cdot 3, \frac{1}{3} \cdot 3)$ or $A'(1, 1)$

$B(6, 0) \longrightarrow B'(\frac{1}{3} \cdot 6, \frac{1}{3} \cdot 0)$ or $B'(2, 0)$

$C(6, 6) \longrightarrow C'(\frac{1}{3} \cdot 6, \frac{1}{3} \cdot 6)$ or $C'(2, 2)$

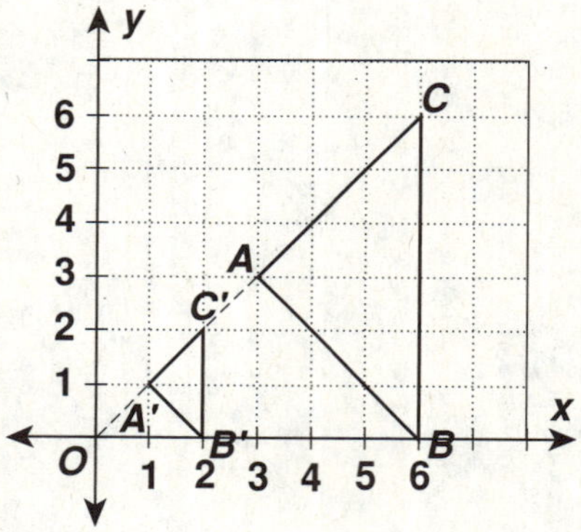

Using the origin as the center of dilation, complete to dilate $\triangle ABC$ by a scale factor of $\frac{1}{2}$.

2. $A(8, 0) \longrightarrow A'(\frac{1}{2} \cdot 8, \frac{1}{2} \cdot 0)$ or $A'($, $)$

$B(4, 4) \longrightarrow B'(\ \ \cdot 4,\ \ \cdot 4)$ or $B'($, $)$

$C(6, 8) \longrightarrow C'(\ \ \cdot 6,\ \ \cdot 8)$ or $C'($, $)$

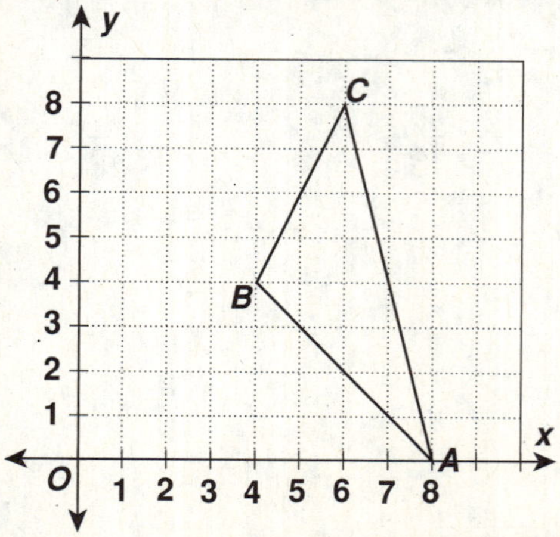

Name ______________________ Date ____________ Class ____________

LESSON 5F

Review for Mastery
Dilations, continued

Dilations in the Coordinate Plane	
For $k > 1$	**For $0 < k < 1$**
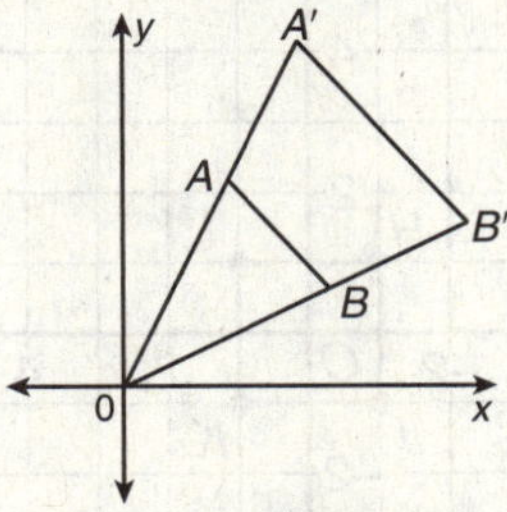 $(x, y) \rightarrow (kx, ky)$	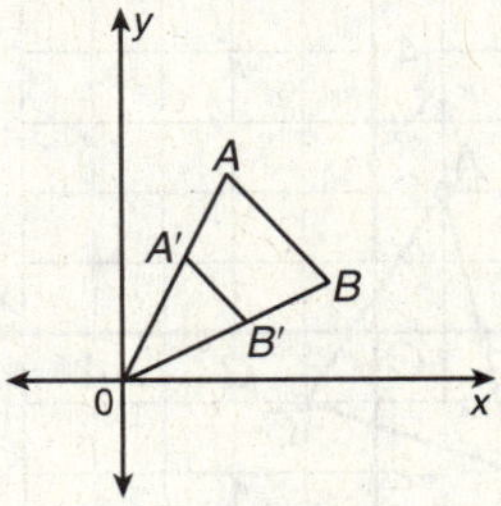 $(x, y) \rightarrow (kx, ky)$

If k has a negative value, the preimage is rotated by 180°.

Draw the image of $\triangle EFG$ with vertices $E(0, 0)$, $F(0, 1)$, and $G(2, 1)$ under a dilation with a scale factor of −3 and centered at the origin.

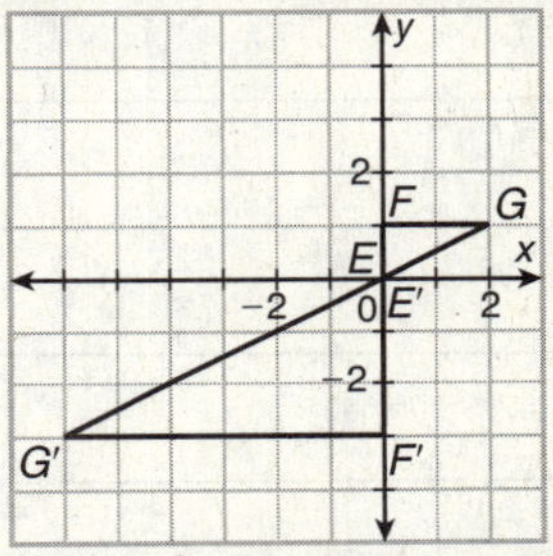

The image of (x, y) is $(-3x, -3y)$.

$E(0, 0) \rightarrow E'(0(-3), 0(-3)) \rightarrow E'(0, 0)$
$F(0, 1) \rightarrow F'(0(-3), 1(-3)) \rightarrow F'(0, -3)$
$G(2, 1) \rightarrow G'(2(-3), 1(-3)) \rightarrow G'(-6, -3)$

Graph the preimage and image.

Draw the image of the figure with the given vertices under a dilation with the given scale factor and centered at the origin.

3. $J(0, 0)$, $K(-1, 2)$, $L(3, 4)$; scale factor: 2

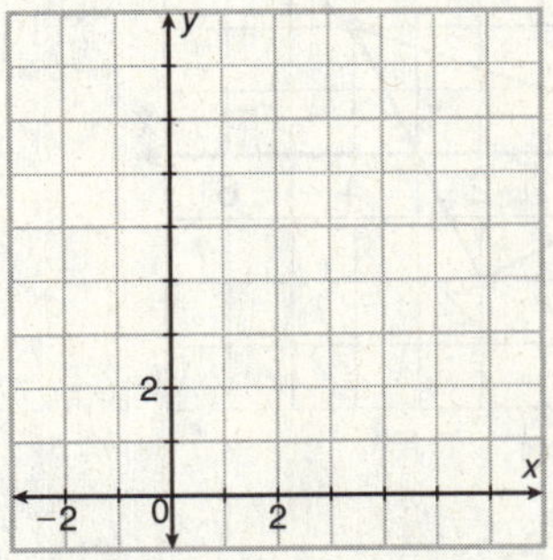

4. $A(0, 0)$, $B(0, 6)$, $C(6, 3)$; scale factor: $\frac{1}{3}$

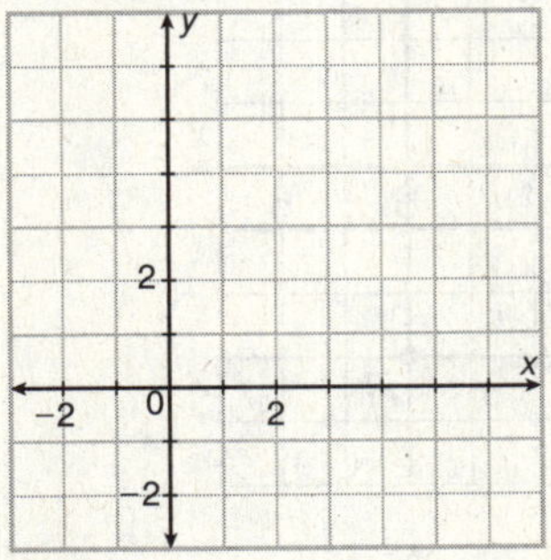

5. $R(1, 0)$, $S(1, -2)$, $T(-1, -2)$; scale factor: −2

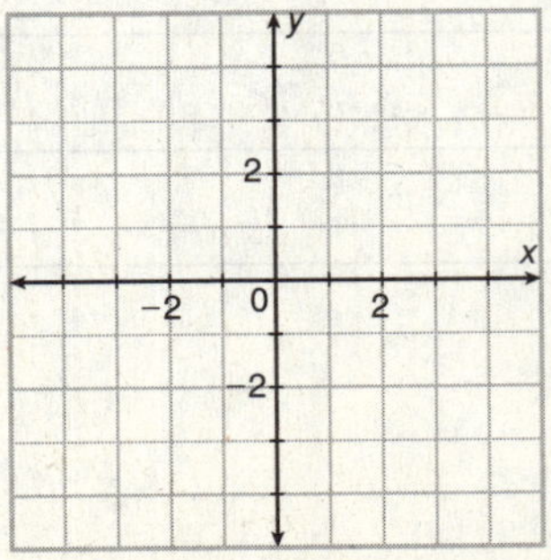

6. $G(2, 0)$, $H(0, 4)$, $I(4, 2)$; scale factor: $-\frac{1}{2}$

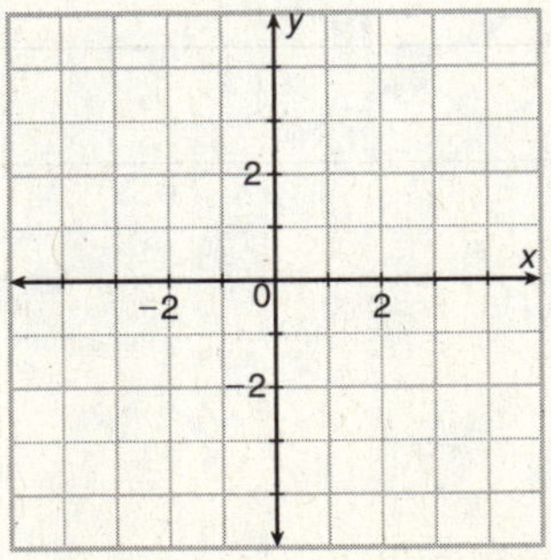

LESSON 5F

Homework and Practice
Dilations

Graph the dilation, using the origin as the center of dilation. Then describe the effect of the dilation on the figure.

1. scale factor: 2

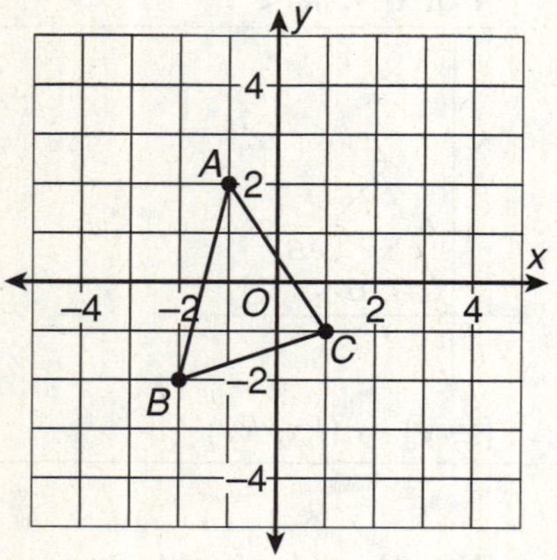

2. scale factor: 4

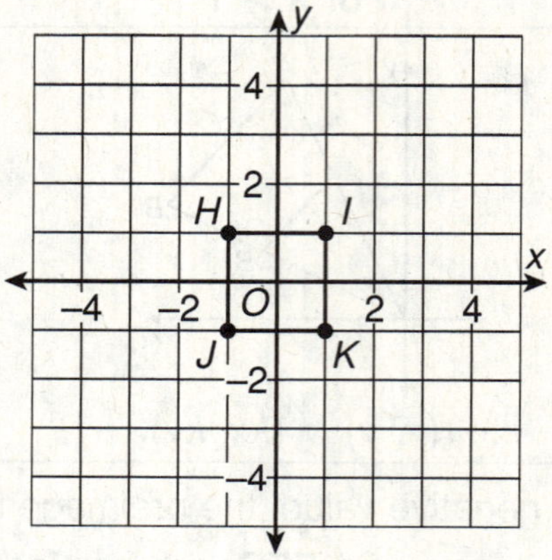

3. scale factor: $\frac{1}{3}$

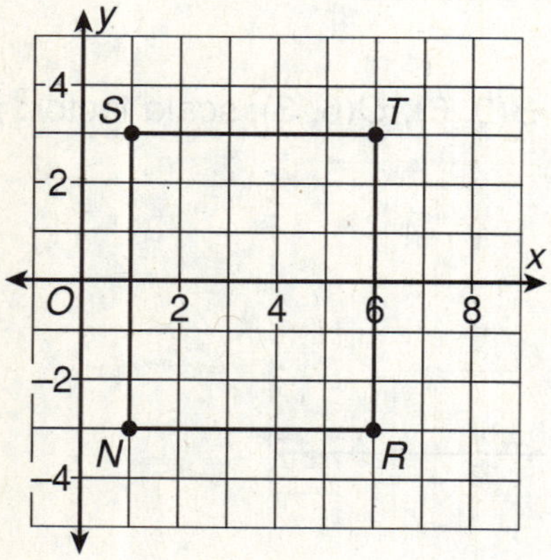

4. scale factor: 1.5

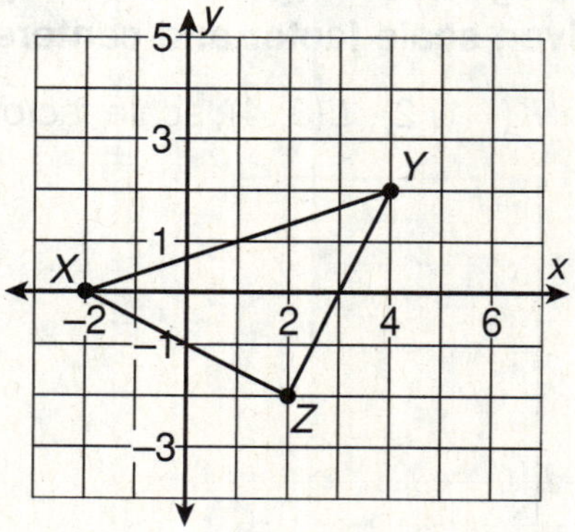

Name ______________________ Date ____________ Class ____________

LESSON 5F

Homework and Practice

Dilations, continued

5. The figure shows a photograph on grid paper. What are the coordinates of C' if the photograph is enlarged with scale factor $\frac{4}{3}$?

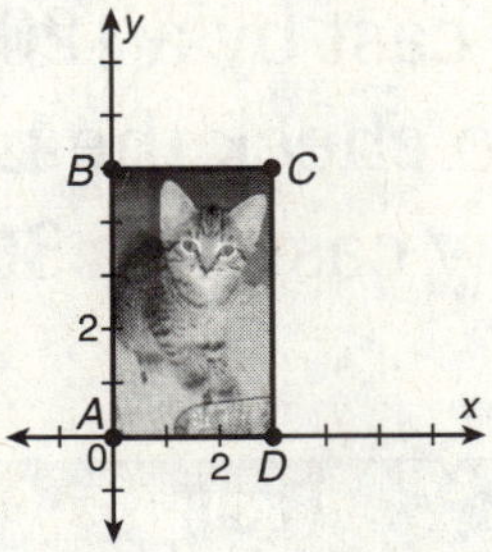

____

6. In the figure, $\triangle HFJ \sim \triangle EFG$. Find the coordinates of G and the scale factor.

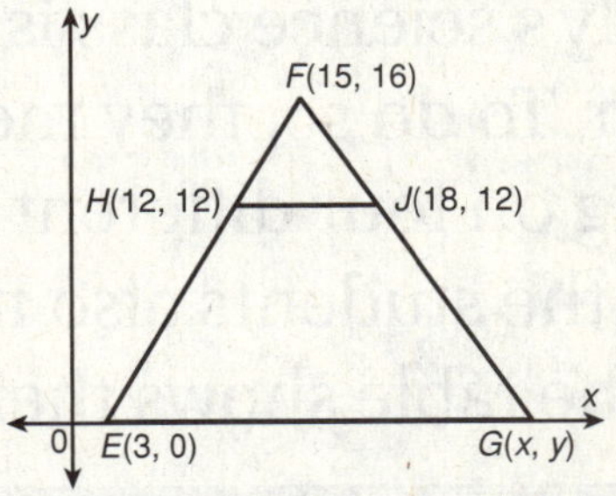

____

7. Triangle LMN has vertices $L(-10, 2)$, $M(-4, 11)$, and $N(6, -6)$. Find the vertices of the image of $\triangle LMN$ after a dilation with scale factor $\frac{5}{2}$.

____

____

8. Triangle HJM has vertices $H(-36, 0)$, $J(0, 20)$, and $M(0, 0)$. Triangle $H'J'M'$ has two vertices at $H'(-27, 0)$ and $M'(0, 0)$, and $\triangle H'J'M'$ is a dilation image of $\triangle HJM$. Find the coordinates of J' and the scale factor.

____

Choose the best answer.

9. The arrow is cut from a logo. The artist needs to make a copy five times as large for a sign. If the coordinates of T are $T(3, 4.5)$, what are the coordinates of T' after the arrow is dilated with scale factor 5?

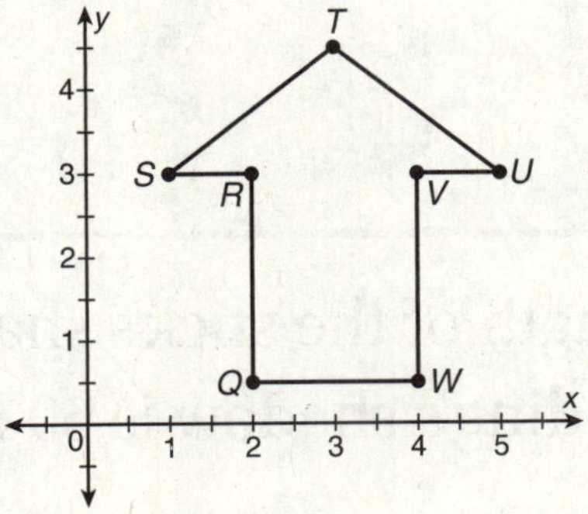

A $T'(15, 22.5)$

B $T'(7.5, 9)$

C $T'(4.5, 6.75)$

D $T'(2.5, 20)$

10. Triangle QRS has vertices $Q(-7, 3)$, $R(9, 8)$, and $S(2, 16)$. What is the scale factor if the vertices after a dilation are $Q'(-10.5, 4.5)$, $R'(13.5, 15)$, and $S'(3, 24)$?

F $\frac{1}{3}$

G $\frac{1}{2}$

H $\frac{2}{3}$

J $\frac{3}{2}$

11. A triangle has vertices $H(-4, 2)$, $J(-8, 6)$, and $K(0, 6)$. If $\triangle ABC \sim \triangle HJK$, what are possible vertices of $\triangle ABC$?

A $A(-4, 3)$, $B(-2, 1)$, $C(0, 3)$

B $A(-2, 1)$, $B(-4, 3)$, $C(0, 3)$

C $A(-2, 4)$, $B(0, 6)$, $C(-2, 8)$

D $A(-2, 4)$, $B(-8, 6)$, $C(-4, 2)$

EXPLORATION

5G Indirect Measurement

Program of Studies and Core Content
MA-8-G-S-SR4, MA-8-G-S-SR5, MA-08-3.1.4

Mr. Kelly's science class is tracking the motion of the sun during the year. To do so, they measure the shadow cast by a 120-foot building on four different days of the year. To check the measurements, the students also measure the shadow cast by a 36-inch stick. The table shows their findings.

Day	Height of Building	Length of Building's Shadow	Height of Stick	Length of Stick's Shadow
1	120 ft	20 ft	36 in.	6 in.
2	120 ft	30 ft	36 in.	9 in.
3	120 ft	40 ft	36 in.	12 in.
4	120 ft	80 ft	36 in.	24 in.

1. a. On Day 1, what was the ratio of the height of the building to the length of the building's shadow?
   b. What was the ratio of the height of the stick to the length of the stick's shadow?
   c. What do you notice about these ratios?
2. Does your observation about the ratios also hold for Day 2, Day 3, and Day 4?

Think and Discuss

3. **Describe** how you could find the length of the stick's shadow if you know that the length of the building's shadow is 60 ft.

4. **Explain** what must be true about the building's shadow on a day when the stick's shadow is 36 in.

Name ______________________ Date ____________ Class ____________

LESSON **5G**

Hands-On Lab

Indirect Measurement

Activity

In similar figures, the ratio of the lengths of two sides of one figure is equal to the ratio of the lengths of the corresponding sides of the other figure. Similar figures have the same shape. They do not always have the same size.

If the day is sunny, you can use shadows to find missing dimensions in similar figures. With a classmate, measure your height. Then go outside and measure the length of your shadow. At the same time, measure the length of the shadow of another object such as a flagpole, or the length of the shadow of your school building. Then follow these instructions.

First, you need to find the ratio formed by the lengths of two sides of one figure. In this case, the two sides are your height and the length of your shadow.

To find the height of the flagpole or school:

1. Measure your height.
2. Divide that measurement by the length of your shadow to get the ratio.
3. Multiply the length of the shadow of the flagpole or school by the ratio to find the actual height of the flagpole or school.

Your height: ______________________

Your shadow's length: ______________________

Ratio: ______________________

Length of flagpole or school's shadow: ______________________

Actual height of flagpole or school: ______________________

Name ______________________ Date __________ Class __________

LESSON 5G

Hands-On Lab

Indirect Measurement, continued

Think and Discuss

1. How could the technique in this activity be useful in real life?

____

____

____

2. How could you use a light bulb to replicate this activity?

____

____

____

____

Try This

Practice using similar figures by measuring the same objects (your height and a flagpole or the school building) at a different time of day, when shadows are a different length.

Your height: _______________________________

Your shadow's length: _______________________________

Ratio: _______________________________

Length of flagpole or school's shadow: _______________________________

Actual height of flagpole or school: _______________________________

What conclusion can you draw about the ratio when the length of the shadow approaches the actual length of the object?

____

____

Name ______________________ Date __________ Class __________

LESSON 5G

Review for Mastery

Indirect Measurement

If two triangles are similar, you can set up and solve a proportion to find a missing side length.

$\triangle ABC \sim \triangle DBE$

$\overline{AB}$ corresponds to $\overline{DB}$.

$\overline{BC}$ corresponds to $\overline{BE}$.

$$\frac{AB}{BC} = \frac{DB}{BE}$$

$$\frac{9}{6} = \frac{27}{x} \quad \text{Find cross products.}$$

$$9x = 6(27) \quad \text{Simplify.}$$

$$\frac{9x}{9} = \frac{162}{9} \quad \text{Divide both sides by 9.}$$

$$x = 18$$

The length of $\overline{EB}$ is 18 feet.

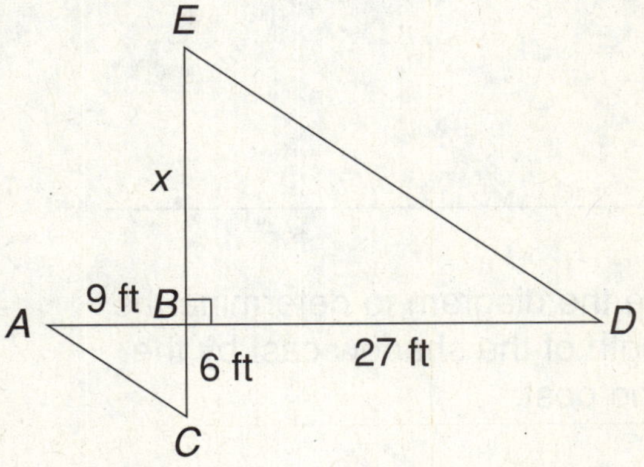

Each pair of triangles is similar. Find each missing length.

1.

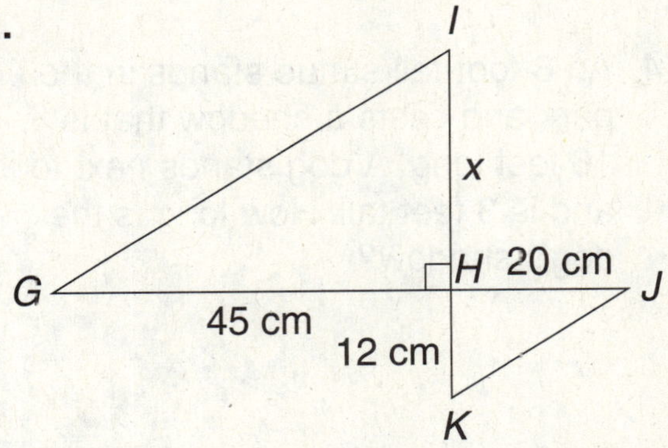

$$\frac{HK}{HJ} = \frac{HI}{HG}$$

$$\frac{12}{20} = \frac{x}{45}$$

$x =$ ______________

$HI =$ ______________

2.

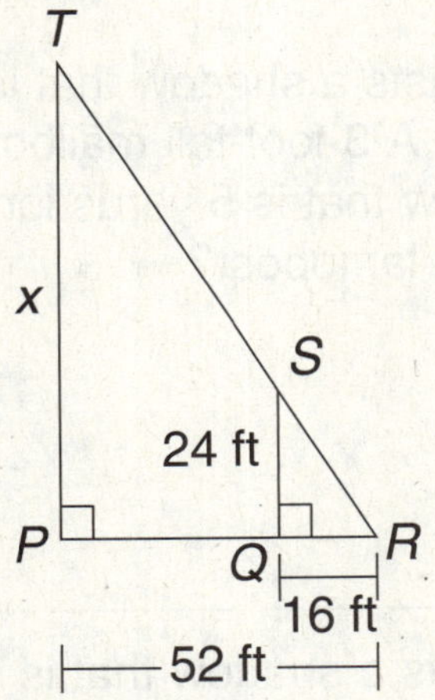

$$\frac{TP}{PR} = \frac{SQ}{QR}$$

$$\frac{x}{52} = \frac{24}{16}$$

$x =$ ______________

$TP =$ ______________

Name ______________________ Date __________ Class __________

LESSON 5G

Homework and Practice

Indirect Measurement

1. Use the diagram to find the height of the building.

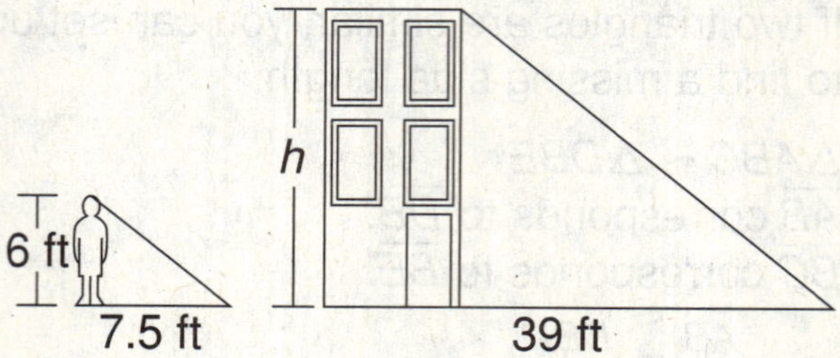

____

2. Use the diagram to determine the length of the shadow cast by the lamp post.

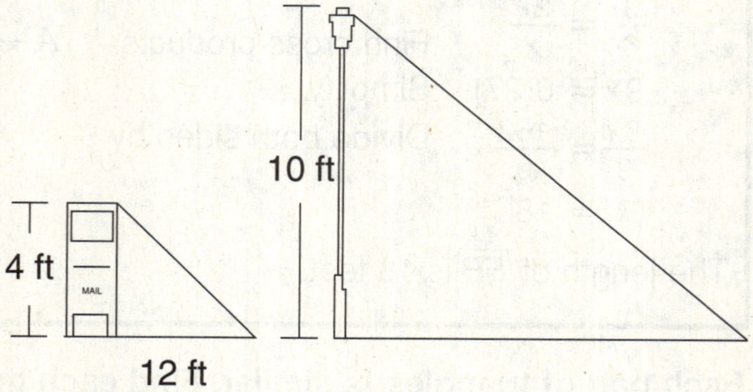

____

3. A lamppost casts a shadow that is 15 yards long. A 3-foot-tall mailbox casts a shadow that is 5 yards long. How tall is the lamppost?

____

4. An 8-foot-tall statue stands in the park and casts a shadow that is 16 feet long. A dog stands next to it and is 3 feet tall. How long is the dog's shadow?

____

5. A building casts a shadow that is 420 meters long. At the same time, a person who is 2 meters tall casts a shadow that is 24 meters long. How tall is the building?

____

6. On a sunny day around noon, a tree casts a shadow that is 12 feet long. At the same time, a person who is 6 feet tall standing beside the tree casts a shadow that is 2 feet long. How tall is the tree?

____

Name ______________________ Date __________ Class __________

LESSON 5G

Homework and Practice

Indirect Measurement, continued

7. Use the diagram to determine the height of the sign.

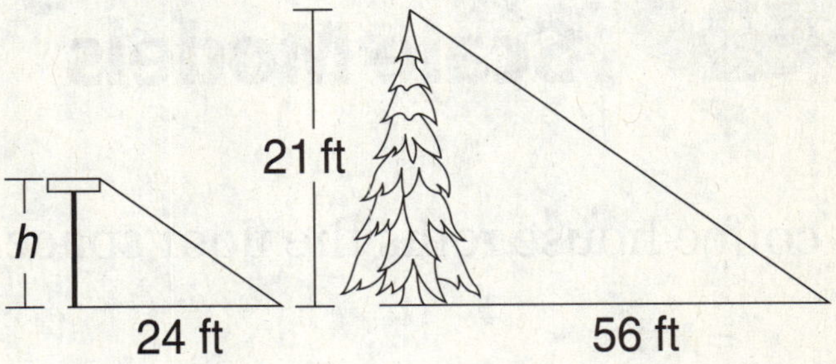

____

8. Use the diagram to determine the height of the tree.

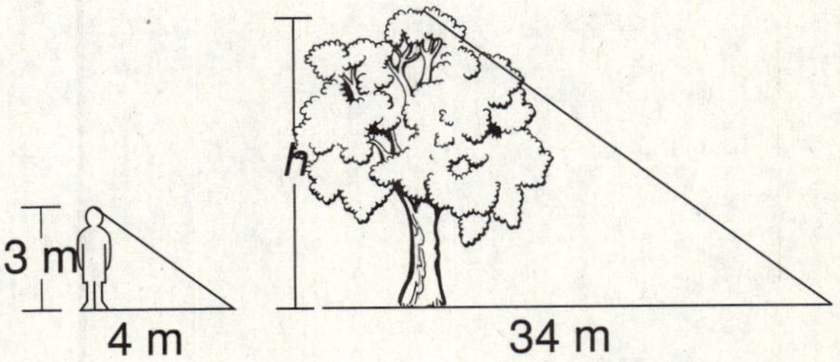

____

9. A pole casts a shadow that is 21 feet long. A 3-feet-tall child standing next to the pole casts a shadow that is 9 feet long. How tall is the pole?

____

10. Jeremy has two trophies next to each other sitting in the window of his room. His football trophy is 7 inches tall and his basketball trophy is 13 inches tall. As the light shines in, the basketball trophy's shadow measures 26 inches. How long is the football trophy's shadow?

____

11. Vladimir wants to know the distance across a busy street. He can take measurements alongside the street, and in the field adjacent to the street. Explain how he can use indirect measurement to estimate the distance across the street. Include a diagram.

____

____

____

EXPLORATION

Program of Studies and Core Content

MA-8-G-S-SR4, MA-8-G-S-SR5, MA-08-3.1.4

5H Scale Drawings and Scale Models

A coffee house rents the floor space modeled by the figure below.

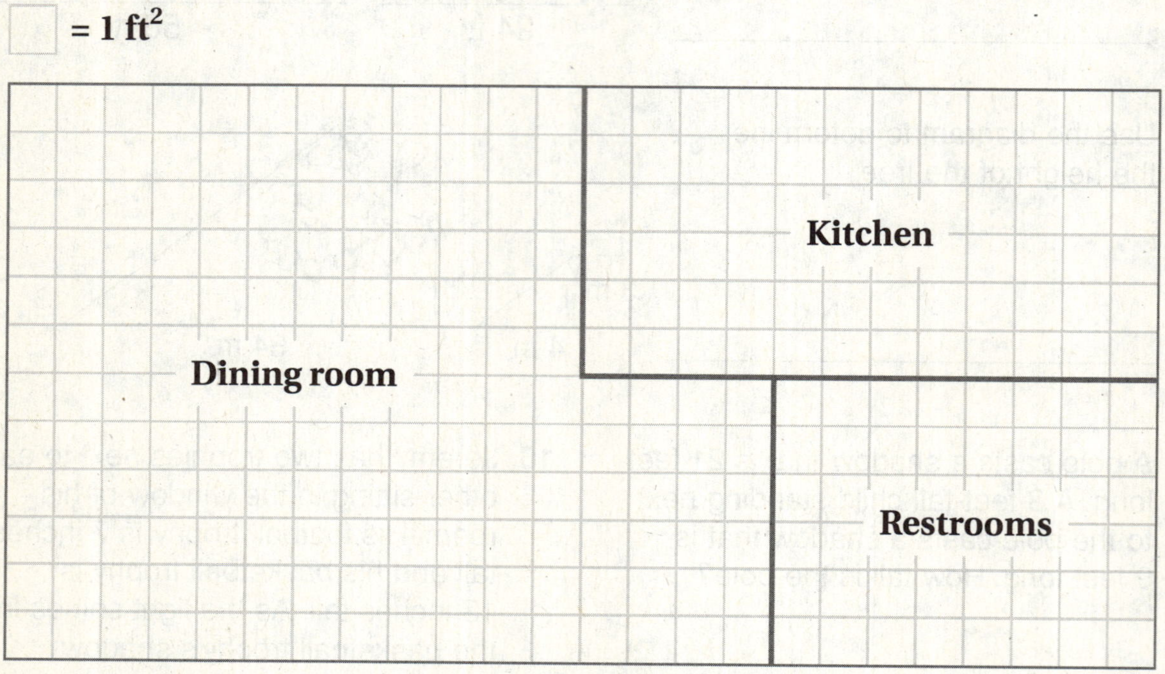

Count the number of squares to answer each question.

1. What is the actual area of the dining room?
2. What is the actual area of the kitchen?
3. What is the actual area of the restrooms?

Think and Discuss

4. **Show** a possible arrangement of square tables in the dining area if the tabletops each measure 2 ft by 2 ft.
5. **Discuss** real-world situations in which scale drawings are used.

Name ______________________ Date __________ Class __________

LESSON 5H

Review for Mastery

Scale Drawings and Scale Models

In a **scale drawing or a scale model,** all the dimensions of the actual object are reduced or enlarged proportionally.

A map is a scale drawing in which actual distance is reduced.

The towns of Ardon and Bacton are on a map with scale 1 cm = 15 km.

If the map distance between Ardon and Bacton is 4.5 cm, what is the actual distance?

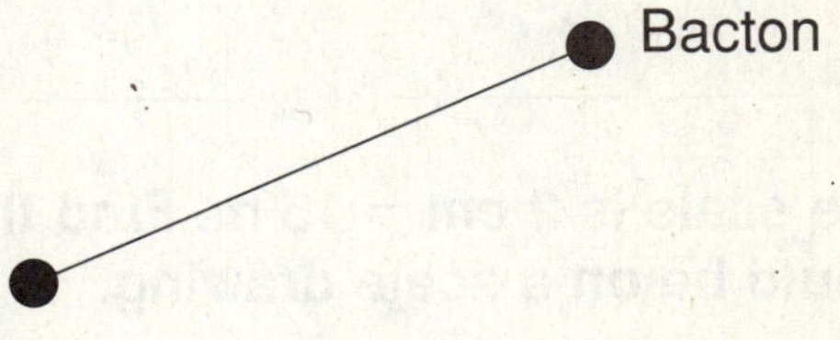

$$\frac{\text{actual distance}}{\text{map distance}} = \frac{\text{actual distance}}{\text{map distance}}$$

$$\frac{15 \text{ km}}{1 \text{ cm}} = \frac{x \text{ km}}{4.5 \text{ cm}}$$

$$1(x) = 15(4.5)$$

$$x = 67.5 \text{ km}$$ ← actual distance between Ardon and Bacton

Complete to find each unknown measure.

1. A map scale is 1 in. = 75 mi. The map distance between two towns is 3.5 in. Find the actual distance x between the towns.

$$\frac{\text{actual distance}}{\text{map distance}} = \frac{\text{actual distance}}{\text{map distance}}$$

$$\frac{75 \text{ mi}}{\_\_\_ \text{ in.}} = \frac{x \text{ mi}}{\_\_\_ \text{ in.}}$$

x = __________

actual distance = __________

2. The actual distance between two towns is 175 km. If the distance between them on a map is 7 cm, what is the map scale?

$$\frac{x \text{ km}}{1 \text{ cm}} = \frac{\_\_\_ \text{ km}}{\_\_\_ \text{ cm}}$$

____

x = __________

map scale: 1 cm = __________

3. An archway in a $\frac{1}{2}$ in. scale drawing is 4.5 in. tall. Find the actual height x.

$$\frac{\text{actual height}}{\text{scale height}} = \frac{\text{actual height}}{\text{scale height}}$$

$$\frac{1 \text{ ft}}{0.5 \text{ in.}} = \frac{\_\_\_ \text{ ft}}{\_\_\_ \text{ in.}}$$

____

____

actual height = __________

4. Under a 7:1 magnification, this letter F appears to be 84 points high. Find the actual height x.

$$\frac{\text{actual height}}{\text{scale height}} = \frac{\text{actual height}}{\text{scale height}}$$

$$\frac{1}{7} = \frac{\_\_\_ \textit{ points}}{\_\_\_ \textit{ points}}$$

____

____

actual height = __________

Name ______________________ Date ____________ Class ____________

LESSON 5H

Homework and Practice

Scale Drawings

The scale of a drawing is $\frac{1}{4}$ in. = 12 ft. Find the actual measurement.

1. 8 in. ________ **2.** 11 in. ________ **3.** 16 in. ________ **4.** 18 in. ________

5. 22 in. ________ **6.** 27 in. ________ **7.** 21.5 in. ________ **8.** 38.5 in. ________

The scale is 2 cm = 15 m. Find the length each measurement would be on a scale drawing.

9. 180 m ________ **10.** 585 m ________ **11.** 330 m ________ **12.** 420 m ________

13. 225 m ________ **14.** 622.5 m ________ **15.** 547.5 m ________ **16.** 682.5 m ________

17. On a map the distance between Charleston and Mt. Pleasant is 3.2 cm. The scale is 1 cm = 25 mi. What is the actual distance in miles between these two towns?

____

18. Blueprints of a building are drawn with a scale of 1 cm = 25 ft. If the base of the building is a square with a perimeter 700 feet, what is the length of one side of the base of the building on the scale drawing?

____

19. If the scale drawing of a room has measurements of 8 cm by 4.5 cm and the scale of the drawing is 1 cm = 8 ft, what are the actual measurements of the room?

____

Name ______________________ Date __________ Class __________

LESSON **5H**

Homework and Practice

Scale Models

Tell whether each scale reduces, enlarges, or preserves the size of the actual object.

20. 1 in.:15 in.

21. 3 ft:1 yd

22. 1 cm:1 in.

23. 40 cm:10 in.

24. 1,760 yd:1 mi

25. 1 mi:1 km

Change both measurements to the same unit of measure, and find the scale factor.

26. 2-in. model of a 5-ft desk

27. 4-ft model of a 120-yd field

28. 30-cm model of a 7.5-m wall

29. 50-in. model of a 7,000-ft volcano

30. The museum has a 2-ft model of a 10-yard shark. What is the scale factor of the model?

31. Marina made a scale model of her yard. Her yard is 40 ft × 28 ft 9 in. Her drawing is 16 in. × 11.5 in. What scale factor did she use?

32. A 17 in. model of a 2002 Buick Rendezvous is made with a scale factor of $\frac{1}{11}$. Estimate the actual length of a 2002 Buick Rendezvous.

33. A miniature fireplace for a dollhouse is 4.5 in. by 6.75 in. If the scale factor is $\frac{1}{12}$, what size fireplace does the miniature represent?

CHAPTER 5

Chapter Summary

Answer these questions to summarize the important concepts from Chapter 5 in your own words.

1. Explain how to find two ratios equivalent to $\frac{12}{6}$.

2. Explain how to use the conversion factor $\frac{1 \text{ year}}{52 \text{ weeks}}$ to convert 221 weeks to years.

3. Explain how to solve the proportion $\frac{40}{15} = \frac{30}{x}$ using equivalent fractions.

4. The length of an object on a scale drawing is 5 in., and its actual length is 50 ft. Explain how to find the scale.

For more review of Chapter 5:

- Complete the Chapter 5 Study Guide and Review on pages 262–264 of your textbook.
- Complete the Ready to Go On quizzes on pages 234 and 258 of your textbook.

Name ______________________ Date ____________ Class ____________

CHAPTER 5

KCCT Prep

Multiple Choice

1. The solution of the equation □ + 2x = –15 is x = –5. What is the missing number in the equation?

(A) –10 (C) 5
(B) –5 (D) 10

2. Pablo writes a fraction in the form $\frac{n}{20}$, where n is a whole number, so that the fraction is greater than 0.5 and less than 0.7. How many possible values are there for n?

(A) 1 (C) 3
(B) 2 (D) 4

3. Jonathan drives 260 feet in 6 seconds. Which is the best estimate of his speed in miles per hour?

(A) 20 mi/h (C) 40 mi/h
(B) 30 mi/h (D) 50 mi/h

4. What is the value of y?

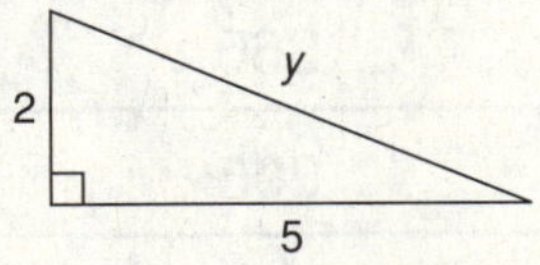

(A) $\sqrt{7}$ (C) $\sqrt{21}$
(B) $\sqrt{14}$ (D) $\sqrt{29}$

5. Notebooks cost \$2.35 each. Regina buys n notebooks and a pen that costs \$1.85. Altogether, she spends more than \$8.90. Which inequality can be used to find the least number of notebooks Regina buys?

(A) $2.35n < 7.05$
(B) $2.35 + n > 7.05$
(C) $2.35n > 7.05$
(D) $2.35 + n < 7.05$

6. Which equation is an example of the Associative Property?

(A) $7(x - 5) = 7x - 35$
(B) $6 + m = m + 6$
(C) $(4 + z) + x = 4 + (z + x)$
(D) $x(y + 1) = xy + x$

7. Which is the best estimate of 98.4 ÷ 0.31?

(A) 3 (C) 300
(B) 30 (B) 3000

8. Which is the best estimate of $\sqrt{80}$?

(A) 7.5 (C) 9
(B) 8 (D) 9.5

9. Which expression has the greatest value when $x = -1$ and $y = 3$?

(A) $5(x + y)$ (C) $5y + x$
(B) $5x + y$ (D) $5 + x + y$

10. A bottle contains $12\frac{1}{3}$ fluid ounces of juice. Kaitlyn uses the juice from 4 such bottles to make punch. She also adds $8\frac{1}{2}$ fluid ounces of mineral water to the juice. What is the total amount of punch that she makes?

(A) $32\frac{2}{3}$ fl oz (C) $49\frac{1}{3}$ fl oz
(B) $40\frac{5}{6}$ fl oz (D) $57\frac{5}{6}$ fl oz

11. What is the value of the expression $(x - y)^{z+1}$ for $x = y + 1$ and $z = 2$?

(A) –3 (C) 1
(B) –1 (D) 3

12. The square shown below is dilated from the origin by a scale factor of 1.5. What are the coordinates of the image of point *R*?

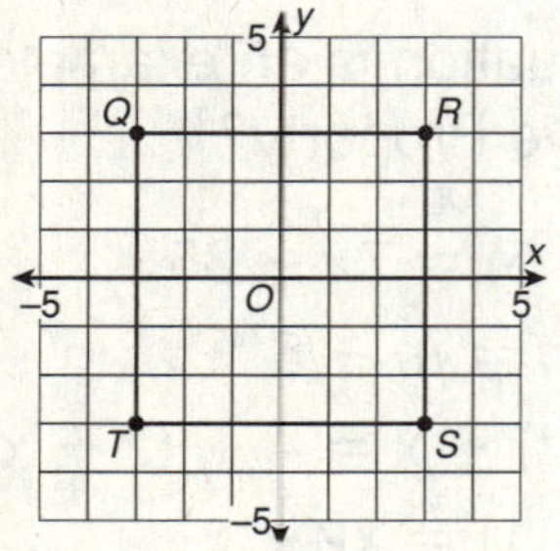

Ⓐ (1.5, 1.5) Ⓒ (3, 3)
Ⓑ (2, 2) Ⓓ (4.5, 4.5)

13. A scale model of an office building is made using the scale 2 in:5 ft. The model is 40.2 inches tall. Which is the best estimate of the height of the actual office building?

Ⓐ 20 ft Ⓒ 100 ft
Ⓑ 60 ft Ⓓ 120 ft

14. The ordered pair (–1, 3) appears in a table of values for the equation $y = 3x + \square$. What is the missing number in the equation?

Ⓐ 0 Ⓒ 3
Ⓑ 2 Ⓓ 6

15. Which equation illustrates the Identity Property of Multiplication?

Ⓐ $(3 \cdot 4)m = 3(4m)$
Ⓑ $8t = t \cdot 8$
Ⓒ $7(d + 3) = 7d + 7 \cdot 3$
Ⓓ $1y = y$

Open Response

16. The cars on an amusement park ride move along a track at a speed of 15 miles per hour.

a. What is the speed of the cars in feet per second? Show your work.

b. A section of the track is 198 feet long. How long does it take a car to move along this section of the track?

17. Holly plots the points (1, 1), (2, 1), and (2, 3). Then she connects the points to form a right triangle.

a. Holly enlarges the triangle using the origin as the center of dilation and a scale factor of 3. What are the coordinates of the vertices of the enlarged triangle?

b. How long is the hypotenuse of the enlarged triangle? Round to the nearest tenth.

18. Three students are comparing their typing speeds. The table gives data on their speeds.

Typing Speeds		
Name	**Number of Words**	**Minutes**
Lee	360	8
Allie	305	5
Brian	312	6

a. What is Lee's unit rate?

b. How long would it take Allie to type 183 words?

c. List the students in order from the slowest typist to the fastest.

d. You want to hire someone to type an essay that is 1000 words long. You need the essay to be typed in less than 20 minutes. Should you hire any of the students? Explain.

CHAPTER 6

Project
Percents: Home Run Derby

Career: Sports Statistician

Statisticians are mathematicians who work with data, creating statistics, graphs, and tables that describe and explain the real world. Sports statisticians combine their love of sports with their ability to use mathematics. Statistics not only explain what has happened, but can help you predict what may happen in the future.

Before you start, work through the Home Run Derby worksheet on the next page.

In this chapter project . . .

You create a fantasy baseball team of past and current players.

What to do:

- Choose six baseball players. They can be current or retired players.
- Research the statistics for each of your six players. Find the number of home runs and at bats for each player, and calculate the percent of hits that are home runs for each player.
- Use the percentages you calculated to estimate how many home runs each player would hit for 200 at bats.
- Compare your team with your classmates' teams. Who has the most home runs?

What to present:

- A table with the names of your six players, their statistics, and the estimated number of home runs for each member of your team
- A bar graph showing the estimated total number of home runs for your team and each of your classmates' teams

Where's the math?

- Use ratios and percents to estimate how many home runs each of your players scores.
- Use a bar graph to compare the performance of your teams.

go.hrw.com
Chapter Project Online
KEYWORD: MT7 Ch6

Name ____________________ Date __________ Class __________

CHAPTER 6

Project Recording Sheet

Percents

Home Run Derby

One of the reasons that people like baseball is that they can spend hours looking at the statistics to compare players and teams. Statistics allow us to see what has happened in the past and try to predict what will happen in the future. Use the data below to compare four great players.

Player	Seasons Played	At-Bats	Hits	Home Runs
Joe DiMaggio	13	6821	2214	361
Mickey Mantle	18	8102	2415	536
Ted Williams	19	7706	2654	521
Willie Mays	22	10881	3283	660

Use the data given above to complete the table.

Player	Average Number of Home Runs per Season	Percentage of Home Runs out of Hits
Joe DiMaggio	$\frac{361}{13} \approx 28$	$\frac{361}{2214} \times 100 \approx 16\%$
Mickey Mantle		
Ted Williams		
Willie Mays		

1. Who has the highest percent of home runs out of hits? __________

2. Who has the lowest average of home runs per season? __________

3. If Mickey Mantle had played for 22 seasons instead of 18, with the same average number of home runs per season, would he have beat Willie Mays total number of home runs? Explain.

____

____

____

EXPLORATION

Program of Studies and Core Content
MA-08-1.1.3

6A Relating Decimals, Fractions, and Percents

A *percent* is a ratio that compares a number to 100. Percents can be modeled on circle graphs. The circle graph below shows the results of a survey in which 25 students were asked which type of music they preferred.

- 16% means 16 per 100.
- 16% as a decimal is 0.16.
- 16% as a fraction is $\frac{16}{100} = \frac{4}{25}$.

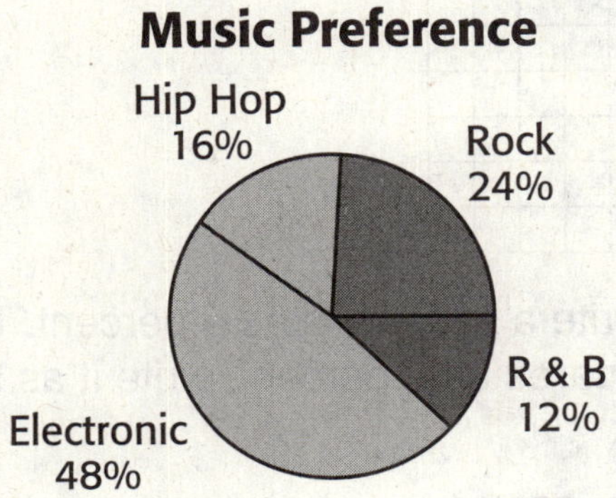

Use the percents in the circle graph to complete the table.

	Music Type	Percent	Decimal	Fraction
1.	Electronic	48%		
2.	Rock	24%		
3.	Hip Hop	16%		
4.	R & B	12%		

Think and Discuss

5. Explain how you wrote each percent as a decimal.

6. Explain how you wrote each decimal as a fraction.

Name ______________________ Date __________ Class __________

LESSON 6A

Hands-On Lab
Relating Decimals, Fractions, and Percents

Activity 1

Draw a grid with 100 squares. Color $\frac{1}{4}$ of the squares red; $\frac{1}{4}$ blue; and $\frac{1}{4}$ green.

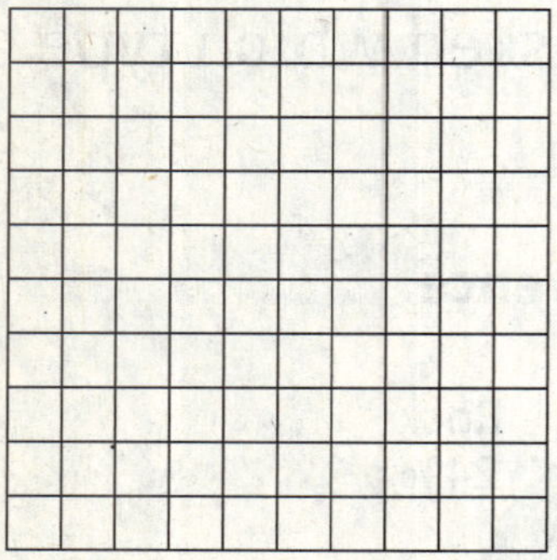

Another way to write a fraction is as a percent. To express the number of red squares as a percent, write it as a fraction with denominator 100:

$\frac{1}{4} = \frac{25}{100} = 0.25 = 25\%.$

Activity 2

Draw a grid with 60 squares and color $\frac{1}{3}$ of the squares.

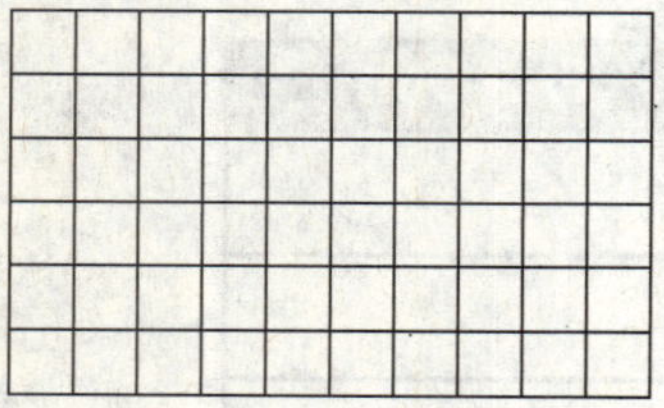

You have colored 20 squares.

Calculate the percent 20 squares represent by using proportions:

$$\frac{y}{100} = \frac{20}{60} \qquad y = \frac{20 \times 100}{60} \qquad \frac{100}{3} \approx 33.33$$

$\approx$ 33.33% of the squares are colored.

Name ______________________ Date ____________ Class ____________

LESSON **6A**

Hands-On Lab

Relating Decimals, Fractions, and Percents, continued

Think and Discuss

If you draw a circle instead of a grid in Activity 1, how much of the circle would you have colored blue? How many degrees would this be?

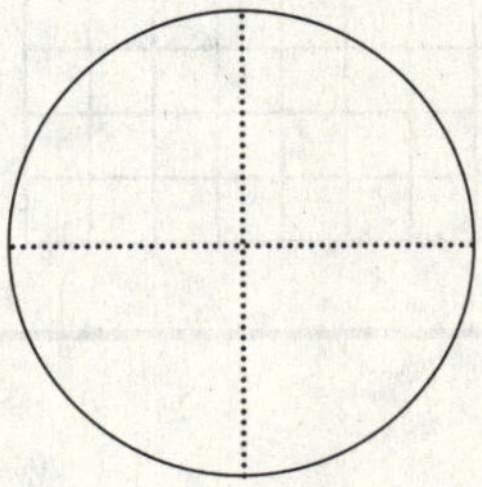

____

Try This

1. Look at the grid below and calculate the percent that is shaded gray and the percent that is white.

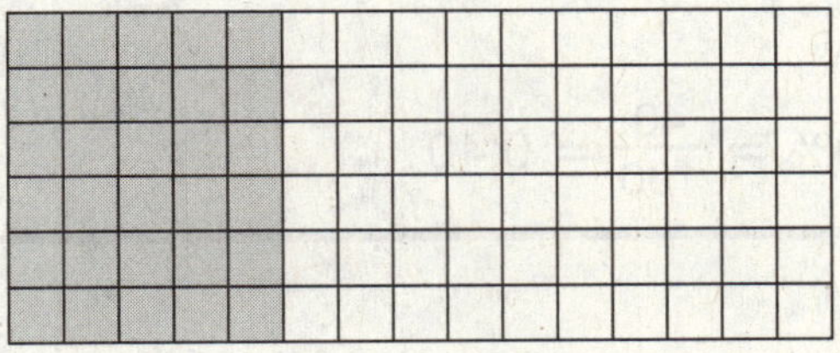

____

____

2. Draw a grid with 55 squares and color 11 squares. What percent of the grid did you color? ____________

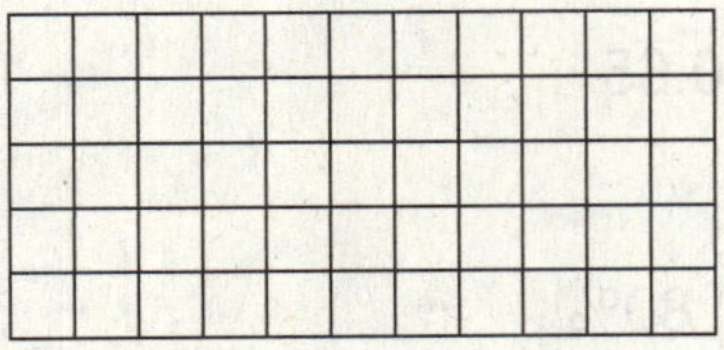

Name ______________________ Date __________ Class __________

LESSON 6A

Review for Mastery

Relating Decimals, Fractions, and Percents

A **percent** (symbol %) is a *ratio* that compares a number to 100.

The ratio is then written in simplest form.

$$40\% = \frac{40}{100} = \frac{40 \div 20}{100 \div 20} = \frac{2}{5}$$

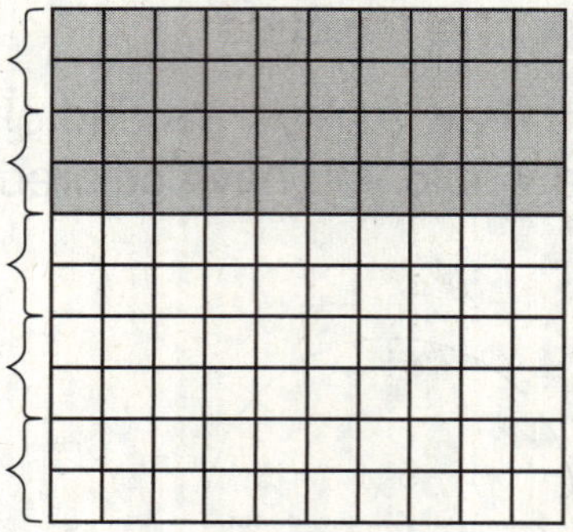

The diagram at the right models 40%. By separating the diagram into five equal sections, you can see that 40 squares make up two of the five sections. $\frac{40}{100} = \frac{2}{5}$.

Write each percent as a ratio in simplest form.

1. $80\% = \frac{\quad}{100}$

   $= \frac{\quad \div \quad}{100 \div 20}$

   $=$ ________

2. $37.5\% = \frac{\quad}{100}$

   $= \frac{\quad}{1000}$

   $= \frac{\quad \div \quad}{1000 \div 125}$

   $=$ ________

3. $65\% = \frac{\quad}{100}$

   $= \frac{\quad \div \quad}{100 \div 5}$

   $=$ ________

Since a percent compares a number to 100, a percent can be written as a decimal. $40\% = \frac{40}{100} = 0.40$

Write each percent as a decimal.

4. $80\% = \frac{\quad}{100}$

   $=$ ________

5. $37.5\% = \frac{\quad}{100}$

   $= \frac{\quad}{1000}$

   $=$ ________

6. $65\% = \frac{\quad}{100}$

   $=$ ________

Use the results of Exercises 1–6 to compare. Write <, >, or =.

7. $\frac{13}{20}$ ☐ 37.5%

8. 80% ☐ 0.65

9. 65% ☐ 0.8

10. 0.375 ☐ 80%

11. 37.5% ☐ 0.65

12. 65% ☐ $\frac{13}{20}$

Name ______________ Date ________ Class ________

LESSON 6A

Homework and Practice

Relating Decimals, Fractions, and Percents

Find the missing fraction or percent equivalent for each letter on the number line.

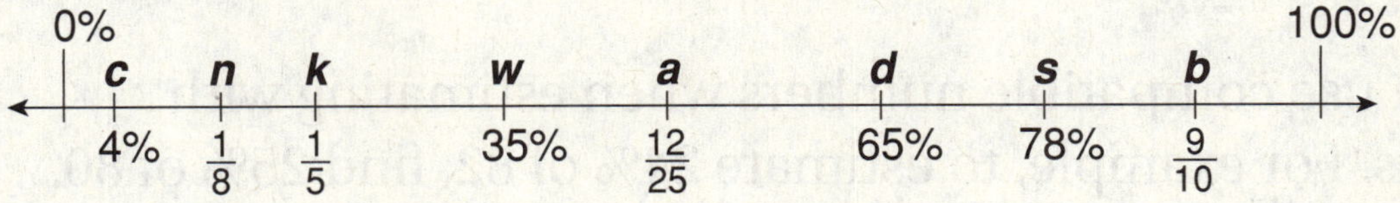

1. *a* ________ **2.** *b* ________ **3.** *c* ________ **4.** *d* ________

5. *k* ________ **6.** *n* ________ **7.** *s* ________ **8.** *w* ________

Find each equivalent value.

9. 44% as a fraction ________

10. $\frac{3}{50}$ as a decimal ________

11. $\frac{13}{40}$ as a percent ________

12. $\frac{3}{5}$ as a percent ________

13. $11\frac{1}{9}\%$ as a fraction ________

14. $\frac{3}{8}$ as a decimal ________

Order the numbers from least to greatest.

15. $\frac{5}{8}$, 0.6, $\frac{63}{100}$, –0.06, $-\frac{2}{3}$

16. 0.125, $-\frac{6}{50}$, –12.5, –0.15, $\frac{3}{8}$

17. At Eduardo's party, 18 friends attended but 6 friends could not come. What percent of the people invited could come to the party?

____

18. On a math quiz, Ande answered 34 out of 40 questions correctly. What percent did she answer correctly?

____

19. There are 300 students and 25 tables in the school cafeteria. If the students are seated equally at the tables, what percent of the students are seated at one table?

____

EXPLORATION

Program of Studies and Core Content

MA-8-NPO-S-E1, MA-8-NPO-S-E2, MA-8-NPO-S-RP1, MA-08-1.2.1, MA-08-1.4.1, MA-08-2.1.2

6B Estimate with Percents

You can use compatible numbers when estimating with percents. For example, to estimate 24% of 82, find 25% of 80.

$0.25 \cdot 80 = 20$
So 24% of 82 is approximately 20.

For each problem, estimate using compatible numbers. Then use a calculator to find the actual answer.

		Estimate	Actual
1.	Of 610 students, 34% prefer domestic cars over foreign cars.		
2.	In a survey, 19% of 152 people selected juice as their favorite drink.		
3.	A family wishes to leave a 15% tip on a $32.15 bill at a restaurant.		
4.	In a country with a population of 10,036,724, 48% are males.		

Think and Discuss

5. **Discuss** your strategies for estimating.

6. **Explain** how to avoid overestimating or underestimating.

Name ______________________________ Date ____________ Class ____________

LESSON 6B

Review for Mastery

Estimate with Percents

You can estimate the solutions to problems involving percents by rounding numbers.

Estimate 38% of 470.

Step 1: Round the given percent to a percent with a simple fractional equivalent.

$38\% \approx 40\% = \frac{40}{100} = \frac{2}{5}$

Step 2: Round the number to a number divisible by the denominator of the fraction.

$470 \approx 500$ 500 is divisible by 5.

Step 3: Use mental math to multiply.

$\frac{2}{5} \cdot 500$ $2 \cdot 500 = 1000$ and $1000 \div 5 = 200$.

So, 38% of 470 is about 200.

Complete to estimate each percent.

1. 32% of 872

Round to a percent with an easy fractional equivalent: 32% ≈ __________% = $\frac{1}{3}$

Round to hundreds place, divisible by 3: 872 ≈ __________

Do mental math to multiply: $\frac{1}{3}$ × __________ = __________

So, 32% of 872 is about __________.

2. 78% of 495

% to easy fraction:

78% ≈ __________% = __________

number divisible by denominator:

495 ≈ __________

Do mental math to multiply.

______ × __________ = __________

So, 78% of 495 is about __________.

3. 73% of 1175

% to easy fraction:

73% ≈ __________% =

number divisible by denominator:

1175 ≈ __________

Do mental math to multiply.

______ × __________ = __________

So, 73% of 1175 is about __________.

Name ______________________ Date ____________ Class ____________

LESSON 6B

Review for Mastery

Estimate with Percents, continued

Compatible numbers (numbers that go well together because they have common factors) are used to estimate the percent one number is of another.

About what percent of 82 is 39?

Step 1: Round to compatible numbers.

$82 \approx 80$ and $39 \approx 40$

Step 2: Reduce the fraction.

$\frac{40}{80} = \frac{1}{2}$

Step 3: Convert the fraction to a percent.

$\frac{1 \cdot 50}{2 \cdot 50} = \frac{50}{100} = 50\%$ or $2\overline{)1.0}$ gives $.5 = 50\%$

So, 39 is about 50% of 82.

Estimate what percent one number is of another.

4. About what percent of 98 is 27?

Use compatible numbers. 98 ≈ ________ 27 ≈ ________

Write fraction. $\frac{\textit{is}\text{ number}}{\textit{of}\text{ number}}$: ________

Reduce the fraction and convert to percent. $\frac{\quad}{\quad} = \frac{\quad}{\quad}$ = ______%

So, 27 is about ________% of 98.

5. About what percent of 19 is 6?

19 ≈ ________ 6 ≈ ________

$\frac{\textit{is}\text{ number}}{\textit{of}\text{ number}}$: $\frac{\quad}{\quad} = \frac{\quad}{\quad}$ = ______%

So, 6 is about ________% of 19.

6. About what percent of 51 is 89?

51 ≈ ________ 89 ≈ 90

$\frac{\textit{is}\text{ number}}{\textit{of}\text{ number}}$: $\frac{\quad}{\quad} = \frac{9}{5} = 1 + \frac{\quad}{\quad}$

= ________%

So, 89 is about ________% of 51.

Name ______________________ Date ____________ Class ____________

Homework and Practice

LESSON **6B**

Relating Decimals, Fractions, and Percents

Estimate.

1. 26 out of 51 ______

2. 24% of 42 ______

3. 19% of 20 ______

4. 30% of 73 ______

5. 68% of 103 ______

6. 61 out of 82 ______

7. 32 out of 98 ______

8. 34 out of 84 ______

9. 50% of 19.6 ______

Estimate each number or percent.

10. 10% of 81.2 is about what number? ______

11. 18 is 26% of about what number? ______

12. About what number is 33% of 46? ______

13. 32 is 39% of about what number? ______

14. On Tuesday, 453 students attend school. This is about 91% of the total enrollment. Estimate the school's total enrollment.

______

15. Suzette bought dinner for her friends. The total cost of the meals was $73.95. Suzette left a tip of $15. Approximate the percent of the tip to the cost of the meals.

______

16. Of the 294 students who took the state test, 31 of them must retake the test. Estimate the percent of students who need to retake the test.

______

17. The usual markup on shoes at Fast Times Department Store is 240%. During a 25%-off sale, the store sold a pair of shoes for $39.95. Estimate the amount of the store's profit on the shoes.

______

EXPLORATION

Program of Studies and Core Content

MA-8-NPO-S-RP1, MA-08-1.4.1

6C Finding Percents

The circle graph shows results from a mock election in which 40 students were polled.

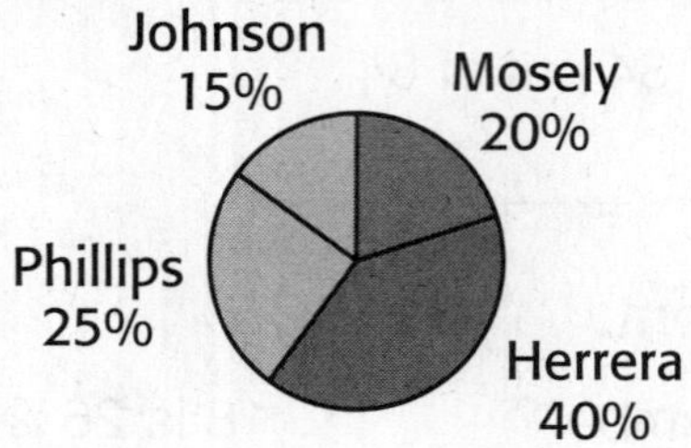

Complete the table to find the number of votes each candidate received.

	Candidate	Percent of Votes	Number of Votes
	Mosely	20%	$0.20 \cdot 40 = 8$
1.	Herrera	40%	
2.	Phillips	25%	
3.	Johnson	15%	

Think and Discuss

4. Discuss whether there is a clear winner in the election.

5. Tell which two candidates combined received the same total votes as another candidate. Did they also get the same percent of the votes as the other candidate?

Name ______________________ Date __________ Class __________

LESSON **6C**

Hands-On Lab
Finding Percents

A percent equation ($\frac{\text{percent}}{100}$ • whole = part) is a linear equation and can be drawn as a line graph.

Activity 1

The whole is 50. Write the percent equation using *y* (percent) and *x* (part).

$\frac{y}{100} \cdot 50 = x$

$y = \frac{100}{50} \cdot x$; $y = \frac{2}{1} \cdot x$

Draw a line graph with a slope $= \frac{2}{1} = 2$ and the point *A* (0, 0) and *B* (50, 100).

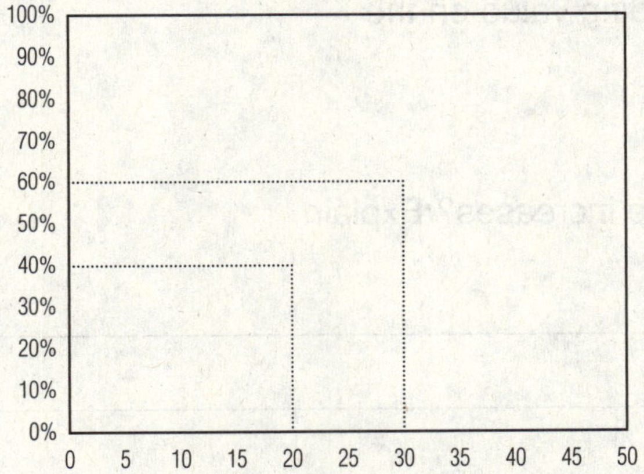

1. What percent of 50 is 30?
   Draw a vertical line from 30 to the graphed line. Then draw a horizontal line to the *y*-axis. The corresponding value on the *y*-axis is 60: 30 is 60% of 50.

2. What percent of 50 is 20?
   The corresponding value on the *y*-axis is 40: 20 is 40% of 50.

Activity 2

Draw a line graph for a whole of 30.

$y = \frac{100}{30} \cdot x$; $y = \frac{10}{3} \cdot x$

The slope $= \frac{10}{3}$.

Name ______________________ Date ____________ Class ____________

LESSON 6C

Hands-On Lab

Finding Percents, continued

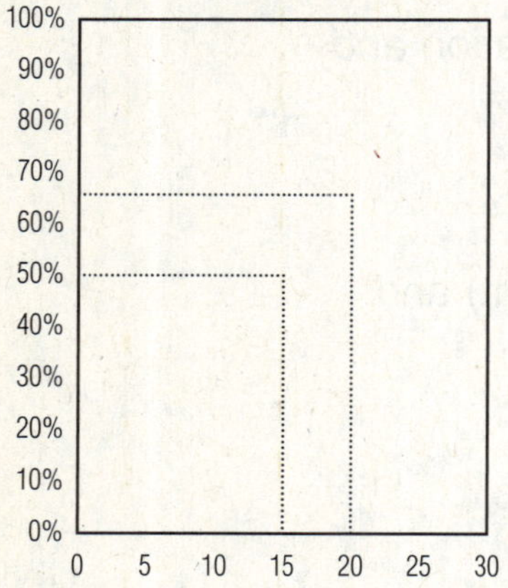

1. What percent of 30 is 15? The corresponding value on the *y*-axis is 50%.

2. What percent of 30 is 20? The corresponding value on the *y*-axis is about 67%.

Think and Discuss

1. How does the line change when the whole increases? Explain.

____

____

Try This

1. What is the slope for a whole of 150? For a point *C* with an *x*-coordinate of 90, what is the *y*-coordinate?

____

2. Draw a line graph using the data above.

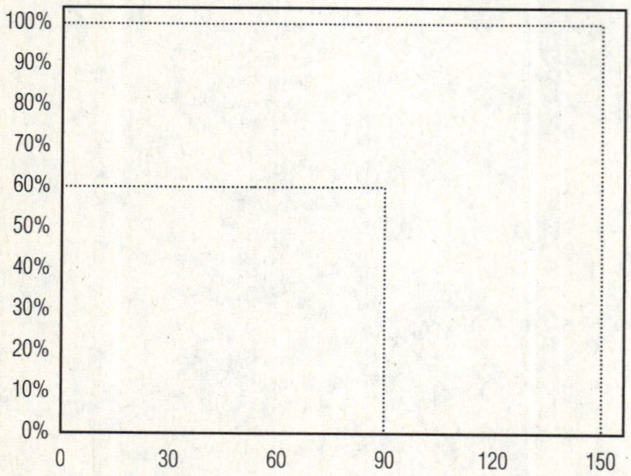

Name ______________________________ Date ____________ Class ____________

Technology Lab

LESSON 6C

Finding Percents

The percent equation is a linear equation and can be drawn as a line graph.

Activity 1

The whole is 50. Write the percent equation using *y* (percent) and *x* (part).

$$y = \frac{100}{50} \cdot x = \frac{2}{1} \cdot x$$

Draw a line graph with a slope $= \frac{2}{1} = 2$ and the point *A*(0, 0) and *B*(50, 100).

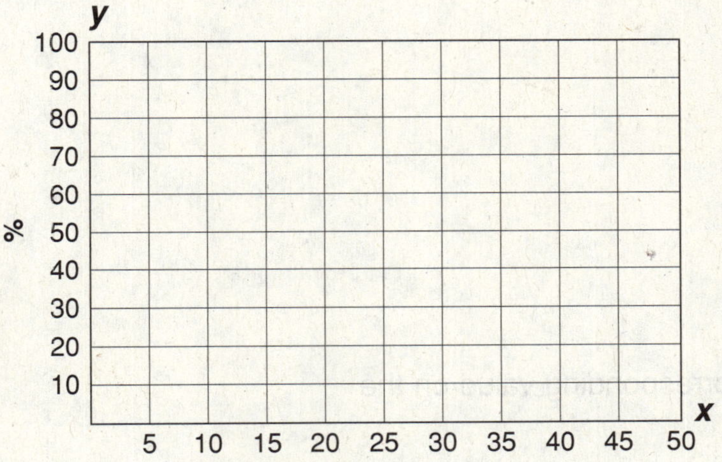

1. What percent of 50 is 30?

Draw a vertical line from 30 on the *x*-axis to the graph of the line. Then draw a horizontal line to the *y*-axis. The corresponding value on the *y*-axis is 60: 30 is 60% of 50.

2. What percent of 50 is 20?

The corresponding value on the y-axis is 40: 20 is 40% of 50.

Name ______________________ Date __________ Class __________

LESSON 6C

Technology Lab
Finding Percents, continued

Activity 2

Draw a line graph of the percent equation for a whole of 30.

$y = \frac{100}{30} \cdot x = \frac{10}{3} \cdot x.$

The slope is $\frac{10}{3}$.

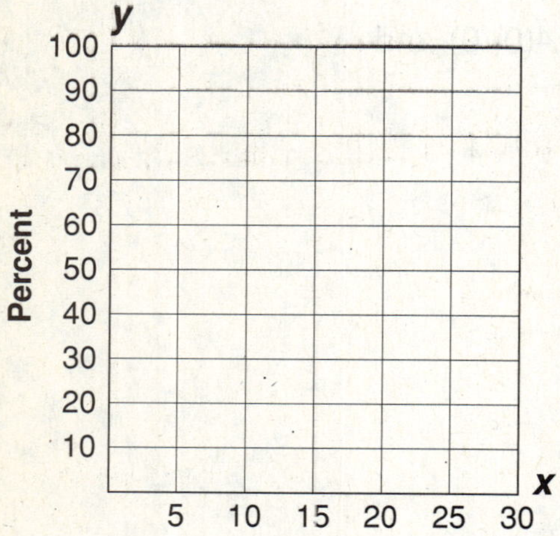

1. What percent of 30 is 15? The corresponding value on the y-axis is 50%.
2. What percent of 30 is 20? The corresponding value on the y-axis is about 67%.

Think and Discuss

1. How does the line change, when the whole increases? Explain.

____

____

____

____

Name ______________________ Date __________ Class __________

LESSON 6C

Technology Lab

Finding Percents, continued

Try This

1. What is the slope of the percent equation for a whole of 150?
For a point *C* with an *x*-coordinate of 90, what is the *y*-coordinate?

____

____

2. Draw a line graph using the data above.

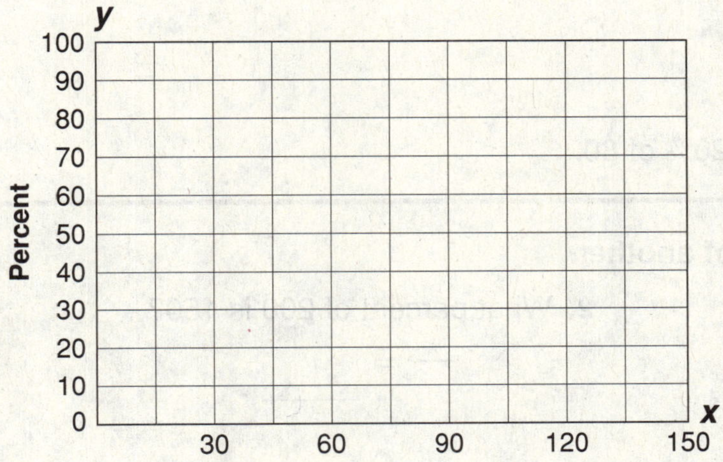

Name ______________________ Date __________ Class __________

LESSON 6C

Review for Mastery

Finding Percents

Since a percent is a ratio, problems involving percent can be solved by using a proportion.

There are different possibilities for an unknown quantity in this proportion.

$$\frac{\text{symbol number}}{100} = \frac{\textit{is}\text{ number}}{\textit{of}\text{ number}}$$

Possibility 1: Find the *symbol number*.

What percent of 80 is 16?

$$\frac{\text{symbol number}}{100} = \frac{\textit{is}\text{ number}}{\textit{of}\text{ number}}$$

$$\frac{x}{100} = \frac{16}{80}$$

$$80 \cdot x = 16 \cdot 100$$

$$\frac{80x}{80} = \frac{1600}{80}$$

$x = 20$ So, 16 is 20% of 80.

Find what percent one number is of another.

1. What percent of 64 is 16?

$\frac{x}{100} = \frac{\_\_\_}{\_\_\_}$

___ $\cdot x =$ ___ $\cdot$ ___

$\frac{\_\_\_ x}{\_\_\_} = \frac{\_\_\_}{\_\_\_}$

$x =$ ___

So, 16 is __________ of 64.

2. What percent of 200 is 150?

$\frac{x}{\_\_\_} = \frac{\_\_\_}{\_\_\_}$

___ $\cdot x =$ ___ $\cdot$ ___

$\frac{\_\_\_ x}{\_\_\_} = \frac{\_\_\_}{\_\_\_}$

$x =$ ___

So, 150 is __________ of 200.

3. What percent of 4 is 6?

$\frac{x}{\_\_\_} = \frac{\_\_\_}{\_\_\_}$

___ $\cdot x =$ ___ $\cdot$ ___

$\frac{\_\_\_ x}{\_\_\_} = \frac{\_\_\_}{\_\_\_}$

$x =$ ___

So, 6 is __________ of 4.

4. About what percent of 115 is 40?

$\frac{\_\_\_}{100} = \frac{\_\_\_}{\_\_\_}$

___ $\cdot x =$ ___ $\cdot$ ___

$\frac{\_\_\_ x}{\_\_\_} = \frac{\_\_\_}{\_\_\_}$

$x =$ ___

So, 40 is about __________ of 115.

Name ______________________ Date __________ Class __________

LESSON 6C

Review for Mastery

Finding Percents, continued

Possibility 2: Find the *is number.*

What is 20% of 80?

$$\frac{\text{symbol number}}{100} = \frac{\textit{is}\text{ number}}{\textit{of}\text{ number}}$$

$$\frac{20}{100} = \frac{x}{80}$$

$$100 \cdot x = 20 \cdot 80$$

$$\frac{100x}{100} = \frac{1600}{100}$$

$$x = 16$$

So, 20% of 80 is 16.

Find the indicated percent of each number.

5. What is 30% of 150?

$\frac{\_\_\_}{100} = \frac{x}{\_\_\_}$

___ • x = ___ • ___

$\frac{\_\_\_ x}{\_\_\_} = \frac{}{\_\_\_}$

x = ___

So, 30% of 150 is ______.

6. What is 75% of 205?

$\frac{\_\_\_}{100} = \frac{x}{\_\_\_}$

___ • x = ___ • ___

$\frac{\_\_\_ x}{\_\_\_} = \frac{}{\_\_\_}$

x = ___

So, 75% of 205 is ______.

7. What is 125% of 300?

$\frac{\_\_\_}{100} = \frac{x}{\_\_\_}$

___ • x = ___ • ___

$\frac{\_\_\_ x}{\_\_\_} = \frac{}{\_\_\_}$

x = ___

So, 125% of 300 is ______.

8. What is $66\frac{2}{3}$% of 81?

$\frac{66\frac{2}{3}}{\_\_\_} = \frac{x}{\_\_\_}$

___ • $x = 66\frac{2}{3}$ • ___

___ • $x = \frac{200}{3}$ • ___

___ • x = ___

$\frac{\_\_\_ x}{\_\_\_} = \frac{}{\_\_\_}$

x = ___

So, $66\frac{2}{3}$% of 81 is ______.

Name ______________________ Date __________ Class __________

LESSON **6C**

Homework and Practice

Finding Percents

Find each percent.

1. What percent of 60 is 45?

____

2. 45 is what percent of 90?

____

3. What percent of 45 is 18?

____

4. What percent of 75 is 15?

____

5. 21 is what percent of 84?

____

6. 51 is what percent of 85?

____

7. What percent of 60 is 27?

____

8. What percent of 90 is 27?

____

9. 18 is what percent of 60?

____

10. 24.2 is what percent of 110?

____

11. What percent of 42 is 35?

____

12. What percent of 64 is 24?

____

13. What percent of 78 is 27.3?

____

14. 105 is what percent of 60?

____

15. An airplane completed 900 miles of a 1200-mile flight. What percent of the trip remains to be completed?

____

16. Cloe had 36 of the 45 questions on the math test correct. What percent of problems did Cloe have correct?

____

17. Lincoln received $55 for his birthday. He puts $22 of it in his savings account. What percent of the money did Lincoln deposit in his savings account?

____

18. Chico bought a jacket for $42. The cost of the jacket including tax, was $44.31. What was the percent of sales tax on the purchase?

____

EXPLORATION

Program of Studies and Core Content

MA-8-NPO-S-RP1, MA-08-1.4.1

6D Finding a Number When the Percent is Known

A CD player is on sale for $15 off the original price. According to the advertisement, this is a 25% discount. You can use what you know about percents to figure out the original price of the CD player.

Think: 25% is $\frac{1}{4}$, so $15 must be $\frac{1}{4}$ of the original price.

You can model the situation as shown.

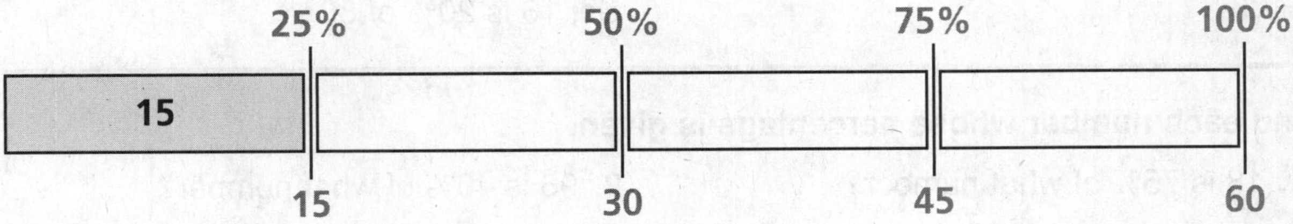

The original price is $60.

Draw a model to find the original price for each of the following.

1. Amount of discount: $8; percent of discount: 20%
2. Amount of discount: $12; percent of discount: $33\frac{1}{3}\%$
3. Amount of discount: $17; percent of discount: 50%

Think and Discuss

4. **Explain** how you could find the value of a number given that 7 is 10% of the number.

5. **Describe** what must be true about a number if you know that 45 is 100% of the number.

Name ____________________ Date __________ Class __________

LESSON 6D

Review for Mastery

Finding a Number When the Percent Is Known

Since a percent is a ratio, problems involving percent can be solved by using a proportion.

$$\frac{\text{symbol number}}{100} = \frac{\textit{is}\text{ number}}{\textit{of}\text{ number}}$$

To find a number when the percent is known, the variable appears in the *of* position in the proportion

16 is 20% of what number?

$$\frac{20}{100} = \frac{16}{x}$$

$$20 \cdot x = 16 \cdot 100$$

$$\frac{20x}{20} = \frac{1600}{20}$$

$$x = 80$$

So, 16 is 20% of 80.

Find each number whose percentage is given.

1. 18 is 75% of what number?

$\frac{\_\_\_}{100} = \frac{\_\_\_}{x}$

____ • x = ____ • ____

$\frac{\_\_\_ x}{\_\_\_} = \frac{\_\_\_}{\_\_\_}$

x = ____

So, 18 is 75% of ______.

2. 96 is 40% of what number?

$\frac{\_\_\_}{100} = \frac{\_\_\_}{x}$

____ • x = ____ • ____

$\frac{\_\_\_ x}{\_\_\_} = \frac{\_\_\_}{\_\_\_}$

x = ____

So, 96 is 40% of ______.

3. 7 is 125% of what number?

$\frac{\_\_\_}{100} = \frac{7}{x}$

____ • x = ____ • ____

$\frac{\_\_\_ x}{\_\_\_} = \frac{\_\_\_}{\_\_\_}$

x = ____

So, 7 is 125% of ____.

4. 40 is about 30% of what number?

$\frac{\_\_\_}{100} = \_\_\_$

____ • x = ____ • ____

$\frac{\_\_\_ x}{\_\_\_} = \frac{\_\_\_}{\_\_\_}$

x = ____

So, 40 is about 30% of ____.

Name ______________________ Date ____________ Class ____________

LESSON 6D

Homework and Practice

Finding a Number When the Percent Is Known

Find each number.

1. 25% of what number is 12?

____

2. 27 is 54% of what number?

____

3. 51 is 60% of what number?

____

4. 32% of what number is 8?

____

5. 80% of what number is 16?

____

6. 76 is 95% of what number?

____

7. 27 is 18% of what number?

____

8. 24% of what number is 72?

____

9. 104 is 65% of what number?

____

10. 12 is 3% of what number?

____

11. 5% of what number is 12.2?

____

12. 16% of what number is 4.8?

____

13. 90 is $66\frac{2}{3}$% of what number?

____

14. 40.5 is 150% of what number?

____

15. Meredith bought a bicycle that cost $385.89. In addition, she had to pay 7% sales tax on her purchase. What was the amount of sales tax Meredith paid? Round the answer to the nearest cent.

____

16. Wilson answered 92% of a 25 question test correctly. How many questions did he answer correctly?

____

17. Rodriquez had $84.60 of his salary withheld for taxes. This was 15% of his gross pay. What is his gross pay?

____

18. A car dealer receives an order of 340 new cars. Thirty percent of the cars are silver in color. How many of the cars received by the dealership are silver?

____

Name ______________________ Date ____________ Class ____________

LESSON 6D

Homework and Practice

Finding a Number When the Percent Is Known, continued

Use the table below, which shows the recommended daily allowance of food components for a 2000-calorie diet, to answer questions 19–26. Round your answers to the nearest tenth.

Component	Recommended Daily Allowance
Toal fat	65 g
Saturated fat	20 g
Cholesterol	300 mg
Sodium	2400 mg
Total Carbohydrates	300 g
Fiber	25 g

19. One serving of oatmeal contains 16% of the recommended daily allowance of fiber. How many grams of fiber are in one serving?

____

20. A certain can of soup contains 30% of the recommended daily allowance of sodium. How many milligrams of sodium are in one can?

____

21. One serving of pure Vermont maple syrup contains 53 grams of total carbohydrates. What percent of the recommended daily allowance is this?

____

22. One serving of pumpkin pie contains 12 total grams of fat. What percent of the recommended daily allowance is this?

____

Select the best answer.

23. A certain nutrition bar contains 15% of the recommended daily allowance of saturated fat. How many grams of saturated fat are in one bar?

A 2.5 g **C** 4 g
B 3 g **D** 5 g

24. One serving of plain yogurt contains 6 mg of cholesterol. What percent of the recommended daily allowance is this?

A 2% **C** 20%
B 5% **D** 50%

25. A cereal contains 90 mg of potassium, and this is 3% of the recommended daily allowance. What is the recommended daily allowance of potassium?

A 27 mg **C** 300 mg
B 270 mg **D** 3000 mg

26. A slice of whole grain bread contains 7 grams of fiber. What percent of the recommended daily allowance is this?

A 18% **C** 28%
B 25% **D** 30%

EXPLORATION

Program of Studies and Core Content

MA-8-NPO-S-RP1, MA-08-1.4.1

6E Percent Increase and Decrease

1. After his first year at a job, Andrew's original hourly wage of $7.95 increased to $9.54.

   a. Subtract the old hourly wage from the new hourly wage to find the amount by which Andrew's hourly wage increased.

   b. Divide the amount in **1a** by Andrew's original hourly wage to compare the amount of increase to the original hourly wage.

   c. Write the decimal in **1b** as a percent to find the percent increase in Andrew's hourly wage.

2. An electronics store offers a $100 discount on everything in the store that is priced between $500 and $900. Complete the table to determine the percent of decrease for each price.

Price	\$500	\$600	\$700	\$800	\$900
Percent of Decrease	$\frac{100}{500}$ = ___%	$\frac{100}{600}$ =	$\frac{100}{700}$ =	$\frac{100}{800}$ =	$\frac{100}{900}$ =

Think and Discuss

3. **Explain** how you wrote the decimal in Problem **1c** as a percent.

4. **Compare** the percents of decrease in the table in Problem **2**.

Name ______________________ Date ___________ Class ___________

LESSON 6E

Review for Mastery

Percent Increase and Decrease

To find the percent increase:

- Find the amount of increase by subtracting the lesser number from the greater.
- Write a fraction: **percent increase** = $\frac{\textbf{amount of increase}}{\textbf{original amount}}$
- If possible, simplify the fraction.
- Rewrite the fraction as a percent.

The fee increased from \$60 to \$75.
Find the percent of increase.

percent of increase = $\frac{75 - 60}{60} = \frac{15}{60} = \frac{1}{4} = 25\%$

Complete to find each percent increase.

1. Membership increased from 80 to 100.

2. Savings increased from \$500 to \$750.

____ − ____   ____ − ____   Find the amount of increase.

= ____   = ____

$\frac{\quad}{80}$ = ____ $\frac{\quad}{500}$ = ____   percent increase = $\frac{\text{amount of increase}}{\text{original amount}}$

= ____ %   = ____ %   Change the fraction to a percent.

3. Price increased from \$20 to \$23.

________   Find the amount of increase.

$\frac{\quad}{20}$ percent increase = $\frac{\text{amount of increase}}{\text{original amount}}$

$\frac{\quad}{20}$ = 20$\overline{)\quad}$ = ______ %

Change the fraction to a percent.

Name ______________________ Date __________ Class __________

LESSON 6E

Review for Mastery

Percent Increase and Decrease, continued

To find the percent decrease:

- Find the amount of decrease by subtracting the lesser number from the greater.
- Write a fraction: **percent decrease** = $\frac{\textbf{amount of decrease}}{\textbf{original amount}}$
- If possible, simplify the fraction.
- Rewrite the fraction as a percent.

Carl's weight decreased from 175 lb to 150 lb.
Find the percent of decrease.

percent of decrease = $\frac{175 - 150}{175} = \frac{25}{175} = \frac{1}{7} \approx 14.3\%$

Complete to find each percent decrease.

4. Enrollment decreased from 1000 to 950.

5. Temperature decreased from 75°F to 60°F.

_____ − _____   _____ − _____   Find the amount of decrease.

= _____   = _____

$\frac{\_\_\_}{1000} = \frac{\_\_\_}{100}$ $\frac{\_\_\_}{75} = \frac{\_\_\_}{15} = \frac{\_\_\_}{100}$ percent decrease = $\frac{\text{amount of decrease}}{\text{original amount}}$

= _____ %   = _____ %   Change the fraction to a percent.

6. Sale price decreased from $22 to $17.

______________________   Find the amount of decrease.

$\frac{\_\_\_}{22}$ percent decrease = $\frac{\text{amount of decrease}}{\text{original amount}}$

$\frac{\_\_\_}{22}$ = 22)‾‾‾‾‾‾ ≈ __________ %   Change the fraction to a percent.

Name ________________ Date ________ Class ________

LESSON 6E

Homework and Practice

Percent Increase and Decrease

Find each percent increase or decrease to the nearest percent.

1. from 16 to 12 ________

2. from 40 to 46 ________

3. from 20 to 26 ________

4. from 50 to 38 ________

5. from 75 to 45 ________

6. from 30 to 45 ________

7. from 27 to 35.1 ________

8. from 35 to 22.75 ________

9. from 84 to 54.6 ________

10. from 35 to 42.7 ________

11. from 39 to 97.5 ________

12. from 108 to 86.4 ________

13. from 63 to 39 ________

14. from 81 to 124 ________

15. from 72 to 98 ________

16. The regular price of track shoes is $49.75. They are on sale for $39.80. What is the percent of decrease for the sale price?

____

17. An electronics store pays a $255.50 for a TV and sells it for $357.70. What is the percent of increase in the price of the TV?

____

18. The Globe Company has laid off 1,220 employees of the company's 24,400 employees. What is the percent of decrease in the company's work force?

____

19. Brooke is given a 4% raise which makes her salary $754 a week. What was Brooke's salary before the salary increase?

____

20. Clyde weighed 200 pounds and lost 15 pounds. What was the percent of decrease in his weight?

____

Name ______________________ Date __________ Class __________

LESSON 6E

Homework and Practice

Percent Increase and Decrease, continued

Write the correct answer.

21. The entrance fee at an art museum is $22. Senior citizens receive a 15% discount. How much do seniors pay?

____

22. Kylie paid $38.62 to fill the gas tank of her Jeep. Two weeks ago, she paid $34.18 to fill the tank. Find the percent increase.

____

23. In 2001, the population of Barbados was approximately 275,000. In 2002, the population increased by 0.5%. What was the approximate population of Barbados in 2002?

____

24. A shoe salesman marked up a pair of sandals in the spring by 75% and advertised them for $42.00. In the fall, he put them on sale at a 30% discount on the spring price. What was the original price of the sandals, before the markup? How much were they selling for after the fall discount?

____

Use the table below to answer questions 25–27. The table shows the U.S. population by region as given in the 1990 and 2000 Census. Select the best answer.

Region	Population (1990)	Population (2000)
Northeast	50,809,229	53,594,378
Midwest	59,668,632	64,392,776
South	85,445,930	100,236,820
West	52,786,082	63,197,932

25. By what percent did the population of the South increase from 1990 to 2000?

A 14.8% **C** 18.6%

B 17.3% **D** 19.8%

26. By what percent did the population of the West increase from 1990 to 2000?

F 16.5% **H** 19.7%

G 18.1% **J** 21.1%

27. Which statement is NOT true?

A The percent increase in the West was twice that in the Northeast.

B The Northeast had the smallest percent change.

C The South had the greatest increase in population.

D All regions had a percent increase.

EXPLORATION

Program of Studies and Core Content

MA-8-NPO-S-RP1, MA-08-1.4.1

6F Applications of Percents

You often need to calculate percents when making a purchase.

Use a calculator to find the tax on each item and the total cost of the item including the tax.

	Item	Cost	Tax Rate	Tax	Total Cost = Cost + Tax
1.	CD	$13.95	8%		
2.	DVD	$24.99	8%		
3.	Headphones	$29.95	8%		

Use a calculator to find the total cost of each item.

	Item	Cost	Tax Rate	Total Cost = 1.08 · Cost
4.	CD	$13.95	8%	
5.	DVD	$24.99	8%	
6.	Headphones	$29.95	8%	

Think and Discuss

7. Explain how you calculated the tax on each item in Problems 1–3.

8. Explain why the total cost is the same whether you use the formula *total cost* = *cost* + *tax* or the formula *total cost* = 1.08 · *cost*.

Name ______________________ Date ____________ Class ____________

LESSON 6F

Technology Lab

Applications of Percents

Percentages are important in financial transactions. Whenever you make a purchase, you may have to add a sales tax to the price of the item. When you get a discount on an item, you need to deduct a certain percentage from the price of the item.

Activity 1

The sales tax is 8.25% in Dallas, Texas. If you want to buy a bicycle that is priced at $220, how much will you pay?

1. First, find the amount of the tax (T) you pay on the bicycle. Use your calculator to compute the following:

$$T = \frac{8.25}{100} \cdot 220 = 18.15$$

You may also use the keystrokes in the order given. 8.25 [×] 220 [%]

2. Then add this amount to the price of the bicycle to find the final amount (A):

$$A = 220 + 18.15 = 238.15$$

You may also use the following keystrokes in the order given. 220 [+] 8.25 [%]

If you have an algebraic calculator, you can input $220 + 0.0825 \times 220 = 238.15$.

Activity 2

The merchant gives you a 15% discount on the $220 bicycle and you live in a state without a sales tax.

How much do you have to pay?

You may also use the following keystrokes in the order given. 220 [−] 15 [%]

The bicycle costs $187.

Name ______________________ Date ____________ Class ____________

LESSON 6F

Technology Lab

Applications of Percents, continued

Think and Discuss

1. A department store in a state without sales tax has advertised its annual sale: 50% off of every item. When you get to the store, an additional 25% discount is offered. You also have a discount coupon for 25%. You buy a coat with an original price tag of $160. Does the coat cost anything? Why or why not?

____

____

____

Try This

1. Use your calculator to complete the following receipt from a department store:

Receipt	
11/5/03	
Coat	110.00
– 15% discount	______
Shirt	22.00
Belt	17.00
Sub Total	______
+ 6% sales tax	
Total	______

Name ______________________ Date __________ Class __________

LESSON 6F

Review for Mastery

Applications of Percents

Salespeople often earn a **commission,** a percent of their total sales.

Find the commission on a real-estate sale of \$125,000 if the commission rate is 4%.

Write the percent as a decimal and multiply.

commission rate × amount of sale = amount of commission
0.04 × \$125,000 = \$5000

If, in addition to the commission, the salesperson earns a salary of \$1000, what is the total pay?

commission + salary = total pay
\$5000 + \$1000 = \$6000

Complete to find each total monthly pay.

1. total monthly sales = \$170,000; commission rate = 3%; salary = \$1500

amount of commission = 0.03 × \$__________ = \$__________

total pay = \$__________ + \$1500 = \$__________

2. total monthly sales = \$16,000; commission rate = 5.5%; salary = \$1750

amount of commission = __________ × \$__________ = \$__________

total pay = \$__________ + \$__________ = \$__________

A **tax** is a charge, usually a percentage, generally imposed by a government.

Sales tax is the tax on the sale of an item or service.

If the sales tax rate is 7%, find the tax on a sale of \$9.49.

Write the tax rate as a decimal and multiply.

tax rate × amount of sale = amount of tax
0.07 × \$9.49 = \$0.6643 ≈ \$0.66

Complete to find each amount of sales tax.

3. item price = \$5.19; sales tax rate = 6%

amount of sales tax = 0.06 × \$__________ = \$__________ ≈ \$__________

4. item price = \$250; sales tax rate = 6.75%

amount of sales tax = __________ × \$__________ = \$__________ ≈ \$__________

Name ______________________ Date __________ Class __________

LESSON 6F

Review for Mastery

Applications of Percents, continued

Use a proportion to find what percent of a person's income goes to a specific expense.

Heather earned \$3,200 last month. She paid \$448 for transportation. To find the percent of her earnings that she put towards transportation, write a proportion.

Think: What percent of 3200 is 448?

$\frac{n}{100} = \frac{448}{3200}$ ← Set up a proportion.

Think: $\frac{\text{part}}{\text{whole}} = \frac{\text{part}}{\text{whole}}$

$3200n = 448 \times 100$ ← Find cross products.

$3200n = 44{,}800$ ← Simplify.

$\frac{3200n}{3200} = \frac{44{,}800}{3200}$ ← Divide both sides by 3200.

$n = 14$ ← Simplify.

Heather put 14% of her earnings towards transportation.

Complete each proportion to find the percent of earnings.

5. Wayne earned \$3,100 last month. He paid \$837 for food. What percent of his earnings went to food?

$\frac{n}{100} = \frac{\_\_\_\_}{3100}$

$3100n = \_\_\_\_ \times 100$

$3100n = \_\_\_\_$

$\frac{3100n}{3100} = \_\_\_\_$

$n = \_\_\_\_$

______ of Wayne's earnings went to food.

6. Leah earned \$1,900 last month. She paid \$304 for utilities. What percent of her earnings went to utilities?

$\frac{n}{100} = \frac{304}{\_\_\_\_}$

$\_\_\_\_ \times n = \_\_\_\_ \times 100$

$\_\_\_\_ = \_\_\_\_$

$\_\_\_\_ = \_\_\_\_$

$n = \_\_\_\_$

______ of Leah's earnings went to utilities.

Name ______________________ Date __________ Class __________

LESSON 6F

Homework and Practice

Applications of Percents

Complete the table to find the amount of sales tax for each sale amount to the nearest cent.

1.

	6% sales tax	7% sales tax	5.5% sales tax
$53.50			
$80.50			
$219.95			
$2,640.00			

Complete the table to find the commission for each sale amount to the nearest cent.

2.

	5% commission	8% commission	7.5% commission
$365.00			
$2,140.00			
$16,300.00			
$94,750.00			

3. Mr. Darney bought a house for $58,000. He made a down payment of 20% and got a loan for the rest. What was the amount of his loan?

____

4. Elijah buys three pens for $1.89 each and a notebook for $3.95. The sales tax rate is 6%. What is Elijah's total cost? Round the answer to the nearest cent.

____

5. Seth is a real estate agent. He receives 8% commission on the selling price of any home he sells. If the last house he sold had a selling price of $145,900, what was Seth's commission?

____

Name ______________________ Date ____________ Class ____________

LESSON 6F

Homework and Practice

Applications of Percents, continued

Write the correct answer.

6. A snack food distributor earns $400 a week plus a 4% commission on sales. What is her total pay for a week in which her sales are $1080?

____

7. Estimate the tax on a digital camera that costs $399 with a 6.25% sales tax.

____

Vira recorded her weekend expenses to start budgeting her money. Use Vira's weekend spending list to answer questions 10–14. Select the best answer.

Weekend Spending

Saturday:
Dinner
$72.40 + 15% tip
Taxi
$22 + $2 tip

Sunday:
Breakfast
$12.75 + 20% tip
Manicure
$13 + 10% tip
New shoes
$89 + 6.25% sales tax

8. What is the tip on Saturday's dinner?

A $3.62　　**C** $10.86
B $7.24　　**D** $14.48

9. Estimate the tax on Vira's new shoes.

F $4.50　　**H** $6.30
G $5.40　　**J** $7.20

10. What is the tip on Sunday's breakfast?

A $0.26　　**C** $1.28
B $0.64　　**D** $2.55

11. What is the tip on Vira's manicure?

F $0.13　　**H** $1.30
G $0.26　　**J** $2.60

12. At what percent rate did Vira tip the taxi driver?

A about 5%　　**C** about 15%
B about 10%　　**D** about 20%

EXPLORATION

Program of Studies and Core Content

MA-8-AT-S-EI2, MA-8-AT-S-VEO2, MA-8-NPO-S-RP1, MA-08-1.4.1

6G Simple Interest

Simple interest is the amount earned on money deposited in some savings accounts.

The interest your account earns is calculated using the formula $I = Prt$. For example, if you start a savings account with $100.00 (***principal P***) and your savings account pays 5% (***interest rate r***), the interest (***I***) you will have earned at the end of 1 year (***time t***) will be $0.05 \cdot 100 = \$5$. Your total balance at the end of 1 year will be $100.00 + $5.00 = $105.00.

Complete the table.

	Savings	Interest Rate	Interest	Total Balance
	$100	5%	$0.05 \cdot 100 = \$5$	$100 + 5 = \$105$
1.	$200	6%		
2.	$300		$24	
3.	$500			$550

Think and Discuss

4. Describe your strategies for completing the table.

5. Explain how to use the formula $I = Prt$ when you need to find an interest rate.

Name ______________________ Date __________ Class __________

LESSON 6G

Review for Mastery

Simple Interest

Interest is money paid on an investment.
A borrower pays the interest. An investor earns the interest.

Simple interest, *I*, is earned when an amount of money, the *principal P*, is borrowed or invested at a *rate of interest r* for a *period of time t*.

Interest = Principal • Rate • Time
$I = P \cdot r \cdot t$

Situation 1: Find *I* given *P*, *r*, and *t*.

Calculate the simple interest on a loan of $3500 for a period of 6 months at a yearly rate of 5%.

Write the interest rate as a decimal. 5% = 0.05
Write the time period in terms of years. 6 months = 0.5 year

$I = P \cdot r \cdot t$

$I = 3500 \cdot 0.05 \cdot 0.5 = \$87.50$ ← interest earned

Find the interest in each case.

1. principal $P = \$5000$; time $t = 2$ years; interest rate $r = 6\%$

$I = P \cdot r \cdot t =$ ________ • 0.06 • ________ = $________

2. principal $P = \$2500$; time $t = 3$ months; interest rate $r = 8\%$

$I = P \cdot r \cdot t =$ ________ • ________ • ________ = $________

Situation 2: Find *t* given *I*, *P*, and *r*.

An investment of $3000 at a yearly rate of 6.5% earned $390 in interest. Find the period of time for which the money was invested.

$I = P \cdot r \cdot t$

$390 = 3000 \cdot 0.065 \cdot t$

$390 = 195t$

$\frac{390}{195} = \frac{195t}{195}$

$2 = t$

The investment was for 2 years.

Find the time in each case.

3. $I = \$1120$; $P = \$4000$; $r = 7\%$

$I = P \cdot r \cdot t$

1120 = ________ • 0.07 • *t*

1120 = ________*t*

________ = ________*t*
(________ / ________)

________ years = *t*

4. $I = \$812.50$; $P = \$5000$; $r = 6.5\%$

$I = P \cdot r \cdot t$

812.50 = ________ • ________ • *t*

812.50 = ________*t*

________ = ________*t*
(________ / ________)

________ years = *t*

Name ______________________ Date ____________ Class ____________

LESSON 6G

Review for Mastery

Simple Interest, continued

Situation 3: Find r given I, P, and t.

$2500 was invested for 3 years and earned $450 in interest. Find the rate of interest.

$$I = P \cdot r \cdot t$$
$$450 = 2500 \cdot r \cdot 3$$
$$450 = 7500r$$
$$\frac{450}{7500} = \frac{7500r}{7500}$$
$$0.06 = r$$

The interest rate was 6%.

Find the interest rate in each case.

5. $I = \$1200$; $P = \$6000$; $t = 4$ years

$I = P \cdot r \cdot t$

1200 = ________ • r • 4

1200 = ________ r

$\frac{\_\_\_\_}{\_\_\_\_\_\_} = \frac{\_\_\_\_ r}{\_\_\_\_\_\_}$

________ = r

The interest rate was ______%.

6. $I = \$325$; $P = \$2000$; $t = 2.5$ years

$I = P \cdot r \cdot t$

325 = ________ • r • ________

325 = ________ r

$\frac{\_\_\_\_}{\_\_\_\_\_\_} = \frac{\_\_\_\_ r}{\_\_\_\_\_\_}$

________ = r

The interest rate was ______%.

The total amount A of money in an account after interest has been earned, is the sum of the principal P and the interest I.

Amount = Principal + Interest
$A = P + I$

Find the amount of money in the account after $3500 has been invested for 3 years at a yearly rate of 6%.

First, find the interest earned.

$I = P \cdot r \cdot t$

$I = 3500 \cdot 0.06 \cdot 3 = \$630$ ⟵ interest earned

Then, add the interest to the principal. $3500 + 630 = 4130$

So, the total amount in the account after 3 years is $4130.

Find the total amount in the account.

7. principal $P = \$4500$; time $t = 2.5$ years; interest rate $r = 5.5\%$

$I = P \cdot r \cdot t =$ ________ • ________ • ________ = $________

Total Amount $= P + I = 4500 +$ ________ = ________

So, after 2.5 years, the total amount in the account was $________.

Name ______________________ Date ____________ Class ____________

LESSON 6G

Homework and Practice

Simple Interest

Find the missing value.

1. principal = $175
   rate = 4%
   time = 5 years
   interest = ?

   ____

2. principal = ?
   rate = 6%
   time = 6 years
   interest = $81

   ____

3. principal = $210
   rate = 5%
   time = ? years
   interest = $31.50

   ____

4. principal = $125
   rate = ?%
   time = 4 years
   interest = $35

   ____

5. principal = ?
   rate = $5\frac{1}{4}$%
   time = 2 years
   interest = $126

   ____

6. principal = ?
   rate = 6.5%
   time = 3 years
   interest = $165.75

   ____

7. Scott deposits $1,000 in an account that earns 5% simple interest. What will the account be worth after two years?

   ____

8. A bank pays 5.5% simple interest per year. Find the total amount at the end of one year on a principal of $3,000.

   ____

9. Mr. Womer borrowed $1,100 to start his own business. National Bank charged him 15% interest per year. Mr. Womer paid $330 in interest. For what period of time did he borrow the money?

   ____

10. Ms. Lang borrows $2,300 for 30 months at 13% interest per year. How much interest will Ms. Lang pay? What is the total amount she will repay?

    ____

CHAPTER 6

Chapter Summary

Answer these questions to summarize the important concepts from Chapter 6 in your own words.

1. Explain how to convert $\frac{3}{8}$ to a percent.

2. Explain how to solve "34 is 25% of what number?".

3. Explain how to find the percent increase or decrease from 200 to 145.

4. Explain how to find the interest and total amount of $400 at 7.5% per year for 8 years.

For more review of Chapter 6:

- Complete the Chapter 6 Study Guide and Review on pages 312–314 of your textbook.
- Complete the Ready to Go On quizzes on pages 292 and 308 of your textbook.

Name ______________________ Date ____________ Class ____________

CHAPTER 6

KCCT Prep

Multiple Choice

1. On a number line, which of these is closest to 0.1?

(A) 0.6% (C) 60%

(B) 6% (D) 600%

2. A family spent \$576.25 on utilities. Of this amount, \$72.90 was for electricity. Which is the best estimate of the percent that was spent on electricity?

(A) 8% (C) 20%

(B) 12% (D) 33%

3. Isaac has 60% more coins in his collection than Reba has in her collection. Reba has 20% more coins in her collection than Matt has in his collection. Isaac has 192 coins in his collection. Which equation can you solve to find the number of coins in Matt's collection?

(A) $(0.6 + 0.2)m = 192$

(B) $(1.6 + 1.2)m = 192$

(C) $(0.6)(0.2)m = 192$

(D) $(1.6)(1.2)m = 192$

4. While training for a race, Kendrick runs 9 miles. This is 60% of the length of the race. How long is the race?

(A) 5.4 mi (C) 13.5 mi

(B) 12 mi (D) 15 mi

5. A helicopter takes off from the roof of a building that is 610 feet tall. The helicopter rises vertically 350 feet each minute. Which equation gives the altitude y of the helicopter after x minutes?

(A) $y = 350 + 610x$

(B) $y = 350 + 610 + x$

(C) $y = 350x + 610$

(D) $y = 350(x + 610)$

6. According to the U.S. Census Bureau, $26\frac{1}{5}$% of all adults visited an art museum in the past year. In a town with a population of 98,956 adults, which is the best estimate of the number of adult residents who visited an art museum in the past year?

(A) 27 (C) 2,650

(B) 265 (D) 26,500

7. Which of these is a rational number?

(A) $-\sqrt{3}$ (C) $2\sqrt{25}$

(B) $\sqrt{5}$ (B) $\sqrt{-10}$

8. The Golden Gate Bridge is 1280 meters long. This is about 91% of the length of the Humber Bridge, in England. Which is the best estimate of the length of the Humber Bridge?

(A) 1100 m (C) 1400 m

(B) 1200 m (D) 1600 m

9. The rectangle shown below is dilated with a scale factor of $\frac{1}{2}$. What are the coordinates of the image of point *K*?

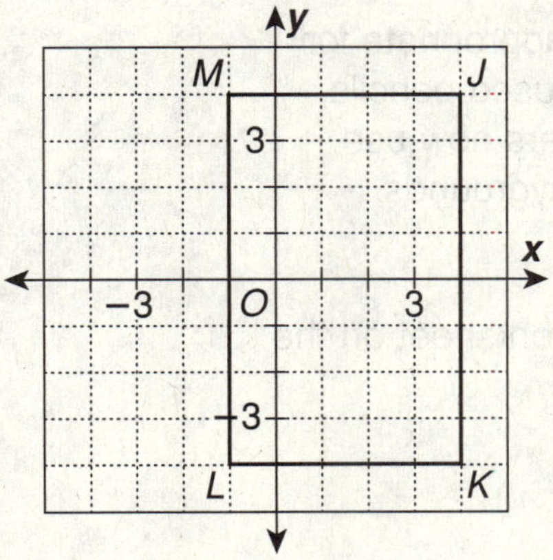

Ⓐ (2, 2) Ⓒ (8, −8)
Ⓑ (2, −2) Ⓓ (8, 8)

10. Which pair of numbers are closest to each other on a number line?

Ⓐ 5% and 0.5 Ⓒ 60% and 0.8
Ⓑ 50% and 0.6 Ⓓ 80% and 0.08

11. A square has an area of 36 cm^2. Which is closest to the length of the diagonal of the square?

Ⓐ 6 cm Ⓒ 10.4 cm
Ⓑ 8.5 cm Ⓓ 12 cm

12. The table shows the rate at which a tank is filling with water. Which equation relates the time *x* and the amount of water *y*?

Time (min)	3	5	6
Water (gal)	25.5	42.5	51

Ⓐ $y = x + 22.5$ Ⓒ $y = 8.5x$
Ⓑ $y = 6x + 7.5$ Ⓓ $y = 28.5 - x$

13. A right triangle has legs of length 7 in. and 9 in. Which is the best estimate of the length of the triangle's hypotenuse?

Ⓐ 11 in. Ⓒ 16 in.
Ⓑ 14 in. Ⓓ 20 in.

14. A bag contains red marbles and blue marbles. There are 16 red marbles in the bag and 60% of the marbles are blue. How many blue marbles are in the bag?

Ⓐ 8 Ⓒ 24
Ⓑ 16 Ⓓ 40

15. Ms. Chu orders 15 calculators for her classroom. There is a shipping fee of $8.75 on the order. The total cost of the order is $380. When she gets the calculators, she finds that two do not work. She asks for a refund of the cost of these two calculators. What is the amount of the refund?

Ⓐ $24.75 Ⓒ $49.50
Ⓑ $25.33 Ⓓ $50.66

Open Response

16. Doug is paid a monthly salary of $3100, plus a commission on his sales. His commission rate is 5%.

a. Last month, Doug's sales were $9500. What was his commission?

b. What was Doug's total salary for last month?

c. Doug would like his total salary for next month to be at least $4000. Explain how he can do this.

17. Sushila is shopping for a new car. The car she wants to buy is advertised in the newspaper for $18,400.

a. When Sushila arrives at the car dealer, the sticker price on the car is $19,136. What is the percent increase compared to the advertised price?

b. The dealer says he will lower the price to match the advertised price. What percent decrease should the dealer apply to the sticker price? Explain your answer.

CHAPTER 7

Project

Geometry: The Ultimate Playground

Career: Playground Equipment Designer

Playground equipment must be attractive, safe, fun, and appropriate for the ages of children who will use it. Years ago, designers used pencils, T-squares, and slide rules to create their designs. Designers now use computers, 3-D programs, and virtual reality to design playgrounds.

Before you start, work through the Ultimate Playground worksheet on the next page.

In this chapter project . . .

You design your own playground.

What to do:

- Research playground equipment. Use the Internet, local parks, or catalogs. Identify as many geometric shapes in each piece of equipment as you can.
- Decide what kind of equipment you would put in your own playground. Make sketches of equipment you've designed yourself, and label the shapes, lengths, and angles. For example, what angle will your slide make with the ground? What shapes would you use for a jungle gym?
- Show your drawings to younger students at your school. Ask each student which piece of equipment they like best.

What to present:

- Drawings of your playground equipment, including measurements.
- A circle graph that shows the percentages of younger students that preferred each piece of equipment.

Where's the math?

- Use geometric shapes, angle measures, and lengths to design playground equipment.
- Use a circle graph to show the percentage of students that preferred each piece of equipment.

go.hrw.com
Chapter Project Online
KEYWORD: MT7 Ch7

Name ______________________ Date __________ Class __________

CHAPTER 7

Project Recording Sheet

Foundations of Geometry

The Ultimate Playground

Do you remember the playground equipment you played on when you were younger? Which kinds did you enjoy? Were some kinds scary? Are all of those pieces of equipment still around? Some kinds of old playground equipment are now considered too dangerous. Below is a list of some popular equipment. How many do you know about?

- Monkey bars
- Chin-up bar
- Tunnel
- Fire pole
- Trapeze
- Merry-go-round
- Bridge
- Slide
- Ladder
- Glider
- Climbing wall
- Teeter-totter
- Spring toy
- Rings
- Track ride
- Swing
- Balance beam
- Rope ladder
- Tire swing
- Platform

Copy and complete the chart below to describe the equipment you are familiar with. The first row is an example.

Equipment	Fun/Not fun	Dangerous/ Not dangerous	Angles and Lines	Activity
Swing	*Fun (if they are tall)*	*Less dangerous with soft seats*	*Parallel lines, right angles*	*Individual swinging*

Use your chart to create a description of the kind of equipment that you prefer.

Create a model of what you think is a perfect piece of playground equipment. Use toothpicks, drinking straws, pipe cleaners, and other materials. Compare your model with those of your classmates. Answer these questions to analyze your findings.

What similarities and differences do you find?

____

____

What geometric shapes are most common? Most unusual?

____

____

____

What kind of activity is the most common?

____

Why do you think the results turned out the way that they did?

____

EXPLORATION

Program of Studies and Core Content

MA-8-G-S-SR1, MA-8-M-S-MPA1, MA-8-M-S-MPA2, MA-8-M-S-MPA8, MA-08-2.1.2, MA-08-2.1.3, MA-08-3.1.1

7A Points, Lines, Planes, and Angles

In each group, one picture is different from the others. Identify the picture that is different and explain why it is different.

1.

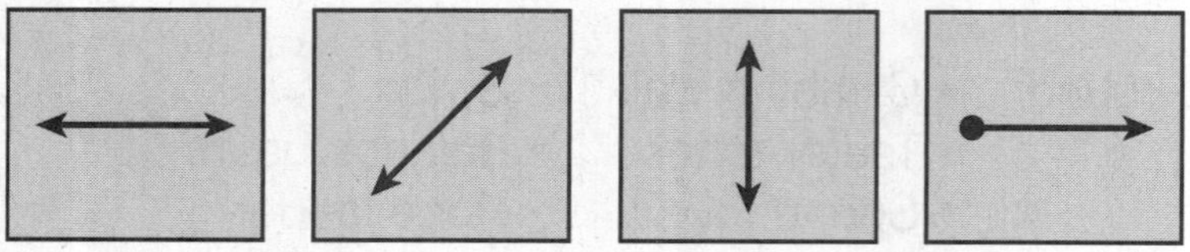

2.

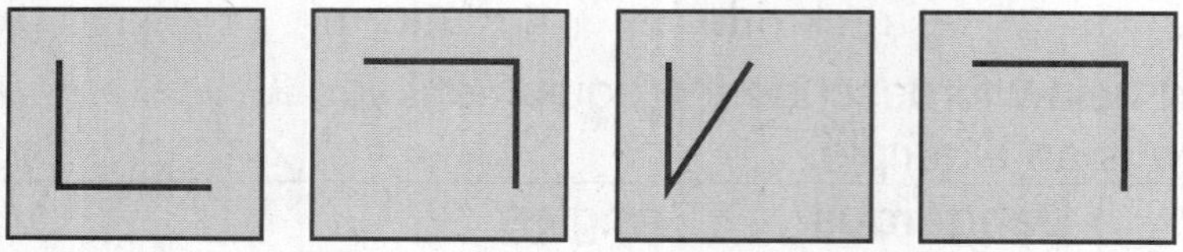

3. Look at the picture for each geometry term and write a real-world example for each term.

Term	Picture	Example
Point		
Segment		
Ray		
Line		
Angle		

Think and Discuss

4. Explain the difference between a *segment* and a *ray.*

5. Explain the difference between a ray and a *line.*

Name ______________________ Date ____________ Class ____________

LESSON 7A

Hands-On Lab 1

Choosing Measuring Tools

Activity

In order to get the most precise measurement, the correct measurement tool must be used. The smallest reasonable measuring tool and unit should always be used for accurate measurements.

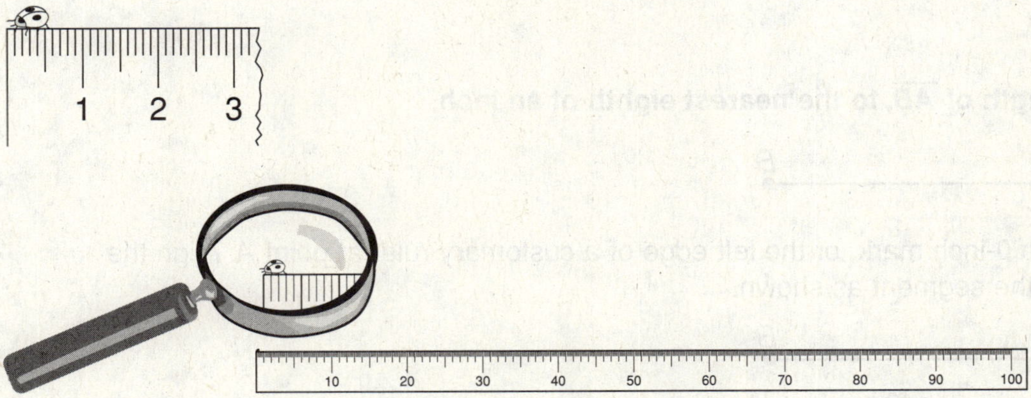

The measurement of the bug is more accurate with the centimeter ruler because the measurement marks are smaller. On the meter stick, the accuracy of measurement is poor because the measurement marks are too large for the bug.

The greatest possible error (GPE) of a measurement is half (0.5) of the measurement unit being used, therefore, the smaller the measurement being used, the less the error in measurement.

Think and Discuss

1. Why is it better to measure a classroom in feet rather than millimeters?

____

____

____

____

Try This

1. Use an appropriate measuring tool to find the length of your desk.

What is the length of your desk? ____________

What is the GPE of your measurement? ____________

Name ______________________ Date __________ Class __________

LESSON **7A**

Hands-On Lab 2
Measuring and Constructing Segments

Line segments can be measured because they have two endpoints. You can use either a customary or metric ruler to measure a segment, depending on the units required. The method to measure a segment is the same no matter which ruler is used, however, the length of a segment is only as precise as the smallest unit on the ruler.

Materials needed: Customary and Metric Ruler

Activity
Find the length of $\overline{AB}$, to the nearest eighth of an inch.

(1) Place the 0-inch mark, or the left edge of a customary ruler at point *A*. Align the ruler on the segment as shown.

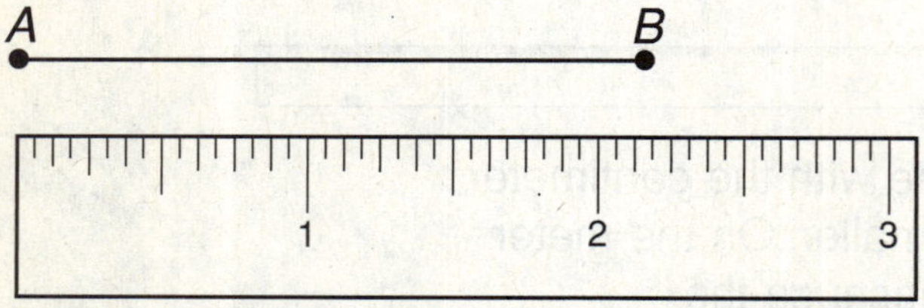

(2) Read the measure on the ruler at point *B*. Point *B* is slightly to the left of the $\frac{3}{16}$-inch mark, so the segment is $2\frac{1}{8}$ inches long, to the nearest eighth of an inch.

Try This
Measure each segment to the nearest eighth of an inch and to the nearest half of a centimeter.

1.  ______

2. ______

3. 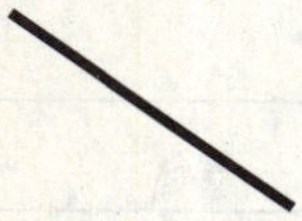 ______

4. 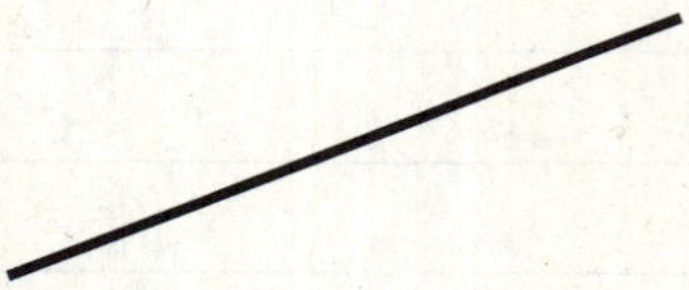 ______

5. ______

6. ______

Name ______________________ Date ____________ Class ____________

LESSON 7A

Hands-On Lab 2

Measuring and Constructing Segments, continued

7.  ____________

8. **Critical Thinking** In Exercises 1–7, which two line segments are congruent? Explain.

____

9. If $\overline{AB}$ is $6\frac{1}{2}$ cm, estimate the length of $\overline{CD}$.

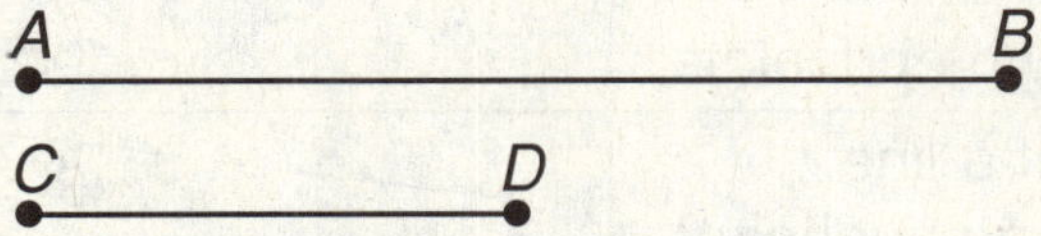

____

10. If $\overline{EF}$ is 32 mm, estimate the length of $\overline{GH}$.

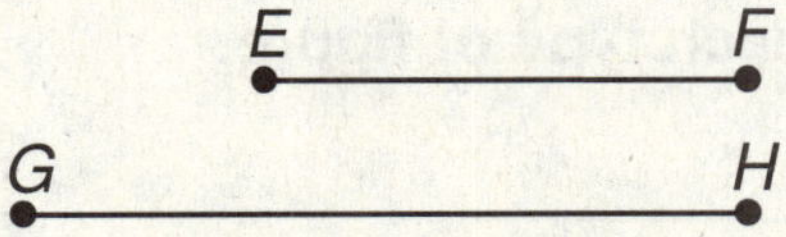

____

11. If $\overline{IJ}$ is $1\frac{1}{4}$ in., estimate the length of $\overline{MN}$.

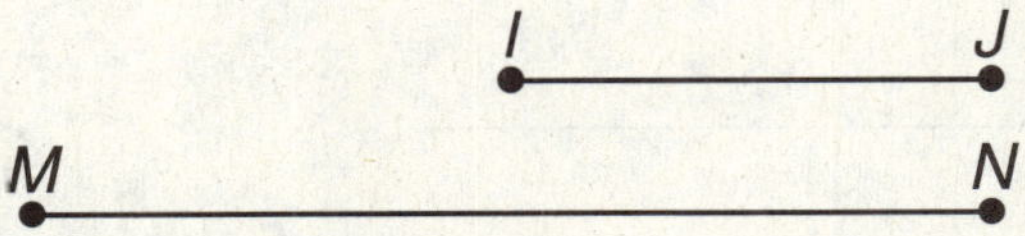

____

Name ______________________ Date ____________ Class ____________

LESSON 7A

Review for Mastery

Points, Lines, Planes, and Angles

Figure	Description	Diagram	Notation: Write	Notation: Read
Line	an infinite collection of points with no beginning and no end	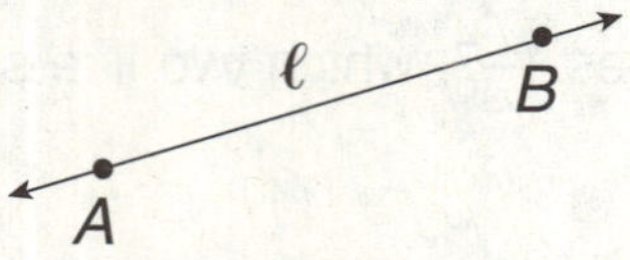	$\overleftrightarrow{AB}$ or $\overleftrightarrow{BA}$ or ℓ	line *AB*, line *BA*, line ℓ
Line Segment	part of a line, with two endpoints	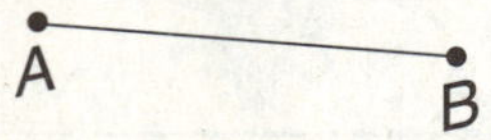	$\overline{AB}$ or $\overline{BA}$	line segment *AB* line segment *BA*
Ray	part of a line, with one endpoint		$\overrightarrow{AB}$	ray *AB*

Use the diagram to name each type of figure.

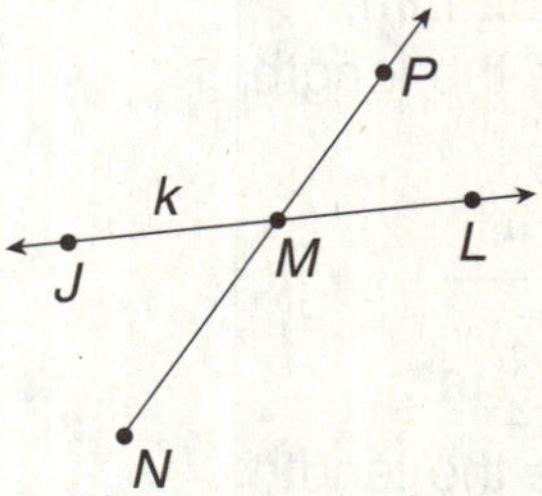

1. $\overrightarrow{MP}$ ______________________

2. k ______________________

3. $\overline{MN}$ ______________________

4. $\overline{LJ}$ ______________________

5. $\overleftrightarrow{JL}$ ______________________

Acute Angle	Right Angle	Obtuse Angle	Straight Angle
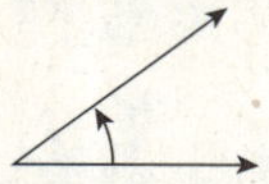	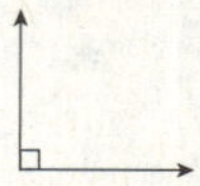	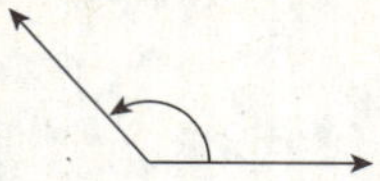	
Measures between 0° and 90°	Measures exactly 90°	Measures between 90° and 180°	Measures exactly 180°.

Use the diagram to name each type of angle.

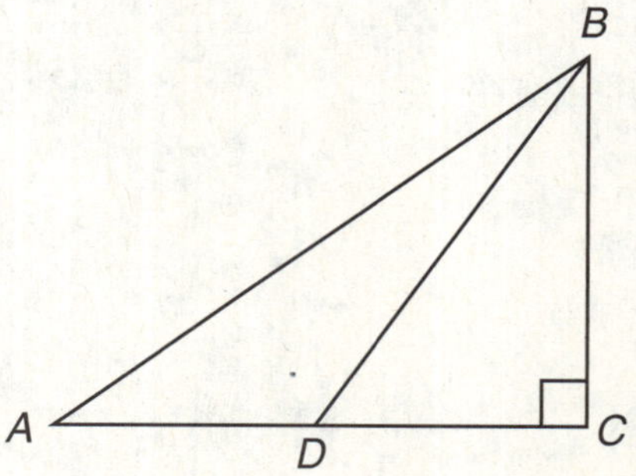

6. ∠*BCD* ______________________

7. ∠*BAD* ______________________

8. ∠*BDA* ______________________

9. ∠*CDA* ______________________

10. ∠*BDC* ______________________

11. ∠*ABC* ______________________

Name ______________________ Date ____________ Class ____________

LESSON 7A

Review for Mastery

Points, Lines, Planes, and Angles, continued

Complementary Angles	Supplementary Angles	Vertical Angles
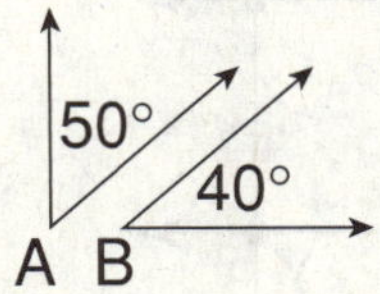	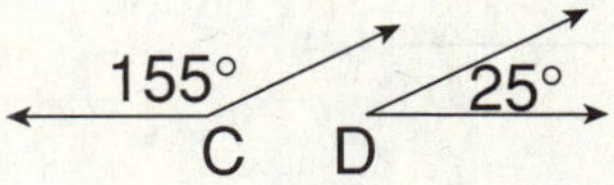	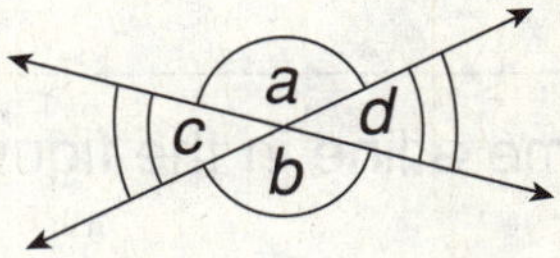
Two angles whose measures have a sum of 90°.	Two angles whose measures have a sum of 180°.	Intersecting lines form two pairs of vertical angles.
∠*A* and ∠*B* are complementary angles.	∠*C* and ∠*D* are supplementary angles.	∠*a* and ∠*b*, ∠*c* and ∠*d* are pairs of vertical angles.

Use the diagram to complete.

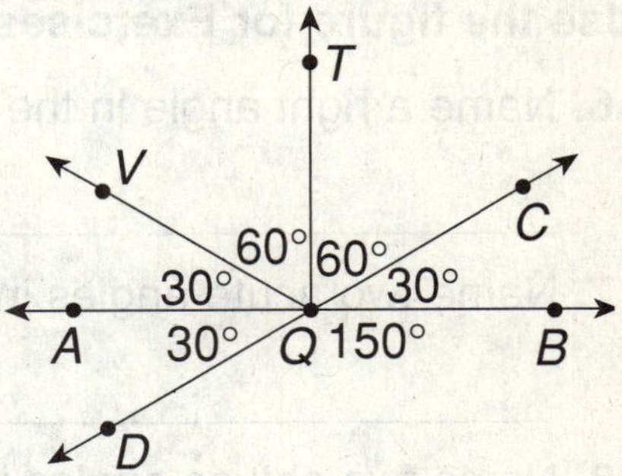

12. Since ∠*AQC* and ∠*DQB* are formed by intersecting lines, $\overleftrightarrow{AQB}$ and $\overleftrightarrow{CQD}$, they are: ____________________

13. The sum of the measures of ∠*AQV* and ∠*VQT* is: ______

So, these angles are: ____________________

14. The sum of the measures of ∠*AQC* and ∠*CQB* is: ______

So, these angles are: ____________________

Congruent figures have the same size and shape.
The symbol ≅ means *is congruent to.*

Complete these statements about congruence.

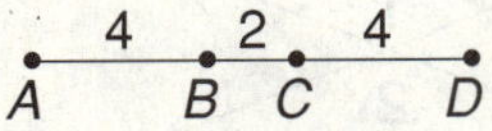

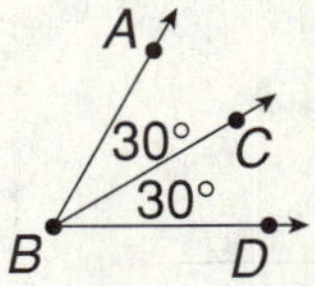

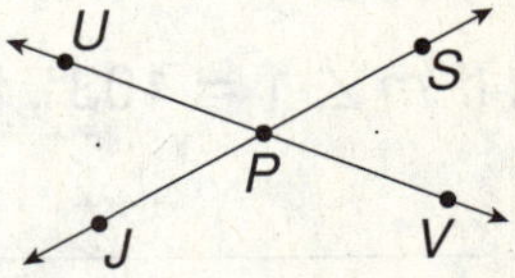

15. Since *AC* = 6 units and *BD* = 6 units,

then $\overline{AC}$ ______ $\overline{BD}$.

16. Since m∠*ABC* = 30° and m∠*CBD* = 30°

then ∠*ABC* __________.

17. Since vertical angles are congruent,

then ∠*UPJ* __________.

Name ______________________ Date __________ Class __________

LESSON 7A

Homework and Practice

Points, Lines, Planes, and Angles

Use the figure for Exercises 1–5.

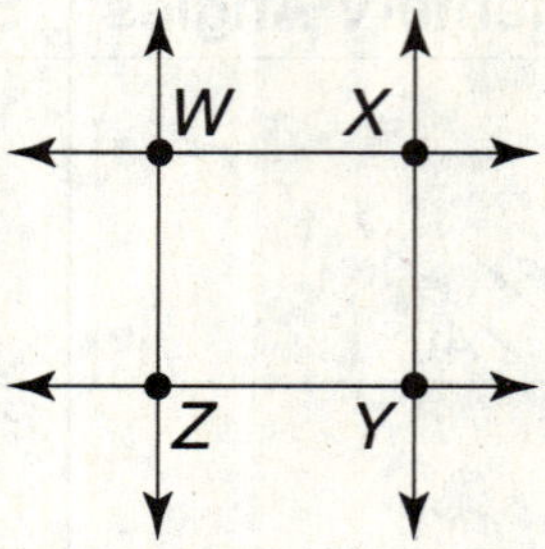

1. Name four points in the figure.

____

2. Name a line in the figure.

____

3. Name the plane in the figure.

____

4. Name three segments in the figure.

____

5. Name four rays in the figure.

____

Use the figure for Exercises 6–11

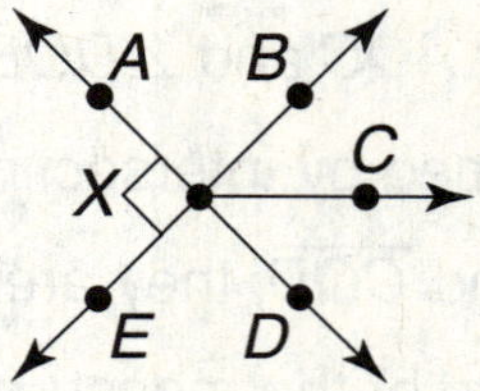

6. Name a right angle in the figure.

____

7. Name two acute angles in the figure

____

8. Name two obtuse angles in the figure.

____

9. Name a pair of complementary angles in the figure.

____

10. Name three pair of supplementary angles in the figure.

____

11. If you know the length of $\overline{BX}$ is approximately 1 cm, estimate the length of $\overline{BE}$.

____

In the figure, ∠1 and ∠2 are vertical angles.

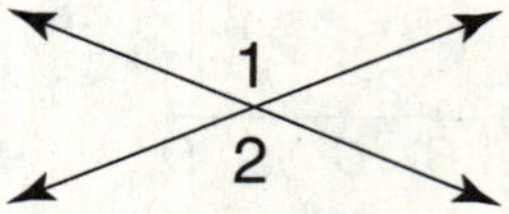

12. If $m\angle 1 = 103°$, find $m\angle 2$.

____

13. If $m\angle 2 = 115°$, find $m\angle 1$.

____

Name ______________________ Date __________ Class __________

LESSON 7A

Homework and Practice

Understanding Points, Lines, and Planes, continued

Use the map of part of San Antonio for Exercises 14 and 15.

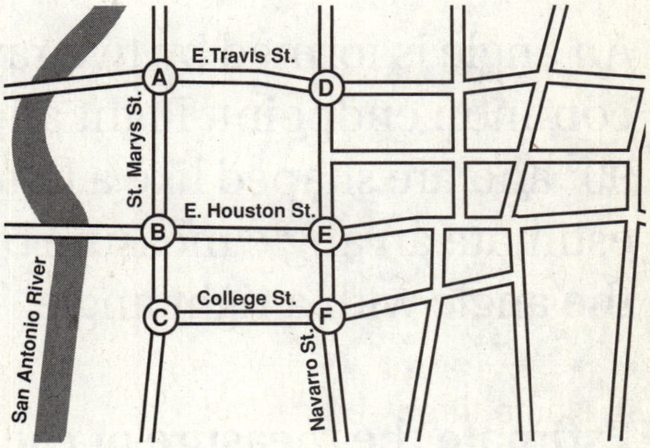

14. Name a point that appears to be on the same line with $\overline{EF}$. Which streets intersect at this point?

____

____

15. Explain why point A is NOT on the same line with $\overline{BE}$.

____

____

16. Suppose $\overline{UV}$ represents the pencil that you are using to do your homework and plane $\mathcal{P}$ represents the paper that you are writing on. Describe the relationship between $\overleftrightarrow{UV}$ and plane $\mathcal{P}$.

____

____

17. Two cyclists start at the same point, but travel along two straight streets in different directions. If they continue, how many times will their paths cross again? Explain.

____

____

Choose the best answer.

18. In a building, planes $\mathcal{W}$, $\mathcal{X}$, and $\mathcal{Y}$ represent each of the three floors; planes $\mathcal{Q}$ and $\mathcal{R}$ represent the front and back of the building; planes $\mathcal{S}$ and $\mathcal{T}$ represent the sides. Which is a true statement?

A Planes $\mathcal{W}$ and $\mathcal{Y}$ intersect in a line.

B Planes $\mathcal{Q}$ and $\mathcal{X}$ intersect in a line.

C Planes $\mathcal{W}$, $\mathcal{X}$, and $\mathcal{T}$ intersect in a point.

D Planes $\mathcal{Q}$, $\mathcal{R}$, and $\mathcal{S}$ intersect in a point.

19. Suppose point G represents a duck flying over a lake, points H and J represent two ducks swimming on the lake, and plane $\mathcal{L}$ represents the lake. Which is a true statement?

F There are two lines through G and J.

G The line containing G and H lies in plane $\mathcal{L}$.

H Plane L does not contain line HJ.

J There is exactly one plane containing points G, H, and J.

Use the figure for Exercise 7.

20. A frame holding two pictures sits on a table. Which is NOT a true statement?

A $\overline{PN}$ and $\overline{NM}$ lie in plane $\mathcal{T}$.

B $\overline{PN}$ and $\overline{NM}$ intersect in a point.

C $\overline{LM}$ and N intersect in a line.

D Points J, K, N, and P lie on the same plane.

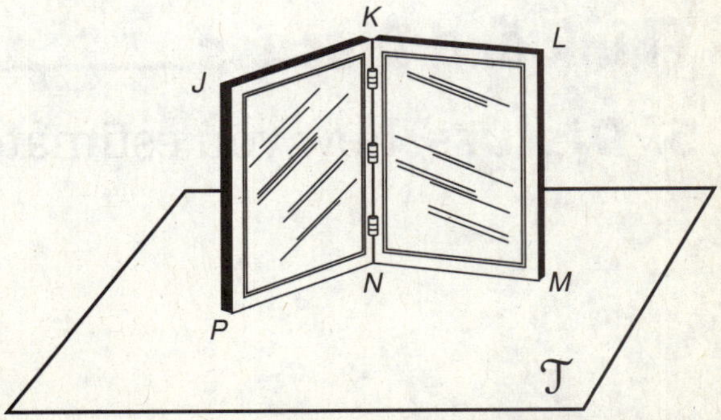

EXPLORATION

Program of Studies and Core Content

MA-8-G-S-SR1, MA-8-M-S-MPA1, MA-8-M-S-MPA2, MA-8-M-S-MPA8, MA-08-2.1.2, MA-08-2.1.3, MA-08-3.1.1

7B Measuring and Constructing Angles

An angle is formed by two rays that have a common endpoint. Right angles measure 90° and are shaped like a letter L. You can estimate an angle measure by comparing the angle with a right angle.

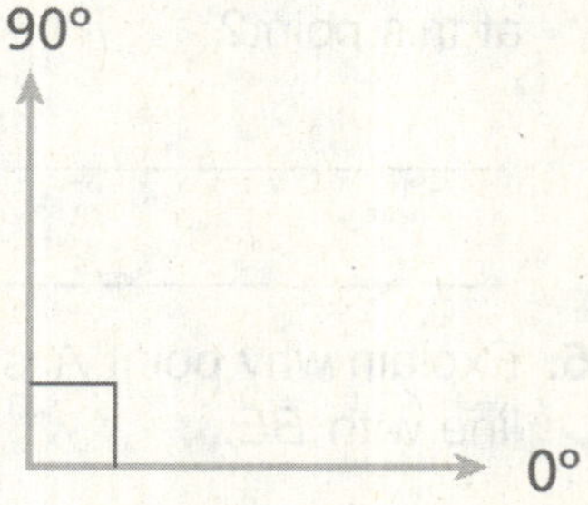

Estimate the measure of each angle. Then measure the angle with a protractor.

		Estimate	Actual
1.	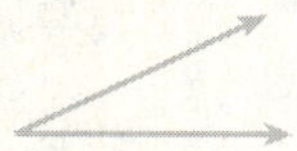		
2.			
3.	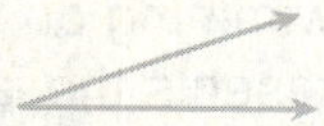		
4.	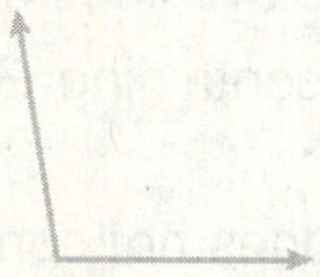		

Think and Discuss

5. Discuss how you estimated the angle measures.

Name ______________________________ Date ___________ Class ___________

LESSON 7B

Hands-On Lab

Measuring and Constructing Angles

When a protractor is not available you can quickly make an angle-measuring device out of an index card. You can use this home-made protractor to estimate angle measurements and to quickly classify an angle.

Materials needed: Index card, or similar

Activity

(1) On the index card, draw the rays forming the right angle at the bottom left corner.

- A right angle has a measure of 90°.
- Any angle greater than this measure is an obtuse angle.
- Any angle smaller than this angle is acute angle.

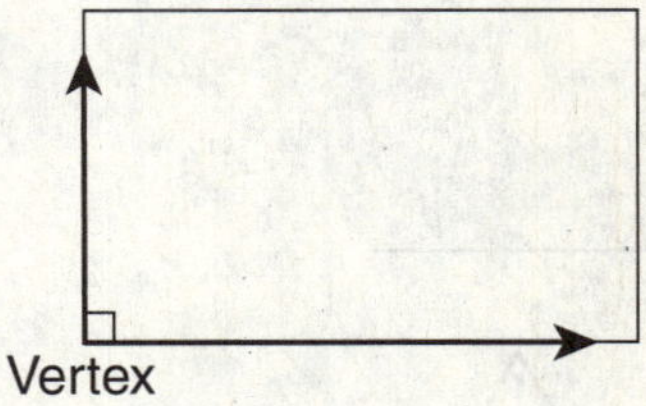

(2) Fold the top left corner of the index card down so that it meets the bottom edge of the card. Press on the fold so that it creases the card. This crease indicates a 45° angle with the bottom edge of the card.

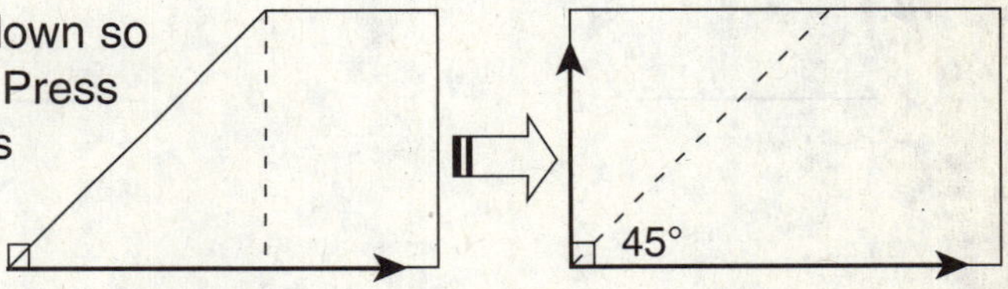

(3) Now fold the top left corner of the index card down so that it meets the crease of the 45° mark. Press on the fold so that it creases the card. This crease indicates a 67.5° angle with the bottom edge of the card.

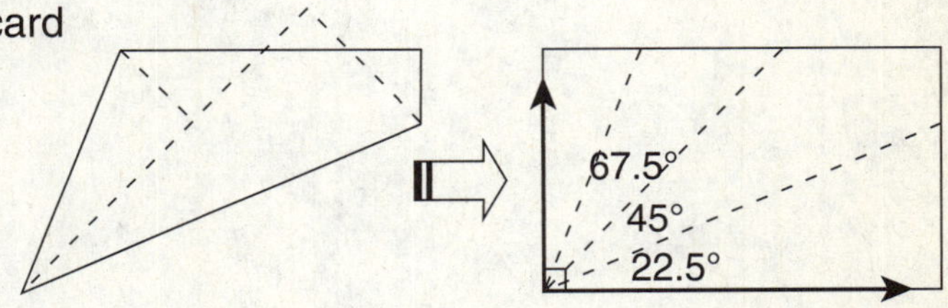

(4) Then fold the bottom left corner up so that it meets the crease of the 45° mark. Press on the fold so that it creases the card. This crease indicates a 22.5° angle with the bottom edge of the card.

Name ______________________ Date __________ Class __________

LESSON **7B**

Hands-On Lab

Measuring and Constructing Angles, continued

Try This

Use your index-card protractor to classify each angle as right, obtuse, or acute and then estimate the approximate measure of each angle.

1.

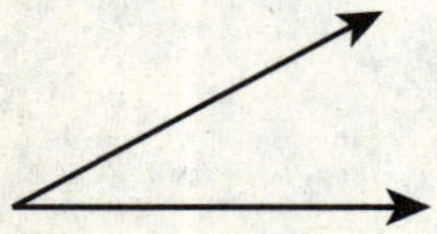

____

2.

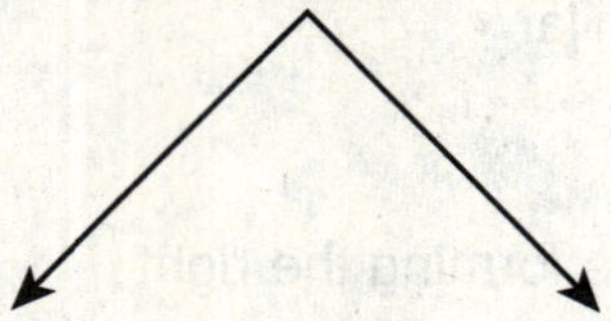

____

3.

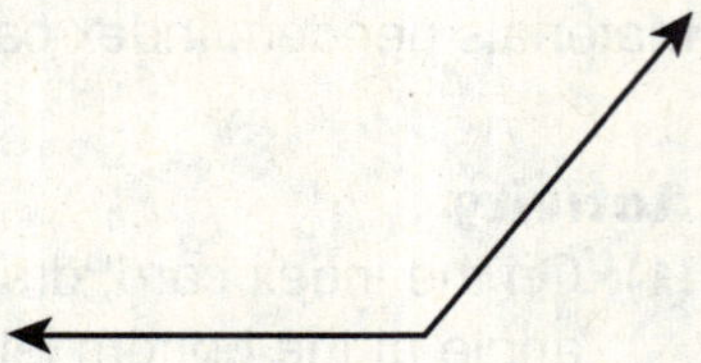

____

4.

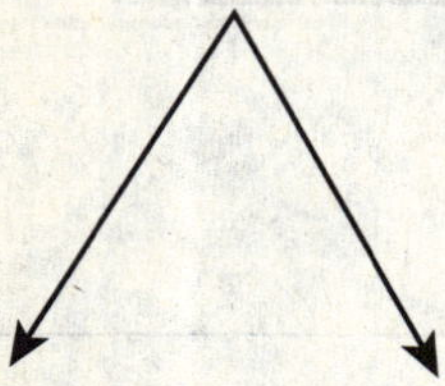

____

5.

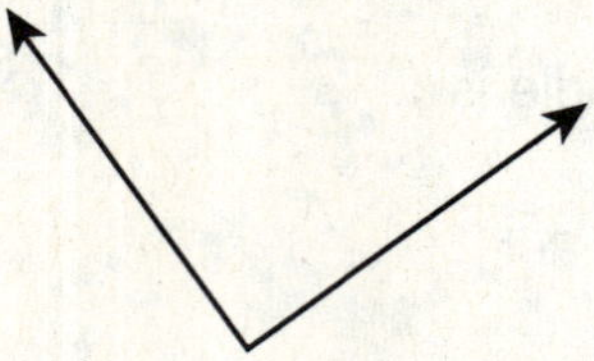

____

6.

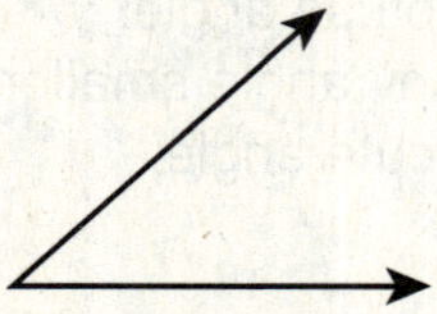

____

Name ____________________ Date __________ Class __________

LESSON 7B

Review for Mastery

Measuring and Constructing Angles

An **angle** is a figure made up of two rays, or **sides,** that have a common endpoint, called the **vertex** of the angle.

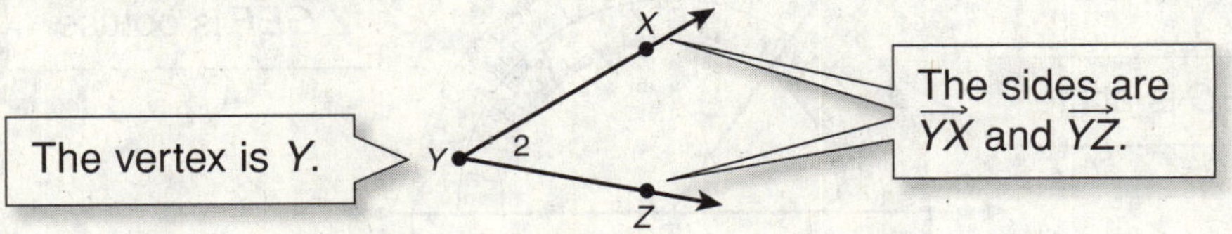

There are four ways to name this angle.

$\angle Y$	*Use the vertex.*
$\angle XYZ$ or $\angle ZYX$	*Use the vertex and a point on each side.*
$\angle 2$	*Use the number.*

Name each angle in three ways.

1.

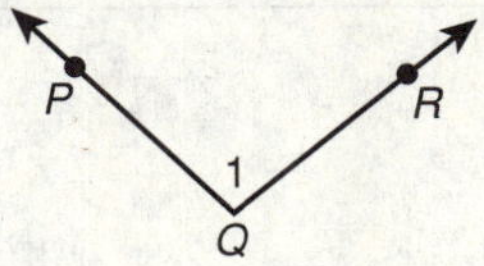

____

2.

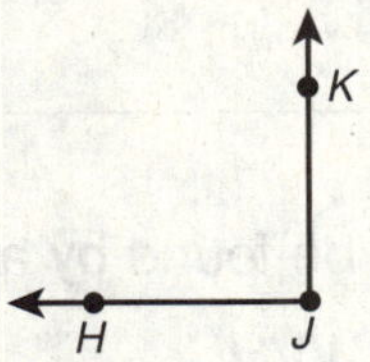

____

3. Name three different angles in the figure.

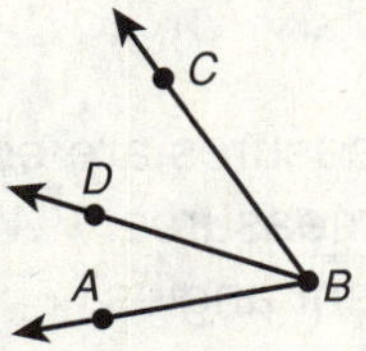

____

Angle	acute	right	obtuse	straight
Model	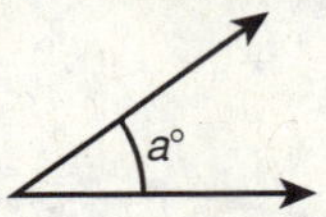	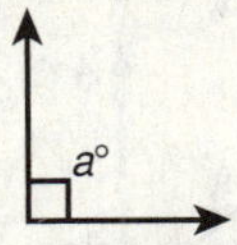	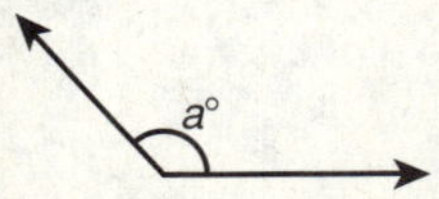	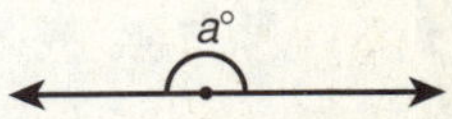
Possible Measures	$0° < a° < 90°$	$a° = 90°$	$90° < a° < 180°$	$a° = 180°$

Classify each angle as acute, right, obtuse, or straight.

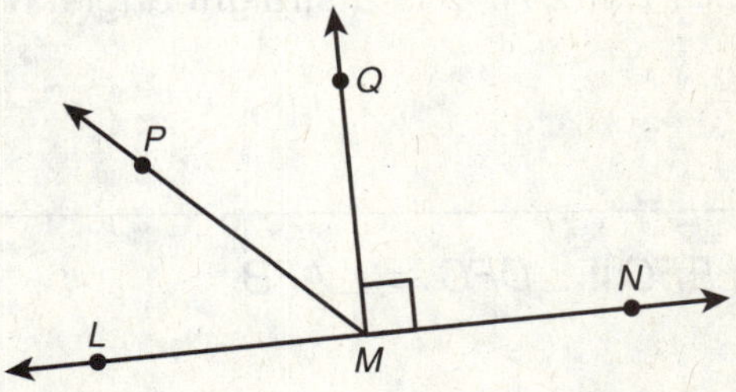

4. $\angle NMP$

____

5. $\angle QMN$

____

6. $\angle PMQ$

____

Name ______________________________ Date ____________ Class ____________

LESSON **7B**

Review for Mastery

Measuring and Constructing Angles, continued

You can use a protractor to find the measure of an angle.

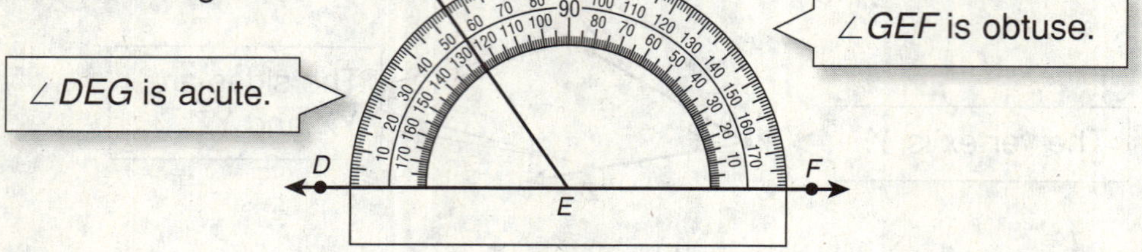

Use the figure above to find the measure of each angle.

7. ∠*DEG* ______________________

8. ∠*GEF* ______________________

The measure of ∠*XVU* can be found by adding.

$m\angle XVU = m\angle XVW + m\angle WVU$

$= 48° + 48°$

$= 96°$

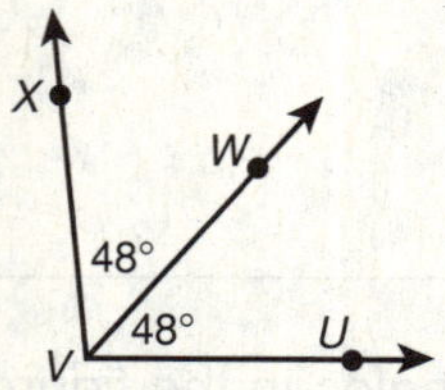

Angles are **congruent** if their measures are equal. In the figure, ∠*XVW* ≅ ∠*WVU* because the angles have equal measures. $\overrightarrow{VW}$ is an **angle bisector** of ∠*XVU* because it divides ∠*XVU* into two congruent angles.

Find each angle measure.

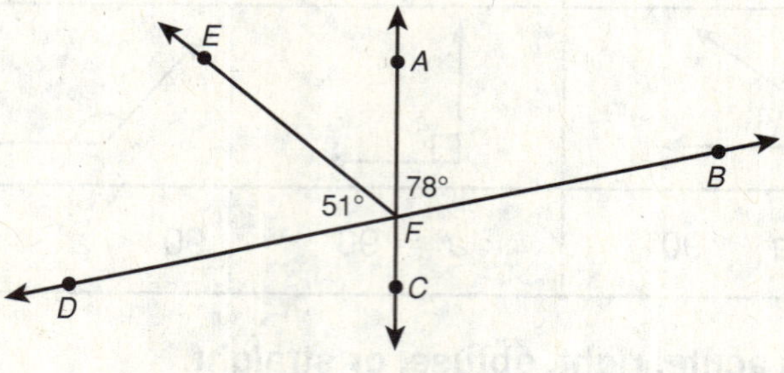

9. m∠*CFB* if ∠*AFC* is a straight angle.

____

10. m∠*EFA* if the angle is congruent to ∠*DFE*.

____

11. m∠*EFC* if ∠*DFC* ≅ ∠*AFB*.

____

12. m∠*CFG* if $\overrightarrow{FG}$ is an angle bisector of ∠*CFB*.

____

Name ______________________________ Date ____________ Class ____________

LESSON 7B

Homework and Practice

Measuring and Constructing Angles

Use the figure for Exercises 1–4.

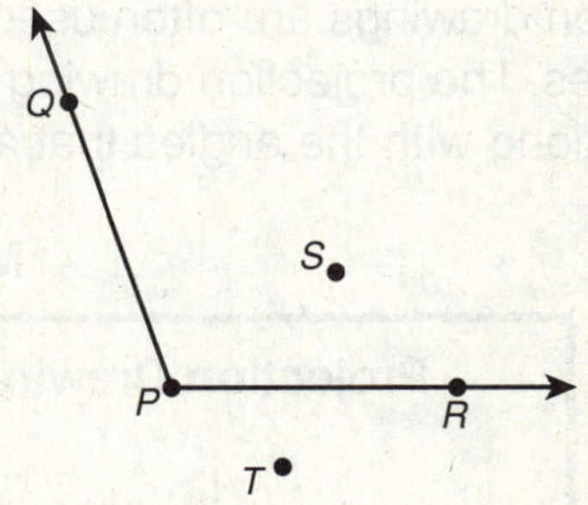

1. An angle is a figure formed by two rays with a common endpoint called the ____________________.

2. Name the two rays that form ∠*P*.

____

3. Use the angle symbol and three letters to name ∠*P* in two ways.

____

4. Name a point that is in the interior of ∠*P*.

____

Complete the statement.

5. A tool used to measure and draw angles is called a ______________________.

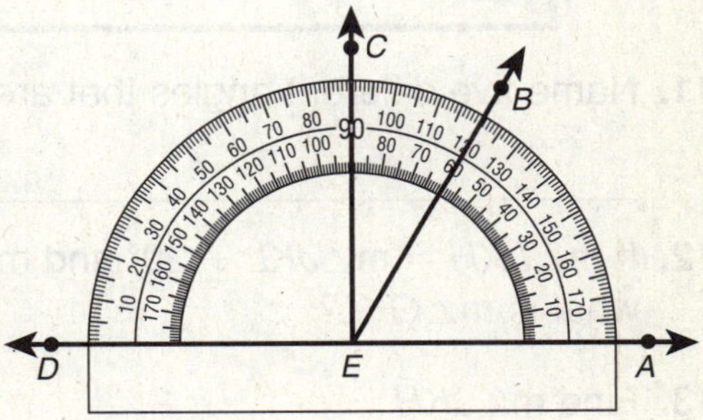

Find the measure of each angle. Then tell whether each is acute, right, obtuse, or straight.

6. ∠*CEA* ______________

7. ∠*AEB* ______________

8. ∠*DEA* ______________

Complete the angle.

9. Use a compass and straightedge to finish constructing ∠*IHJ* congruent to ∠*MLN*.

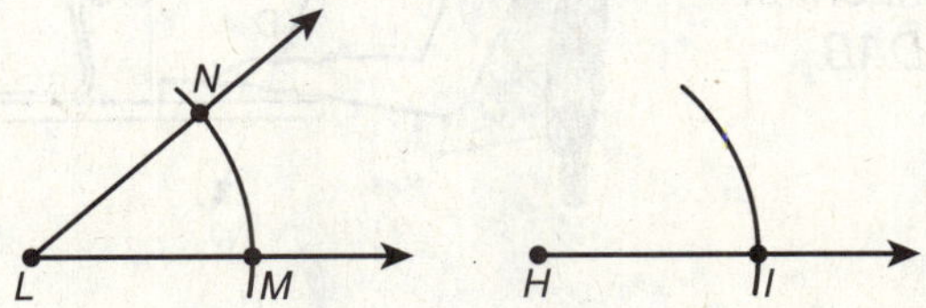

10. Marc doesn't think that the angle of the front seat in his mom's car is very cool, so he tilts the seat back. m∠*ZWY* = 95° and m∠*YWX* = 30°.

Find m∠*ZWX*. ____________

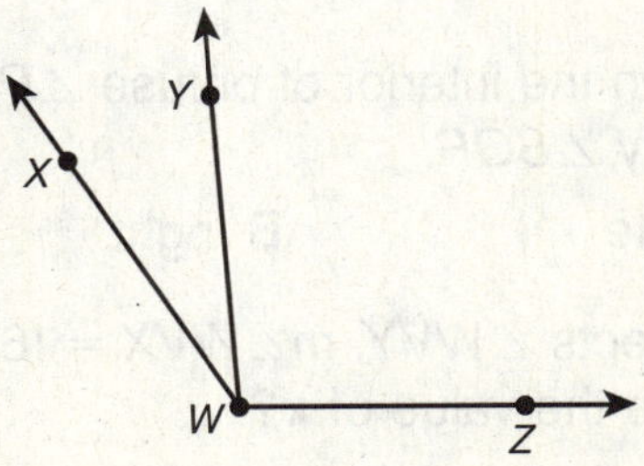

Name ____________________ Date __________ Class __________

LESSON 7B

Homework and Practice

Measuring and Constructing Angles, continued

Projection drawings are often used to represent three-dimensional molecules. The projection drawing of a methane molecule is shown below, along with the angles that are formed in the drawing.

Methane Molecule

Projection Drawing

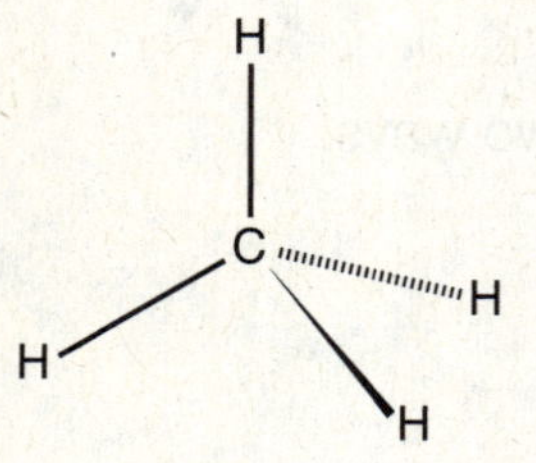

Angles Formed

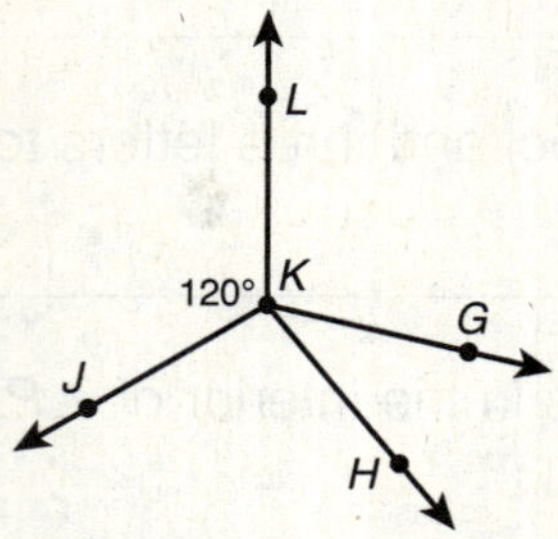

11. Name five different angles that are formed in the drawing.

____

12. If m∠*LKH* = m∠*JKL* + 20° and m∠*HKG* = 37°, what is m∠*GKL*?

____

13. Find m∠*JKH*.

____

The figure shows the proper way to sit at a computer to avoid straining your back or eyes. Use the figure for Exercises 14 and 15.

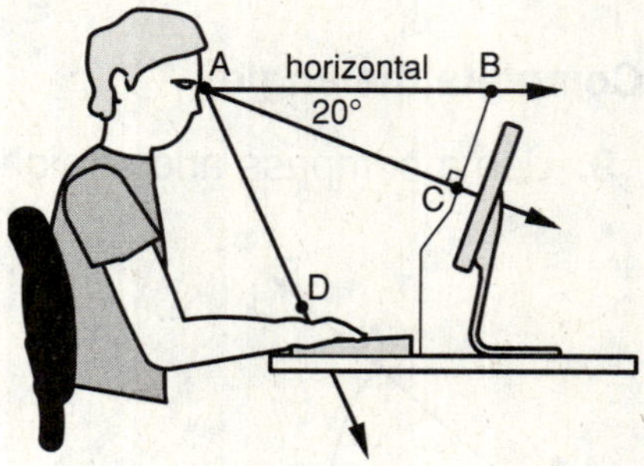

14. The *total viewing angle* is ∠*DAB*. If m∠*DAC* = $\frac{1}{2}$(m∠*ACB*), what is the measure of the total viewing angle?

____

15. The *optimum viewing angle* is 38° below the horizontal. If $\overrightarrow{AV}$ is drawn to form this angle with $\overrightarrow{AB}$ and ∠*DAB* measures 65°, what is the measure of ∠*DAV*?

____

Choose the best answer.

16. $\overrightarrow{QR}$ is in the interior of obtuse ∠*PQS*, and ∠*PQR* is a right angle. Classify ∠*SQR*.

A acute **B** right **C** obtuse **D** straight

17. $\overrightarrow{VX}$ bisects ∠*WVY*, m∠*WVX* = (6*x*)°, and m∠*WVY* = (16*x* − 42)°. What is the value of *x*?

F $\frac{21}{11}$ **G** $\frac{42}{13}$ **H** 4.2 **J** 10.5

EXPLORATION

Program of Studies and Core Content

MA-8-G-S-SR1,
MA-08-3.1.1

7C Pairs of Angles

The line segments in some letters and symbols form angles.

Numbers 1–5 describe types of angle pairs. Determine which marked angle pairs in the figures apply to each description, and then check the appropriate boxes.

	W	X	T	↗
1. Congruent angles: same measure				
2. Vertical angles: opposite each other when two lines intersect				
3. Adjacent angles: side by side with a common vertex and ray				
4. Complementary angles: sum equals 90°				
5. Supplementary angles: sum equals 180°				

Think and Discuss

6. Discuss other examples of vertical angles in the real world.

Name ______________________________ Date ____________ Class ____________

LESSON **7C**

Review for Mastery

Pairs of Angles

Angle Pairs		
Adjacent Angles	**Linear Pairs**	**Vertical Angles**
have the same vertex and share a common side	adjacent angles whose noncommon sides are opposite rays	nonadjacent angles formed by two intersecting lines
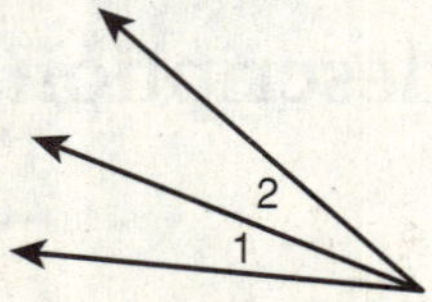 ∠1 and ∠2 are adjacent.	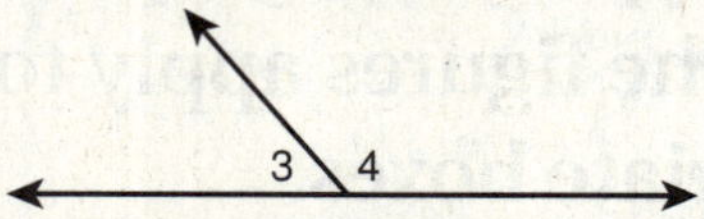 ∠3 and ∠4 are adjacent and form a linear pair.	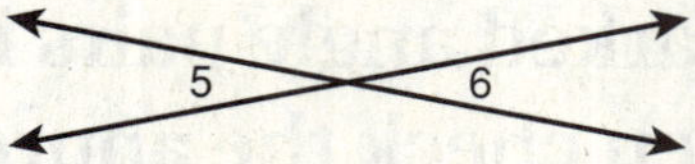 ∠5 and ∠6 are vertical angles.

Tell whether ∠7 and ∠8 in each figure are only adjacent, are adjacent and form a linear pair, or are not adjacent.

1.

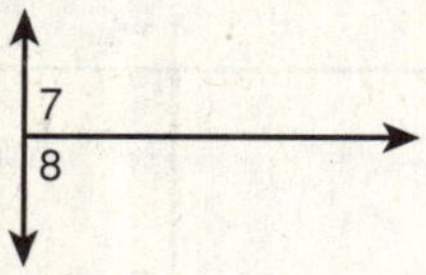

2.

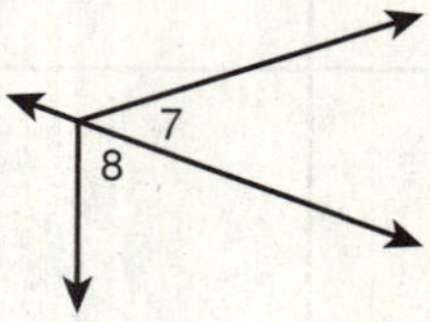

3.

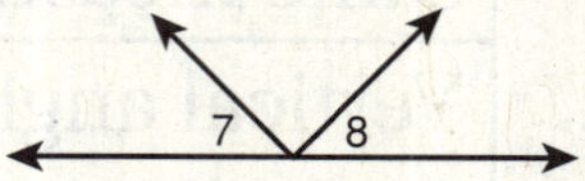

______________ ______________ ______________

Tell whether the indicated angles are only adjacent, are adjacent and form a linear pair, or are not adjacent.

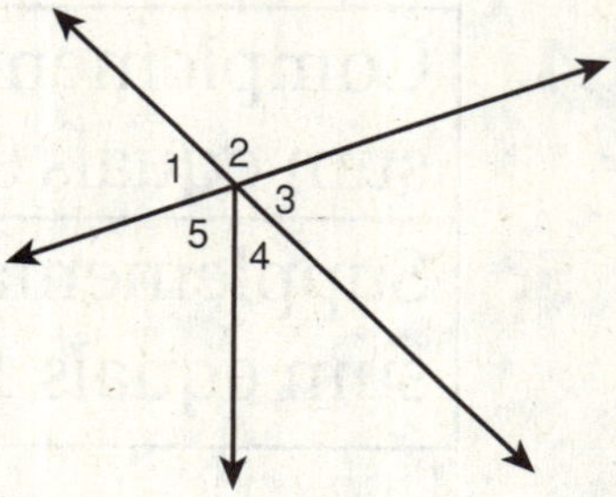

4. ∠5 and ∠4 ______________________________

5. ∠1 and ∠4 ______________________________

6. ∠2 and ∠3 ______________________________

Name each of the following.

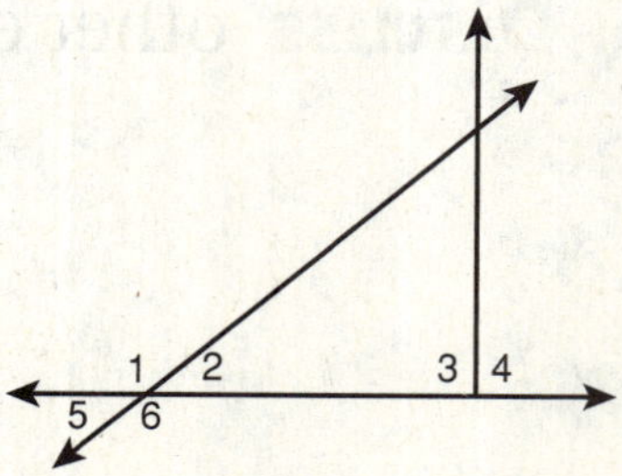

7. a pair of vertical angles ______________________

8. a linear pair ______________________

9. an angle adjacent to ∠4 ______________________

Name ______________________ Date __________ Class __________

LESSON **7C**

Review for Mastery

Pairs of Angles, continued

Angle Pairs	
Complementary Angles	**Supplementary Angles**
sum of angle measures is 90°	sum of angle measures is 180°
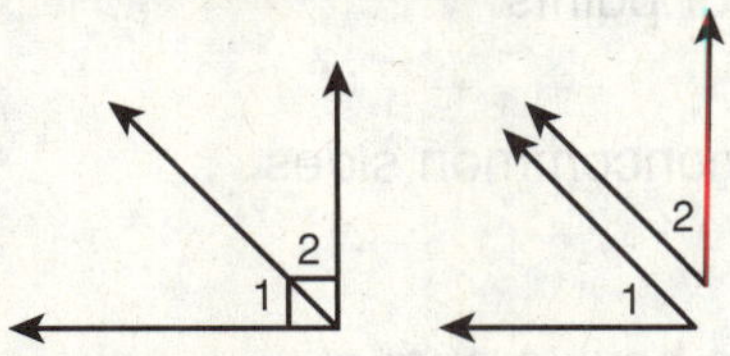 $m\angle 1 + m\angle 2 = 90°$ In each pair, ∠1 and ∠2 are complementary.	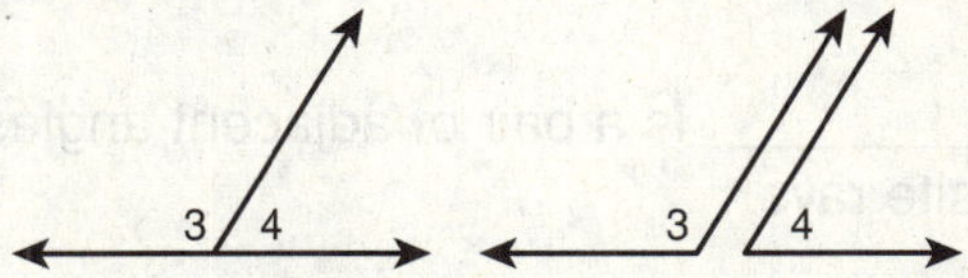 $m\angle 3 + m\angle 4 = 180°$ In each pair, ∠3 and ∠4 are supplementary.

Tell whether each pair of labeled angles is complementary, supplementary, or neither.

10.

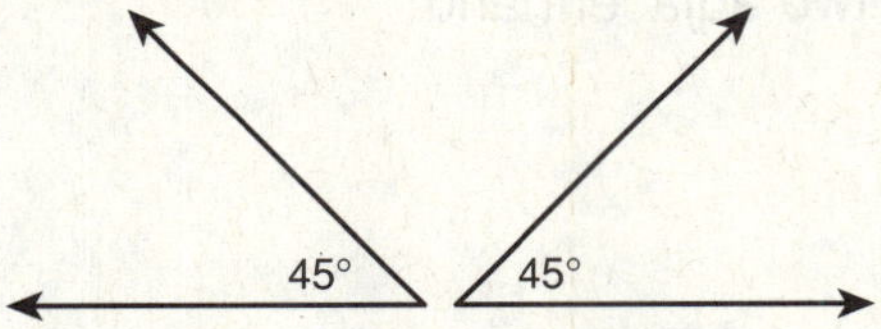

____

11.

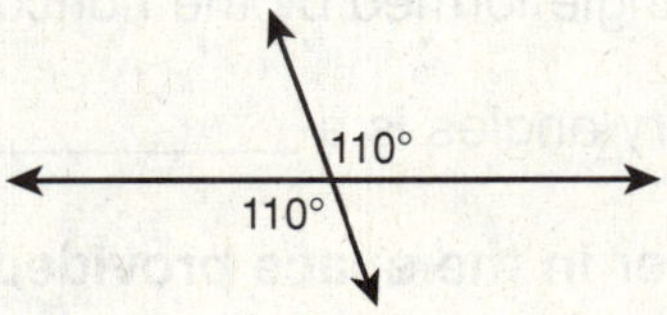

____

Find the measure of each of the following angles.

12. complement of ∠*S* ____________

13. supplement of ∠*S* ____________

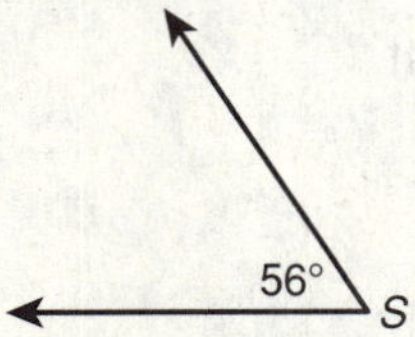

14. complement of ∠*R* ____________

15. supplement of ∠*R* ____________

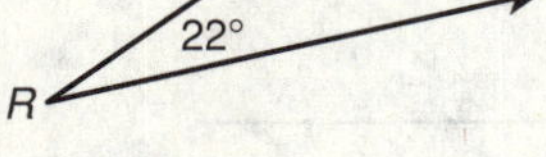

16. ∠*LMN* and ∠*UVW* are complementary. Find the measure of each angle if $m\angle LMN = (3x + 5)°$ and $m\angle UVW = 2x°$.

____

Name ______________________ Date ____________ Class ____________

LESSON 7C

Homework and Practice

Pairs of Angles

Complete the statements.

1. Adjacent angles are two angles in the same plane with a common ____________ and a common ____________, but no common interior points.

2. A ____________ is a pair of adjacent angles whose noncommon sides are opposite rays.

3. Complementary angles are two angles whose measures have a sum of ____________.

4. The kind of angle formed by the noncommon sides of two adjacent and complementary angles is a ____________.

5. ____________ angles are two angles whose measures have a sum of 180°.

6. The kind of angle formed by the noncommon sides of two adjacent and supplementary angles is a ____________.

Draw your answer in the space provided.

7. Sketch ∠1 and ∠2 so that they are adjacent angles.

8. Sketch ∠1 and ∠2 so that they form a linear pair.

In an equilateral triangle, all three sides have equal lengths and all three angles have equal measures. Find the measure of the following angles.

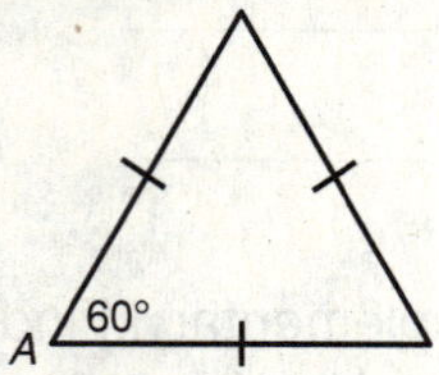

9. supplement of ∠*A* ____________

10. complement of ∠*A* ____________

Draw your answer in the space provided.

11. Sketch ∠1 and ∠2 so that they are vertical angles.

Name ______________________ Date __________ Class __________

LESSON 7C

Homework and Practice

Pairs of Angles, continued

Use the drawing of part of the Eiffel Tower for Exercises 12–16.

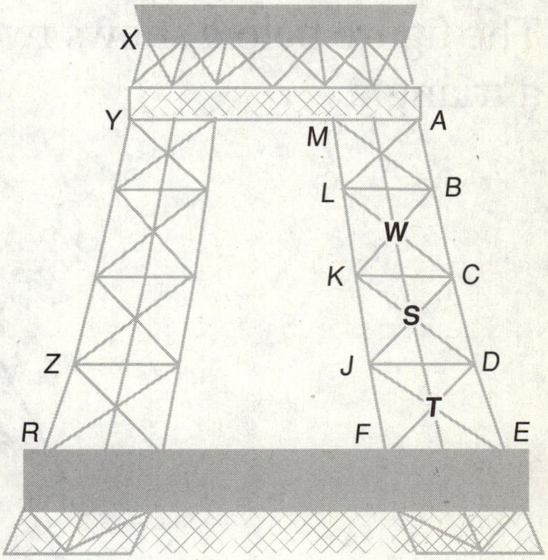

12. Name a pair of angles that appear to be complementary.

____

13. Name a pair of supplementary angles.

____

14. If $m\angle CSW = 45°$, what is $m\angle JST$? How do you know?

____

____

15. If $m\angle FKB = 135°$, what is $m\angle BKL$? How do you know?

____

____

16. Name three angles whose measures sum to 180°.

____

Choose the best answer.

17. A landscaper uses paving stones for a walkway. Which are possible angle measures for $a°$ and $b°$ so that the stones do not have space between them?

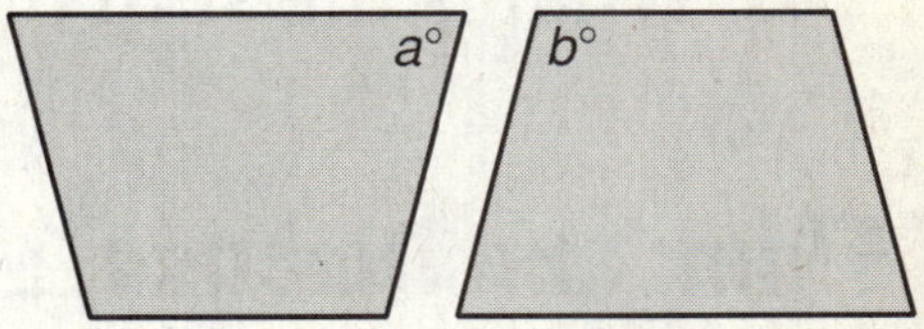

A 50°, 100° **C** 75°, 105°

B 45°, 45° **D** 90°, 80°

18. The angle formed by a tree branch and the part of the trunk above it is 68°. What is the measure of the angle that is formed by the branch and the part of the trunk below it?

F 22° **H** 158°

G 112° **J** 180°

19. $\angle R$ and $\angle S$ are complementary. If $m\angle R = (7 + 3x)°$ and $m\angle S = (2x + 13)°$, which is a true statement?

A $\angle R$ is acute. **C** $\angle R$ and $\angle S$ are right angles.

B $\angle R$ is obtuse. **D** $m\angle S > m\angle R$

EXPLORATION

Program of Studies and Core Content

MA-8-G-S-SR2, MA-8-M-S-MPA2, MA-08-2.1.3, MA-08-3.1.2

7D Angles in Triangles

The figure below shows two parallel lines *m* and *n*, and a triangle.

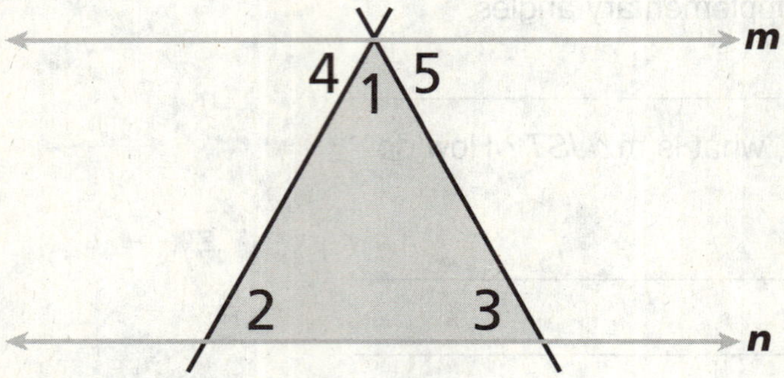

1. What is the sum of the measures of ∠4, ∠1, and ∠5?

2. Name an angle with the same measure as ∠4.

3. Name an angle with the same measure as ∠5.

4. What is the sum of the measures of ∠1, ∠2, and ∠3?

Think and Discuss

5. Discuss what would happen to the sum of the measures of ∠1, ∠2, and ∠3 if the triangle above were bigger.

6. Explain why each angle measure in an ***equilateral triangle*** is 60°. (In an equilateral triangle, the three sides are equal.)

Name ______________________ Date __________ Class __________

LESSON 7D

Hands-On Lab

Angles in Triangles

Activity

1. The sum of the measures of the angles of a triangle equals 180°. You can prove this. Use a straight edge to construct a triangle on a sheet of paper. Measure each of the angles with a protractor and write each angle measure here:

$\angle A =$ ______________________

$\angle B =$ ______________________

$\angle C =$ ______________________

What is the total of the angles?

____

Now cut off two of the corners of the triangle. Paste them to the sides of the other corner. Together, they form a straight line that equals 180°.

2. In the space below, draw an equilateral triangle. Write the measurements of each angle. Why do you know this is an equilateral triangle?

____

____

____

Name ______________________ Date ___________ Class ___________

LESSON 7D

Hands-On Lab

Angles in Triangles, continued

Think and Discuss

1. What is a fact you know about the angles of every triangle?

____

____

2. If you know one angle in a triangle is 90°, can that triangle be an acute triangle?

____

____

Try This

In the space below, draw an acute triangle, an obtuse triangle, and a right triangle. Label each triangle and include each angle measurement.

____

Name ______________________ Date __________ Class __________

LESSON 7D

Review for Mastery

Angles in Triangles

You can classify triangles by their angle measures. An **equiangular triangle,** for example, is a triangle with three congruent angles.

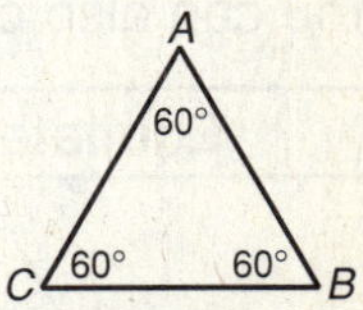

$\angle A \cong \angle B \cong \angle C$

$\triangle ABC$ is equiangular.

Examples of three other triangle classifications are shown in the table.

Acute Triangle	Right Triangle	Obtuse Triangle
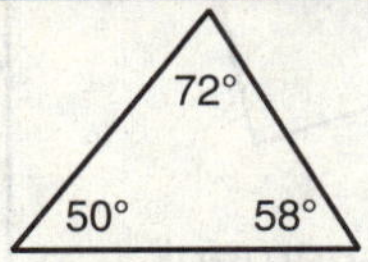	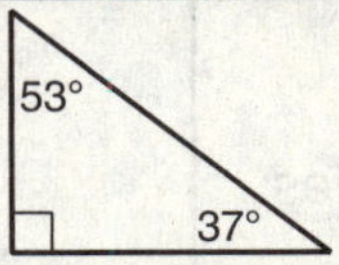	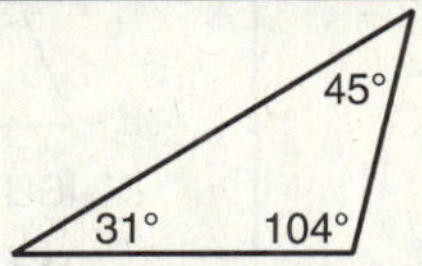
all acute angles	one right angle	one obtuse angle

You can use angle measures to classify $\triangle JML$ at right.

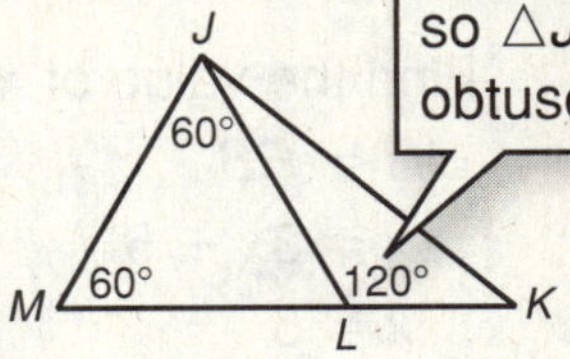

$\angle JKL$ is obtuse so $\triangle JLK$ is an obtuse triangle.

$\angle JLM$ and $\angle JLK$ form a linear pair, so they are supplementary.

$m\angle JLM + m\angle JLK = 180°$	Def. of supp. ∠s
$m\angle JLM + 120° = 180°$	Substitution
$m\angle JLM = 60°$	Subtract.

Since all the angles in $\triangle JLM$ are congruent, $\triangle JLM$ is an equiangular triangle.

Classify each triangle by its angle measures.

1.

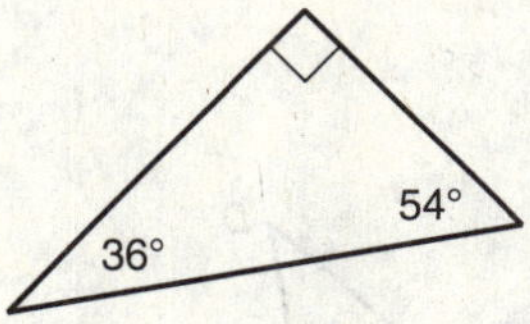

____

2.

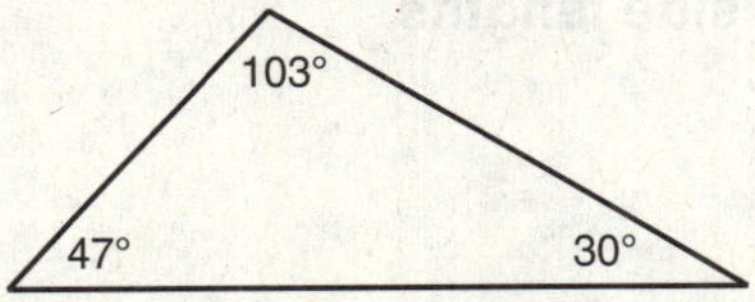

____

3.

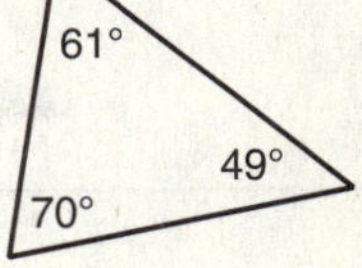

____

Use the figure to classify each triangle by its angle measures.

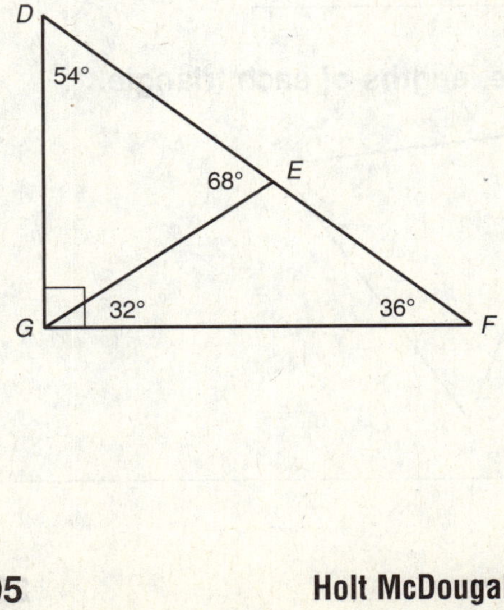

4. $\triangle DFG$

____

5. $\triangle DEG$

____

6. $\triangle EFG$

____

Name ______________________ Date __________ Class __________

LESSON 7D

Review for Mastery

Angles in Triangles, continued

You can also classify triangles by their side lengths.

Equilateral Triangle	Isosceles Triangle	Scalene Triangle
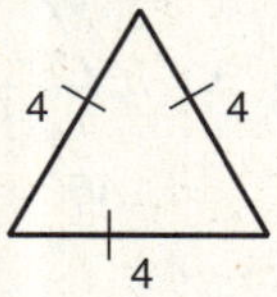 all sides congruent	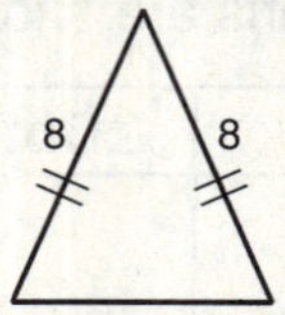 at least two sides congruent	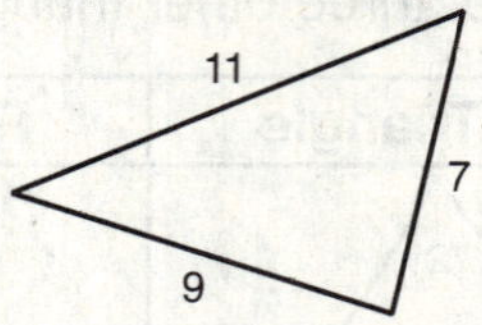 no sides congruent

You can use triangle classification to find the side lengths of a triangle.

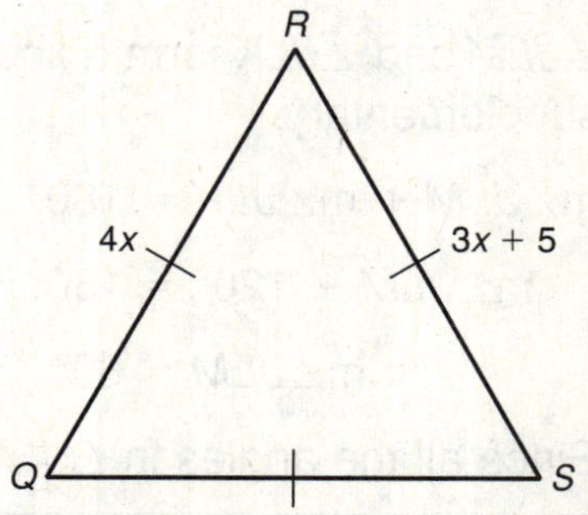

Step 1 Find the value of x.

$QR = RS$	Def. of ≅ segs.
$4x = 3x + 5$	Substitution
$x = 5$	Simplify.

Step 2 Use substitution to find the length of a side.

$4x = 4(5)$	Substitute 5 for x.
$= 20$	Simplify.

Each side length of $\triangle QRS$ is 20.

Classify each triangle by its side lengths.

7. $\triangle EGF$

____

8. $\triangle DEF$

____

9. $\triangle DFG$

____

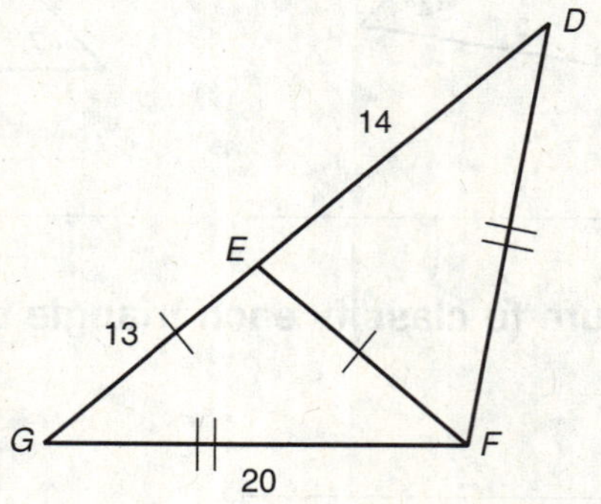

Find the side lengths of each triangle.

10.

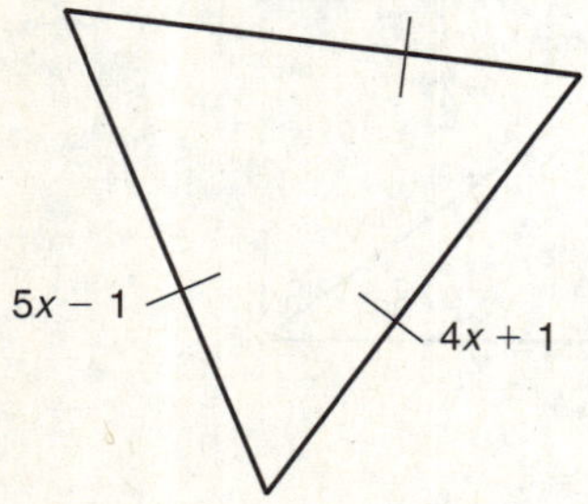

____

11.

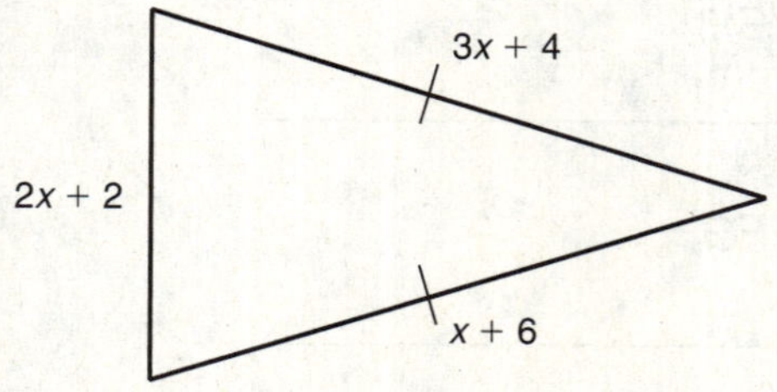

____

Name ______________________ Date ____________ Class ____________

LESSON 7D

Homework and Practice

Angles in Triangles

Match the letter of the figure to the correct vocabulary word in Exercises 1–4.

1. right triangle ________
2. obtuse triangle ________
3. acute triangle ________
4. equiangular triangle ________

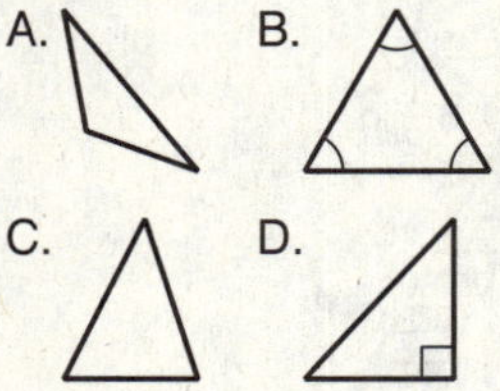

Classify each triangle by its angle measures as acute, equiangular, right, or obtuse. (*Note:* Give two classifications for Exercise 7.)

5. 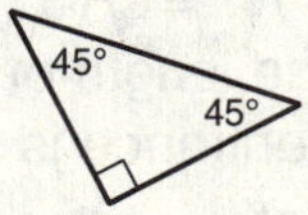

____

6. 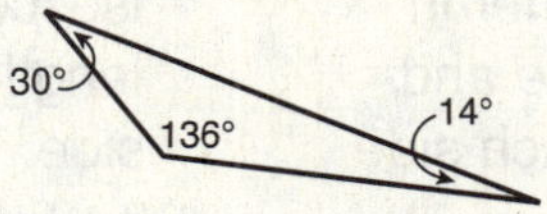

____

7. 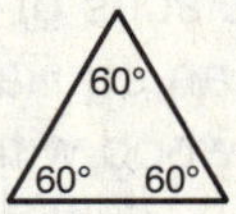

____

For Exercises 8–10, fill in the blanks to complete each definition.

8. An isosceles triangle has ________ congruent sides.
9. An ________ triangle has three congruent sides.
10. A ________ triangle has no congruent sides.

Classify each triangle by its side lengths as equilateral, isosceles, or scalene. (*Note:* Give two classifications in Exercise 13.)

11. 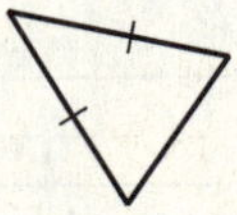

____

12. 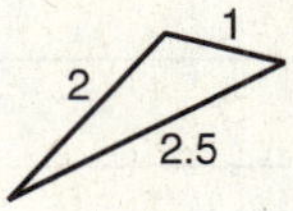

____

13. 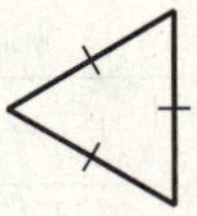

____

Find the side lengths of the triangle.

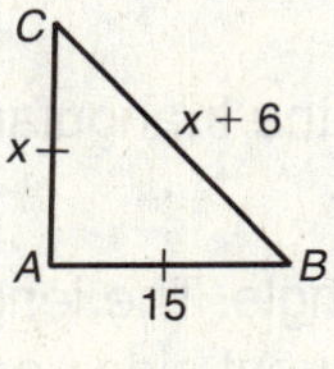

14. AB = ________ AC = ________ BC = ________

15. The New York City subway is known for its crowded cars. If all the seats in a car are taken, passengers must stand and steady themselves with railings or handholds. The last subway cars designed with steel hand straps were the "Redbirds" made in the late 1950s and early 1960s. The figure gives the dimensions of one of these triangular hand straps. How many hand straps could have been made from 99 inches of steel?

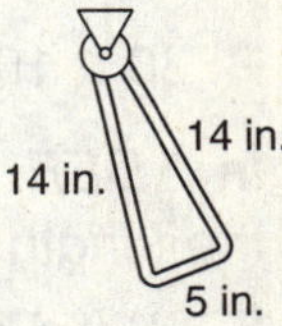

____

Name ________________________________ Date ____________ Class ____________

LESSON **7D**

Homework and Practice

Angles in Triangles, continued

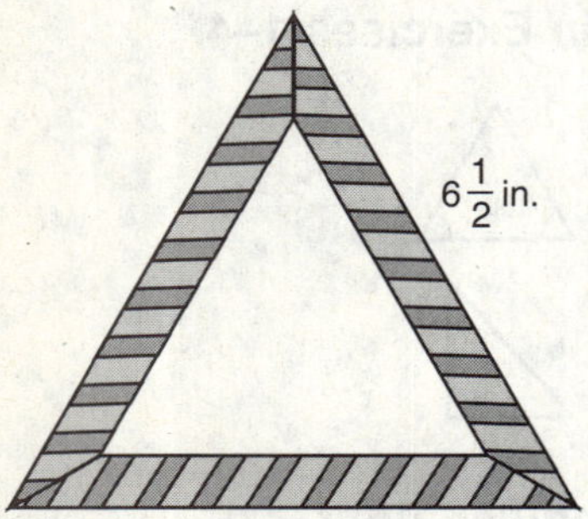

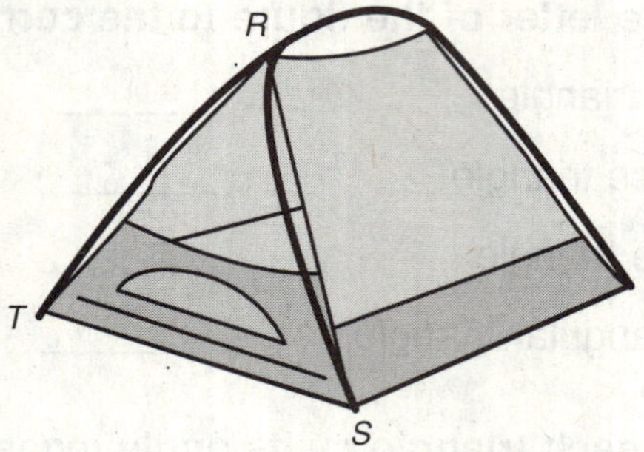

16. Aisha makes triangular picture frames by gluing three pieces of wood together in the shape of an equilateral triangle and covering the wood with ribbon. Each side of a frame is $6\frac{1}{2}$ inches long. How many frames can she cover with 2 yards of ribbon?

17. A tent's entrance is in the shape of an isosceles triangle in which $\overline{RT} \cong \overline{RS}$. The length of $\overline{TS}$ is 1.2 times the length of a side. The perimeter of the entrance is 14 feet. Find each side length.

____

Use the figure and the following information for Exercises 18 and 19.

The distance "as the crow flies" between Santa Fe and Phoenix is 609 kilometers. This is 245 kilometers less than twice the distance between Santa Fe and El Paso. Phoenix is 48 kilometers closer to El Paso than it is to Santa Fe.

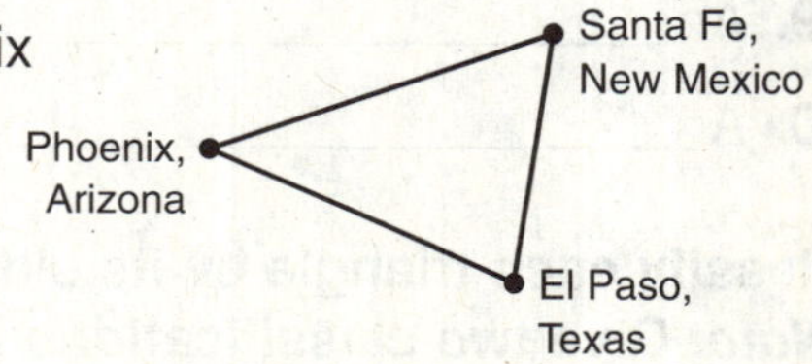

18. What is the distance between each pair of cities?

____

____

19. Classify the triangle that connects the cities by its side lengths. ____________________

Choose the best answer.

A *gable,* as shown in the diagram, is the triangular portion of a wall between a sloping roof.

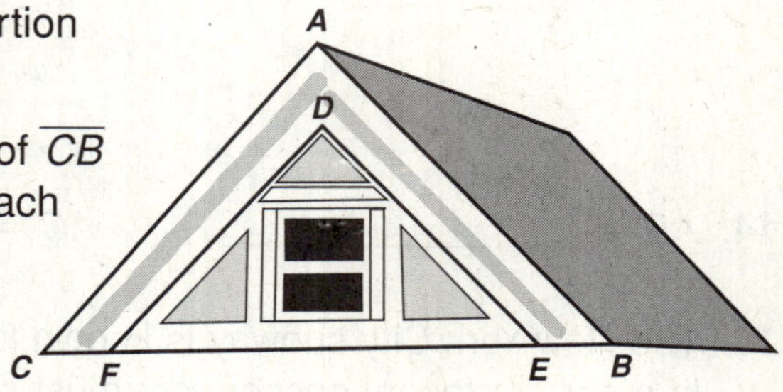

20. Triangle *ABC* is an isosceles triangle. The length of $\overline{CB}$ is 12 feet 4 inches and the congruent sides are each $\frac{3}{4}$ this length. What is the perimeter of $\triangle ABC$?

A 31 ft 4 in. **C** 21 ft 7 in.

B 30 ft 10 in. **D** 18 ft 6 in.

21. In $\triangle DEF$, $\overline{DE}$ and $\overline{DF}$ are each 6 feet 3 inches long. This length is 0.75 times the length of $\overline{FE}$. What is the perimeter of $\triangle DEF$?

F 12 ft 4 in. **H** 17 ft 2 in.

G 14 ft 7 in. **J** 20 ft 10 in.

EXPLORATION

Program of Studies and Core Content

MA-8-G-S-SR2, MA-8-M-S-MPA1, MA-08-3.1.2

7E Classifying Polygons

Use a protractor to measure each angle in each figure. Then find the sum of the angle measures.

	Polygon	Picture	Sum of Angles
1.	Triangle (3 sides)	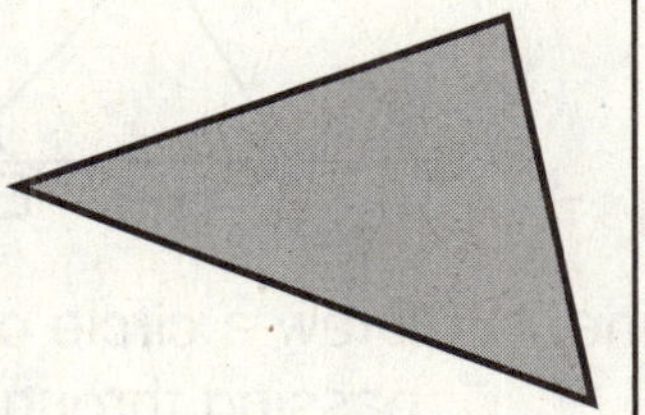	
2.	Pentagon (5 sides)	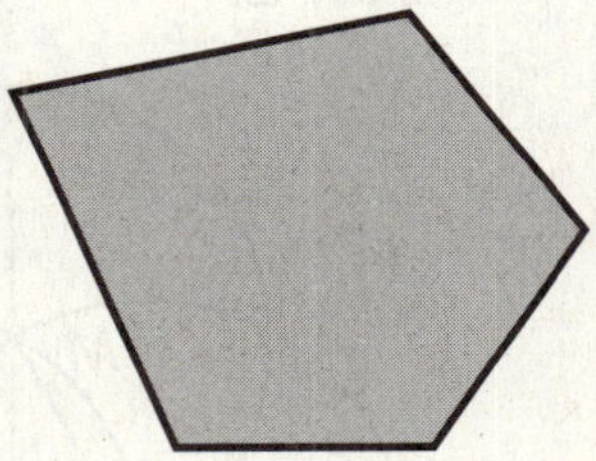	
3.	Hexagon (6 sides)	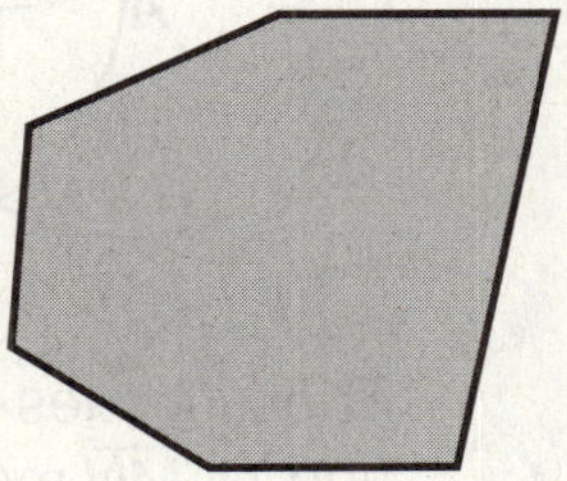	

Think and Discuss

4. Explain how you know the sum of the angle measures in a square.

5. Discuss how you can use the two triangles to show that the sum of the angle measures in a four-sided figure is 360°.

Name ______________________________ Date ____________ Class ____________

LESSON 7E

Hands-On Lab 1

Constructing Altitudes and Diagonals of Triangles

Activity

You can construct the altitude of a triangle using a compass and straightedge.

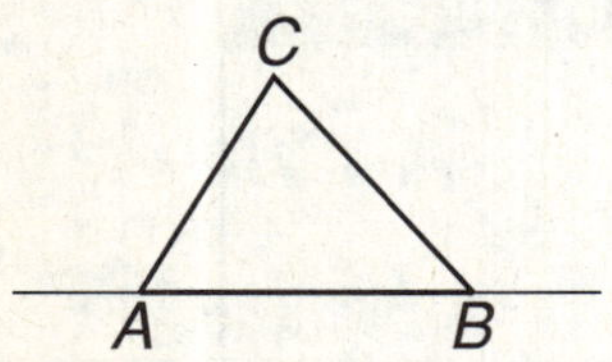

Make line segment $\overline{AB}$ into a line.

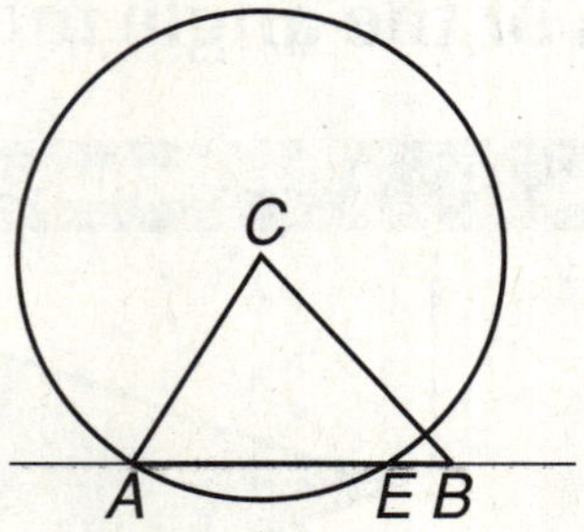

Draw a circle centered at *C* and passing through *A*. Label the other point of intersection through $\overline{AB}$, *E*.

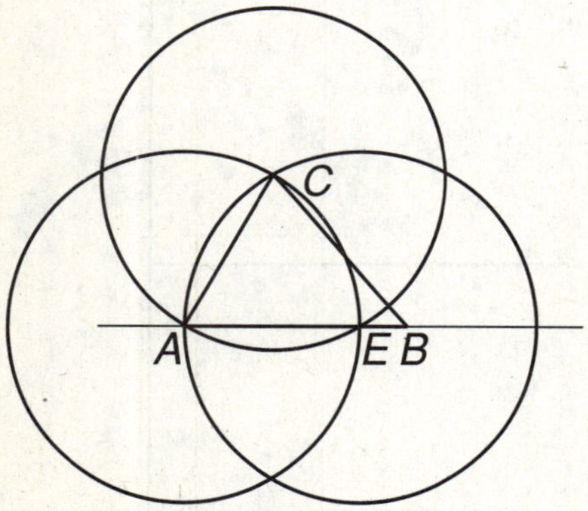

Draw two more circles.One centered at *A* and passing through *E*. The other centered at *E* and passing through *A*.

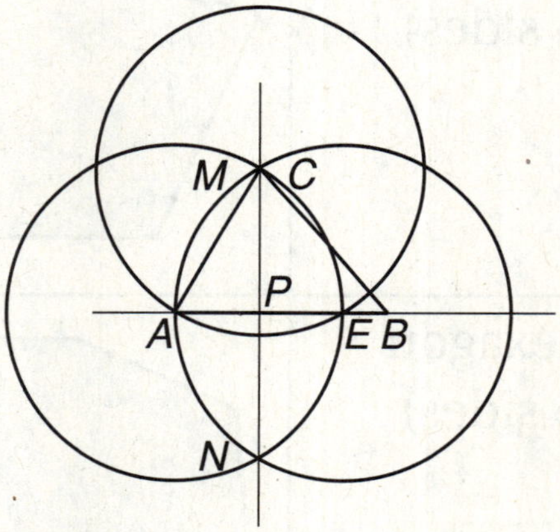

These circles intersect at *M* and *N*. $\overline{MN}$ meets $\overline{AB}$ at point *P*. $\overline{CP}$ is the altitude from *C*.

Name ______________________ Date __________ Class __________

LESSON 7E

Hands-On Lab 1

Constructing Altitudes and Diagonals of Triangles, continued

You can construct an angle bisector of a triangle using a compass and straightedge.

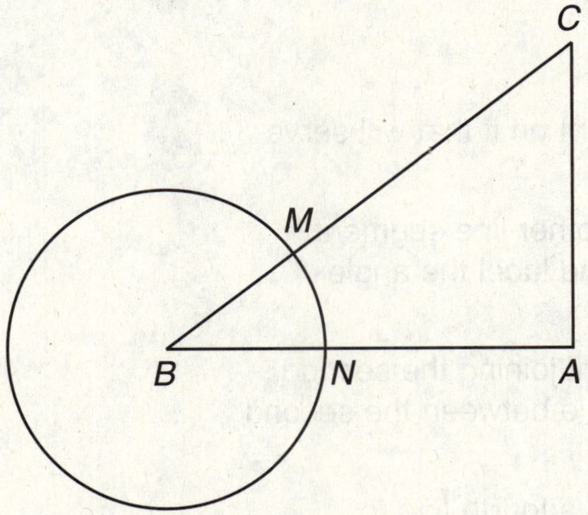

First draw a circle centered at *B*. Label intersection points *M* and *N*.

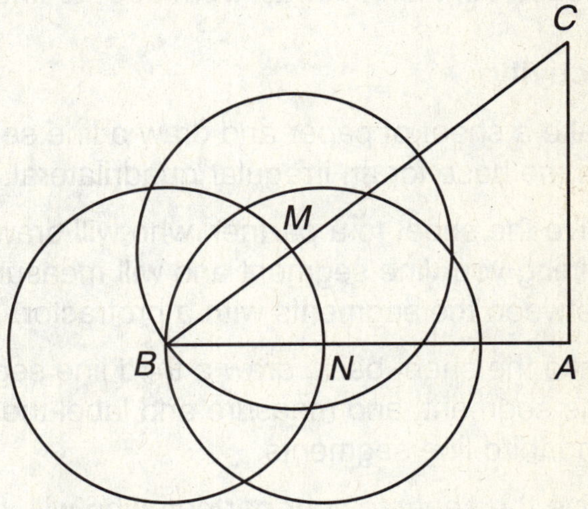

Draw circles through *B* centered at *M* and *N*.

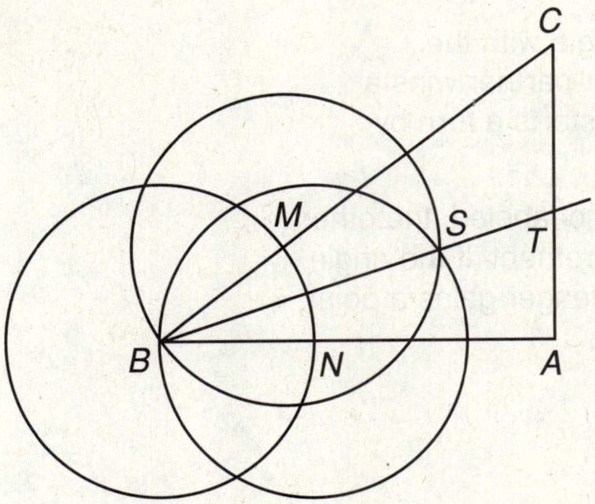

Find intersection point *S* of the two circles. Draw a line from point *B*, through point *S* and intersecting $\overline{CA}$. Label the point *T*. $\overrightarrow{BT}$ is the angle bisector.

Try This

1. Find an angle bisector of triangle *ABC*. Label it $\overrightarrow{BK}$.

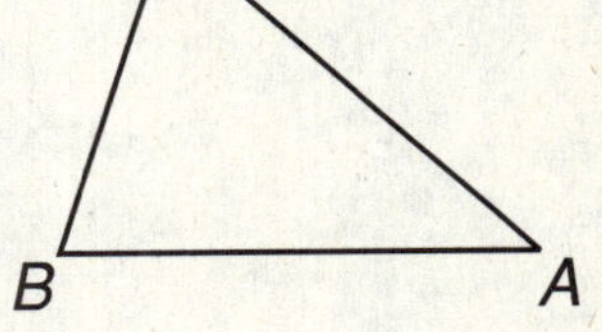

Name ______________________ Date ____________ Class ____________

LESSON 7E

Hands-On Lab 2
Polygons

The sum of the angles of a quadrilateral is always 360°. That means that if you know three of the angles, you can combine the angle measures, then subtract from 360° to find the fourth angle.

Activity

Take a sheet of paper and draw a line segment on it that will serve as the base for an irregular quadrilateral.

Give the sheet to a partner, who will draw another line segment joining your line segment and will measure and label the angle between the segments with a protractor.

Take the sheet back, draw a third line segment joining the second line segment, and measure and label the angle between the second and third line segments.

Give the sheet to your partner, who will draw a fourth line to complete the quadrilateral. The partner must measure and label the third angle created, using the protractor, and then label the fourth angle by subtracting the sum of the other angles from 360°.

Take the sheet back and measure the fourth angle with the protractor. If it matches the label within 10°, your partner wins a point. If not, you win a point. Then your partner starts a turn by drawing another baseline for another polygon.

At any point where an angle has been drawn and labeled, the other party may challenge the accuracy of the measurement. If the angle was not correctly measured within 10°, the challenger gains a point. Otherwise, the challenger loses a point.

See who gets to 5 points first.

Name ____________________ Date __________ Class __________

LESSON 7E

Hands-On Lab 2
Polygons, continued

Think and Discuss

1. Suppose player 1 draws a baseline, player 2 draws a line segment at an angle of 135°, player 1 continues with another line segment at an angle of 135°, and player 2 adds another line segment at an angle of 135°. What will be the fourth angle to complete the quadrilateral?

____

____

2. If player 1 creates an angle of 70° and labels it 60° without the mistake being discovered, how will player 2's calculations be wrong at the end?

____

____

Try This

Extend the game to polygons from 3 to 8 sides. Before starting each game, roll a number cube and add 2 to the result to get a number from 3 to 8. Follow the game as before, creating a polygon with a number of sides determined by rolling the number cube.

Name ______________________________ Date ____________ Class ____________

LESSON 7E

Review for Mastery

Classifying Polygons

The parts of a polygon are named on the quadrilateral below.

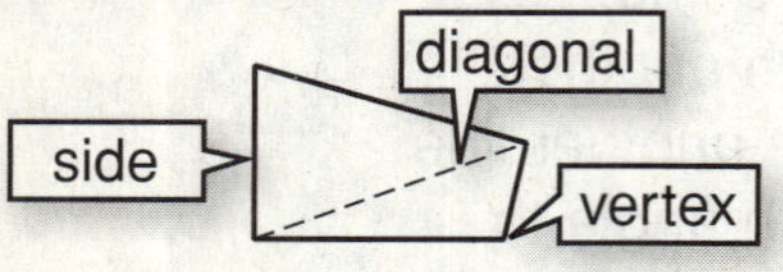

Number of Sides	Polygon
3	triangle
4	quadrilateral
5	pentagon
6	hexagon
7	heptagon
8	octagon
9	nonagon
10	decagon
n	*n*-gon

You can name a polygon by the number of its sides.

A **regular polygon** has all sides congruent and all angles congruent. A polygon is **convex** if all its diagonals lie in the interior of the polygon. A polygon is **concave** if all or part of at least one diagonal lies outside the polygon.

Types of Polygons		
regular, convex	**irregular, convex**	**irregular, concave**
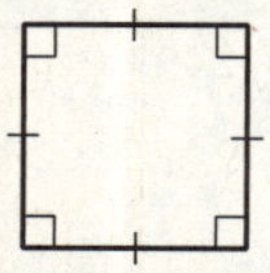	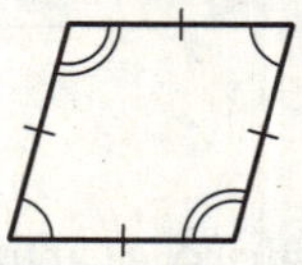	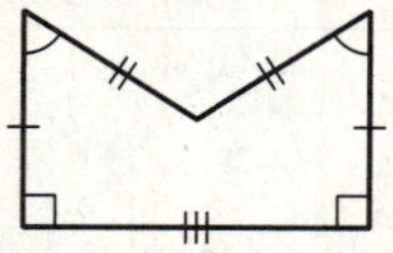

Tell whether each figure is a polygon. If it is a polygon, name it by the number of sides.

1. 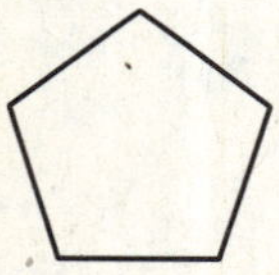

____

2. 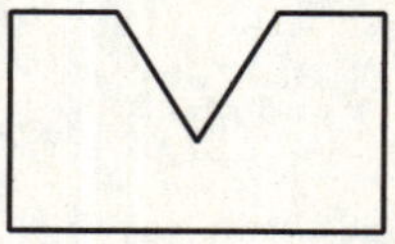

____

3. 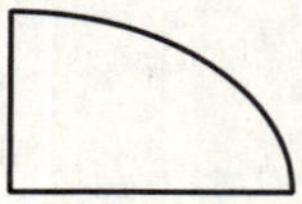

____

Tell whether each polygon is regular or irregular. Then tell whether it is concave or convex.

4. 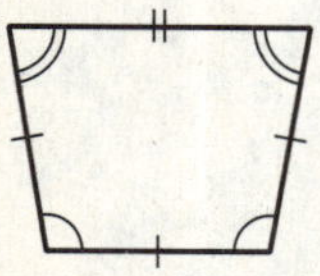

____

5. 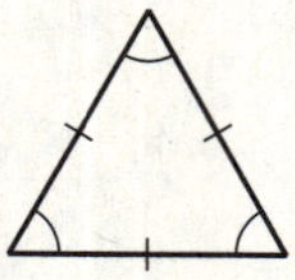

____

6. 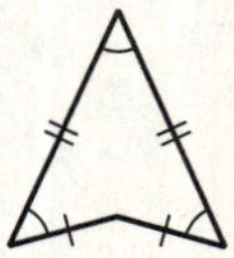

____

Name ______________________ Date ____________ Class ____________

LESSON 7E

Review for Mastery

Classifying Polygons, continued

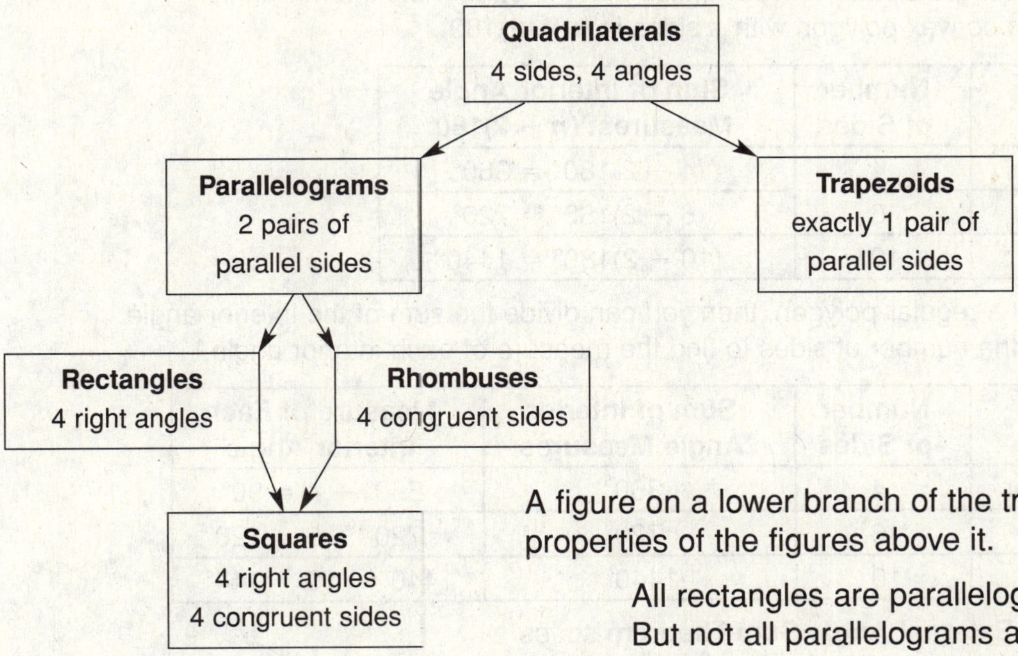

A figure on a lower branch of the tree has the properties of the figures above it.

All rectangles are parallelograms.
But not all parallelograms are rectangles.

Write all the names that apply to each figure.

7. 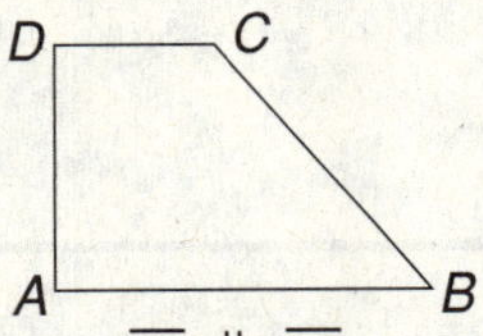

$\overline{AB} \parallel \overline{DC}$

four-sided polygon ____________

1 pair of parallel sides ____________

8. 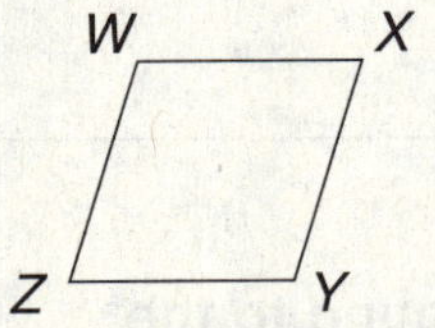

$\overline{WX} \parallel \overline{ZY}$, $\overline{WZ} \parallel \overline{XY}$
$\overline{WX} \cong \overline{XY} \cong \overline{ZY} \cong \overline{WZ}$

four-sided polygon ____________

2 pairs of parallel sides ____________

4 congruent sides ____________

9. 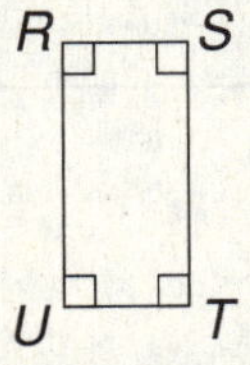

$\overline{RS} \parallel \overline{UT}$, $\overline{RU} \parallel \overline{ST}$

four-sided polygon ____________

2 pairs of parallel sides ____________

4 right angles ____________

Name ______________________ Date __________ Class __________

LESSON 7E

Review for Mastery

Classifying Polygons, continued

The **Polygon Angle Sum Theorem** states that the sum of the interior angle measures of a convex polygon with n sides is $(n - 2)180°$.

Convex Polygon	Number of Sides	Sum of Interior Angle Measures: $(n - 2)180°$
quadrilateral	4	$(4 - 2)180° = 360°$
hexagon	6	$(6 - 2)180° = 720°$
decagon	10	$(10 - 2)180° = 1440°$

If a polygon is a regular polygon, then you can divide the sum of the interior angle measures by the number of sides to find the measure of each interior angle.

Regular Polygon	Number of Sides	Sum of Interior Angle Measures	Measure of Each Interior Angle
quadrilateral	4	360°	$360° \div 4 = 90°$
hexagon	6	720°	$720° \div 6 = 120°$
decagon	10	1440°	$1440° \div 10 = 144°$

The **Polygon External Angle Sum Theorem** states that the sum of the exterior angle measures, one angle at each vertex, of a convex polygon is 360°.

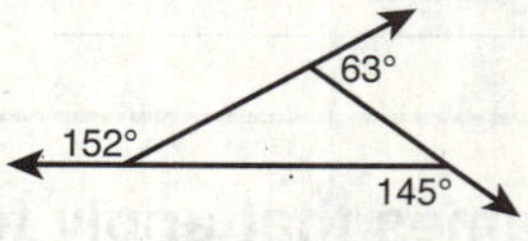

$152° + 63° + 145° = 360°$

The measure of each exterior angle of a regular polygon with n exterior angles is $360° \div n$. So the measure of each exterior angle of a regular decagon is $360° \div 10 = 36°$.

Find the sum of the interior angle measures of each convex polygon.

10. pentagon ____________

11. octagon ____________

12. nonagon ____________

Find the measure of each interior angle of each regular polygon. Round to the nearest tenth if necessary.

13. pentagon ____________

14. heptagon ____________

15. 15-gon ____________

Find the measure of each exterior angle of each regular polygon.

16. quadrilateral ____________

17. octagon ____________

Name ______________________ Date ____________ Class ____________

LESSON 7E

Homework and Practice

Classifying Polygons

Match each vocabulary term on the left with a part of polygon *ABCDE* on the right.

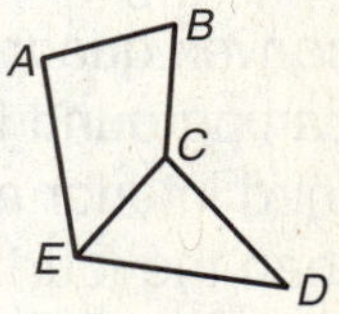

1. a diagonal ________ **A.** point *D*
2. a side of the polygon ________ **B.** $\overline{CE}$
3. a vertex of the polygon ________ **C.** $\overline{CD}$

A polygon is a closed flat figure made of straight segments that do not cross each other. Tell whether each figure is a polygon. If it is a polygon, name it by the number of its sides.

4. 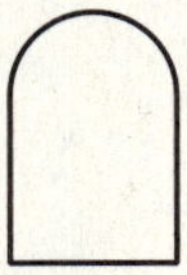 **5.**  **6.** 

______________________ ______________________ ______________________

A regular polygon has all sides congruent and all angles congruent. Tell whether each polygon is regular or irregular. A concave polygon has a pair of sides that make a "cave" in the polygon. Tell whether each polygon is concave or convex.

7. 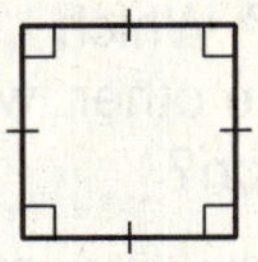 **8.** 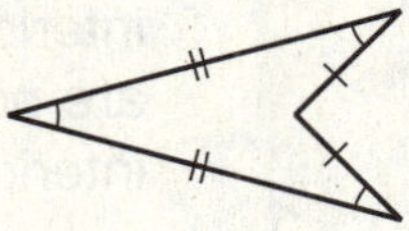 **9.** 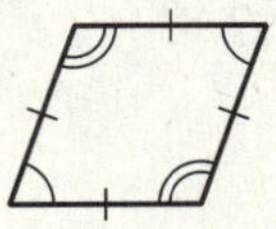

______________________ ______________________ ______________________

Honeybees store their honey in honeycombs. The honeycomb is made of many small wax compartments that are perfect regular hexagons.

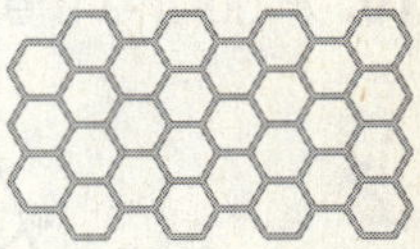

10. Use the Polygon Angle Sum Theorem to find the sum of the interior angle measures of a regular hexagon. ________

11. Find the measure of one interior angle of a regular hexagon. (*Hint:* Divide the answer to Exercise 10 by the number of sides.) ________

12. Use the Polygon Exterior Angle Sum Theorem to find the sum of the exterior angle measures, one exterior angle at each vertex, of a regular hexagon. ________

13. Find the measure of one exterior angle of a regular hexagon. (*Hint:* Divide the answer to Exercise 12 by the number of sides.) ________

Name ______________________________ Date ____________ Class ____________

LESSON 7E

Homework and Practice

Classifying Polygons, continued

14. A campground site is in the shape of a convex quadrilateral. Three sides of the campground form two right angles. The third interior angle measures 10° less than the fourth angle. Find the measure of each interior angle.

____

15. A pentagon has two exterior angles that measure $(3x)°$, two exterior angles that measure $(2x + 22)°$, and an exterior angle that measures $(x + 41)°$. If all of these angles have different vertices, what are the measures of the exterior angles of the pentagon?

____

16. The top view of a hexagonal greenhouse is shown at the right. What is the measure of $\angle PQR$, the acute angle formed by the house and the greenhouse?

____

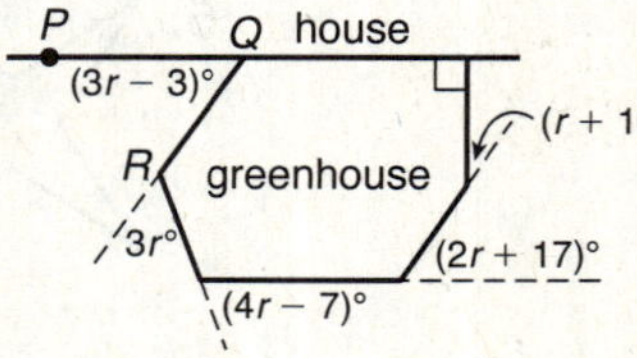

Choose the best answer.

17. A figure is an equiangular 18-gon. What is the measure of each exterior angle of the polygon?

A 10°

B 18°

C 20°

D 36°

18. Three interior angles of a convex heptagon measure 125°, and two of the interior angles measure 143°. Which are possible measures for the other two interior angles of the heptagon?

F 48° and 48°

G 39° and 100°

H 100° and 116°

J 89° and 150°

19. Find the measure of $\angle RKL$.

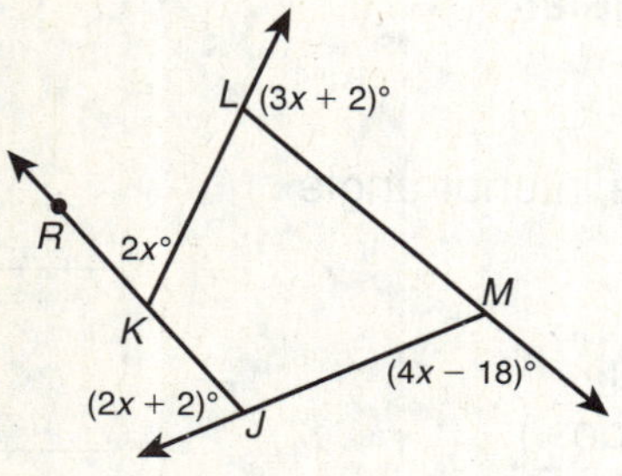

A 34°

B 68°

C 86°

D 148°

20. What is the measure of $\angle GCD$?

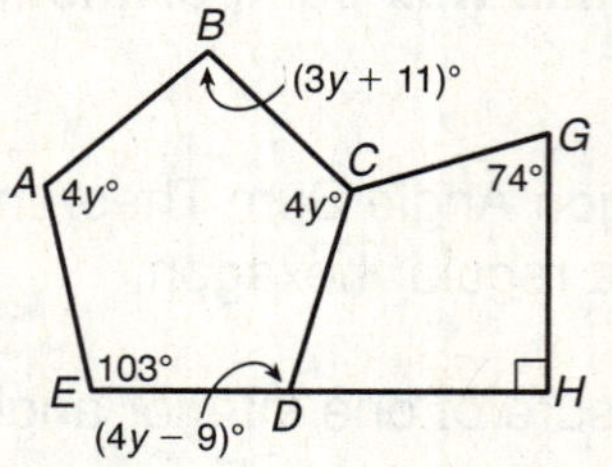

F 123°

G 116°

H 73°

J 29°

EXPLORATION

Program of Studies and Core Content
MA-8-G-S-CG2,
MA-08-3.3.1

7F Coordinate Geometry

Quadrilaterals are figures with four sides. Use the coordinate grid to draw each of the following quadrilaterals. Label vertices with *A, B, C,* and *D*.

	Quadrilateral	Graph
1.	Square *A* (3, 2), *B* (3, 8), *C* (−3, 8), *D* (−3, 2)	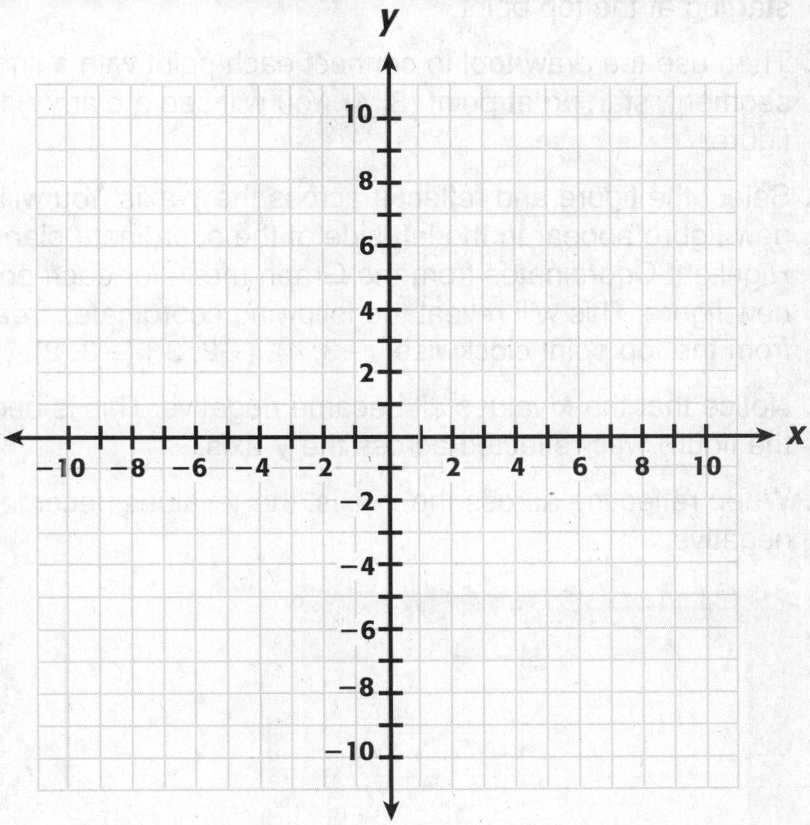
2.	Rectangle *A* (5, −1), *B* (5, −7), *C* (−5, −7), *D* (−5, −1)	

Think and Discuss

3. Describe the relationship between the number of sides and the number of vertices of quadrilaterals.

4. Discuss whether the relationship you described in Problem **3** is also true for other polygons.

Name ______________________ Date __________ Class __________

LESSON 7F

Technology Lab
Coordinate Geometry

Using Geometer's Sketchpad you can automatically plot points on a coordinate plane. These points will help you construct, manipulate and analyze geometric figures.

Activity

1. In Geometer's Sketchpad use the Graph menu to Show Grid. Then plot the following points using the Plot Points function: (3, 4), (4, 3), (3, 2), (2, 3). Label the points *A, B, C,* and *D* starting at the top point.
2. Then use the draw tool to connect each point with a line segment, starting at point (3, 4). You will see a diamond-shaped figure.
3. Select the figure and reflect it across the *y*-axis. You will see a new figure appear in the left side of the coordinate plane. Highlight Coordinates from the Graph menu for each point in the new figure. This will reveal the following coordinates, reading from the top point clockwise: (−3, 4), (−2, 3), (−3, 2), (−4, 3).
4. Notice that the *x*-values all became negative. This is because the figure was reflected across the *y*-axis.
5. When reflecting across the *x*-axis, the *y*-values become negative.

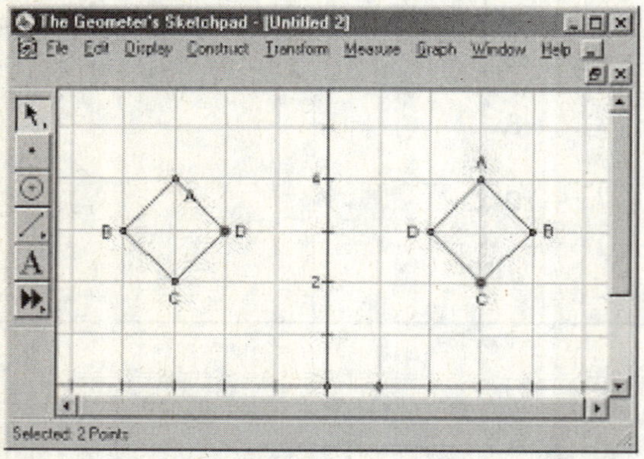

Name ______________________ Date ____________ Class ____________

LESSON 7F

Technology Lab

Coordinate Geometry, continued

Think and Discuss

1. What happens to the coordinates of the original figure if you plot points to reflect your second diamond across the *x*-axis into the lower left quadrant?

____

____

2. What happens if you plot points to reflect the original figure across line segment *AB*?

____

____

Name ______________________ Date ____________ Class ____________

LESSON 7F

Technology Lab

Coordinate Geometry, continued

Try This

1. Plot the following points on a coordinate grid in the upper right quadrant. Connect the points to form a square. Then reflect the figure across each axis. What are the coordinates of each reflection?

A(3, 4) *B*(6, 4) *C*(6, 1) *D*(3, 1)

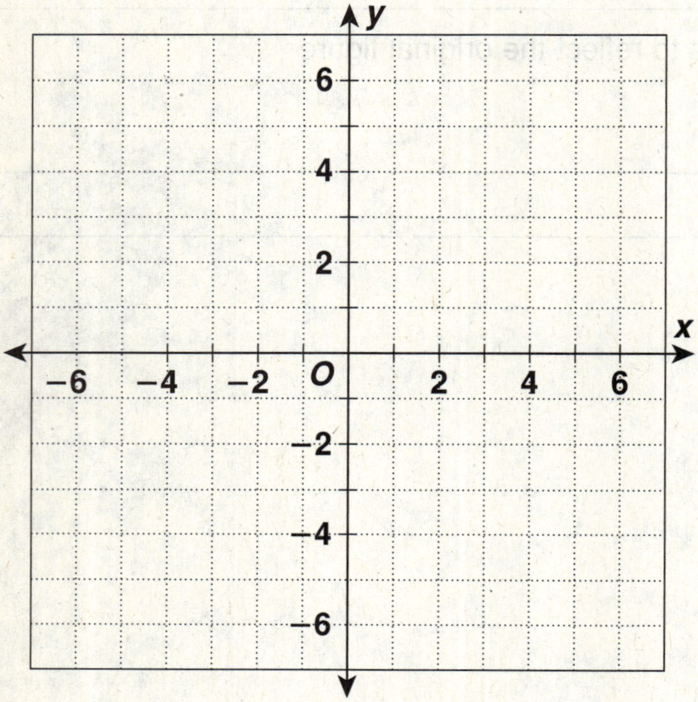

____

____

Name ______________________ Date ____________ Class ____________

LESSON 7F

Review for Mastery

Coordinate Geometry

The slope of a line is a number that describes the steepness of the line.

Possible Values for Slope

Slope is Positive	**Slope is Negative**	**Slope = 0**	**Slope is Undefined**
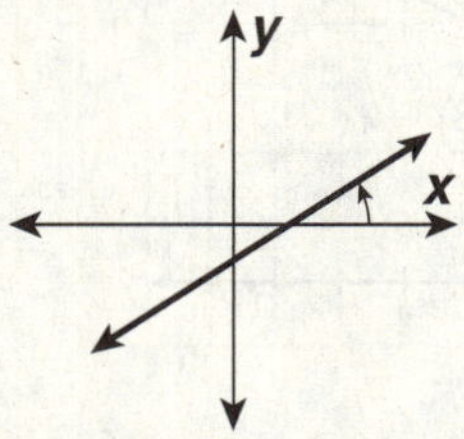	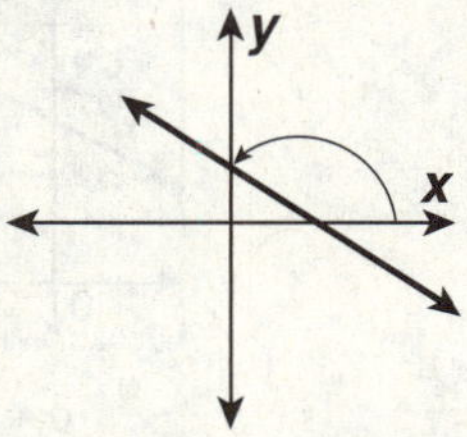	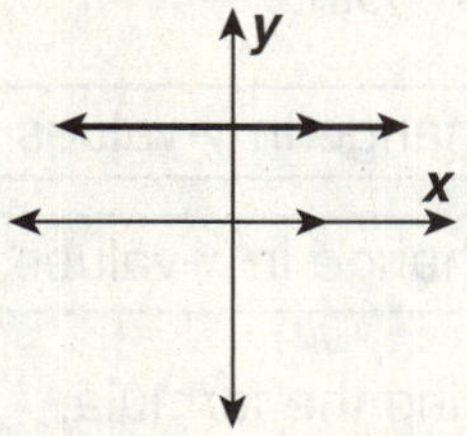	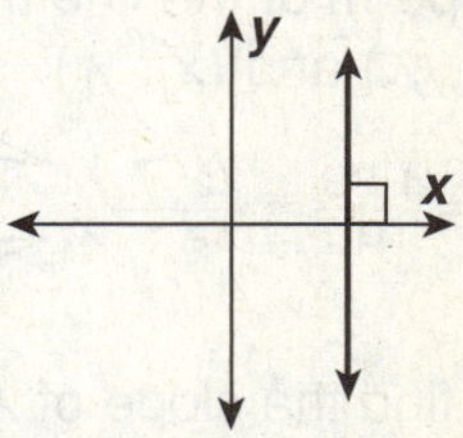
Line slants up. Forms acute angle with the positive direction of *x*-axis.	*Line slants down.* Forms obtuse angle with positive direction of *x*-axis.	*Horizonal Line* Parallel to *x*-axis.	*Vertical Line* Perpendicular to *x*-axis

Plot the given points. Describe the slope of the line that joins them.

1. (−2, 2) and (2, 5)

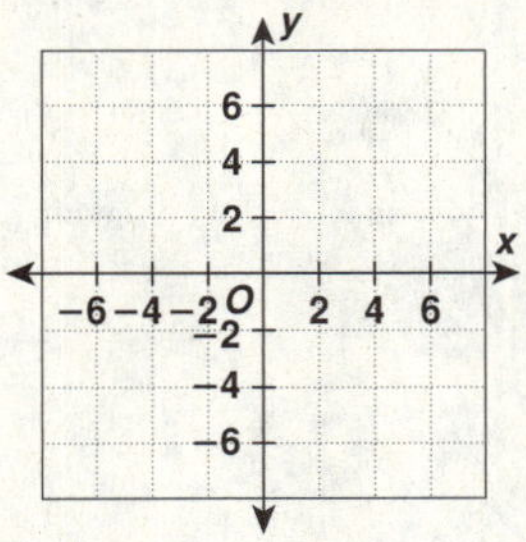

slope is: ______________

2. (−2, −5) and (−2, 2)

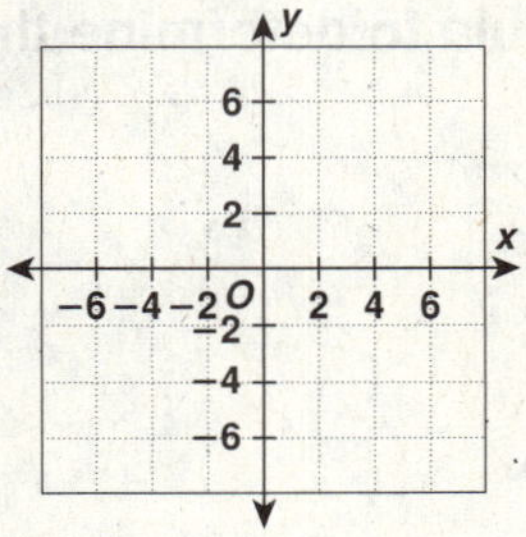

slope is: ______________

3. (1, 2) and (5, −2)

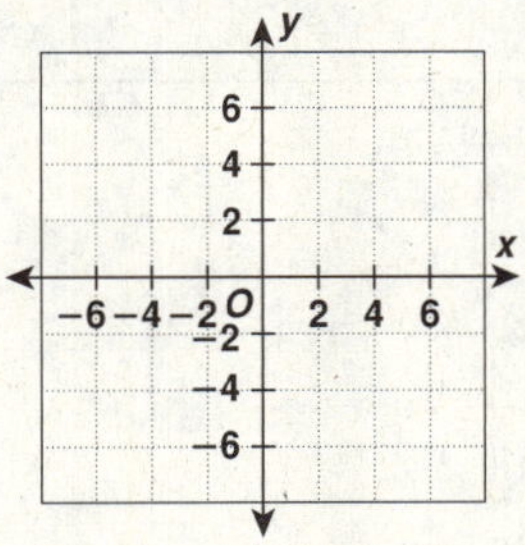

slope is: ______________

4. (−2, −2) and (4, −2)

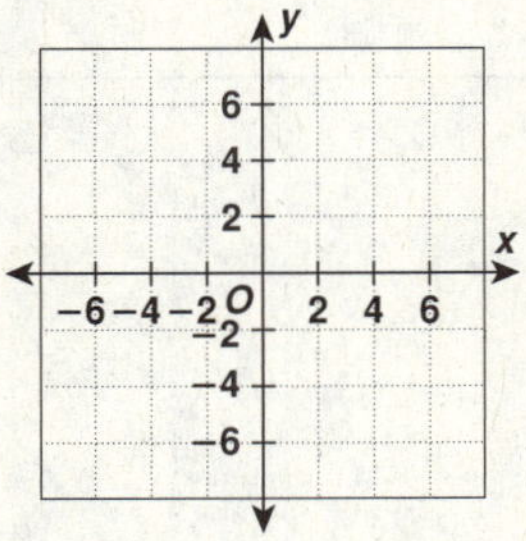

slope is: ______________

Name ______________________________ Date ____________ Class ____________

LESSON 7F

Review for Mastery

Coordinate Geometry, continued

The **slope** of a line describes how steep the line is. You can find the slope by writing the ratio of the **rise** to the **run**.

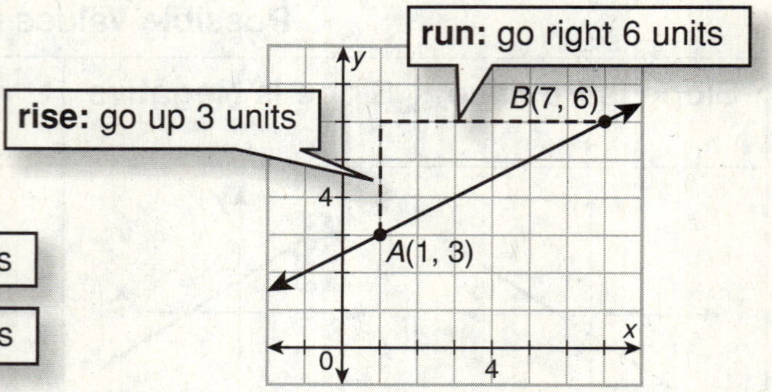

$$\text{slope} = \frac{\text{rise}}{\text{run}} = \frac{3}{6} = \frac{1}{2}$$

You can use a formula to calculate the slope *m* of the line through points (x_1, y_1) and (x_2, y_2).

$$m = \frac{\text{rise}}{\text{run}} = \frac{y_2 - y_1}{x_2 - x_1}$$

Change in *y*-values

Change in *x*-values

To find the slope of $\overleftrightarrow{AB}$ using the formula, substitute (1, 3) for (x_1, y_1) and (7, 6) for (x_2, y_2).

$m = \frac{y_2 - y_1}{x_2 - x_1}$	Slope formula
$= \frac{6 - 3}{7 - 1}$	Substitution
$= \frac{3}{6}$	Simplify.
$= \frac{1}{2}$	Simplify.

Use the slope formula to determine the slope of each line.

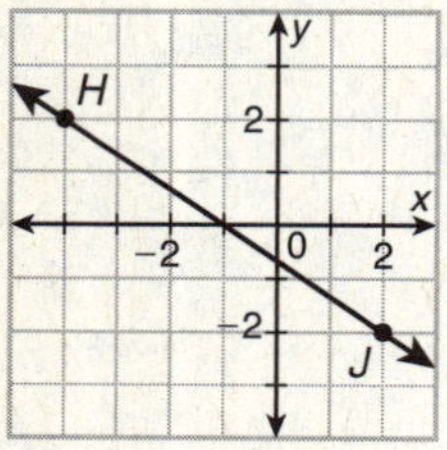

5. $\overleftrightarrow{HJ}$

____

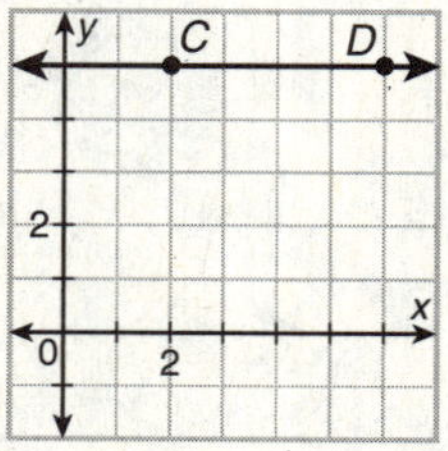

6. $\overleftrightarrow{CD}$

____

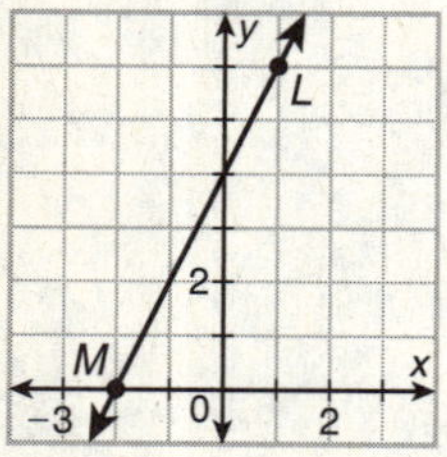

7. $\overleftrightarrow{LM}$

____

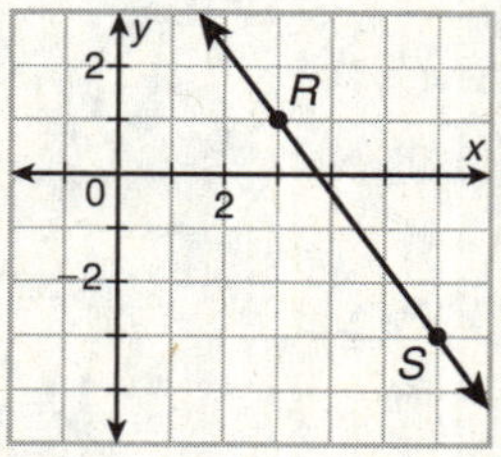

8. $\overleftrightarrow{RS}$

____

Name ______________________ Date __________ Class __________

LESSON **7F**

Review for Mastery

Coordinate Geometry, continued

Slopes of Parallel and Perpendicular Lines

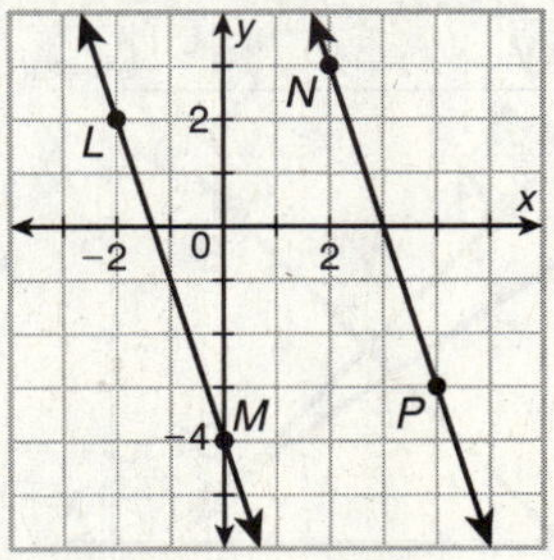

slope of $\overleftrightarrow{LM} = -3$

slope of $\overleftrightarrow{NP} = -3$

Parallel lines have the same slope.

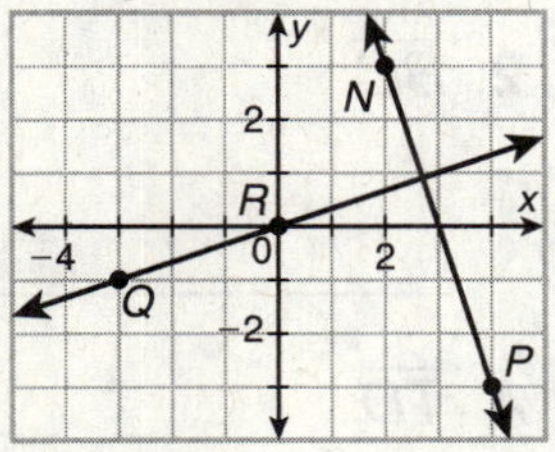

slope of $\overleftrightarrow{NP} = -3$

slope of $\overleftrightarrow{QR} = \frac{1}{3}$

product of slopes:

$-3\left(\frac{1}{3}\right) = -1$

Perpendicular lines have slopes that are *opposite reciprocals*. The product of the slopes is −1.

Use slopes to determine whether each pair of distinct lines is parallel, perpendicular, or neither.

9. slope of $\overleftrightarrow{PQ} = 5$

slope of $\overleftrightarrow{JK} = -\frac{1}{5}$

____

10. slope of $\overleftrightarrow{EF} = -\frac{3}{4}$

slope of $\overleftrightarrow{CD} = -\frac{3}{4}$

____

11. slope of $\overleftrightarrow{BC} = -\frac{5}{3}$

slope of $\overleftrightarrow{ST} = \frac{3}{5}$

____

12. slope of $\overleftrightarrow{WX} = \frac{1}{2}$

slope of $\overleftrightarrow{YZ} = -\frac{1}{2}$

____

Graph each pair of lines. Use slopes to determine whether the lines are parallel, perpendicular, or neither.

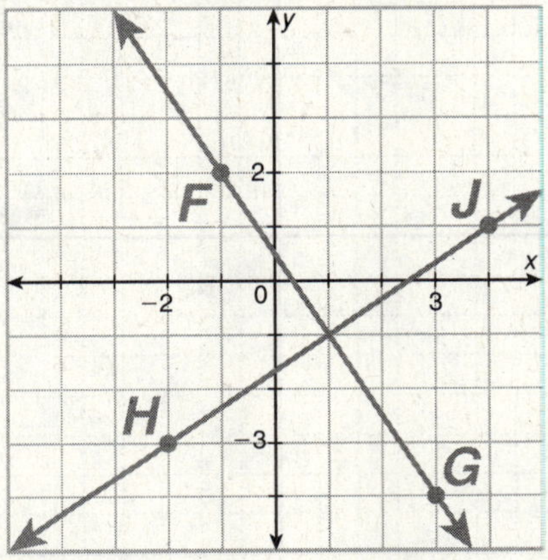

13. $\overleftrightarrow{FG}$ and $\overleftrightarrow{HJ}$ for $F(-1, 2)$, $G(3, -4)$, $H(-2, -3)$, and $J(4, 1)$

____

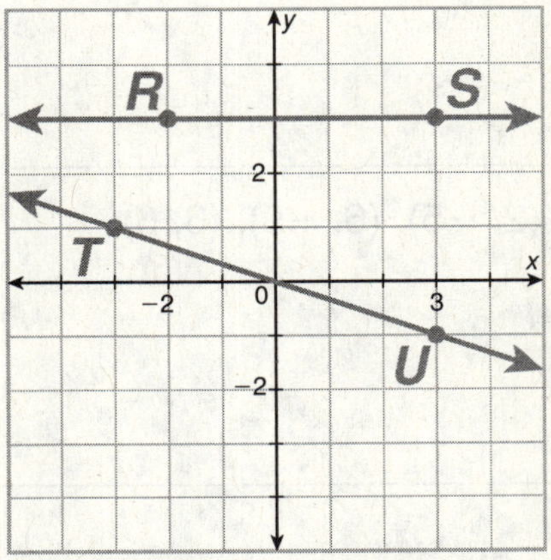

14. $\overleftrightarrow{RS}$ and $\overleftrightarrow{TU}$ for $R(-2, 3)$, $S(3, 3)$, $T(-3, 1)$, and $U(3, -1)$

____

Name ______________________ Date __________ Class __________

LESSON **7F**

Homework and Practice

Coordinate Geometry

Determine if the slope of each line is positive, negative, 0, or undefined. Then find the slope of each line.

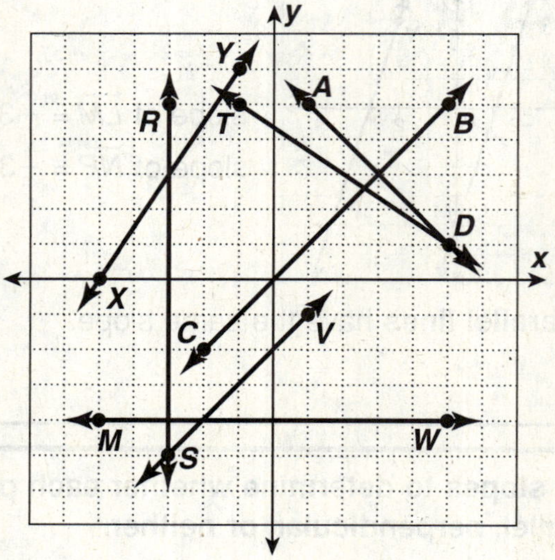

1. $\overleftrightarrow{AD}$ __________ 2. $\overleftrightarrow{BC}$ __________

3. $\overleftrightarrow{MW}$ __________ 4. $\overleftrightarrow{TD}$ __________

5. $\overleftrightarrow{SV}$ __________ 6. $\overleftrightarrow{RS}$ __________

7. Which lines are parallel?

____

8. Which lines are perpendicular?

____

____

Graph the quadrilaterals with the given vertices. Write all the names that apply to each quadrilateral.

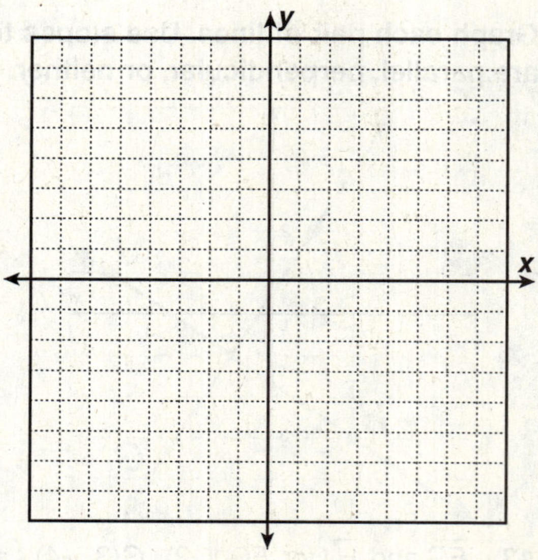

9. (−3, 1), (−6, 4), (4, 2), (6, −4)

____

10. (2, 6), (2, −5), (6, −5), (6, 6)

____

Name ______________________ Date __________ Class __________

LESSON 7F

Homework and Practice

Coordinate Geometry, continued

Tell whether each pair of lines is parallel, perpendicular, or neither.

11. $\overleftrightarrow{EF}$ with slope = 3 and $\overleftrightarrow{GH}$ with slope = −1 ____________

12. $\overleftrightarrow{PQ}$ with slope = $\frac{2}{3}$ and $\overleftrightarrow{RS}$ with slope = $-\frac{3}{2}$ ____________

Graph each pair of lines. Use slopes to determine whether the lines are parallel, perpendicular, or neither.

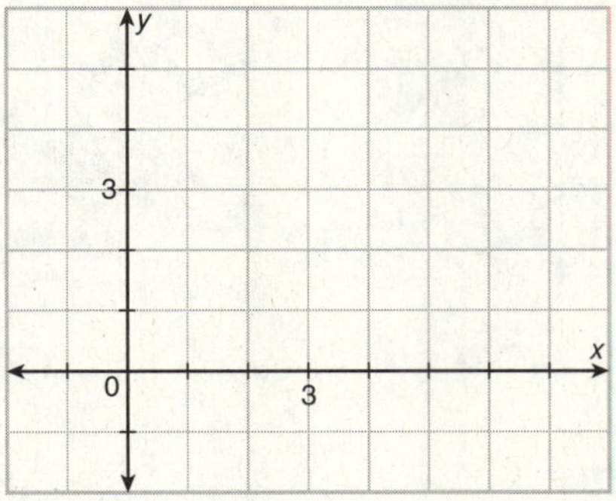

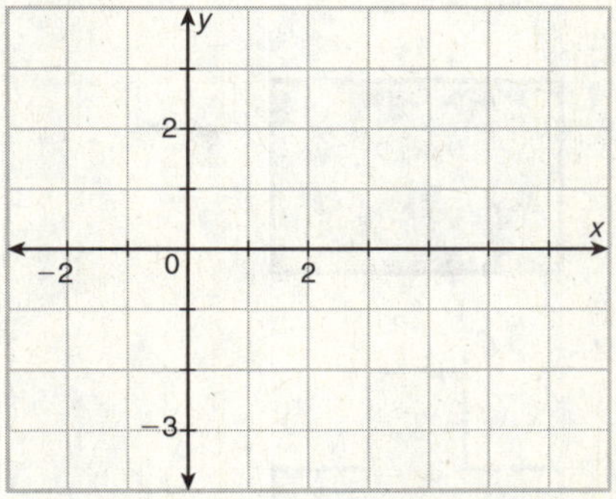

13. $\overleftrightarrow{IJ}$ and $\overleftrightarrow{KL}$ for $I(1, 0)$, $J(5, 3)$, $K(6, -1)$, and $L(0, 2)$ ____________

14. $\overleftrightarrow{PQ}$ and $\overleftrightarrow{RS}$ for $P(5, 1)$, $Q(-1, -1)$, $R(2, 1)$, and $S(3, -2)$ ____________

15. At a ski resort, the different ski runs down the mountain are color-coded according to difficulty. Green is easy, blue is medium, and black is hard. Assume that the ski runs below are rated only according to their slope (steeper is harder) and that there is one green, one blue, and one black run. Assign a color to each ski run.

Emerald $\left(m = \frac{4}{7}\right)$ ____________

Diamond $\left(m = \frac{5}{4}\right)$ ____________

Ruby $\left(m = \frac{5}{8}\right)$ ____________

16. Mohammed earns money tutoring math students on Saturdays. He starts at noon. By 2:00 P.M., he has earned $16. At 5:00 P.M., Mohammed has earned $40. Graph a line to show Mohammed's earnings and find the slope of the line to tell how many dollars Mohammed earns per hour.

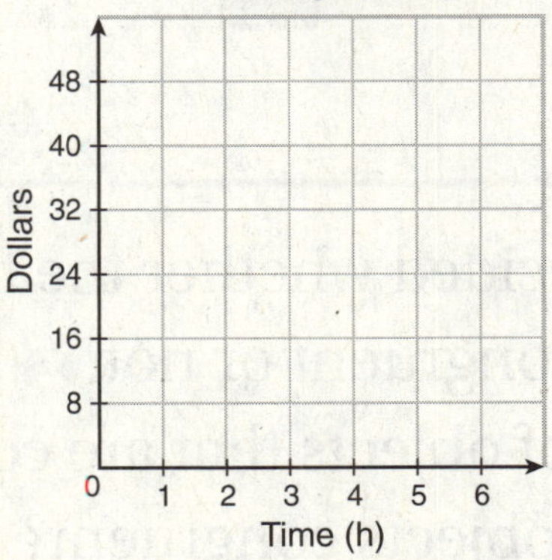

____

EXPLORATION

7G Congruence

Program of Studies and Core Content

MA-8-G-S-SR4, MA-08-3.1.2, MA-08-3.1.4

Congruent figures have the same size and shape.

Determine whether the figures in each pair are congruent.

1.

2.

3.

4.

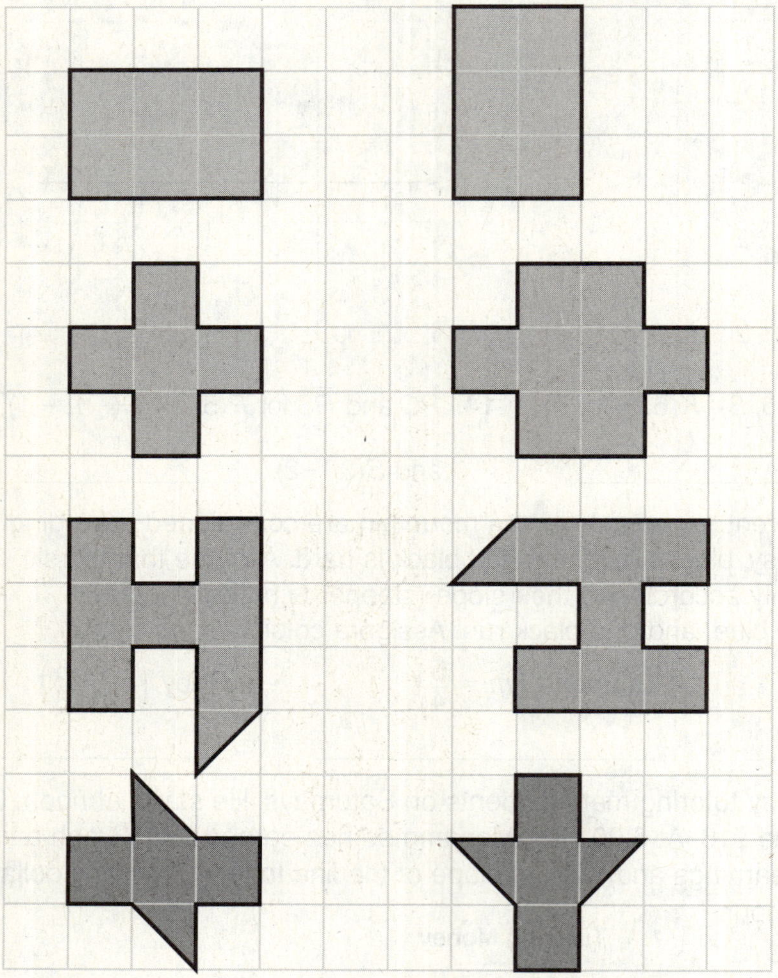

Think and Discuss

5. **Explain** how you decided whether the figures in Problems **1–4** were congruent or not.

6. **Find** two examples of objects that are congruent. What makes the two objects congruent?

Name ______________________ Date __________ Class __________

LESSON **7G**

Hands-On Lab
Congruence

Activity

When two polygons are congruent, their corresponding sides and angles are congruent.

In the space below, construct a polygon congruent to this polygon but in a different position on the page. Label the angles with different letters, and write in the degrees of each angle:

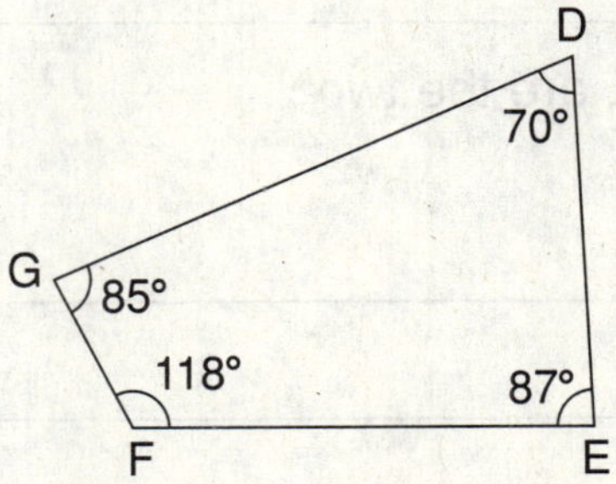

Write a congruence statement for the pair of polygons. In a congruence statement, the vertices in the second polygon are written in the order of correspondence. For example, you might write $\angle A \cong \angle G$, $\angle B \cong \angle D$, and so on.

____

____

____

____

____

Name ______________________ Date __________ Class __________

LESSON 7G

Hands-On Lab

Congruence, continued

Think and Discuss

1. How can you tell if two polygons are congruent?

____

____

____

2. If one polygon is a dilation of another, are the two polygons congruent?

____

____

Try This

In the space below, draw two congruent polygons in different positions. Write in the lengths of the sides and measurements of the angles. Write a congruence statement to prove the polygons are congruent.

____

____

____

____

Name ______________________ Date __________ Class __________

LESSON 7G

Review for Mastery

Congruence

Congruent polygons have the same size and shape.

Corresponding angles are congruent.

$\angle J \cong \angle J'$ $\angle K \cong \angle K'$ $\angle L \cong \angle L'$

(Read *J'* as *J prime.*)

Corresponding sides are congruent.

$\overline{JK} \cong \overline{J'K'}$ $\overline{KL} \cong \overline{K'L'}$ $\overline{LJ} \cong \overline{L'J'}$

In a congruence statement, the vertices of the second polygon are written in order of correspondence with the first polygon.

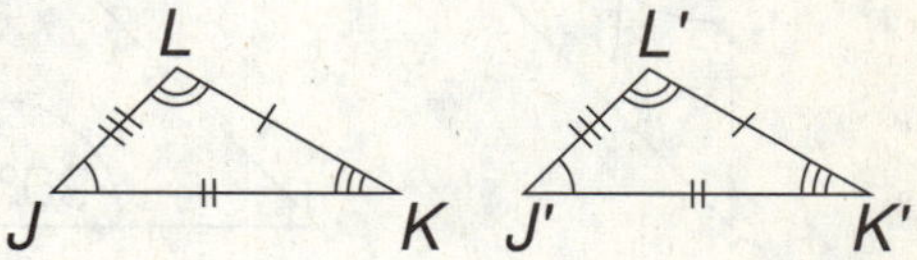

$\triangle JKL \cong \triangle J'K'L'$

Use the markings in each diagram.
Complete to write each congruence statement.

1.

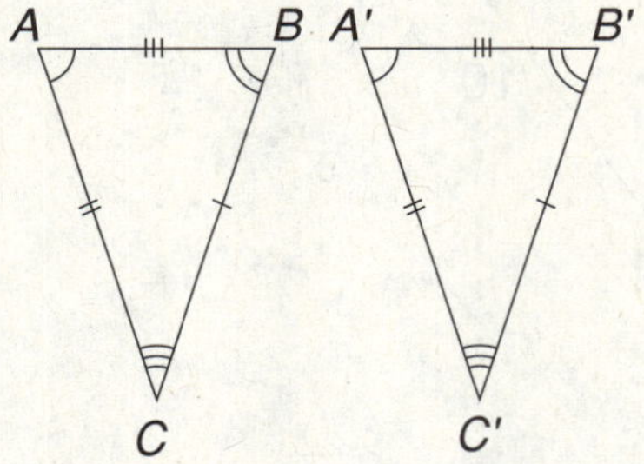

$\angle A \cong$ ______, $\angle B \cong$ ______,

$\angle C \cong$ ______, $\overline{AB} \cong$ ______,

$\overline{BC} \cong$ ______, $\overline{AC} \cong$ ______,

$\triangle ABC \cong$ ____________

2.

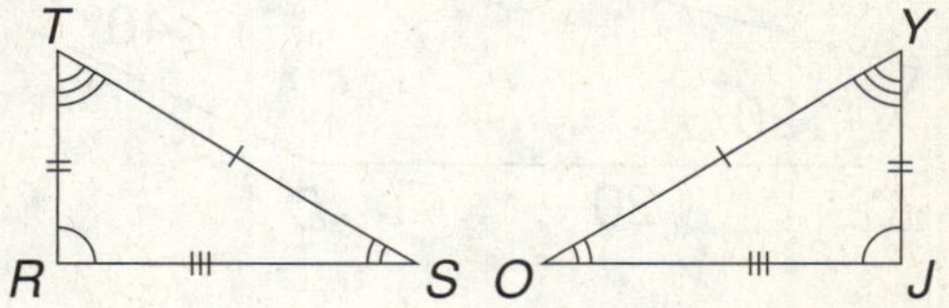

$\angle R \cong$ ______, $\angle S \cong$ ______,

$\angle T \cong$ ______, $\overline{RS} \cong$ ______,

$\overline{RT} \cong$ ______, $\overline{TS} \cong$ ______,

$\triangle RST \cong$ ____________

3.

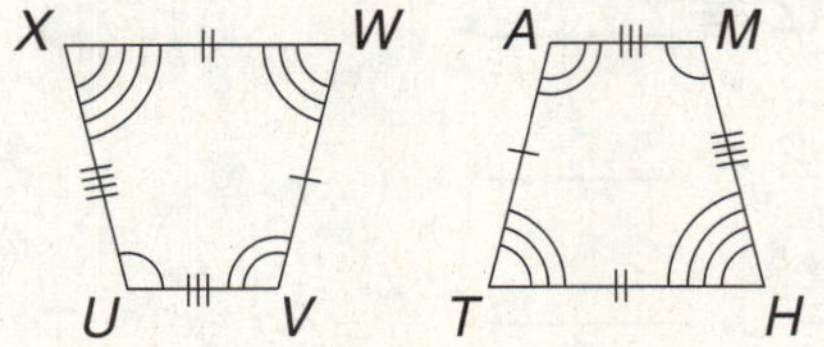

$\angle U \cong$ ______, $\angle V \cong$ ______, $\angle W \cong$ ______, $\angle X \cong$ ______,

$\overline{UV} \cong$ ______, $\overline{VW} \cong$ ______, $\overline{WX} \cong$ ______, $\overline{XU} \cong$ ______

quad. $UVWX \cong$ ______________

Name ______________________ Date ____________ Class ____________

LESSON **7G**

Review for Mastery
Congruence, continued

Congruence relationships can be used to find unknown values.

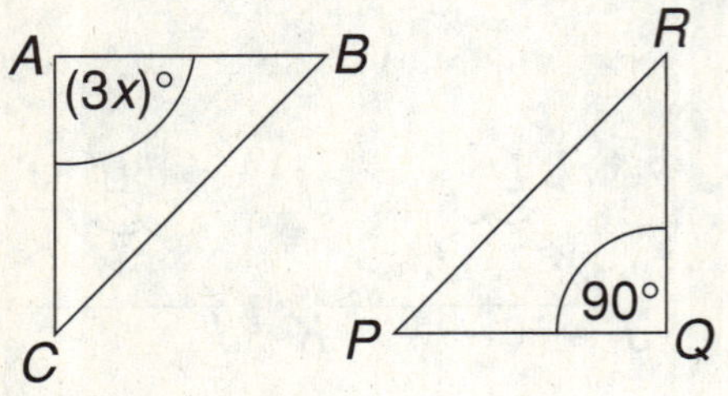

$\angle A \cong \angle Q$

$3x = 90$

$\frac{3x}{3} = \frac{90}{3}$

$x = 30$

$\triangle ABC \cong \triangle QPR$

Using the congruence relationship, complete to find each unknown value.

CDEFG ≅ SMZAR

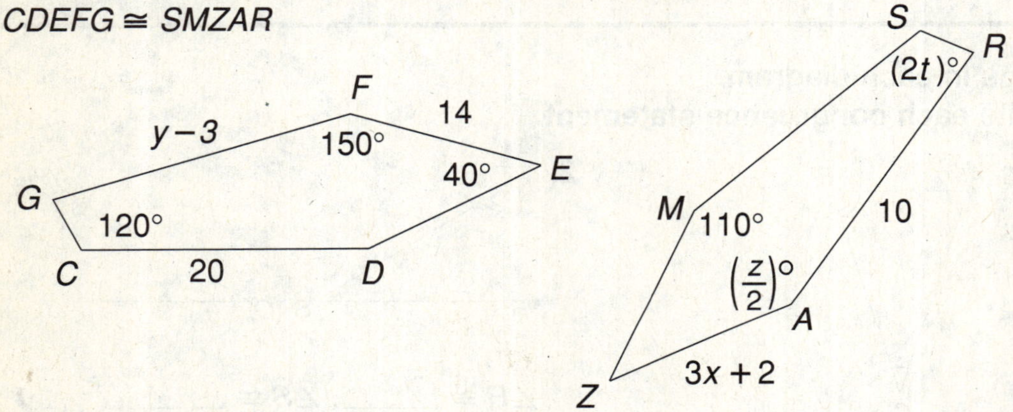

4. $\overline{GF} \cong \overline{RA}$

$y - 3 =$ ______

______ ______

$y =$ ______

5. $\angle C \cong$ ______

______ $= 2t$

______ $= \frac{2t}{}$

______ $= t$

6. $\angle A \cong$ ______

$\frac{z}{2} =$ ______

______ $\times \frac{z}{2} =$ ______ $\times$ ______

$z =$ ______

7. $\overline{AZ} \cong$ ______

$3x + 2 =$ ______

______ ______

$3x =$ ______

$\frac{3x}{} =$ ______

$x =$ ______

Name ______________________ Date __________ Class __________

LESSON 7G

Homework and Practice

Congruence

Write a congruence statement for each pair of congruent polygons.

1.

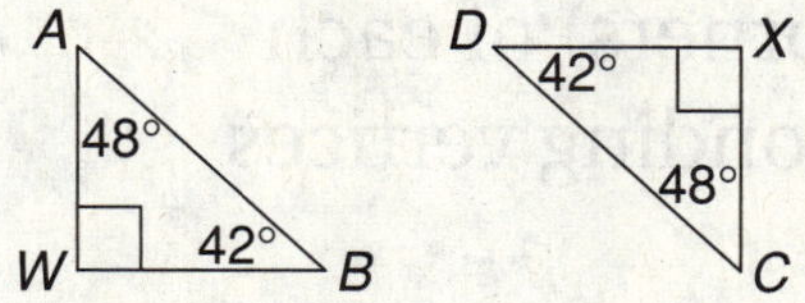

____

2.

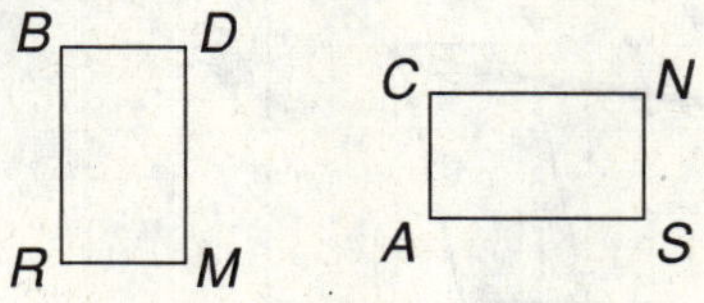

____

3.

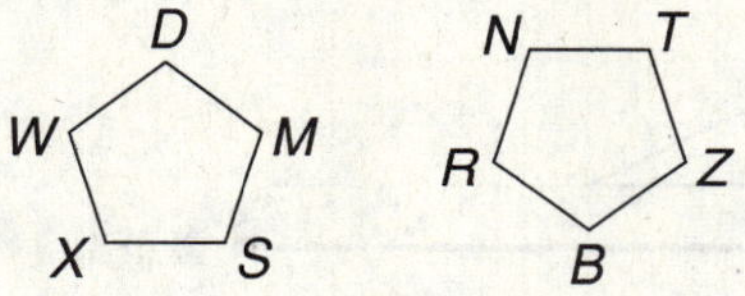

____

4.

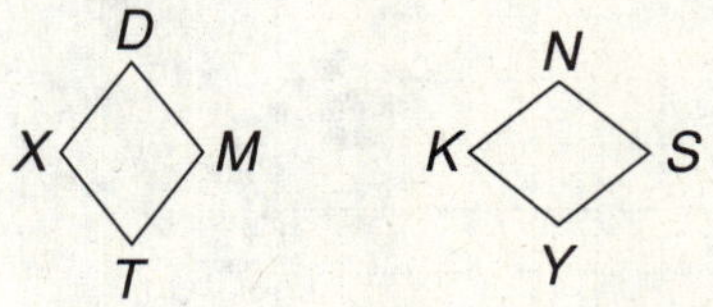

____

Find the value of the variable if triangle *ABC* is congruent to triangle *XYZ*.

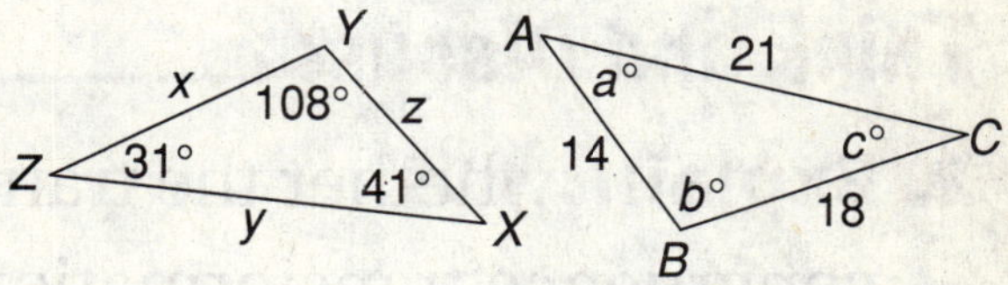

5. Find *a*. __________

6. Find *b*. __________

7. Find *c*. __________

8. Find *x*. __________

9. Find *y*. __________

10. Find *z*. __________

Program of Studies and Core Content

MA-8-G-S-TS2, MA-8-G-S-TS3, MA-8-G-S-TS4, MA-08-3.1.4, MA-08-3.2.2, MA-08-3.2.3

7H Transformations

1. Draw arrows from all the vertices (corners) of each original figure (white) to the corresponding vertices of its *image* (gray).

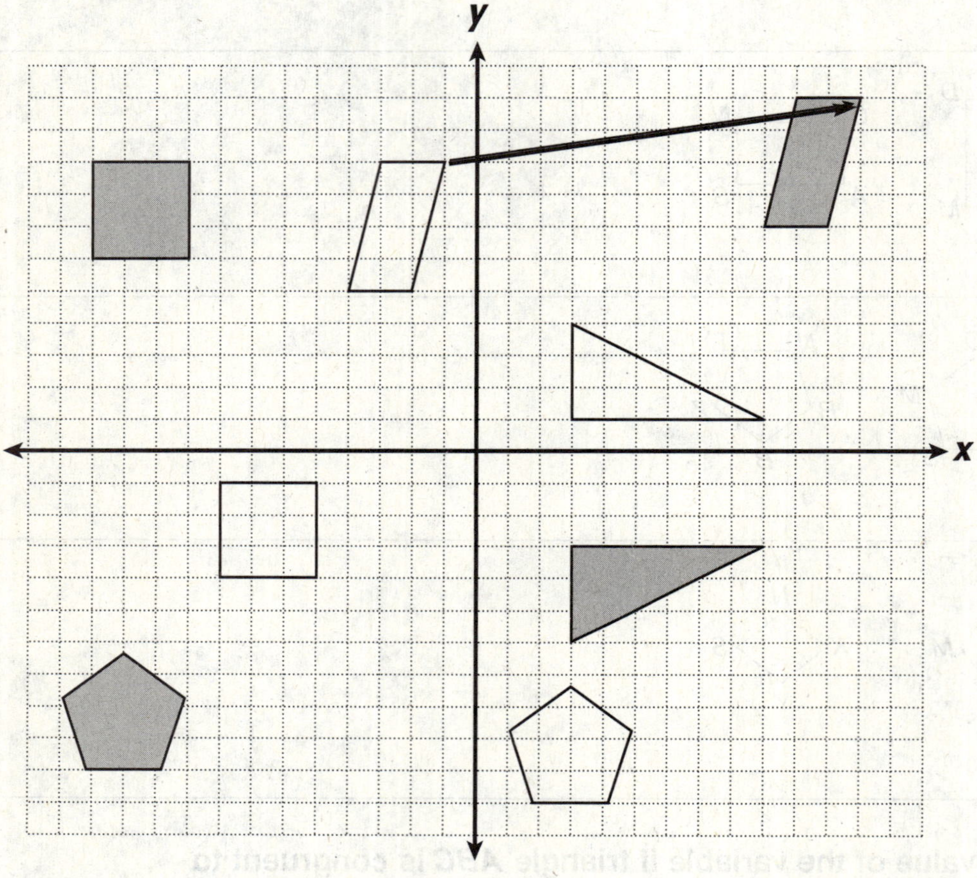

Think and Discuss

2. Explain whether the transformations in Problem **1** are congruence transformations, which means that the image is the same shape and size as the original figure.

3. Compare the transformation of the triangle with the transformations of the other figures. How is it different?

Name ______________________ Date __________ Class __________

LESSON 7H

Review for Mastery

Transformations

A *transformation* is a change in the position, size, or shape of a figure. The figure that results from a transformation is called the *image*.

A *translation* (or *slide*) is a transformation in which all the points of a figure move the same distance in the same direction.

Find the coordinates for the image of $\triangle ABC$ after the translation $(x, y) \rightarrow (x + 5, y - 5)$. Draw the image.

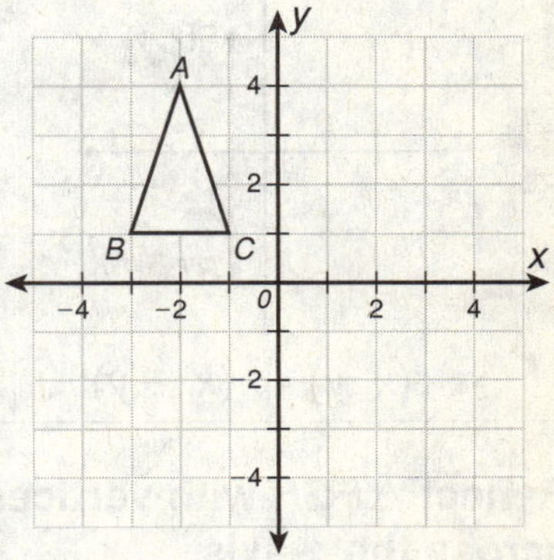

Step 1 Find the coordinates of $\triangle ABC$. The vertices of $\triangle ABC$ are $A(-2, 4)$, $B(-3, 1)$, and $C(-1, 1)$.

Step 2 The rule for a translation tells you how much to add or subtract from the *x*- and *y*-coordinates of a point.

$(x, y) \rightarrow (x + 5, y - 5)$ means add 5 to the *x*-coordinates and subtract 5 from the *y*-coordinates. Apply the rule to find the vertices of the image.

$A'(-2 + 5, 4 - 5) = A'(3, -1)$

$B'(-3 + 5, 1 - 5) = B'(2, -4)$

$C'(-1 + 5, 1 - 5) = C'(4, -4)$

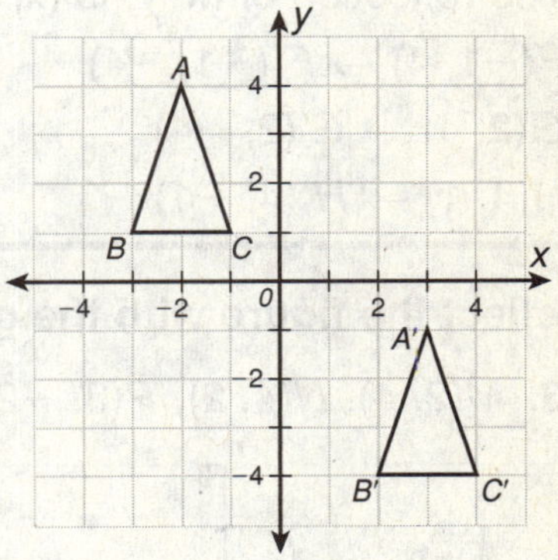

Step 3 Plot the points. Then finish drawing the image by using a straightedge to connect the vertices.

Find the coordinates for each image after the given translation.

1. $D(2, 4)$, $E(4, 1)$, $F(4, -2)$
$(x, y) \rightarrow (x - 2, y - 2)$

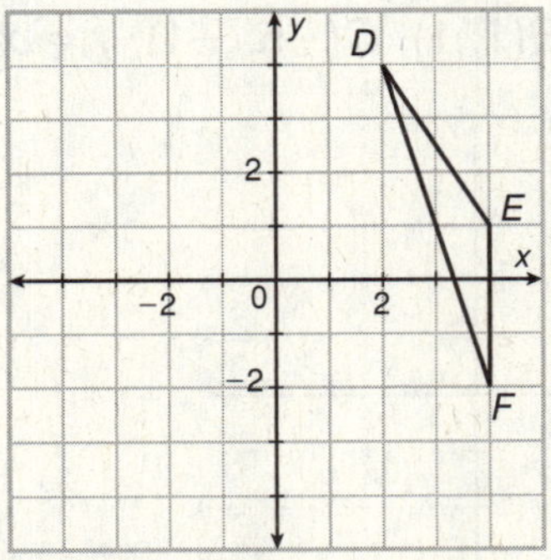

2. $J(-3, -1)$, $K(-2, -3)$, $L(1, -3)$, $M(2, -1)$
$(x, y) \rightarrow (x + 1, y + 4)$

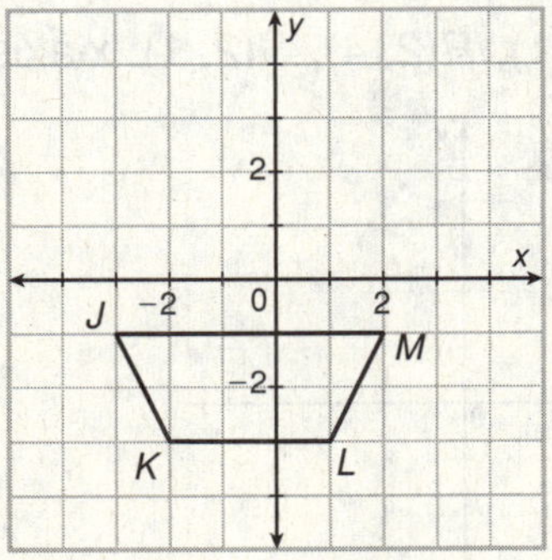

Name ______________________________ Date ____________ Class ____________

LESSON 7H

Review for Mastery

Transformations, continued

A *reflection* is a transformation that flips a figure over a line.

Reflections in the Coordinate Plane		
Across the *x*-axis	**Across the *y*-axis**	**Across the line *y* = *x***
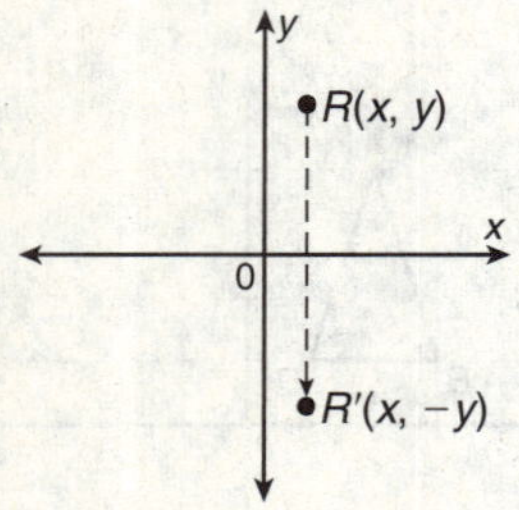	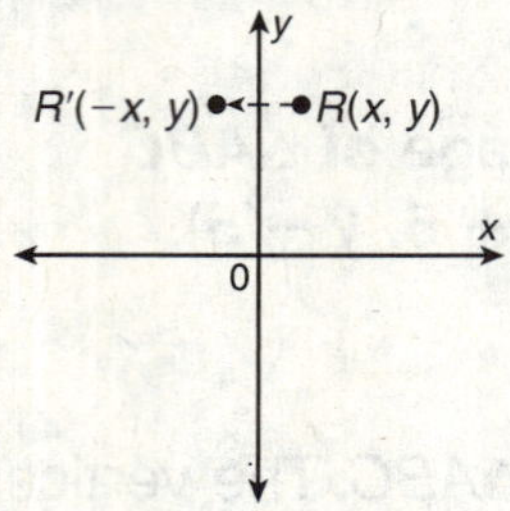	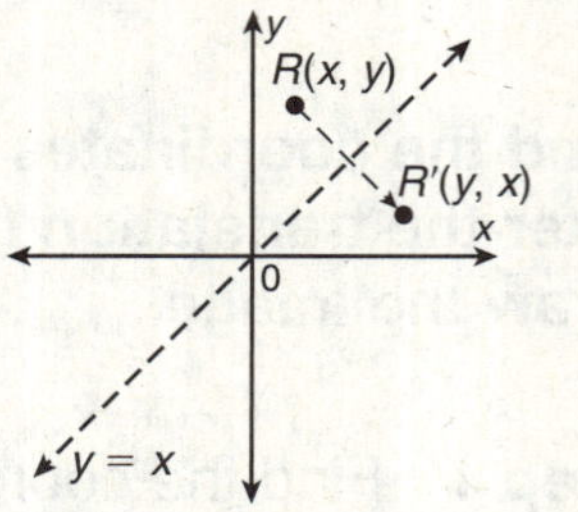
$(x, y) \rightarrow (x, -y)$	$(x, y) \rightarrow (-x, y)$	$(x, y) \rightarrow (y, x)$

Reflect △*FGH* with vertices *F*(−1, 4), *G*(2, 4), and *H*(4, 1) across the *x*-axis.

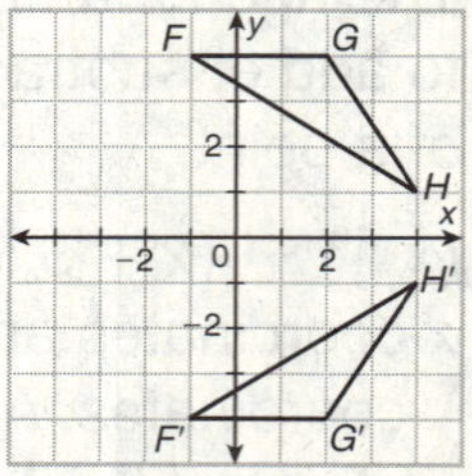

The reflection of (*x*, *y*) is (*x*, −*y*).

F(−1, 4) → *F*′(−1, −4)

G(2, 4) → *G*′(2, −4)

H(4, 1) → *H*′(4, −1)

Reflect the figure with the given vertices across the given line.

3. *M*(2, 4), *N*(4, 2), *P*(3, −2); *y*-axis

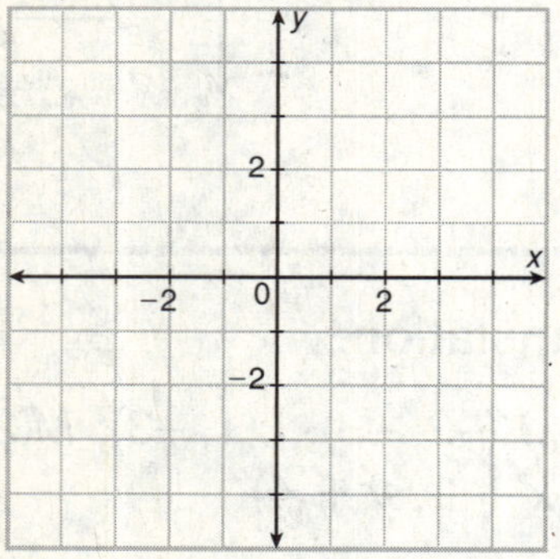

4. *T*(−4, 1), *U*(−3, 4), *V*(2, 3), *W*(0, 1); *x*-axis

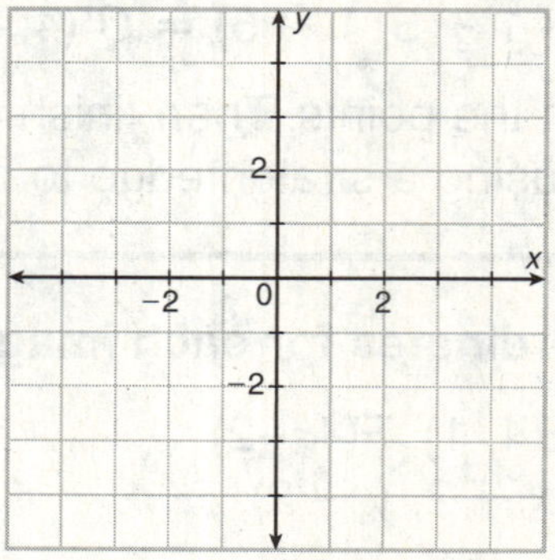

5. *Q*(−3, −1), *R*(2, 4), *S*(2, 1); *x*-axis

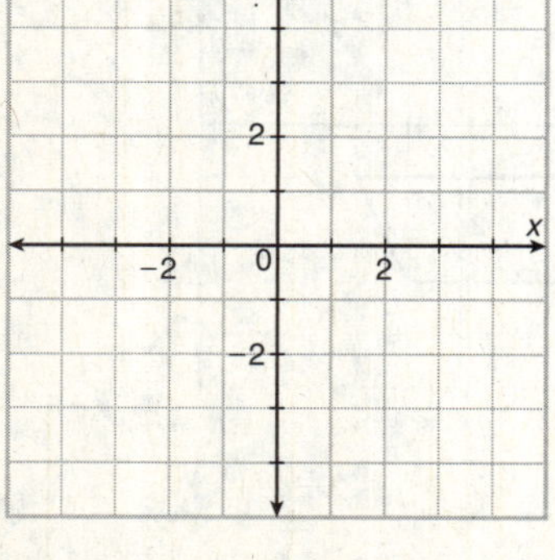

6. *A*(−2, 4), *B*(1, 1), *C*(−5, −1); *y* = *x*

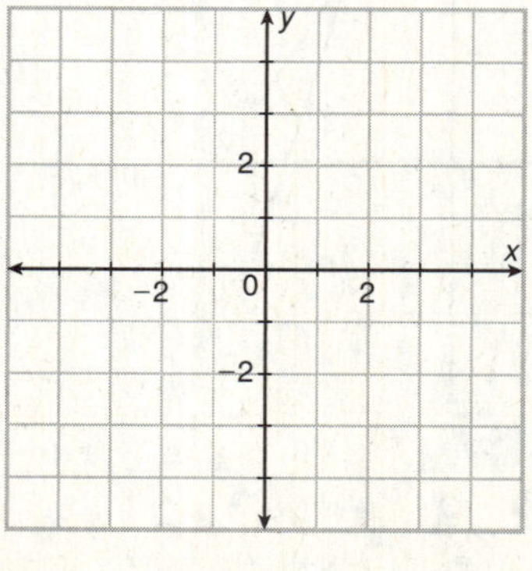

Name ______________________ Date ____________ Class ____________

LESSON 7H

Review for Mastery

Transformations, continued

A *rotation* is a transformation that turns a figure around a fixed point.

Rotations in the Coordinate Plane	
By 90° About the Origin	**By 180° About the Origin**
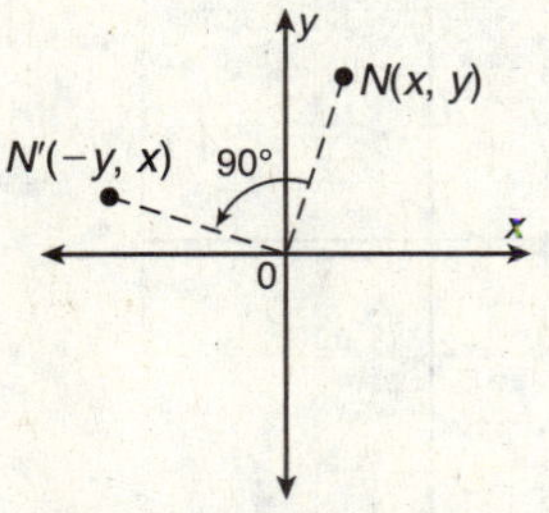 $(x, y) \rightarrow (-y, x)$	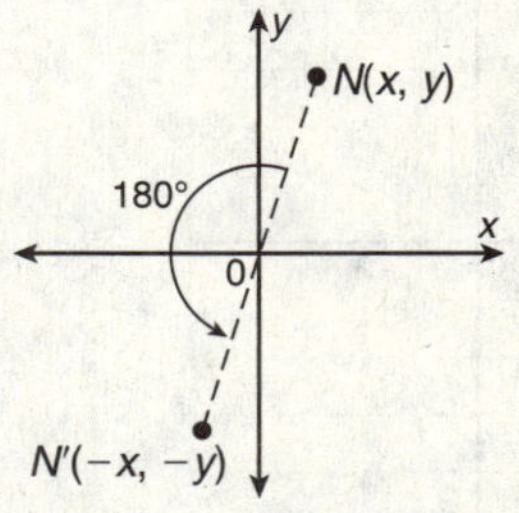 $(x, y) \rightarrow (-x, -y)$

Rotate △*MNP* with vertices *M*(1, 1), *N*(2, 4), and *P*(4, 3) by 180° about the origin.

The image of (x, y) is $(-x, -y)$.

$M(1, 1) \rightarrow M'(-1, -1)$

$N(2, 4) \rightarrow N'(-2, -4)$

$P(4, 3) \rightarrow P'(-4, -3)$

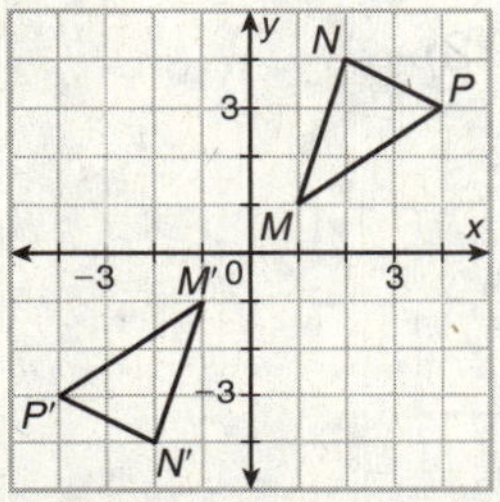

Rotate the figure with the given vertices about the origin using the given angle.

7. $R(0, 0)$, $S(3, 1)$, $T(2, 4)$; 90°

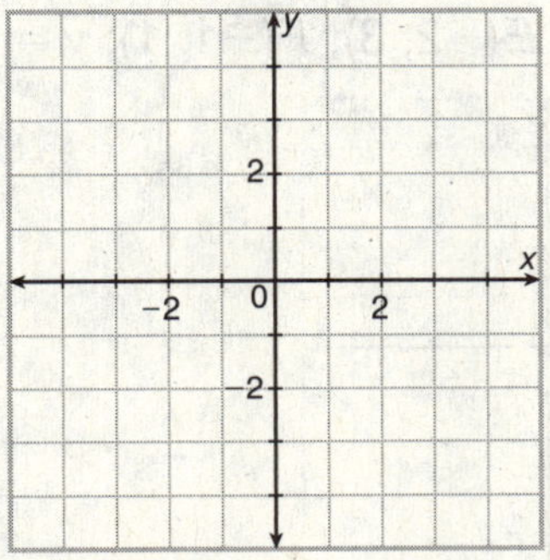

8. $A(0, 0)$, $B(-4, 2)$, $C(-1, 4)$; 180°

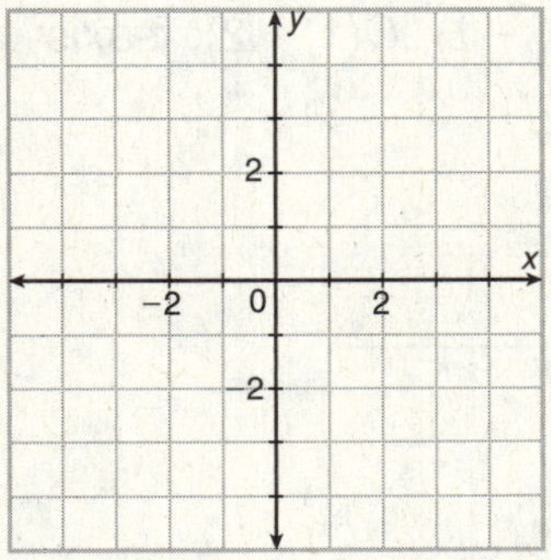

9. $E(0, 3)$, $F(3, 5)$, $G(4, 0)$; 180°

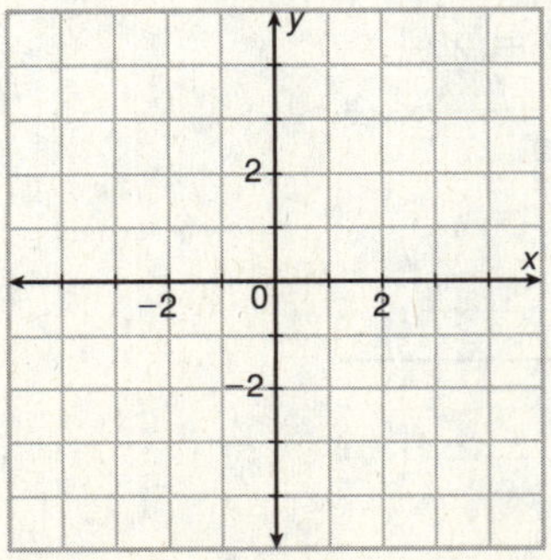

10. $U(1, -1)$, $V(4, -2)$, $W(3, -4)$; 90°

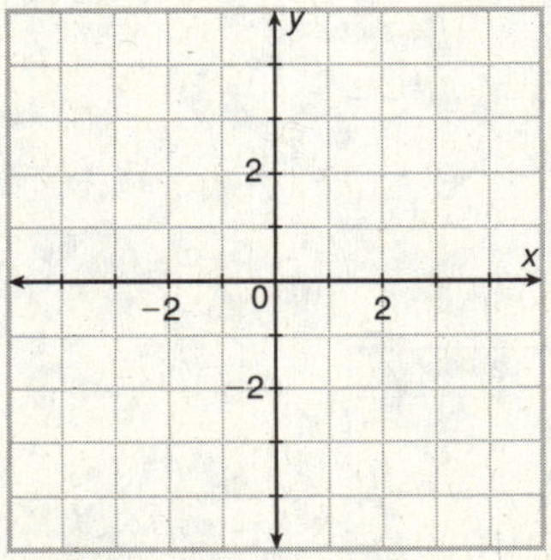

Name ______________________ Date ____________ Class ____________

LESSON **7H**

Homework and Practice

Transformations

Translate the figure with the given vertices using the given rule.

1. $A(-1, 3)$, $B(1, 1)$, $C(4, 4)$; $(x, y) \rightarrow (x, y - 5)$

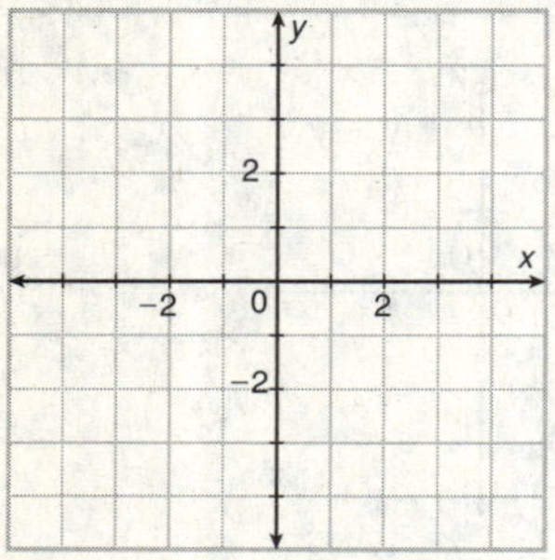

2. $P(-1, 2)$, $Q(0, 3)$, $R(1, 2)$, $S(0, 1)$; $(x, y) \rightarrow (x + 1, y)$

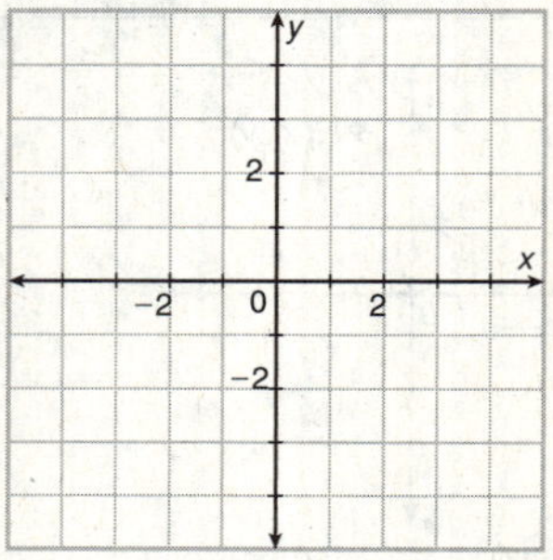

3. $L(3, 2)$, $M(1, -3)$, $N(-2, -2)$; $(x, y) \rightarrow (x - 2, y + 3)$

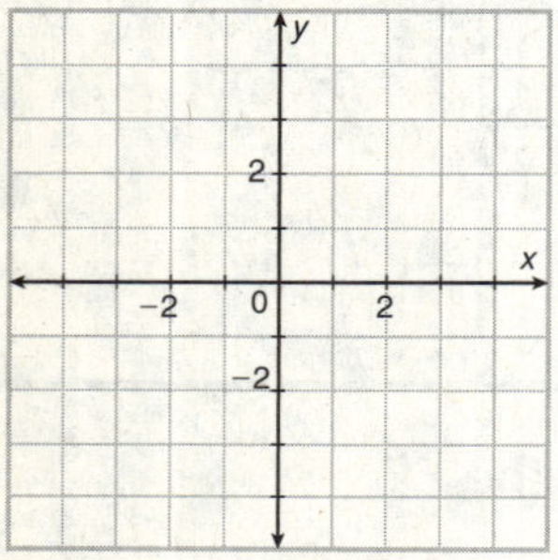

4. $D(2, -2)$, $E(2, -4)$, $F(1, -4)$, $G(-2, -2)$; $(x, y) \rightarrow (x + 2, y + 5)$

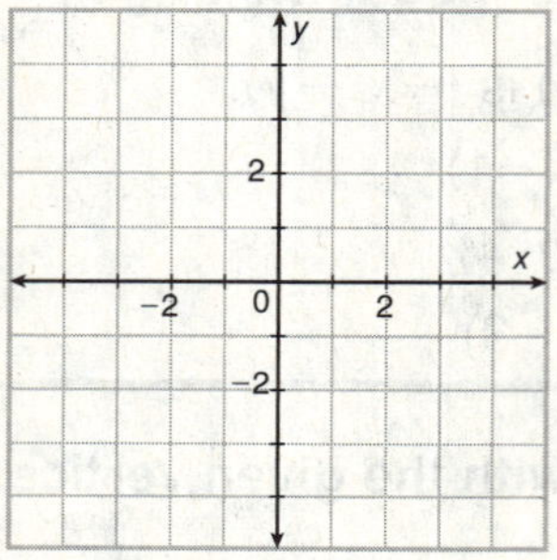

Reflect the figure with the given vertices across the given line.

5. $A(4, 4)$, $B(3, -1)$, $C(1, -2)$; y-axis

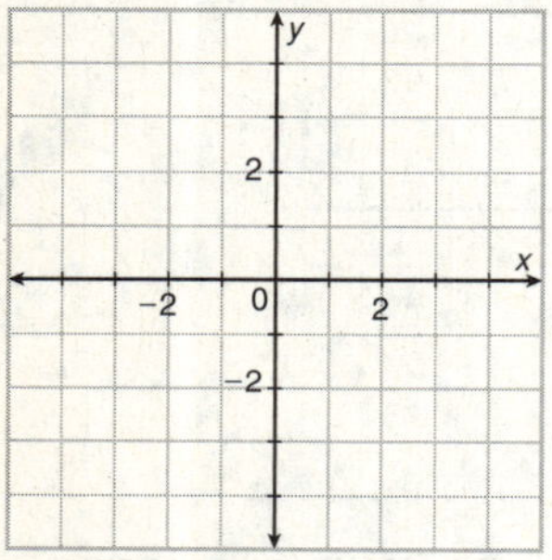

6. $D(-4, -1)$, $E(-2, 3)$, $F(-1, 1)$; $y = x$

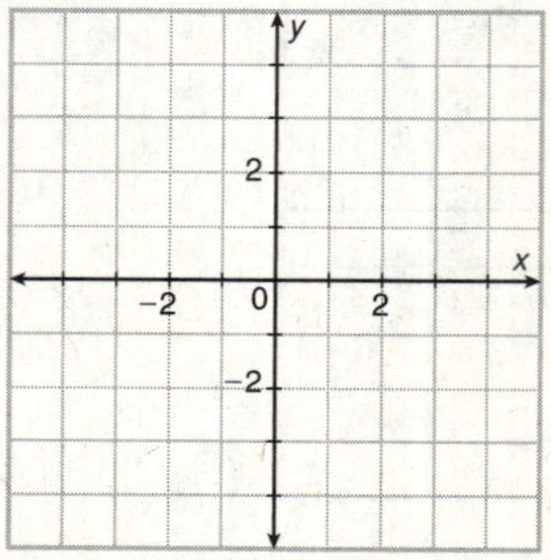

7. $P(1, 3)$, $Q(-2, 3)$, $R(-2, 1)$, $S(1, 0)$; x-axis

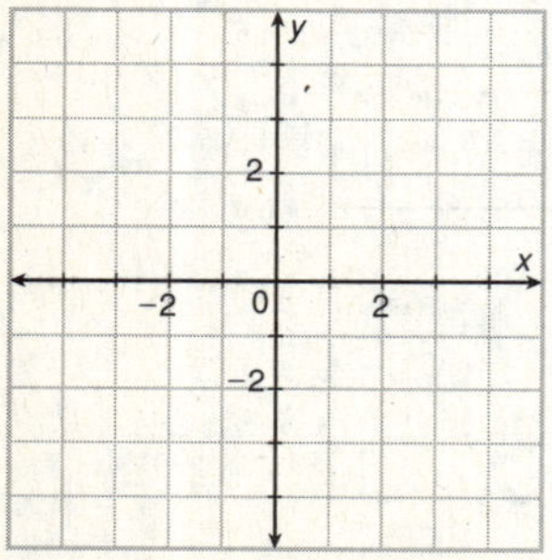

8. $J(3, -4)$, $K(1, -1)$, $L(-1, -1)$, $M(-2, -4)$; $y = x$

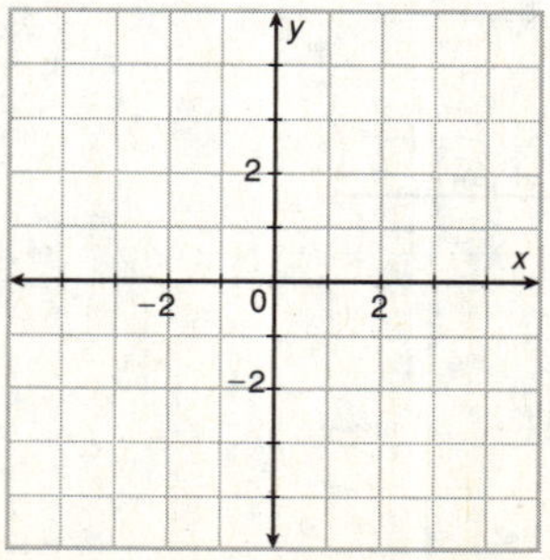

Name ______________________ Date ___________ Class ___________

LESSON 7H

Homework and Practice

Transformations, continued

Rotate the figure with the given vertices about the origin using the given angle of rotation.

9. $A(-2, 3)$, $B(3, 4)$, $C(0, 1)$; 90°

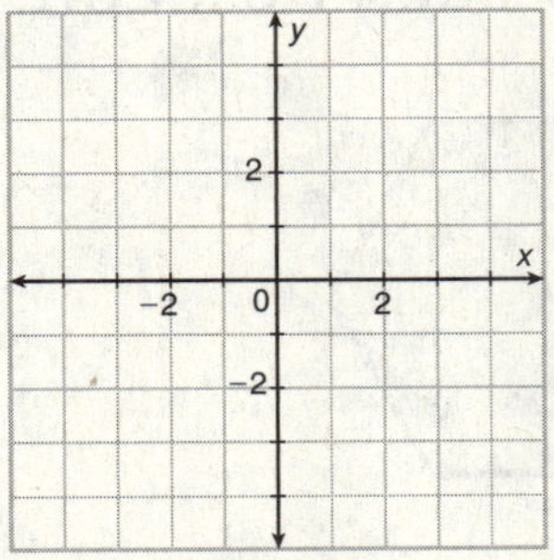

10. $D(-3, 2)$, $E(-4, 1)$, $F(-2, -2)$, $G(-1, -1)$; 90°

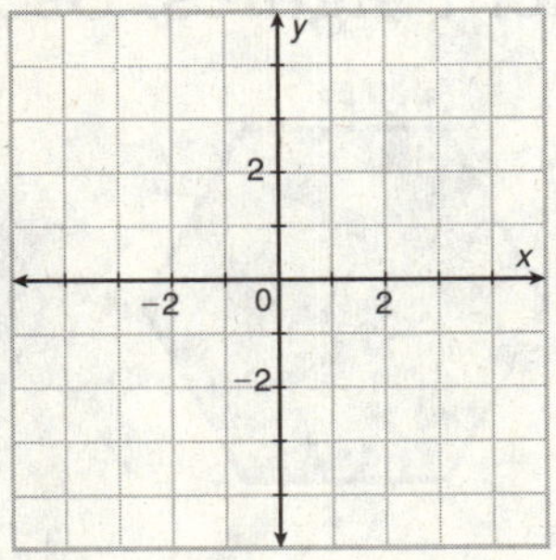

11. $J(2, 3)$, $K(3, 3)$, $L(1, -2)$; 180°

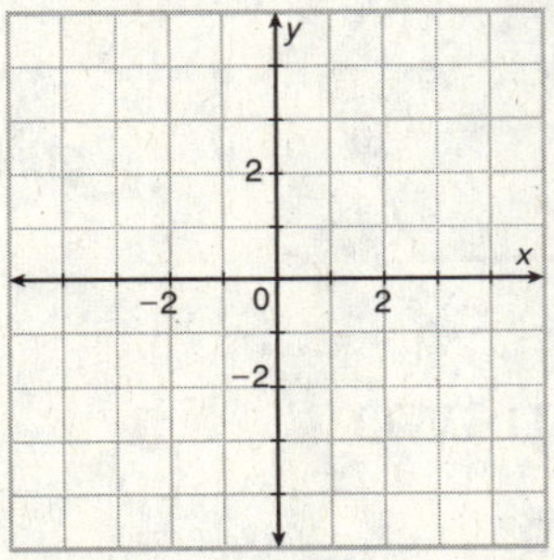

12. $P(0, 4)$, $Q(0, 1)$, $R(-2, 2)$, $S(-2, 3)$; 180°

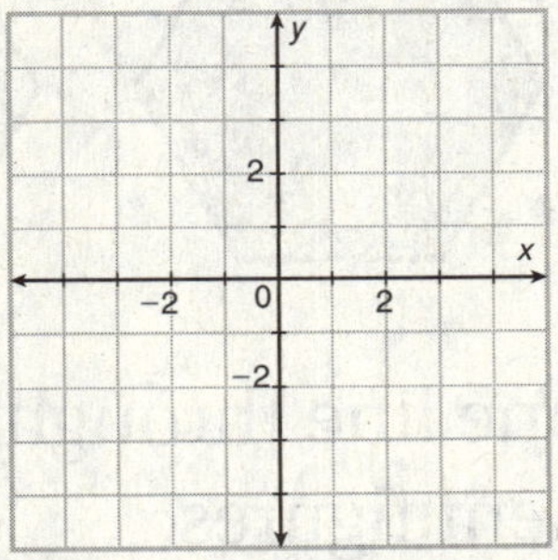

13. Fareek needs to put air in a tire on his bike. The wheel has a 26-inch diameter, and its center is at (0, 0). The valve stem, where the pump attaches to the tire, is at (5, 12). Fareek rotates the wheel 180° to get the valve stem closer to the ground. Find the coordinates of the valve stem after the rotation.

____

Choose the best answer.

14. Point *K* has coordinates (6, 8). After a counterclockwise rotation about the origin, the image of point *K* lies on the *y*-axis. What are the coordinates of *K′*?

A (0, 5) **C** (0, 8)

B (0, 6) **D** (0, 10)

15. $\triangle NPQ$ has vertices $N(-6, -4)$, $P(-3, 4)$, and $Q(1, 1)$. If the triangle is rotated 90° counterclockwise about the origin, what are the coordinates of P'?

F (−4, −3) **H** (3, 4)

G (−4, 3) **J** (3, −4)

EXPLORATION

Program of Studies and Core Content
MA-8-G-S-TS2,
MA-08-3.2.1

7I Symmetry

1. Draw one line through each hexagon to form two congruent figures. Use a different line for each hexagon.

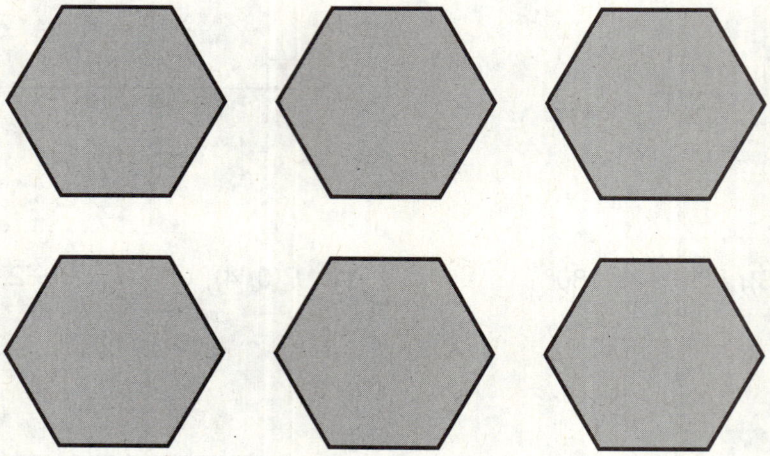

2. Draw one line through each figure to form two congruent figures.

Think and Discuss

3. Draw a figure that cannot be cut into two congruent figures with one line.

4. Describe the characteristics of a figure that can be cut into two congruent figures.

Name ______________________ Date ____________ Class ____________

LESSON 71

Technology Lab

Symmetry

Sketching and manipulating figures in Geometer's Sketchpad will help you understand the principle of symmetry. A figure has symmetry if a line through its center makes two figures that are reflections. A reflection folds a figure backwards along one side.

Activity

Using Geometer's Sketchpad, draw a vertical line segment approximately 2 inches long. Select the segment and use the Construct menu to select Perpendicular Lines. The program will draw two lines across the screen, each line intersecting an endpoint of your line segment.

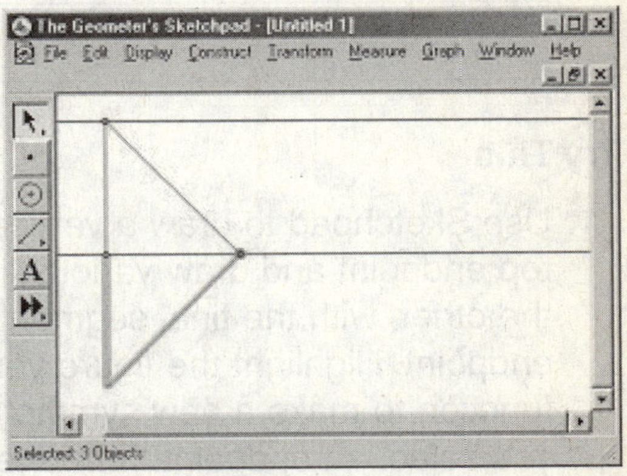

Draw a line segment from the top point to a point along the bottom line. You will now have a right triangle. Hold down the Shift key and select each of the three line segments in the triangle. Then use the Transform menu to Reflect the triangle.

You will now see a new triangle created by the reflection. This new triangle is said to have symmetry because it contains a reflection on either side of the bisecting line.

You can also use Sketchpad to explore symmetry in circles. Draw a horizontal line segment approximately 2 inches long in the center of your screen. Then use the Construct menu to mark the Midpoint of the segment. Use the circle tool to select the midpoint of the segment and draw a circle. Stretch the figure so that the borders pass through the endpoints of the line segment.

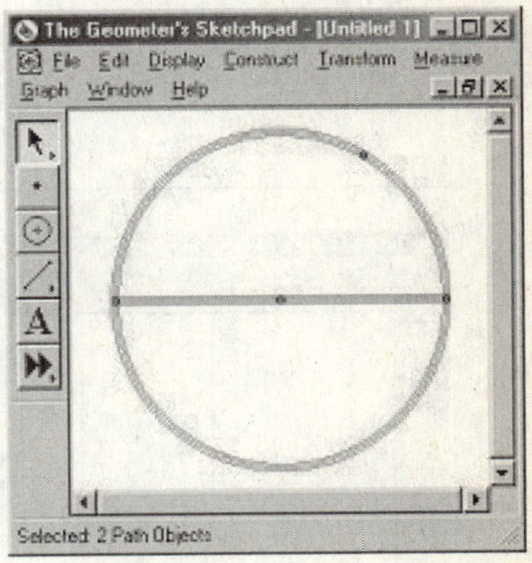

The segment follows what is called the circle's line of symmetry. The semicircles on either side of the line are perfect reflections.

Name ______________________ Date ____________ Class ____________

LESSON 7I

Technology Lab

Symmetry, continued

Think and Discuss

1. Select the righthand endpoint of the triangle you created, and move the point farther to the right. How is the symmetry of the triangle affected?

____

____

Try This

1. Use Sketchpad to draw a vertical line segment. Then start at the top endpoint and draw various line segments, each connected to the other, with the final segment connected to the lower endpoint. Highlight the figure you create and use the Reflect function to make a new symmetrical figure. What happens when you select an endpoint from one of the line segments and move the point? When you place the two figures back to back along the straight line segment?

____

____

____

____

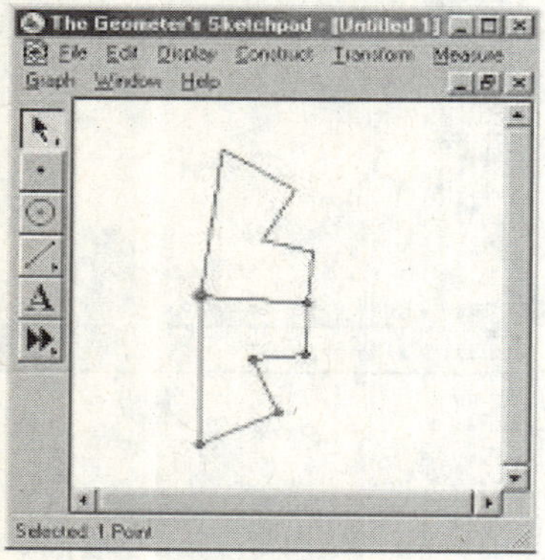

Name ______________________ Date ____________ Class ____________

LESSON 7I

Review for Mastery

Symmetry

A figure has **symmetry** if there is a transformation of the figure such that the image and preimage are identical. There are two kinds of symmetry.

Line Symmetry	The figure has a **line of symmetry** that divides the figure into two congruent halves. 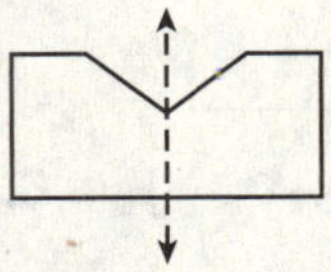 one line of symmetry 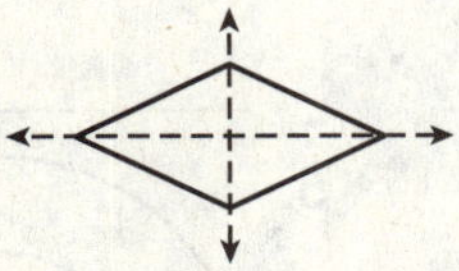 two lines of symmetry 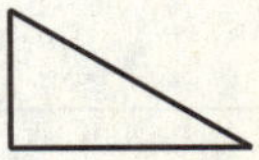 no line symmetry
Rotational Symmetry	When a figure is rotated between 0° and 360°, the resulting figure coincides with the original. • The smallest angle through which the figure is rotated to coincide with itself is called the *angle of rotational symmetry.* • The number of times that you can get an identical figure when repeating the degree of rotation is called the *order* of the rotational symmetry.  angle: 180° order: 2 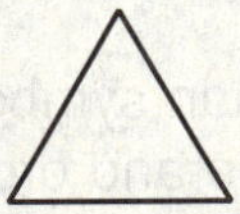 angle: 120° order: 3 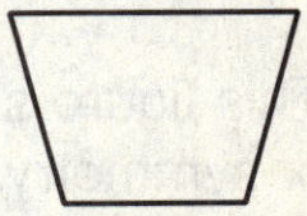 no rotational symmetry

Tell whether each figure has line symmetry. If so, draw all lines of symmetry.

1.

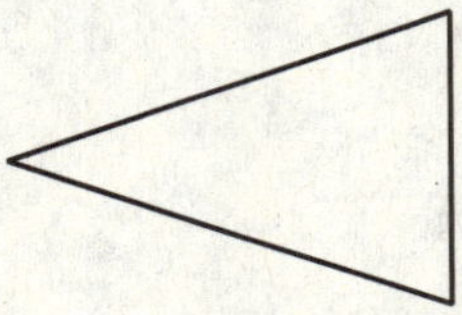

____

2.

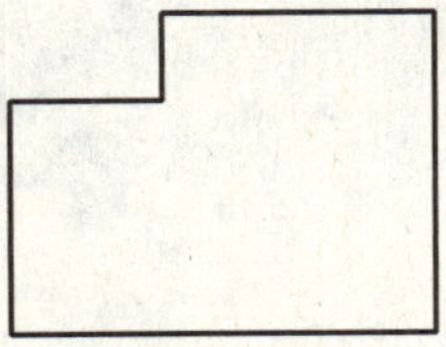

____

Tell whether each figure has rotational symmetry. If so, give the angle of rotational symmetry and the order of the symmetry.

3.

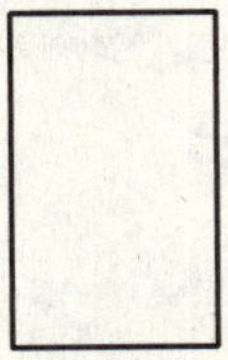

____

4.

____

Name ______________________ Date ____________ Class ____________

LESSON 7-1

Homework and Practice

Symmetry

Tell whether each figure has rotational symmetry. If so, give the angle of rotational symmetry and the order of the symmetry.

1. 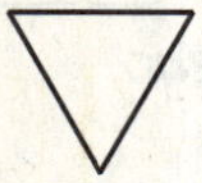

____

2. 

____

3. 

____

4. 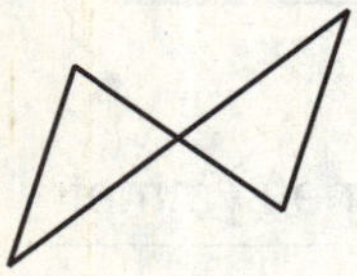

____

5. 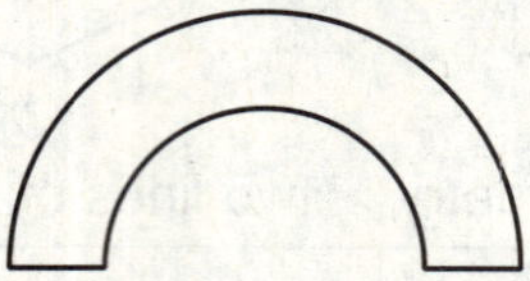

____

6. 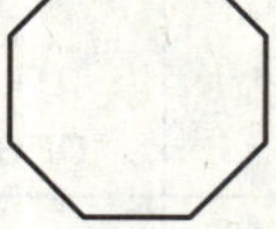

____

7. This figure shows the zodiac symbol for Pisces. Draw all lines of symmetry. Give the angle and the order of any rotational symmetry.

____

8. This figure shows the Roman symbol for Earth. Draw all lines of symmetry. Give the angle and order of any rotational symmetry.

____

9. Tell whether a figure can have rotational symmetry that only occurs at 270° and 360°. If so, draw a figure for which this is true. If not, explain your answer.

Choose the best answer.

10. Which is a true statement about the figure with vertices $Q(-2, -4)$, $R(0, 1)$, $S(8, 1)$, and $T(5, -4)$?

A *QRST* has line symmetry only.

B *QRST* has rotational symmetry only.

C *QRST* has both line symmetry and rotational symmetry.

D *QRST* has neither line symmetry nor rotational symmetry.

11. What is the order of rotational symmetry for the figure shown?

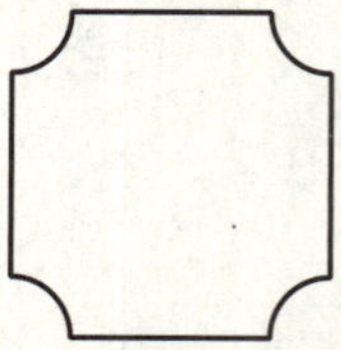

F 2

G 3

H 4

J 6

CHAPTER 7

Chapter Summary

Answer these questions to summarize the important concepts from Chapter 9 in your own words.

1. Explain the difference between complementary angles and supplementary angles.

2. Explain the difference between a convex polygon and a concave polygon.

3. Describe the image of a figure after a 360° rotation about the origin.

For more review of Chapter 7:

- Complete the Chapter 9 Study Guide and Review on pages 510–512 of your textbook.
- Complete the Ready to Go On quizzes on pages 482 and 506 of your textbook.

Name ______________________ Date ____________ Class ____________

CHAPTER 7

KCCT Prep

Multiple Choice

1. In the figure, triangle $ABC \cong$ triangle DEF. What is the value of x?

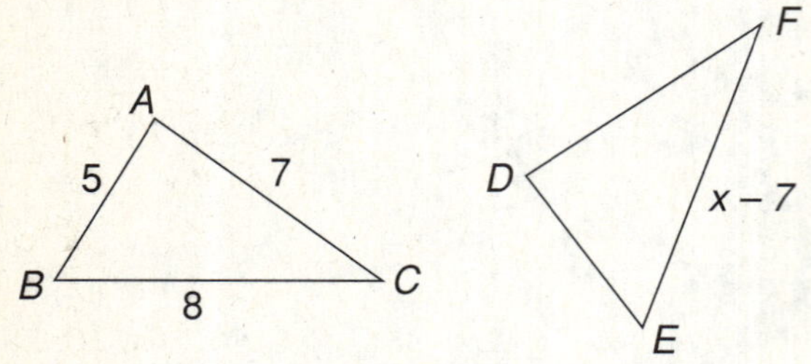

Ⓐ 1 Ⓑ 12 Ⓒ 14 Ⓓ 15

2. The measure of $\angle A$ is 143° and $\angle A$ is supplementary to $\angle B$. $\angle C$ is complementary to $\angle B$. What is $m\angle C$?

Ⓐ 37° Ⓑ 53° Ⓒ 74° Ⓓ 143°

3. Two angles of a triangle measure 33.7° and 15.2°. Which is the best estimate of the measure of the third angle?

Ⓐ 40° Ⓑ 50° Ⓒ 130° Ⓓ 150°

4. A pair of shoes normally costs \$60. The shoes are on sale for \$50. What is the percent decrease in the price?

Ⓐ 10% Ⓑ $16\frac{2}{3}$% Ⓒ 20% Ⓓ $23\frac{1}{3}$%

5. The three sides of triangle MNP all measure 5 cm. Which is the best way to classify the triangle?

Ⓐ Equilateral Ⓑ Isosceles Ⓒ Obtuse Ⓓ Scalene

6. What is the value of m in the triangle shown below?

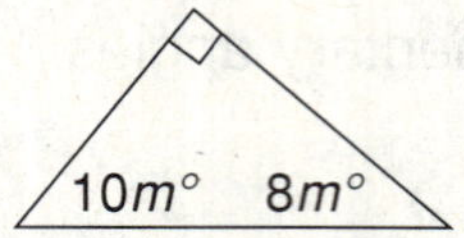

Ⓐ 2 Ⓑ 4 Ⓒ 5 Ⓑ 10

7. Judith rides her bike at 17.1 mi/h. Which is the best estimate of her speed in feet per second?

Ⓐ 10 ft/s Ⓑ 25 ft/s Ⓒ 30 ft/s Ⓑ 35 ft/s

8. The equation $-56 = \square\, z + 8$ has the solution $z = 16$. What is the missing number in the equation?

Ⓐ −64 Ⓑ −16 Ⓒ −12 Ⓓ −4

9. At a restaurant, Manuel wants to leave a 15% tip on his bill of \$24.37. Which is the best estimate for the tip he should leave?

Ⓐ \$2.40 Ⓑ \$3.00 Ⓒ \$3.20 Ⓓ \$3.60

10. Carrie draws a quadrilateral with exactly one pair of parallel sides. Which of these is the best to way describe the quadrilateral?

Ⓐ Parallelogram Ⓑ Rhombus Ⓒ Square Ⓓ Trapezoid

11. What is the mean of the angle measures in an octagon?

Ⓐ 45° Ⓑ 60° Ⓒ 120° Ⓓ 135°

12. $\angle A$ is supplementary to the angle whose measure is $y°$ in the figure. What is $m\angle A$?

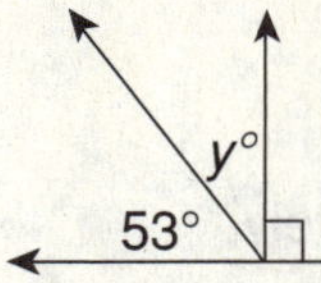

Ⓐ 17° Ⓒ 53°
Ⓑ 37° Ⓓ 143°

13. What is the least possible value of the expression $(a + b)^2 - c$ when $a = 3$, $b = 1$, and $c \leq 3$?

Ⓐ –16 Ⓒ 13
Ⓑ –3 Ⓓ 16

14. What are the coordinates of the image of point A after a reflection across the x-axis?

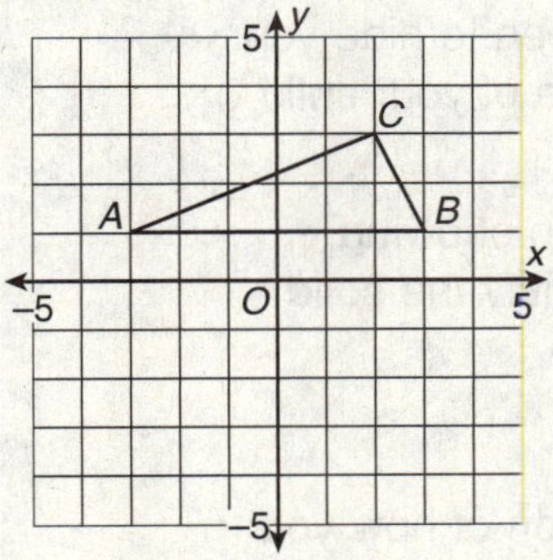

Ⓐ (–3, –1) Ⓒ (3, –1)
Ⓑ (–3, 1) Ⓓ (3, 10)

15. $\angle J$ and $\angle K$ are vertical angles. $\angle K$ and $\angle L$ are complementary angles. If $m\angle J = 28°$, what is $m\angle L$?

Ⓐ 28° Ⓒ 62°
Ⓑ 56° Ⓓ 152°

Open Response

16. Brodie plots the points (0, 3), (2, 2), (2, –1), and (0, –3). Then he connects the points to form a quadrilateral.

a. What type of quadrilateral is formed?

b. He reflects the quadrilateral across the y-axis. Sketch the image.

c. What type of polygon is formed by the original quadrilateral and its image?

d. Is it a regular polygon? Explain?

17. In $\triangle ABC$, $\angle B$ is six times as large as $\angle A$ and $\angle C$ is twice as large as $\angle A$.

a. Let x represent the measure of $\angle A$. Write an equation you can use to find the value of x.

b. Solve the equation. What are the measures of the angles of the triangle?

c. Classify $\triangle ABC$ based on its angles.

18. Latrisha is sketching a map of her county. The table shows some actual distances and the corresponding distances on her map.

Actual Distance (mi)	18	27	45
Map Distance (in.)	4	6	10

a. What is the scale of the map?

b. Two towns in the county are 3 miles apart. How far apart should Latrisha draw them on her map?

c. The county resembles a rectangle with a length of 45 miles and a width of 40 miles. Will Latrisha be able to fit the map on a $8\frac{1}{2}$ in. by 11 in. sheet of paper? Explain.

d. Revise the scale so the map fits on the sheet of paper. Justify your choice of scale.

Project

Perimeter, Area, and Volume: Solid Guessing

Career: Surgeon

Today, some surgeons perform specialized operations known as laser surgery. With many laser surgeries, surgeons cannot actually see the three-dimensional area where they are operating; instead, they must rely on what they can see in two-dimensional images projected onto a screen to guide them.

Before you start, work through the Mystery Solids worksheet on the next page.

In this chapter project . . .

You try to identify three-dimensional shapes from their shadows.

What to do:

- Create a paper model of a three-dimensional figure, but don't let your classmates see it. You can look through Chapter 8 of your textbook for ideas or use the Internet to find more examples. For example, look for "Platonic solids."
- Your teacher will set up an overhead projector and a screen to hide your solid. Hold your model in front of the light so that one side of your solid is projected as a shadow against the wall.
- Repeat the last step for a different side of your solid. Keep showing different sides until one of your classmates is able to identify the solid and make a sketch.

What to present:

- The name of your solid, your model, and a brief description of how you assembled it.

Where's the math?

- Use nets to construct models of three-dimensional solids.
- Use orthographic projections to describe and identify solids.

go.hrw.com
Chapter Project Online
KEYWORD: MT7 Ch8

Name ______________________ Date ____________ Class ____________

CHAPTER 8

Project Recording Sheet

Perimeter, Area, and Volume

Mystery Solids

There are many opportunities to identify three-dimensional objects by looking at their surfaces. Have you ever been able to recognize a bird, a car, a tool or even a person by seeing the shadow? What are the aspects of the object that make recognition possible? Use this worksheet to enhance your skills.

Here are four more mystery geometric solids. Figure out what they are and check your decisions by using an overhead projector, flashlight or other light source to produce shadows of the surfaces.

Mystery Solid	Front View	Side View	Top View
	Triangle	Triangle	Square
	Square	Square	Square
	Rectangle	Rectangle	Triangle
	Rectangle	Rectangle	Rectangle

Create some mystery solids. Draw the various surface views. Exchange your mystery solids with classmates. Check your answers with the light sources when possible.

Mystery Solid	Front View	Side View	Top View

Make some shadow drawings of common objects. Take turns identifying the objects with your classmates.

Shadow	Mystery Object

EXPLORATION

8A Perimeter and Area of Rectangles and Parallelograms

Program of Studies and Core Content

MA-8-AT-S-EI2, MA-8-AT-S-VE02, MA-8-G-S-SR2, MA-8-M-S-MPA9, MA-8-M-S-MPA3, MA-08-2.1.1, MA-08-3.1.2

Recall that the area of a rectangle is the product of its length times its width, or the product of its base times its height, *bh*. You can use this fact to develop the formula for the area of a parallelogram.

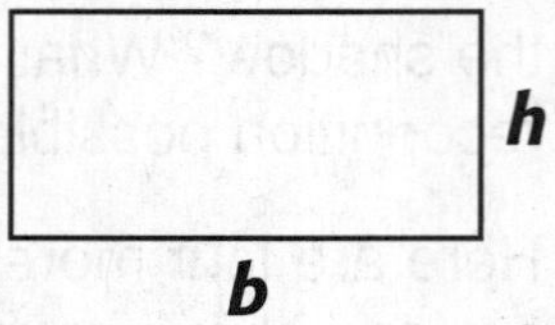

1. Draw a parallelogram on a sheet of graph paper. Label the base and height as shown.

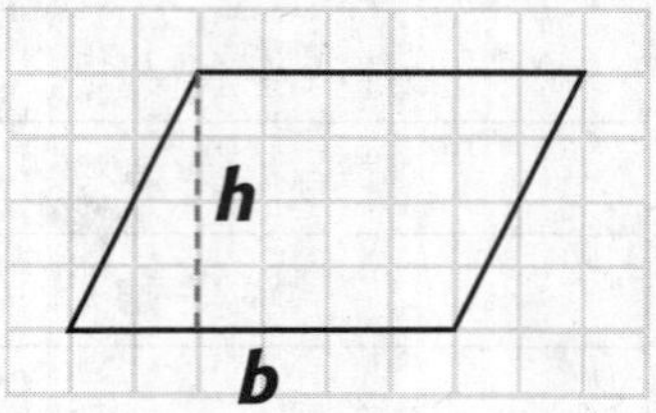

2. Cut out the parallelogram.
3. Now cut a right triangle off the end of the parallelogram as shown.

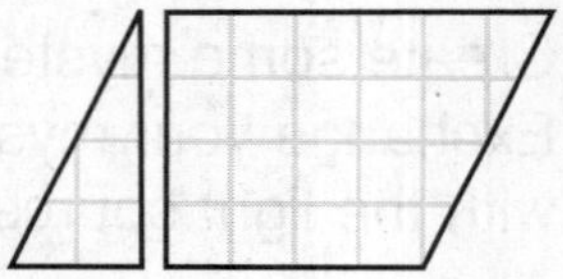

4. Arrange the two pieces to form a rectangle.
5. What are the base and height of the rectangle you made? What is the rectangle's area?
6. How is the area of the parallelogram related to the area of the rectangle?

Think and Discuss

7. **Explain** how to write a formula for the area of a parallelogram with base *b* and height *h*.
8. **Describe** how you can use your formula to find the area of this parallelogram.

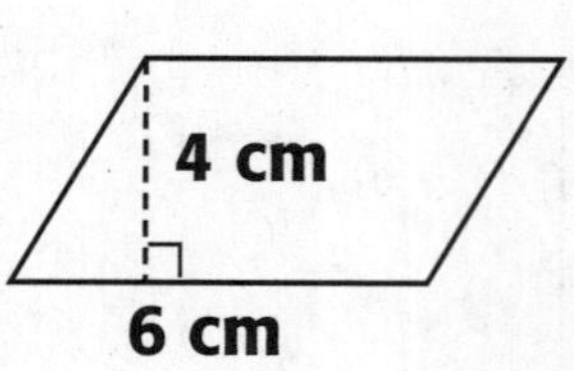

Name ______________________ Date __________ Class __________

LESSON **8A**

Hands-On Lab

Perimeter & Area of Rectangles & Parallelograms

Activity 1

To find the perimeter of a rectangle, use the formula $P = 2\ell + 2w$ where P equals the perimeter, ℓ equals the length, and w equals the width.

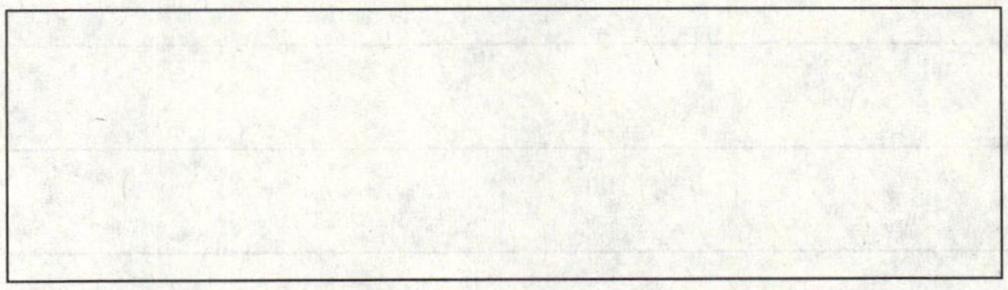

1. Determine the perimeter of this rectangle in centimeters. What is the perimeter?

____

2. Outline an index card in the area below. Measure and label the length and width. Then use the formula to determine the perimeter in centimeters.

____

Name ______________________ Date ____________ Class ____________

LESSON 8A

Hands-On Lab

Perimeter & Area of Rectangles & Parallelograms, continued

Think and Discuss

1. Determine the formula to use for finding the perimeter of a square with one side 15.5 inches long.

____

____

____

____

2. Why does the formula $P = 2l + 2w$ work for parallelograms?

____

____

Try This

Work in pairs. Use masking tape to mark off an area in the classroom that forms a rectangle. Then switch rectangles with another pair. Use the formula to find the perimeter. Write your work here:

____

____

____

Now use the formula to measure the perimeter of the first rectangle you marked off. Write your work here:

____

____

____

Compare the answers your pair got with the other pair. If you got different answers, remeasure and do the calculations again.

Name ______________________ Date ____________ Class ____________

LESSON 8A

Hands-On Lab

Perimeter & Area of Rectangles & Parallelograms, continued

Activity 2

1. Using graph paper, draw a rectangle that is 20 cm long and 12 cm wide with each square representing 1 cm. Then cut out enough squares to completely fill the rectangle. Count the number of squares it takes to fill the rectangle. Then check your answer by calculating the area of the rectangle using the formula: $A = l \times w$.
   Write the formula and the calculations you used to find the area.

   ____

   ____

   ____

   Remember: area is the number of square units in a figure, and is expressed in square units.

2. Find the number of square units in this parallelogram. Trace the figure. Then cut out the figure and reassemble it so it forms a rectangle. Use the formula to find the area.

   ____

   ____

   ____

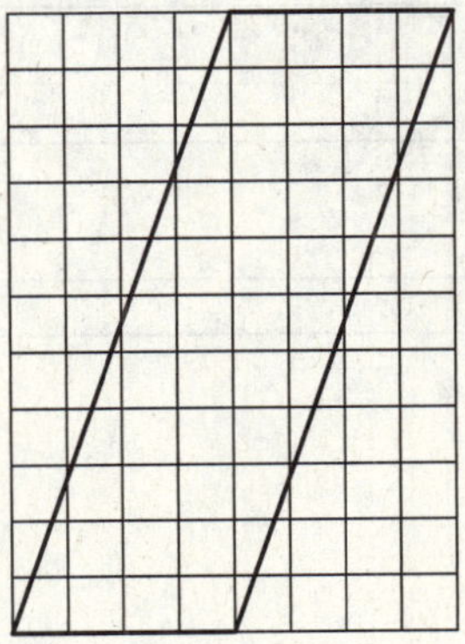

Name ______________________ Date ____________ Class ____________

LESSON 8A

Hands-On Lab

Perimeter & Area of Rectangles & Parallelograms, continued

Think and Discuss

1. How is finding the area of rectangles and parallelograms alike and different?

____

____

____

Try This

1. Find the perimeter of a room with a length of 28 feet and a width of 13.5 feet. Then find the area of the room. Show your work below.

____

____

____

____

2. Find an object that forms a rectangle or a parallelogram and that will fit in the area below. Trace around the object. Then find both the perimeter and area of the object.

____

____

____

Name ______________________ Date ____________ Class ____________

LESSON 8A

Review for Mastery

Perimeter and Area of Rectangles and Parallelograms

Perimeter = distance around the outside of a figure.
To find the perimeter of a figure, add the lengths of all its sides.

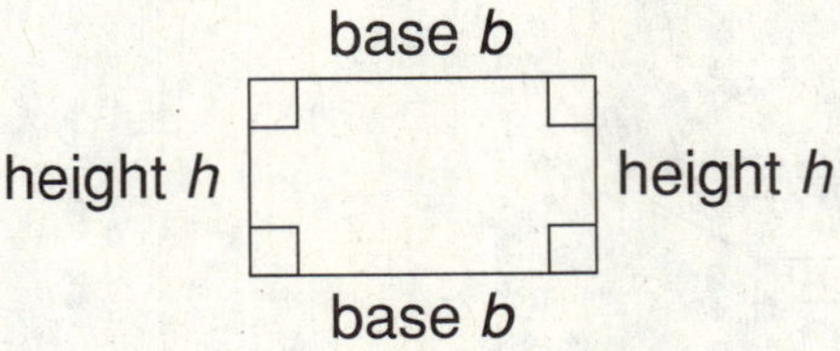

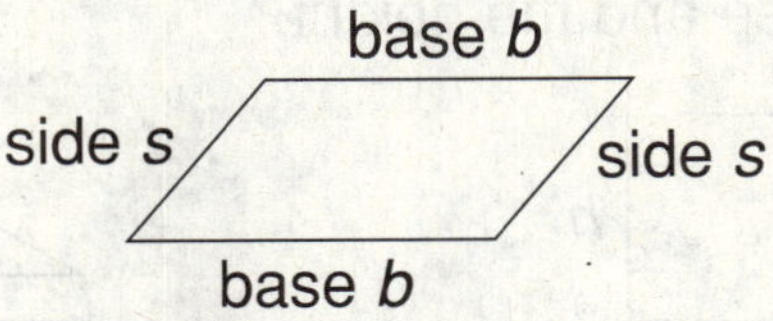

Perimeter of Rectangle

$P = b + h + b + h$

$P = \mathbf{2b + 2h}$

Perimeter of Parallelogram

$P = b + s + b + s$

$P = \mathbf{2b + 2s}$

Complete to find the perimeter of each figure.

1.

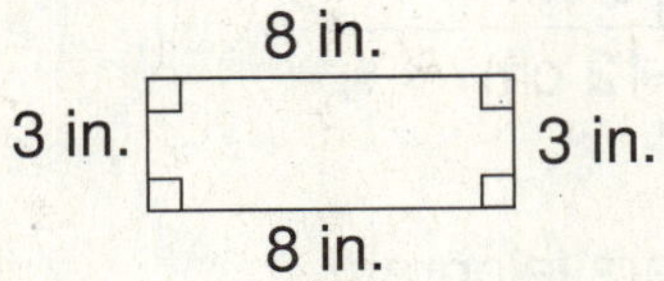

Perimeter of rectangle

$P = 2b + 2h$

$P = 2(\_\_\_\_) + 2(\_\_\_\_)$

$P = \_\_\_\_ + \_\_\_\_$

$P = \_\_\_\_$ in.

2.

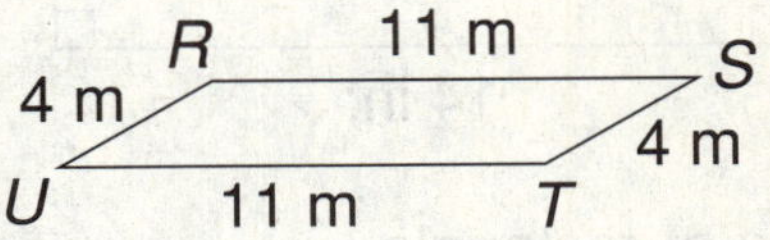

Perimeter of parallelogram

$P = 2b + 2s$

$P = 2(\_\_\_\_) + 2(\_\_\_\_)$

$P = \_\_\_\_ + \_\_\_\_$

$P = \_\_\_\_$ m

Find the perimeter of each.

3. Large rectangle

$P = \_\_\_\_ + \_\_\_\_ + \_\_\_\_ + \_\_\_\_ = \_\_\_\_$

4. Small rectangle

$P = \_\_\_\_ + \_\_\_\_ + \_\_\_\_ + \_\_\_\_ = \_\_\_\_$

5. The combined rectangles as shown in the figure.

$P = \_\_\_\_ + \_\_\_\_ + \_\_\_\_ + \_\_\_\_ + \_\_\_\_ + \_\_\_\_ = \_\_\_\_$

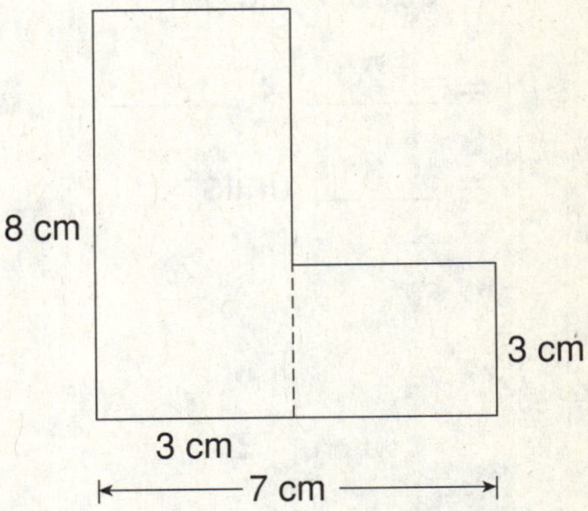

Name ______________________ Date __________ Class __________

LESSON 8A

Review for Mastery

Perimeter and Area of Rectangles and Parallelograms, cont.

Area = number of square units contained inside a figure.

To find the area of a rectangle or a parallelogram, multiply the base and the height.

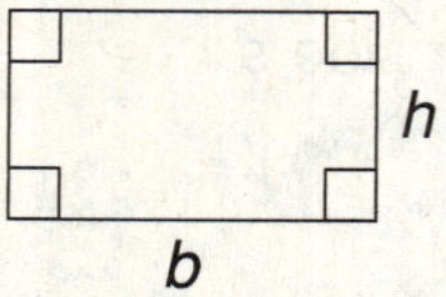

Area of Rectangle = $b \times h$ **Area of Parallelogram = $b \times h$**

Complete to find the area of each figure.

6.

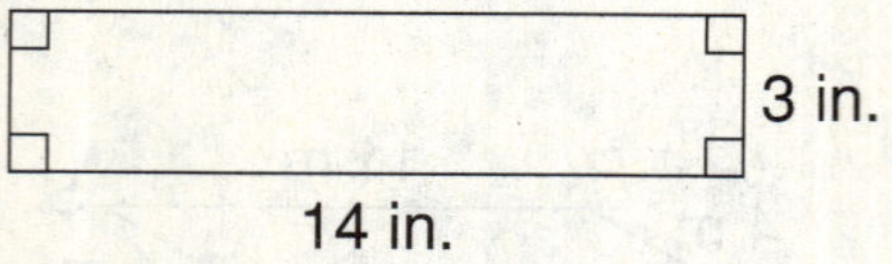

Area of rectangle
$= b \times h$
= 14 × 3 = 42 in²

7.

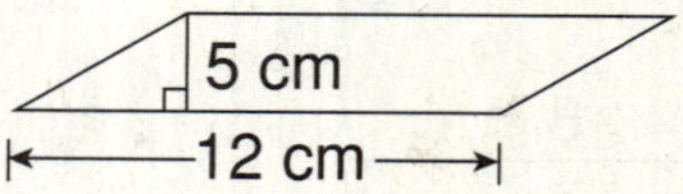

Area of parallelogram
$= b \times h$
= 12 × 5 = 60 cm²

8. In the rectangle graphed on the coordinate plane:

base = 2 units

height = 4 units.

Area of rectangle
= base × height

= 2 × 4

= 8 units²

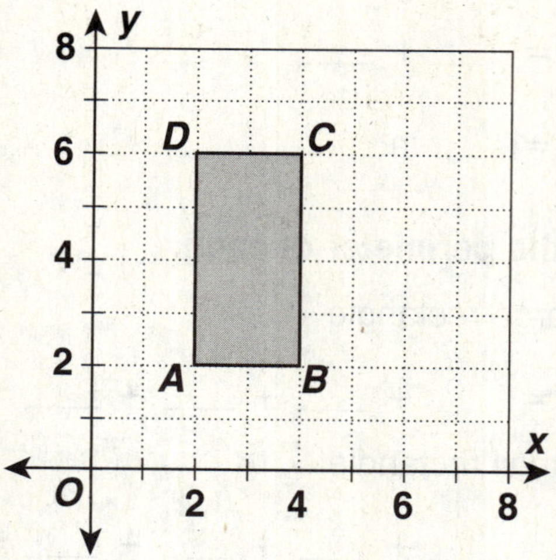

Name ______________________ Date ____________ Class ____________

LESSON **8A**

Homework and Practice

Perimeter & Area of Rectangles & Parallelograms

Find the perimeter of each figure.

1.

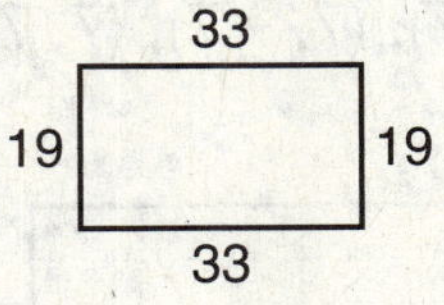

____

2.

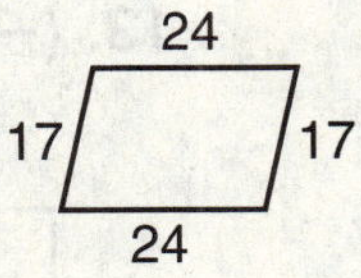

____

3.

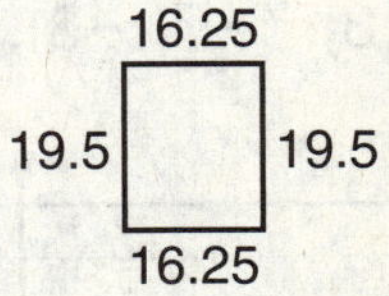

____

Use a metric ruler to measure the dimensions of each figure to the nearest millimeter. Then find the perimeter of each figure.

4.

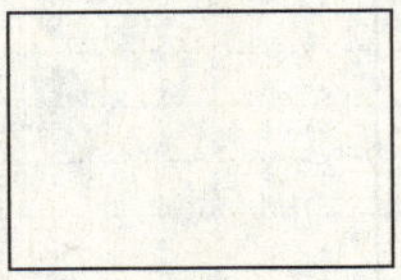

____

5.

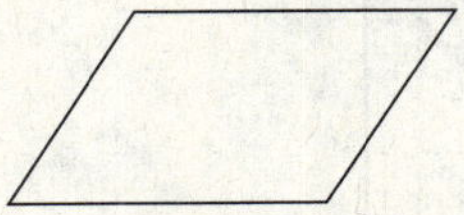

____

6.

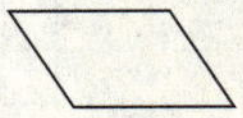

____

7.

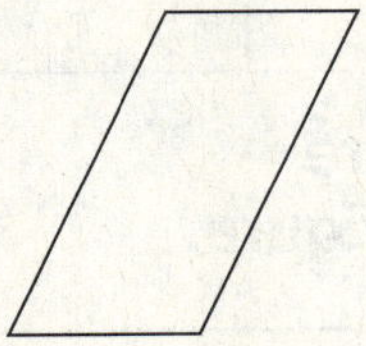

____

8.

____

9.

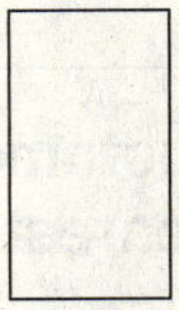

____

Find the perimeter of each figure.

10.

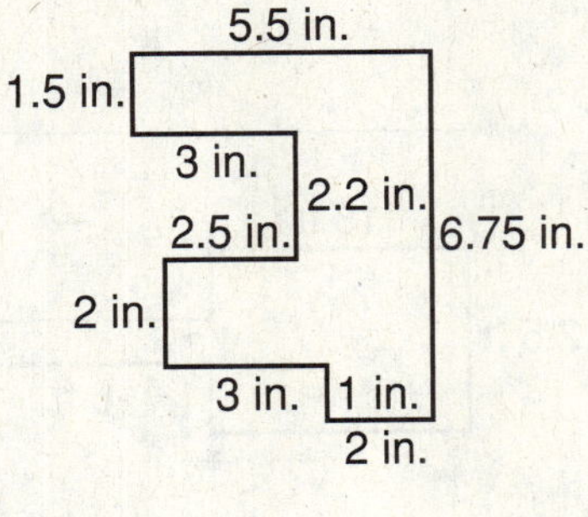

____

11.

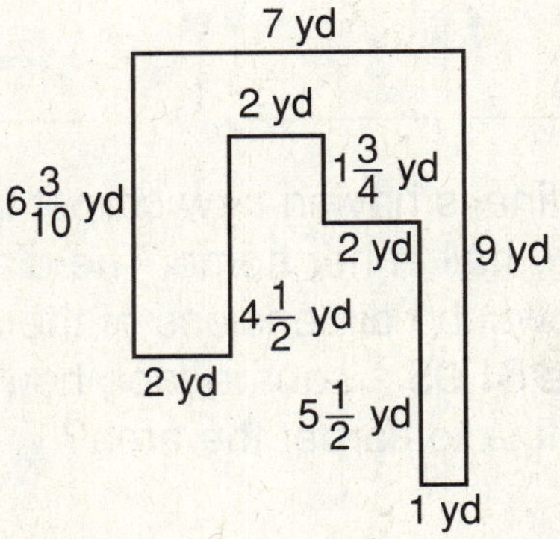

____

Name ______________________ Date ____________ Class ____________

LESSON **8A**

Homework and Practice

Perimeter & Area of Rectangles & Parallelograms, continued

Graph each figure with the given vertices. Then find the area of each figure.

12. (−4, 3), (−7, −3), (4, −3), (7, 3)

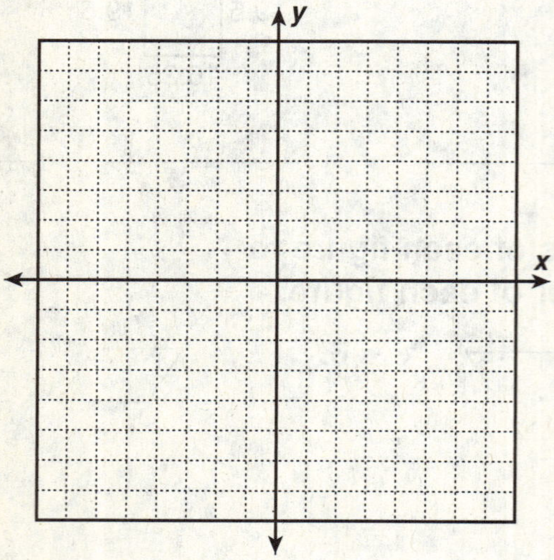

13. (−7, 7), (−7, −7), (7, −7), (7, 7)

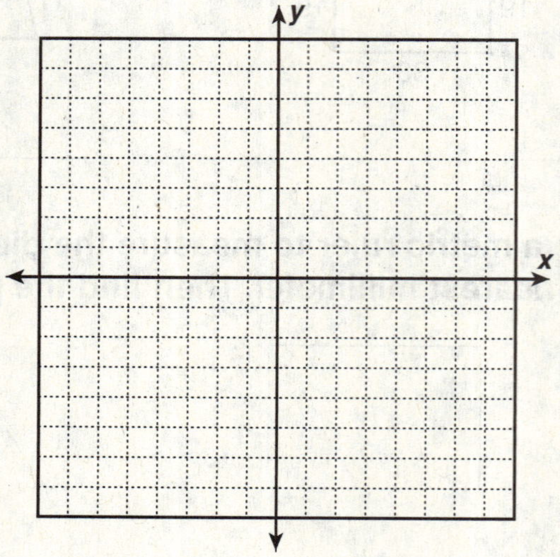

______________________ ______________________

Use a customary ruler to measure the dimensions of each figure to the nearest sixteenth of an inch. Then find the area of each figure.

14.

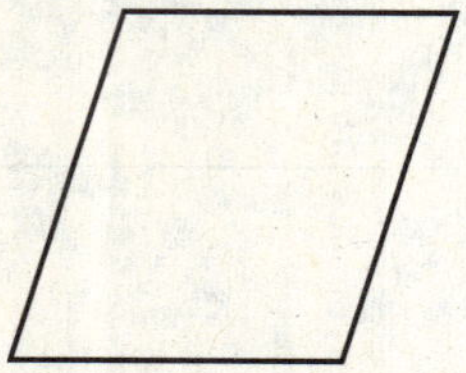

15.

16.

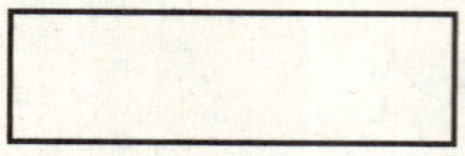

______________ ______________ ______________

17. Mrs. Delfina is having new carpet installed in a room and hall in her home. The diagram at the right shows the dimensions of the area. If the carpet costs $1.95 a square foot, how much will it cost Mrs. Delfina to carpet the area?

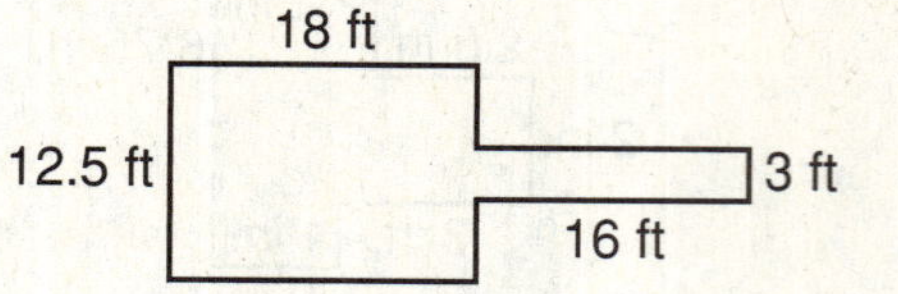

____

18. Explain how the units for the perimeter and the area of the same square are different.

____

EXPLORATION

Program of Studies and Core Content

MA-8-AT-S-EI2, MA-8-AT-S-VE02, MA-8-G-S-SR2, MA-8-M-S-MPA4, MA-8-M-S-MPA9, MA-8-M-S-MPA3, MA-08-2.1.1, MA-08-3.1.2

8B Perimeter and Area of Trapezoids

You can find the area of a trapezoid by drawing a diagonal as shown below. Then use the formula for the area of a triangle.

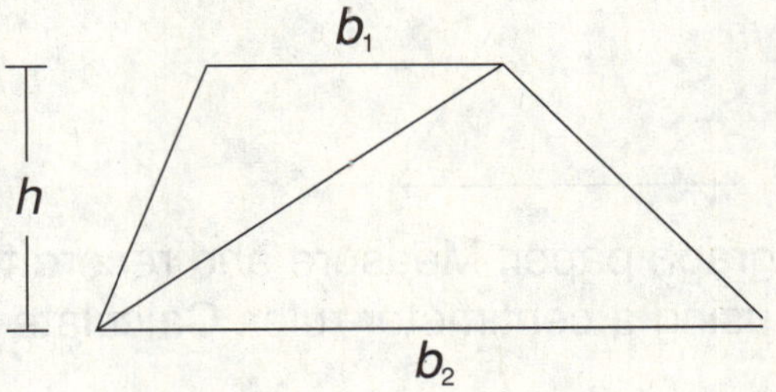

1. Explain why the height of either triangle in the figure is h. What is the base of each triangle?

2. Write an expression for the area of each triangle. Then use those expressions to write a formula for the area of a trapezoid.

3. The formula for the area of a trapezoid is $A = \frac{1}{2}h(b_1 + b_2)$. If necessary, rewrite your answer from number 2 to agree with this formula.

4. Suppose the other diagonal of the trapezoid was used to form two triangles. Would the triangles have different areas?

Think and Discuss

5. Discuss how the diagram and the formula would change if $b_1 = b_2$. Would the figure still be a trapezoid?

Name ______________________ Date ____________ Class ____________

LESSON **8B**

Hands-On Lab

Perimeter and Area of Trapezoids

The area of a trapezoid is equal to the height times half the sum of the bases or $A = h\frac{b_1 + b_2}{2}$. In this activity you will see why this is true.

Activity

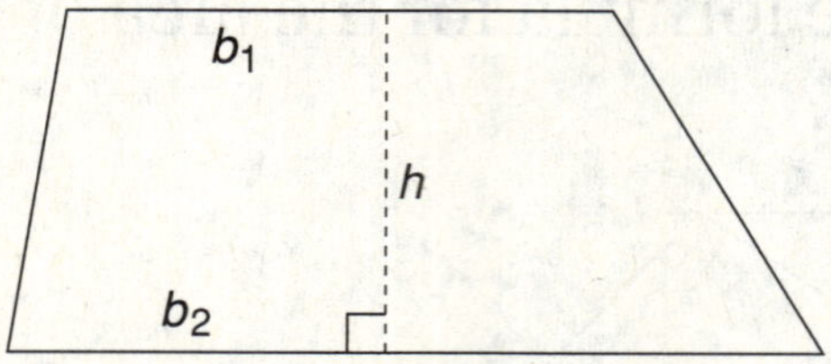

Trace the trapezoid shown here on graph paper. Measure and record the height and the length of each base using a centimeter ruler. Calculate the area and fill in the table below.

	Height (h)	Base #1 (b_1)	Base #2 (b_2)	$A = h\frac{b_1 + b_2}{2}$
Trapezoid				

Use scissors to cut out this trapezoid. Fold it in half so that two parallel bases line up. Crease the fold to make a parallel line halfway between the two bases. With pencil and ruler, draw a line along the length of the fold. The length of this line equals the mean average of the two bases: $\frac{b_1 + b_2}{2}$.

With pencil and ruler, draw vertical lines downward from the ends of the pencil line you have drawn. With the scissors, clip one of these triangular pieces off the lower left corner of the trapezoid, rotate it 180°, and tape it to the upper left corner of the trapezoid. Clip the other triangular piece off the lower right corner, rotate it 180°, and tape it to the upper right corner. The shape should now be a rectangle.

What is the length of the base of this rectangle? What is the height?

____

What is the area of this rectangle?

____

Name ______________________ Date ____________ Class ____________

LESSON 8B

Hands-On Lab

Perimeter and Area of Trapezoids, continued

Think and Discuss

1. Does the rectangle you have formed have the same area as the trapezoid? Justify your response. (Hint: Compare the length of the rectangle base with the value of $\frac{b_1 + b_2}{2}$.)

____

____

____

____

2. Does the rectangle have the same perimeter as the trapezoid?

____

Try This

1. Use the steps you took in the Activity to complete the table below and find the area of this trapezoid.

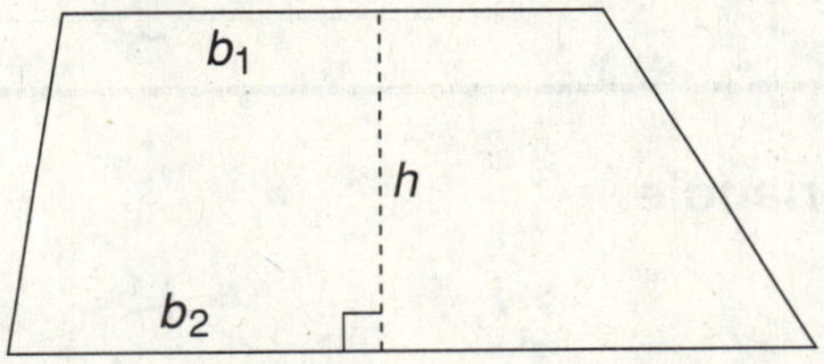

	Height (h)	Base #1 (b_1)	Base #2 (b_2)	$A = h\frac{b_1 + b_2}{2}$
Trapezoid				

Name ______________________ Date __________ Class __________

LESSON **8B**

Review for Mastery

Perimeter and Area of Triangles and Trapezoids

To find the perimeter of a figure, add the lengths of all its sides.

Complete to find the perimeter of each figure.

1.

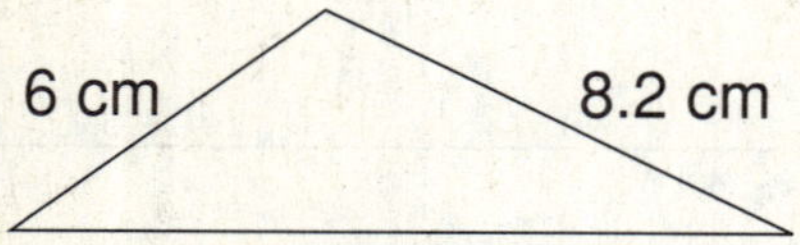

Perimeter of triangle

= ____ + ____ + ____

= ____ cm

2.

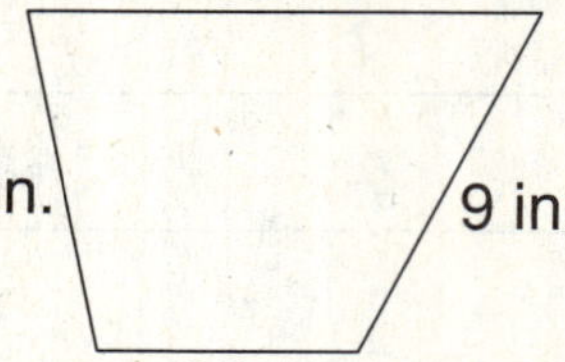

Perimeter of trapezoid

= ____ + ____ + ____ + ____

= ____ in.

Area of Triangle = $\frac{1}{2}bh$

The area of a triangle is one-half the product of a base length *b* and the height *h* drawn to that base.

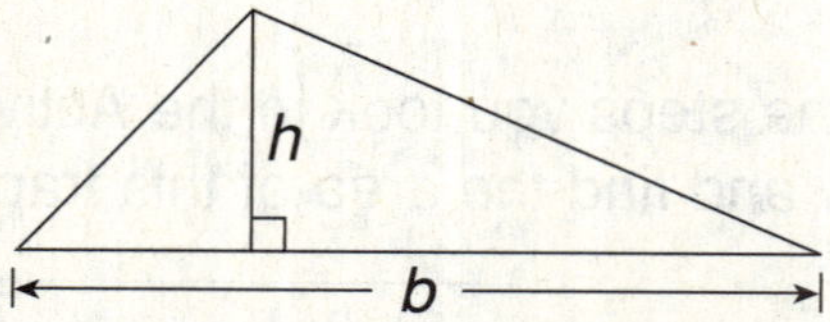

Complete to find the area of each triangle.

3. Area of triangle

$= \frac{1}{2}bh$

$= \frac{1}{2} \times$ ____ $\times$ ____

$= \frac{1}{2} \times$ ____ $=$ ____ in^2

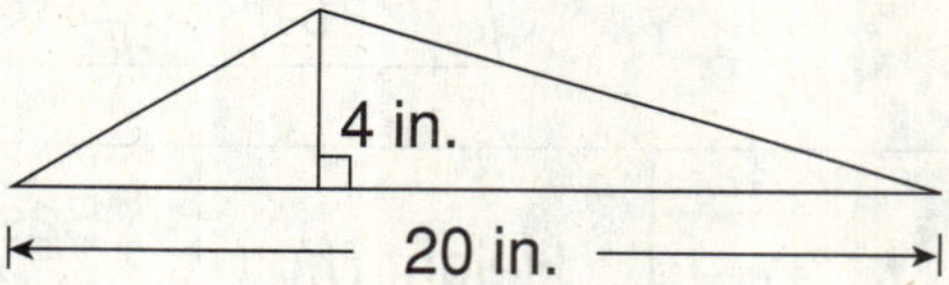

4. In the triangle graphed on the coordinate plane:

base = 10 − 3 = ____ units

height = 4 − (−2) = ____ units.

Area of triangle

$= \frac{1}{2} \times \text{base} \times \text{height}$

$= \frac{1}{2} \times$ ____ $\times$ ____

$= \frac{1}{2} \times$ ____ $=$ ____ units^2

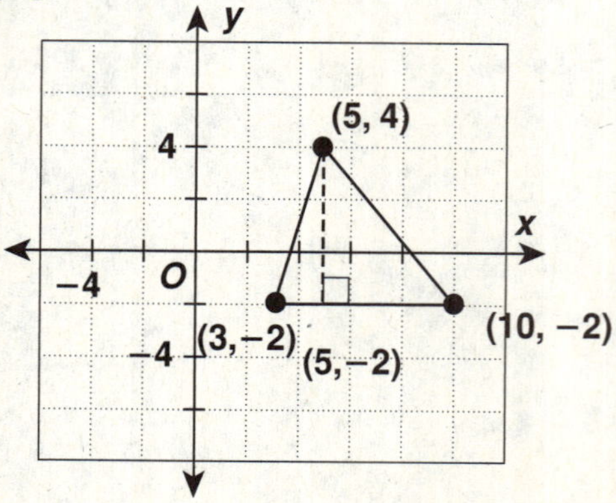

Name ______________________________ Date ____________ Class ____________

LESSON 8B

Review for Mastery

Perimeter and Area of Triangles and Trapezoids, continued

Area of Trapezoid $= \frac{1}{2}h(b_1 + b_2)$

The area of a trapezoid is one-half the height h times the sum of the base lengths b_1 and b_2.

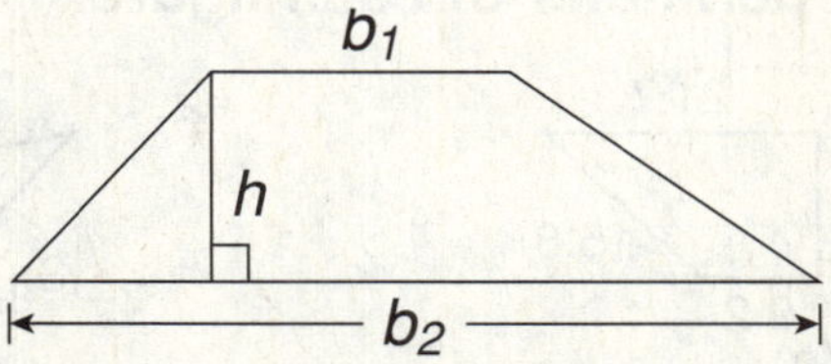

Complete to find the area of each trapezoid.

5.

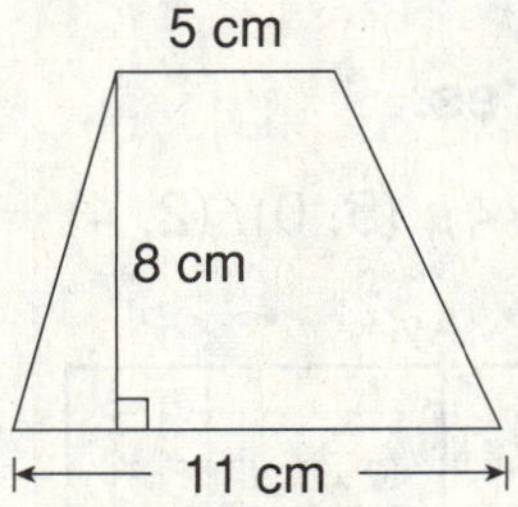

Area of trapezoid

$= \frac{1}{2}h(b_1 + b_2)$

$= \frac{1}{2} \times$ _____ $\times$ (_____ + _____)

$=$ _____ $\times$ (_____) $=$ _____ cm^2

6.

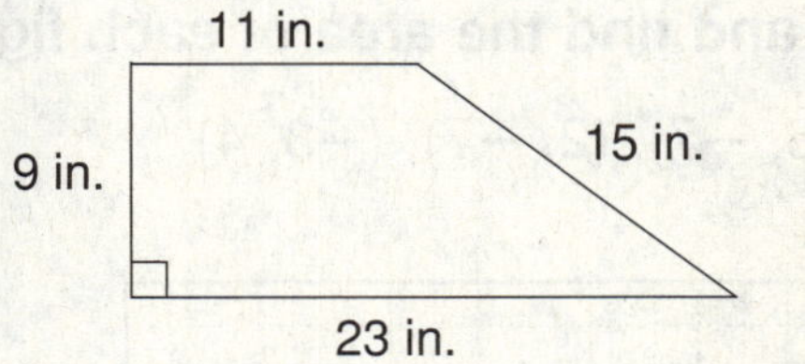

Area of trapezoid

$= \frac{1}{2}h(b_1 + b_2)$

$= \frac{1}{2} \times$ _____ $\times$ (_____ + _____)

$=$ _____ $\times$ (_____) $=$ _____ in^2

7. In the trapezoid graphed on coordinate plane:

$base_1 = 8 - 4 =$ _____ units

$base_2 = 11 - 2 =$ _____ units

height $= 6 - 2 =$ _____ units.

Area of trapezoid

$= \frac{1}{2} \times$ height $\times$ $(base_1 + base_2)$

$= \frac{1}{2} \times$ _____ $\times$ (_____ + _____)

$=$ _____ $\times$ (_____)

$=$ _____ $units^2$

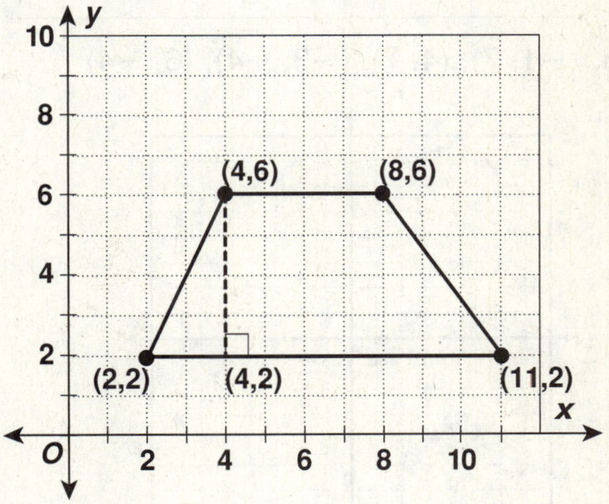

Name ______________________ Date ____________ Class ____________

LESSON 8B

Homework and Practice

Perimeter and Area of Triangles and Trapezoids

Find the perimeter of each figure.

1.

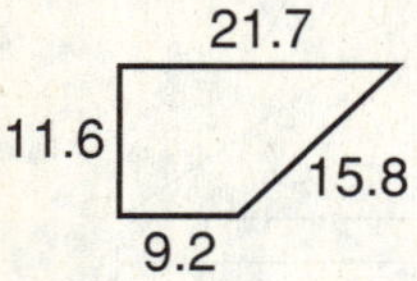

2.

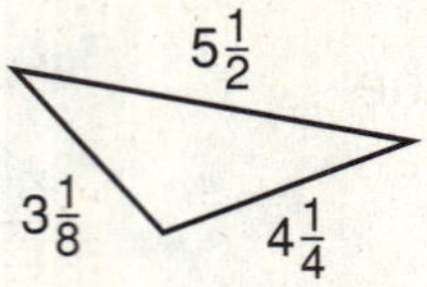

3.

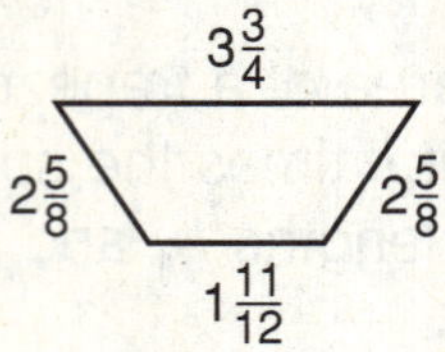

______________ ______________ ______________

Graph and find the area of each figure with the given vertices.

4. (−6, −7), (2, −7), (−3, 4)

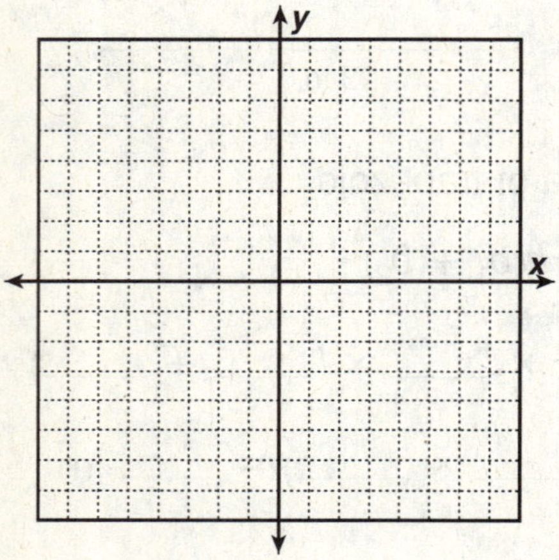

____

5. (−5, 0), (−1, 4), (5, 0), (2, 4)

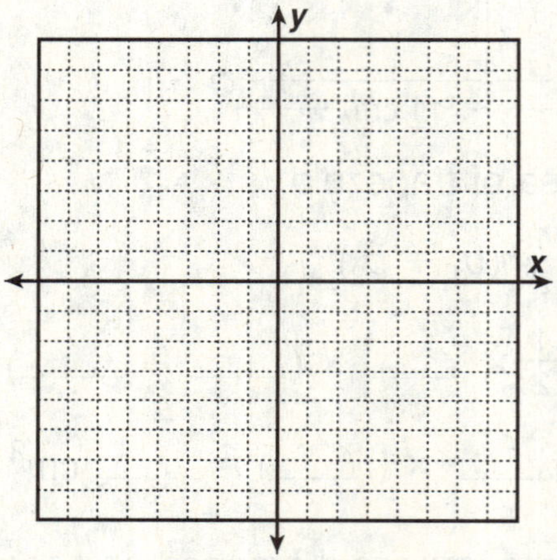

____

6. (−1, 7), (4, 7), (−3, −4), (6, −4)

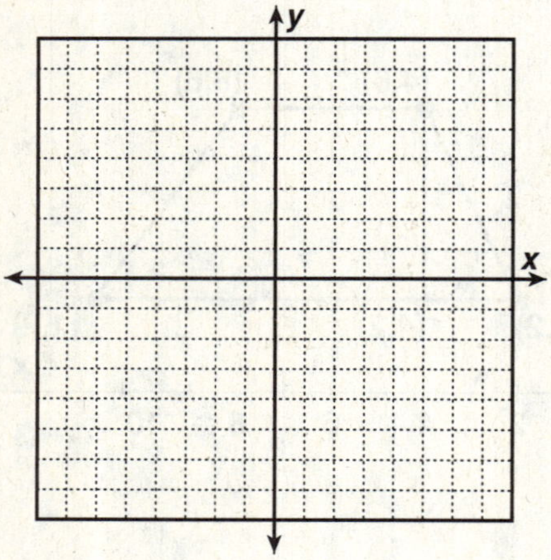

____

7. (−6, 2), (6, 2), (6, −4)

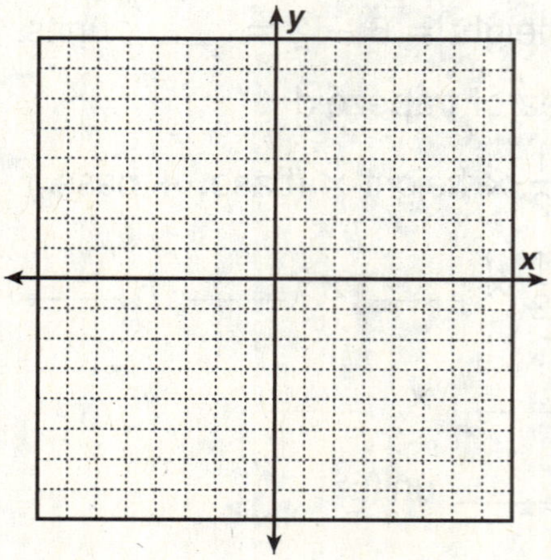

____

Name ______________________ Date ____________ Class ____________

LESSON 8B

Homework and Practice

Perimeter and Area of Triangles and Trapezoids, continued

Use a metric ruler to measure the dimensions of each figure to the nearest millimeter. Then find the perimeter of each figure.

8.

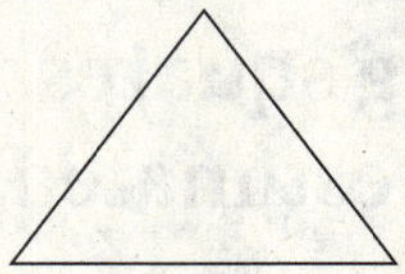

9.

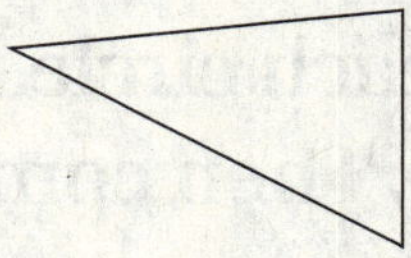

10.

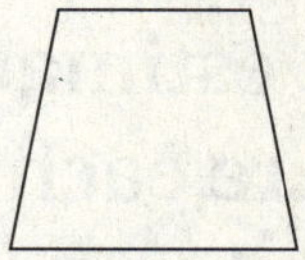

11.

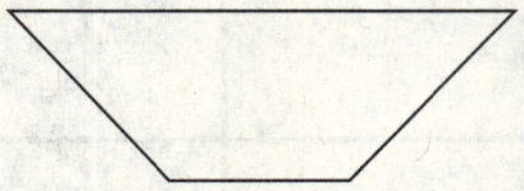

12.

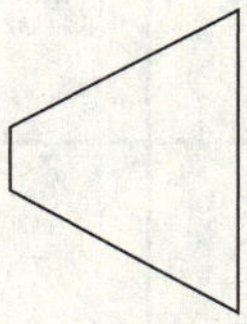

13.

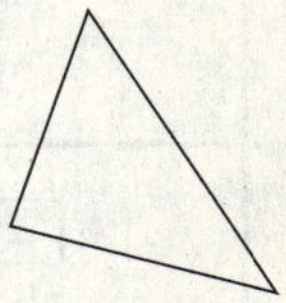

Find the perimeter and area of each figure.

14.

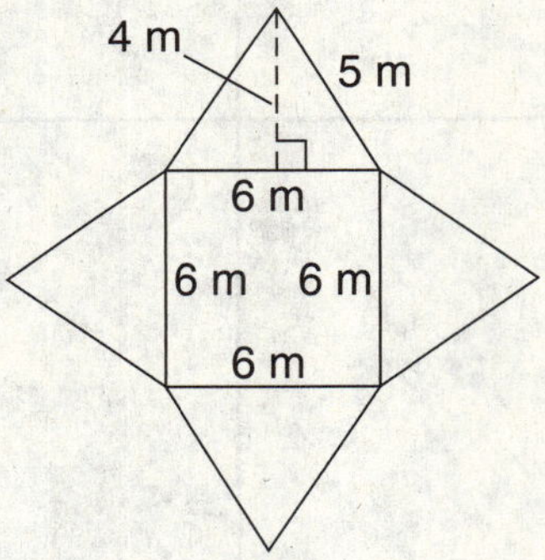

15.

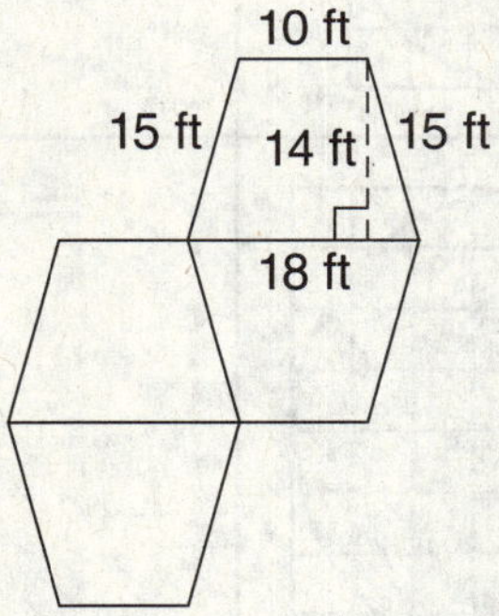

EXPLORATION

8C Circles

Program of Studies and Core Content

MA-8-AT-S-EI2, MA-8-AT-S-VEO2, MA-8-G-S-SR2, MA-8-M-S-MPA5, MA-8-NPO-S-NS1, MA-8-NPO-S-NS3, MA-8-M-S-MPA9, MA-08-1.1.1, MA-08-1.1.2, MA-08-2.1.1, MA-08-3.1.2, MA-08-2.1.2

First estimate the area of each circle by counting squares. Then square each circle's radius. Then compare each estimated area with the square of the radius by computing $\frac{A}{r^2}$.

		Radius	Estimated Area	r^2	$\frac{A}{r^2}$
1.		$r = 1$			
2.	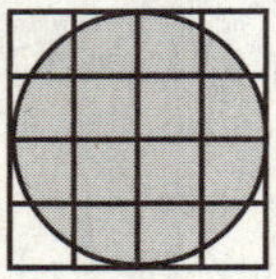	$r = 2$			
3.	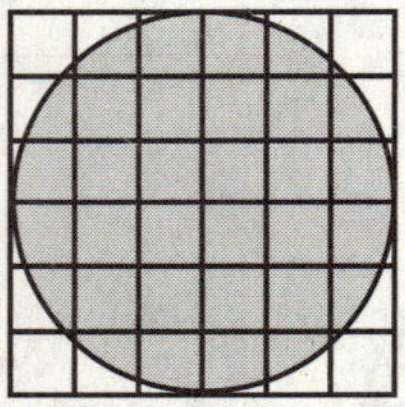	$r = 3$			
4.	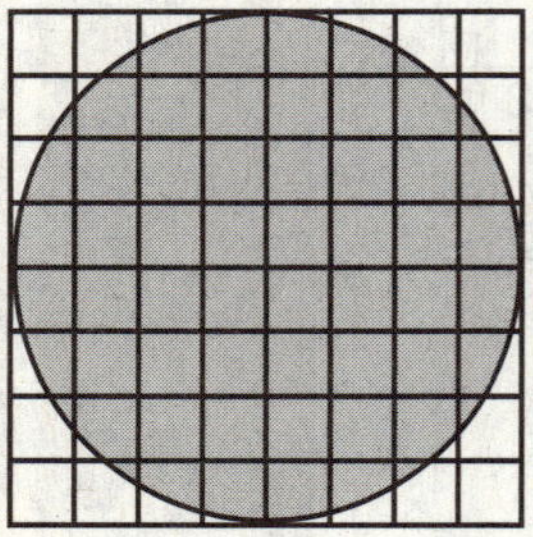	$r = 4$			

Think and Discuss

5. Discuss how you can make a generalization about how to estimate the area of a circle if you know the radius.

Name ______________________________ Date __________ Class __________

LESSON 8C

Hands-On Lab

Area of a Circle

You have learned that the area of a circle is πr^2 when the radius is r. In this activity you will prove it by showing that the area of a circle is equal to the area of a rectangle with a length of πr and a width of r.

Activity

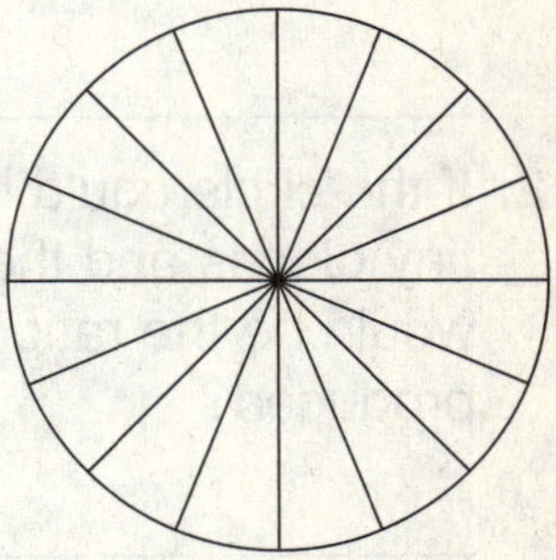

Take a sheet of paper and use a compass to draw a circle with a radius of about 5 cm or trace the circle given. Use a centimeter ruler to measure the radius of your circle. Use a calculator and the formula πr^2 to calculate the area of your circle in square centimeters.

With a ruler and protractor, draw lines to divide the circle into sixteenths as shown. How large will each angle be at the center?

____

Cut out the circle and cut it apart into fourths. Arrange the pieces on a piece of graph paper to make a sort of parallelogram with scalloped edges as shown. Measure the base and height of the parallelogram as accurately as you can with the centimeter ruler and enter the results in the table.

Now cut the fourths into eighths. Arrange the pieces again to make a parallelogram with scalloped edges. Measure the base and height of the parallelogram as accurately as you can and enter the results in the table. Then cut the eighths into sixteenths and repeat the process.

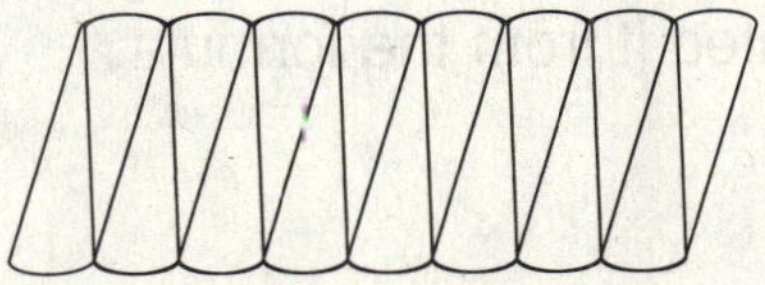

Radius = _______, Area ≈ _______

	Base	Height	Base/Height	Base * Height
Parallelogram #1				
Parallelogram #2				
Parallelogram #3				

Name ____________________ Date __________ Class __________

LESSON 8C

Hands-On Lab

Area of a Circle, continued

Think and Discuss

1. As the circle is divided into a larger number of pieces, what happens to the general shape of the scalloped parallelogram?

____

2. If the circle could be divided into an indefinitely large number of tiny pieces and the pieces were arranged as in this activity, what would be the ratio of the base to the height of the figure produced?

____

____

____

Try This

Draw a circle with a radius of about 3 inches or 7 cm. Measure the radius with a centimeter ruler and calculate the area using a calculator and the formula for the area of a circle. Use a protractor and ruler to divide the circle into 24 segments of 15° apiece. Cut the segments apart and arrange as in the previous example, using glue or rubber cement if needed to stick the segments in place on top of another sheet of paper. Measure the base and height of the approximate parallelogram created. Compare the ratio of the base to the height with π. Compare the product of the base and height with the area of the circle as you calculated it from the formula.

Name	Date	Class

LESSON 8C Review for Mastery
Circles

Circles		
	Circumference	**Area**
Models	distance around the circle 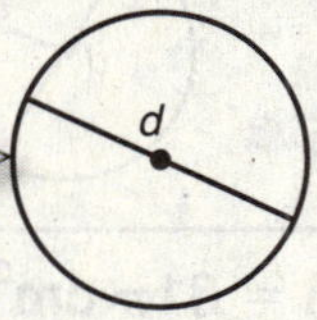	space inside the circle 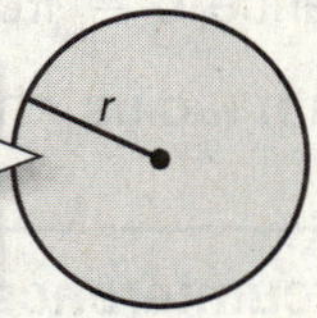
Words	pi times the diameter or 2 times pi times the radius	pi times the square of the radius
Formulas	$C = \pi d$ or $C = 2\pi r$	$A = \pi r^2$

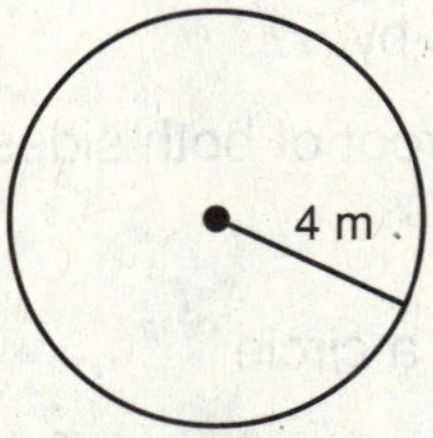

$C = 2\pi r$ | $A = \pi r^2$

$C = 2\pi(4)$ | $A = \pi(4)^2$

$C = 8\pi$ | $A = 16\pi$

$C \approx 25.1$ m | $A \approx 50.3$ m^2

Find the circumference and area of each circle. Use the π key on your calculator. Round to the nearest tenth.

1. circle with a radius of 11 inches

____

2. circle with a diameter of 15 millimeters

____

3. 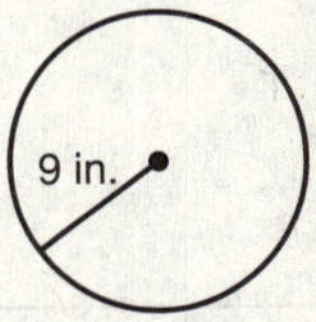

____

4. 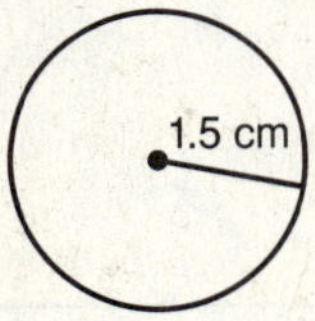

____

5. 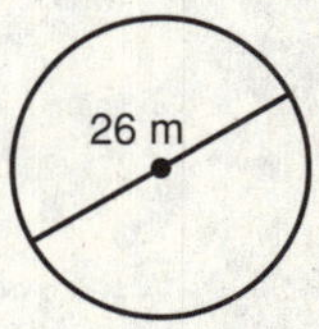

____

6. 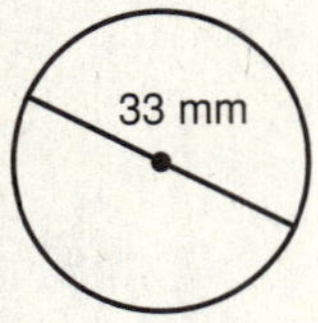

____

Name ______________________ Date ____________ Class ____________

LESSON 8C

Review for Mastery

Circles, continued

Circumference and Area of Circles

A circle with diameter d and radius r has circumference $C = \pi d$ or $C = 2\pi r$.

A circle with radius r has area $A = \pi r^2$.

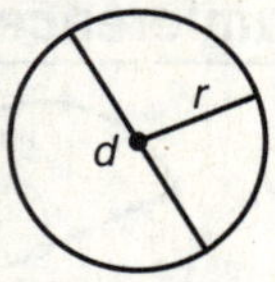

Find the circumference of circle S in which $A = 81\pi$ cm^2.

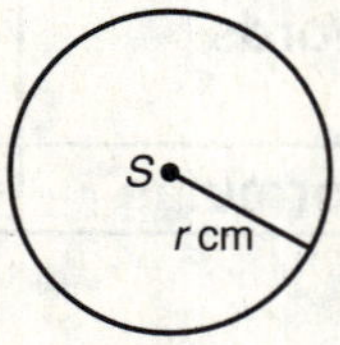

Step 1 Use the given area to solve for r.

$A = \pi r^2$	Area of a circle
$81\pi \text{ cm}^2 = \pi r^2$	Substitute 81π for A.
$81 \text{ cm}^2 = r^2$	Divide both sides by π.
$9 \text{ cm} = r$	Take the square root of both sides.

Step 2 Use the value of r to find the circumference.

$C = 2\pi r$	Circumference of a circle
$C = 2\pi(9 \text{ cm}) = 18\pi \text{ cm}$	Substitute 9 cm for r and simplify.

Find each measurement.

7. the circumference of circle B

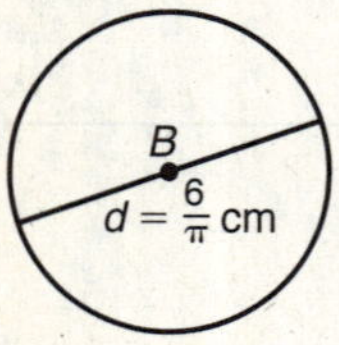

____

8. the area of circle R in terms of π

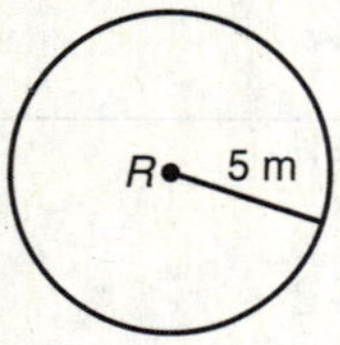

____

9. the area of circle Z in terms of π

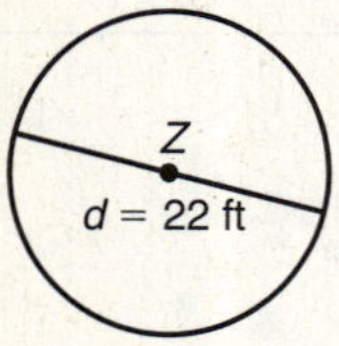

____

10. the circumference of circle T in terms of π

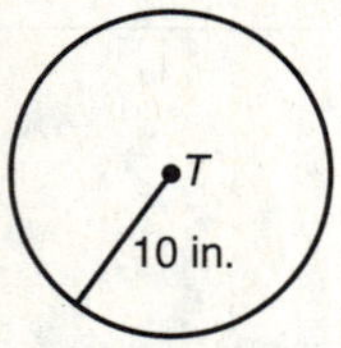

____

11. the circumference of circle X in which $A = 49\pi$ in^2

____

12. the radius of circle Y in which $C = 18\pi$ cm

____

Name ______________________ Date __________ Class __________

LESSON 8C

Homework and Practice

Circles

Use the area and circumference formulas for circles to find each measurement. Give your answers in terms of π.

1.

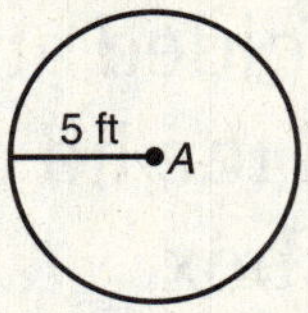

the area of circle A

____

2.

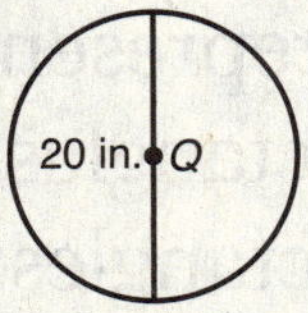

the area of circle Q

____

3.

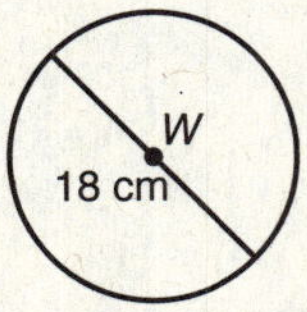

the circumference of circle W

____

4.

the circumference of circle N

____

5.

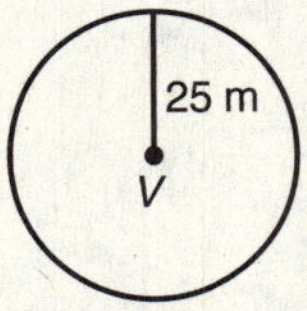

the area of circle V

____

6.

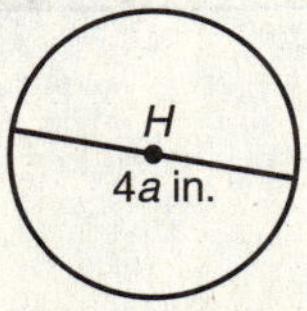

the area of circle circle H

____

7.

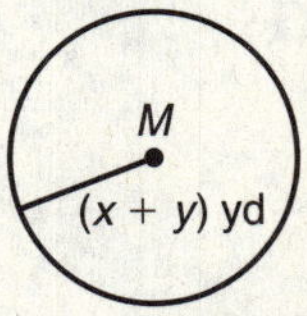

the circumference of circle M

____

8.

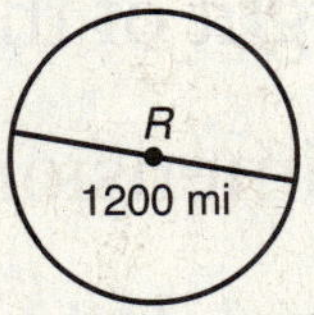

the circumference of circle R

____

9. the radius of $\odot I$ in which $A = 144\pi$ meters^2 ______________________

10. the diameter of $\odot L$ in which $C = 2\pi$ kilometers ______________________

11. the area of $\odot P$ in which $C = 32\pi$ yards ______________________

12. Emile is shopping for a new bicycle. He sees that a trick bike has 20-inch-diameter wheels, a mountain bike has 26-inch-diameter wheels, and a racing bike has 27-inch-diameter wheels. Find the area of each wheel. Round to the nearest tenth.

____

EXPLORATION

8D Volume of Prisms and Cylinders

Program of Studies and Core Content
MA-8-AT-S-EI2, MA-8-AT-S-VE02, MA-8-G-S-SR3, MA-8-M-S-MPA7, MA-08-2.1.4, MA-08-3.1.3

The drawing below represents a box that has been unfolded and laid flat. The dark rectangles represent the bottom and the lid of the box. The light rectangles represent the sides of the box.

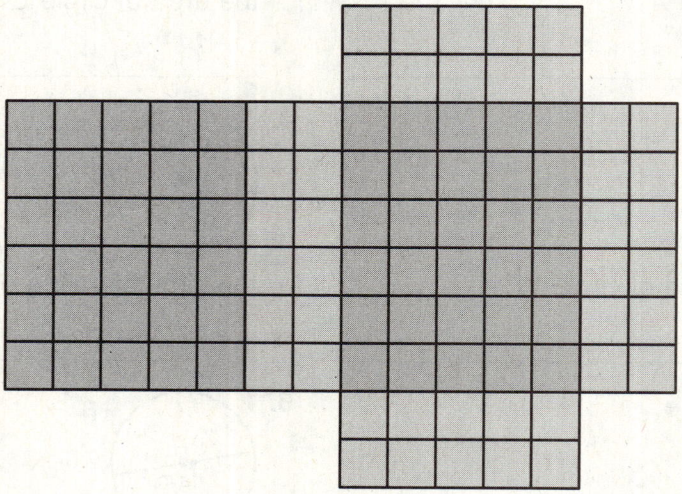

1. What are the dimensions of the bottom and the lid of the box?
2. What is the height of the box when it is assembled?
3. What are the dimensions of each side of the box?
4. How many 1-by-1-by-1 cubes will it take to fill the box when it is assembled? (*Hint:* This is the ***volume*** of the box.)

Think and Discuss

5. **Explain** how you can determine the volume of a box if you know the dimensions.
6. **Discuss** whether you could figure out the height of a box if you know the volume and the dimensions of the base.

Name ______________________________ Date ____________ Class ____________

LESSON **8D**

Hands-On Lab 1

Volume of Prisms and Cylinders

Activity

Use a drinking glass (shape of the cylinder below) and fill it with water using a measuring cup. The volume will be measured in milliliters (a glass holding 8 oz. of water is equal to 236.59 ml of water). Use a metric ruler to find the radius and the height of the glass.

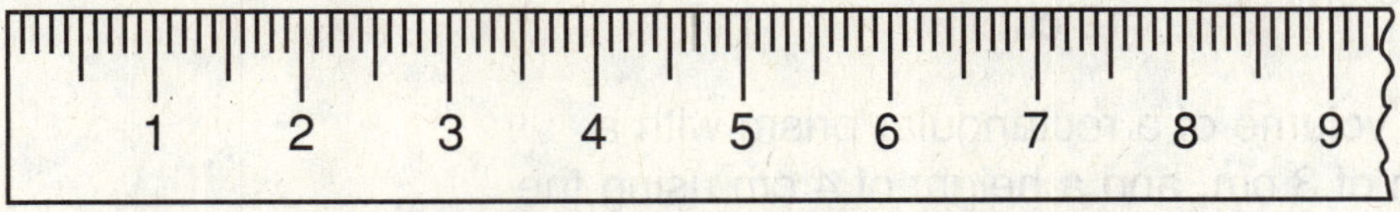

Centimeters and Millimeters

If a cylinder has a diameter of 5 cm and a height of 10 cm, find the volume of the cylinder by using these steps.

Step 1 If a cylinder has a base with a radius of r units and a height of h units, the volume, V, is $\pi r^2 h$ cubic units, or $V = \pi r^2 h$. ($\pi \approx 3.14$)

Step 2 $V = \pi(2.5)^2(10)$

Step 3 $V = \pi(6.25)(10)$

Step 4 $V \approx 3.14(6.25)(10)$

Step 5 $V \approx 196.25 \text{ cm}^3 \approx 196.25 \text{ ml}$

Cylinder

Name ______________________ Date __________ Class __________

LESSON 8D

Hands-On Lab 1

Volume of Prisms and Cylinders, continued

Think and Discuss

1.

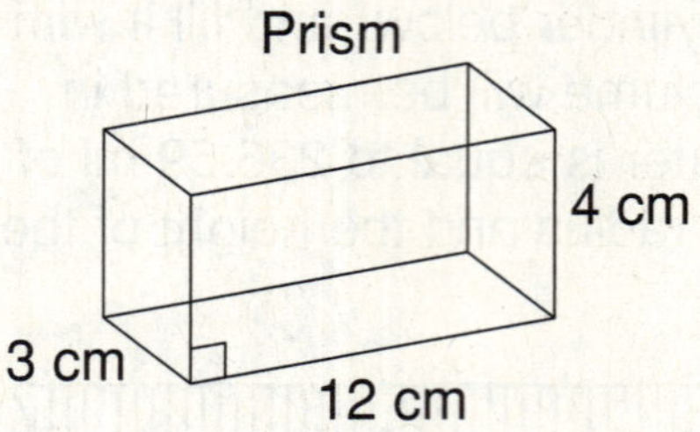

How would you find the volume of a rectangular prism with a length of 12 cm, a width of 3 cm, and a height of 4 cm using the formula below?

Formula: $V = Bh$

Since the base of the prism is a rectangle, then $B = \ell w$

Step 1 $V = (\ell w) \times h$

Step 2 $V = (12)(3)(4)$

Step 3 $V = 144 \text{ cm}^3$

____

Try This

1. Write in words the formula for finding the volume of a cylinder.

____

____

2. Write in words the formula for finding the volume of a prism.

____

____

Name ______________________________ Date ____________ Class ____________

LESSON **8D**

Hands-On Lab 2

Volume of Prisms and Cylinders

What kind of container could be made from a sheet of paper? How much would it hold? In this activity we will use a sheet of paper to form two different cylinders and find out which cylinder has the greater volume.

Activity

1. Tape the two long edges of a piece of paper together to make a tall cylinder. (Roll the paper up tightly before using it.) Create a bottom for the cylinder by placing one end of the cylinder on another piece of paper, tracing around the cylinder, cutting out the traced circular shape, and taping it to an open end of the cylinder.

Measure the cylinder's volume by counting the number of cups of packing peanuts the tall cylinder will hold.

____

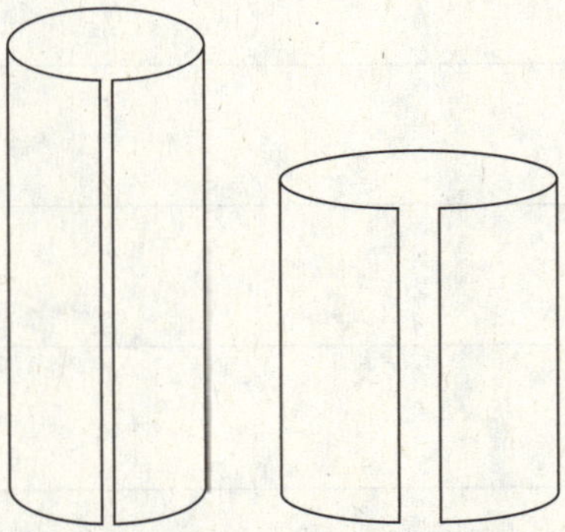

Cylinder #1 Cylinder #2

2. Now roll up another piece of paper from top to bottom. (Again, it helps to roll it up tightly before taping it.) Use tape to join the top and bottom of the sheet to make a shorter, wider cylinder than the first. Create a circular bottom for it as you did for the first cylinder.

Measure its volume in cups of packing peanuts.

____

3. Which cylinder has the greater volume?

____

Name ______________________ Date ____________ Class ____________

LESSON 8D

Hands-On Lab 2

Volume of Prisms and Cylinders, continued

Think and Discuss

1. What are the height and circumference of the two cylinders? How do they compare?

____

____

____

____

2. Kristen says the volumes of the cylinders must be the same since the same size pieces of paper are used to make them. Use the appropriate formula to calculate each cylinder's volume to the nearest 0.1. Use your calculations to show why the volumes are different.

____

____

____

____

____

Try This

1. Empty the packing peanuts and pour them back in, one cup at a time. As you pour each cup in, use the felt-tip pen to mark the level on the outside of the cylinder.

Name ______________________ Date ____________ Class ____________

LESSON 8D

Hands-On Lab 3

Volume of Prisms and Cylinders

If one three-dimensional figure has a greater surface area than another, does it also have a greater volume? In this activity we will find out if that is necessarily so.

Activity

Take a sheet of 8.5 in. × 11 in. paper. Use a ruler to draw lines across the paper 1 in. from the top and 1 in. from the bottom. Also, draw lines from top to bottom, 1 in. from the left and 1 in. from the right. Cut away the squares at the corners and fold up the remaining four sides and tape them together to make a box. Measure the capacity of the box using plastic packing peanuts measured with a coffee mug.

____

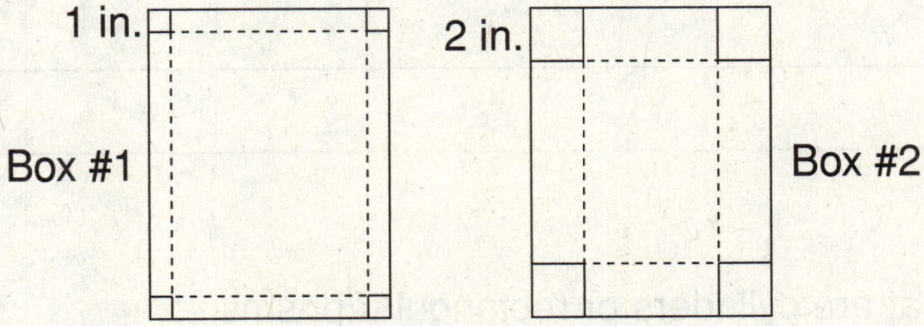

Take another sheet of paper. This time draw lines 2 in. from top and bottom and from left and right. Cut away the squares at the corners and fold up to make a box as before. Measure the capacity of the new box.

____

Name ________________________ Date ____________ Class ____________

LESSON 8D

Hands-On Lab 3

Volume of Prisms and Cylinders, continued

Think and Discuss

1. Which box has a greater surface area? Can you say without calculating?

____

____

____

2. Box #2, which is 2 in. tall, had a greater volume than Box #1, which is 1 in. tall. If you drew the lines 3 in. from every side of the paper and folded up the paper to make a box 3 in. tall, do you think it would have an even greater volume? Explain.

____

____

____

Try This

Arrange several containers that are cylinders or rectangular prisms. Assign each container a letter. Invite a student to estimate which container holds the most, second most, and so on and to write the letters in sequence on the blackboard: "C < A < B," for example. Then measure the capacities with plastic packing peanuts to see which container really holds more.

Name ______________________ Date ____________ Class ____________

LESSON **8D**

Review for Mastery

Volume of Prisms and Cylinders

Solids are made up of flat surfaces and curved surfaces. A flat surface is called a **face.** The intersection of two faces is a segment called an **edge.** Three or more faces intersect at a **vertex.** Use the graphic aid below to understand solids.

Solids

Prism

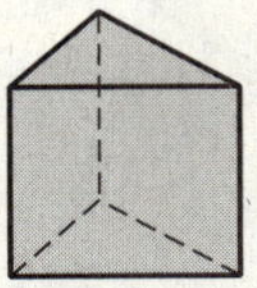

- Two parallel congruent polygonal bases
- The bases are connected by faces that are parallelograms.
- Named for the shape of its base

Pyramid

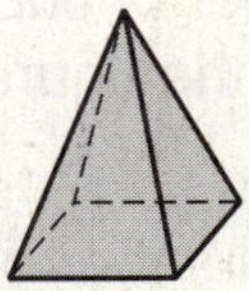

- One polygonal base
- The triangular faces meet at a common vertex.
- Named for the shape of its base

Cylinder

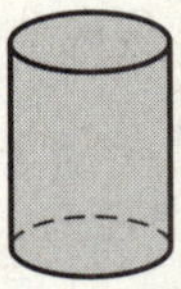

- Two parallel congruent circular bases
- A curved surface connects the bases.

Cone

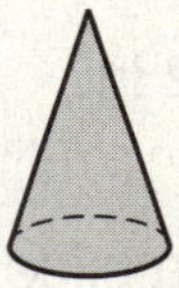

- One circular base
- A curved surface connects the base to a vertex.

Answer the following.

1. Name two solids that have circular bases.

   ____________________ ____________________

2. A(n) ____________________ has two parallel congruent polygonal bases.

3. The intersection of two faces is called a(n) ____________________.

Identify each solid.

4. 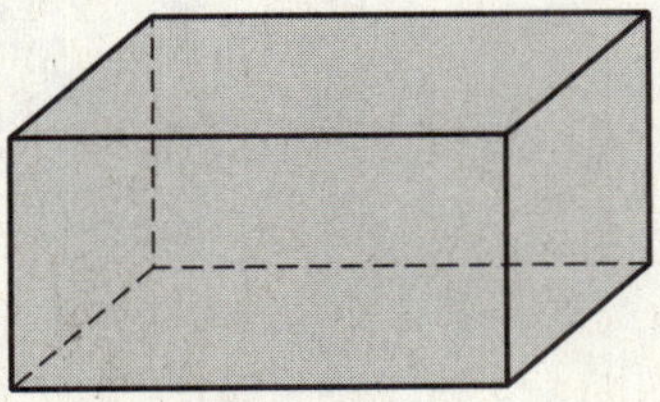

   ____

5. 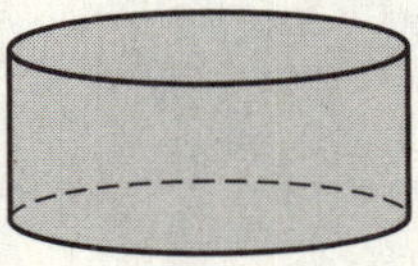

   ____

6. 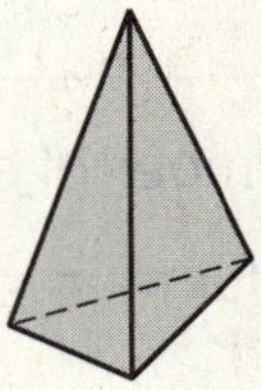

   ____

Name ______________________________ Date ___________ Class ___________

LESSON 8D

Review for Mastery

Volume of Prisms and Cylinders, continued

Volume = number of cubic units inside a solid figure

To find the volume of this solid figure:

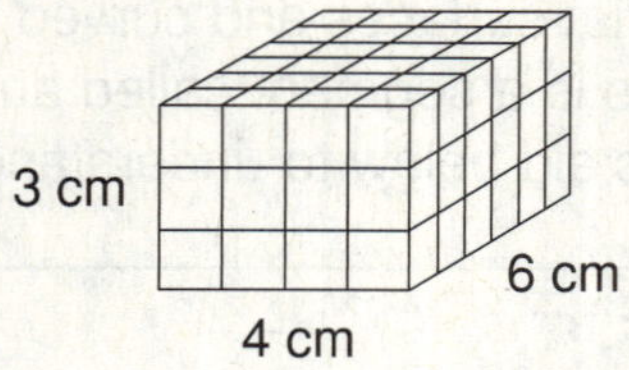

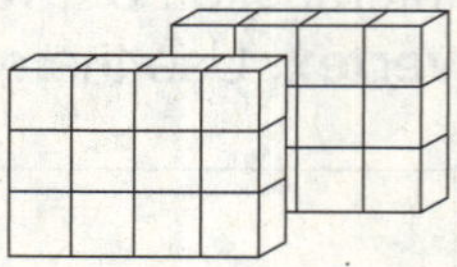

Count the number of cubic centimeters in one "slice" of the figure.

$4 \times 3 = 12$

Multiply by the number of "slices." $12 \times 6 = 72$ cm^3

Complete to find the volume of each solid figure.

7.

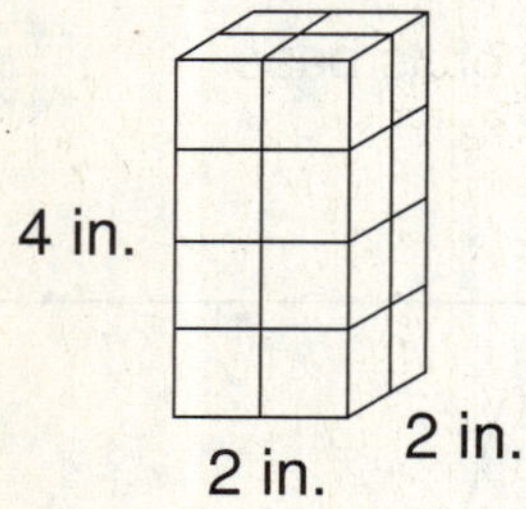

number in^3 in a slice

= ______ × ______ = ______

number of slices = ______

volume = ______ in^3

8.

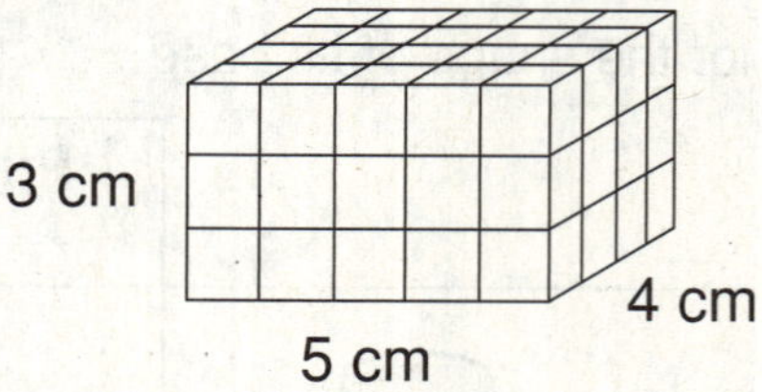

number cm^3 in a slice

= ____________ = ______

number of slices = ______

volume = ______ cm^3

9.

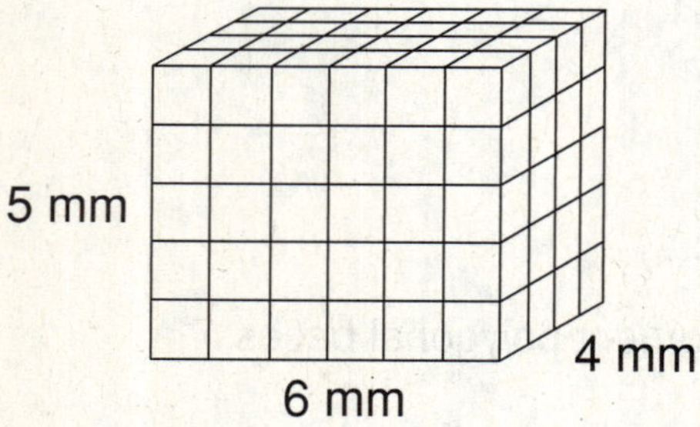

number mm^3 in a slice

= ____________ = ______

number of slices = ______

volume = ______ mm^3

Name ______________________________ Date ____________ Class ____________

LESSON 8D

Review for Mastery

Volume of Prisms and Cylinders, continued

Prism: solid figure named for the shape of its two congruent bases

Volume V of a prism = area of base B × height h

$$V = Bh$$
$$V = (6 \times 4)5$$
$$V = 24(5)$$
$$V = 120 \text{ in}^3$$

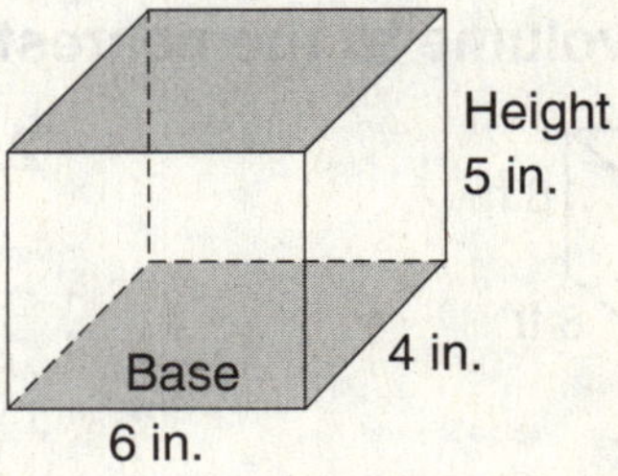

Rectangular Prism

Complete to find the volume of each prism.

10. rectangular prism

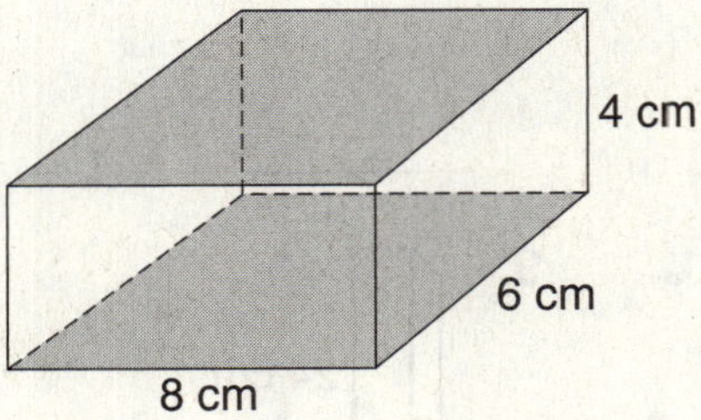

base is a rectangle

$V = Bh$

$V = (\_\_\_ \times \_\_\_) \times \_\_\_$

$= \_\_\_ \text{ cm}^3$

11. cube

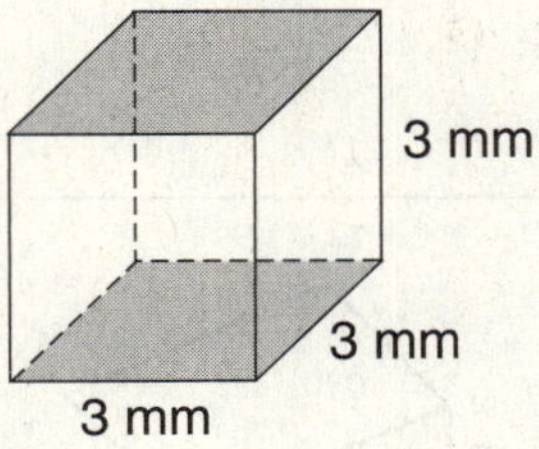

base is a ____________

$V = Bh$

$V = (\_\_\_ \times \_\_\_) \times \_\_\_$

$= \_\_\_ \text{ mm}^3$

12. triangular prism

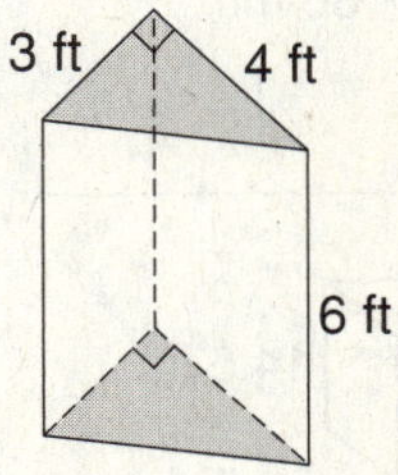

base is a ____________

$V = Bh$

$V = \frac{1}{2}(\_\_\_ \times \_\_\_)$

$\times \_\_\_ = \_\_\_ \text{ ft}^3$

Cylinder: solid figure with a circular base

Volume V of a cylinder = area of base B × height h

$$V = Bh$$
$$V = (\pi \times 4^2)7 = (16\pi)7$$
$$V = 112\pi$$
$$V \approx 112(3.14) \approx 351.7 \text{ units}^3$$

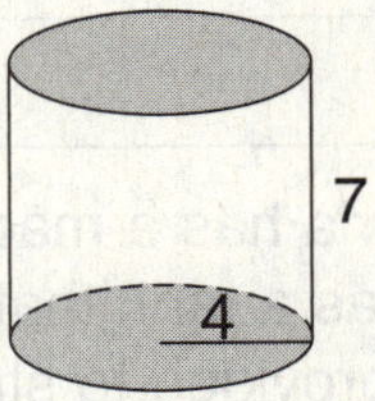

Complete to find the volume of the cylinder.

13.

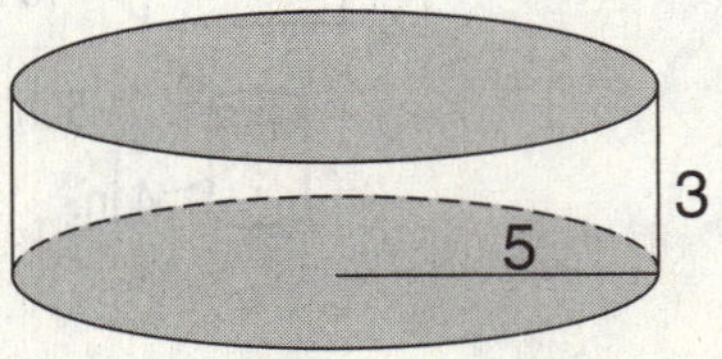

$V = Bh$

$V = (\pi \times \_\_\_\_\_) \times \_\_\_\_\_$

$V = \_\_\_\_\_ \approx \_\_\_\_\_ \text{ units}^3$

Name ______________________________ Date ____________ Class ____________

LESSON 8D

Homework and Practice

Volume of Prisms and Cylinders

Find the volume to the nearest tenth of a unit. Use 3.14 for π.

1.

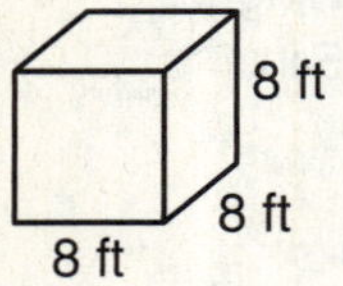

____

2.

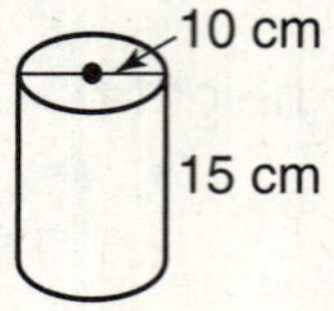

____

3.

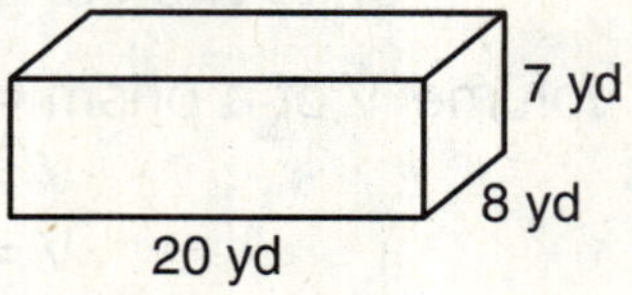

____

4.

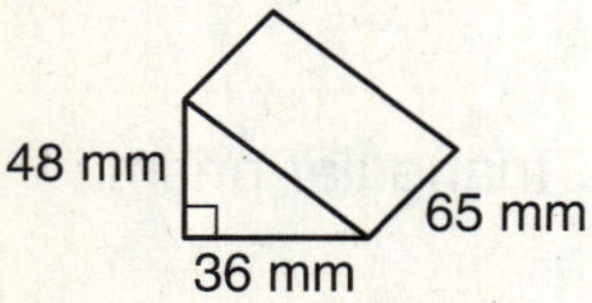

____

5.

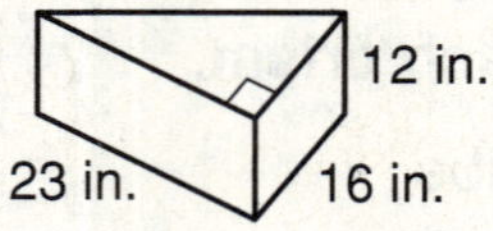

____

6.

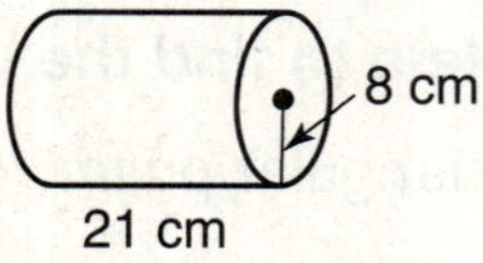

____

7.

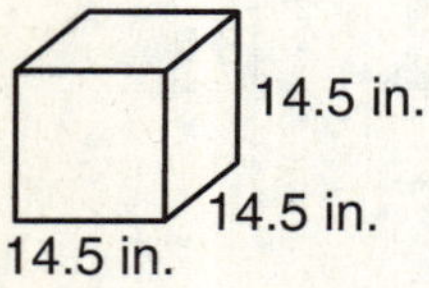

____

8.

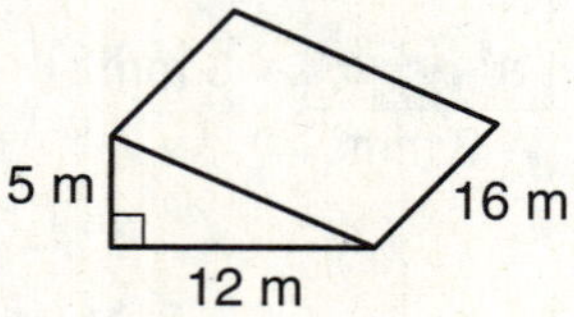

____

9.

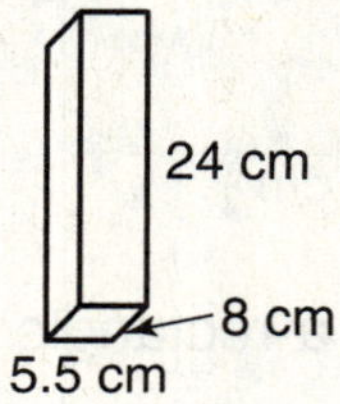

____

10. A cylinder has a radius of 8 cm and a height of 20 cm. Explain whether tripling the height will triple the volume of the cylinder.

____

____

11. The cafeteria has a machine that makes fruit smoothies. The machine has a tank that measures 18 in. by 18 in. by 24 in. The cups provided to students have a radius of 1.5 in. and a height of 4 in.

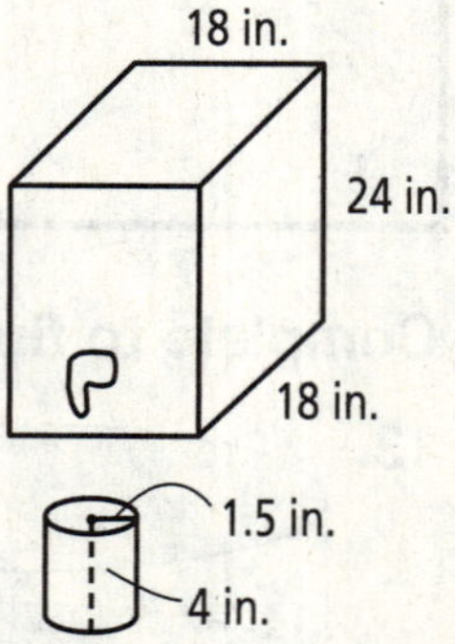

a. What is the volume of the tank?

____

b. What is the volume of a cup?

____

c. How many cups can one tank fill? ______________________________

Name ______________________ Date __________ Class __________

LESSON 8D

Homework and Practice

Volume of Prisms and Cylinders, continued

12. A landscape designer created a fountain with a rectangular prism that drains into a tray below. The water is then pumped out of the tray and back into the prism. The dimensions of the prism are shown.

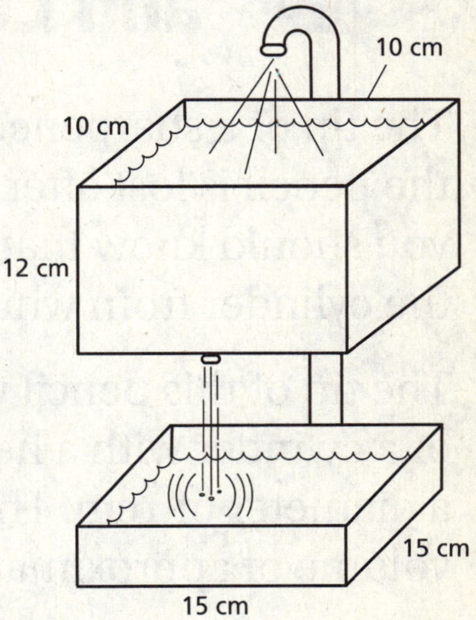

a. Find the volume of the prism.

____

b. The base of the tray is 15 cm long by 15 cm wide. How high should the tray be so that it holds the same amount of water as the prism?

____

13. A local candy store is having a contest. The person who comes closest to guessing the number of rolls of fruit candy in a large jar wins the candy. The dimensions of the jar and the rolls of fruit candy are shown.

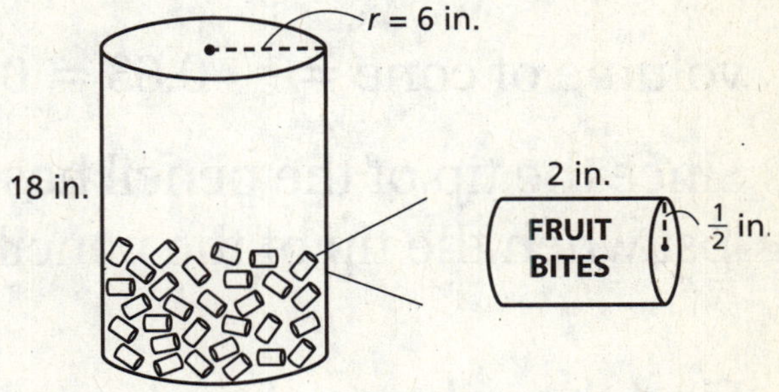

a. Find the volume of the jar.

Round to the nearest tenth. ______________________

b. Find the volume of one candy roll. Round to the nearest tenth. __________

c. Divide your answer in part **a** by part **b** to estimate how many candy rolls will fit in the jar. ______________________

d. Is your answer an overestimate or underestimate? ______________________

Program of Studies and Core Content

MA-8-AT-S-EI2, MA-8-AT-S-VE02, MA-8-M-S-MPA7, MA-08-3.1.3

8E Volume of Pyramids and Cones

The tip of a sharpened pencil is shaped like a cone. How much of the pencil is lost after the tip is formed? To answer this question, you should know that the volume of a cone is $\frac{1}{3}$ the volume of the cylinder from which it was formed.

The tip of this pencil was formed out of a cylinder with a height of 0.8 cm and a diameter of 1cm. The cylinder had a volume of approximately 0.63 cm^3.

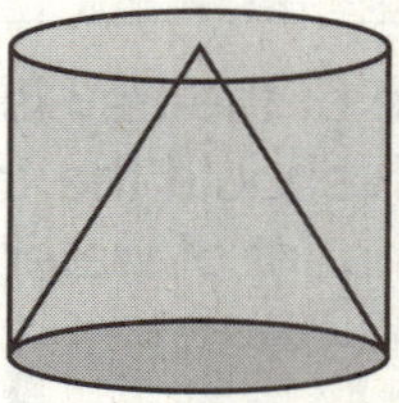

volume of cone = $\frac{1}{3}$ · volume of cylinder

volume of cone = $\frac{1}{3} \cdot 0.63 = 0.21$

Since the tip of the pencil has a volume of 0.21 cm^3, 0.42 cm^3 was lost when the tip of the pencil was formed.

Find the volume of each cone.

	Volume of Cylinder	Volume of Cone = $\frac{1}{3}$ · Volume of Cylinder
1.	66.9 $in.^3$	
2.	99 cm^3	
3.	108 $in.^3$	

Think and Discuss

4. Name familiar objects that are shaped like a cone.

5. Explain how to estimate the volume of one of the cone-shaped objects you named in Problem 4.

Name ______________________ Date ____________ Class ____________

LESSON 8E

Hands-On Lab 1

Volume of Pyramids and Cones

Activity

Roll a single sheet of paper into the shape of a cone. Tape it together. Draw a line around the cone 15 cm from the point of the cone. Cut around the line. You should have a cone with a height of 15 cm. Use a metric ruler to find the diameter of the cone.

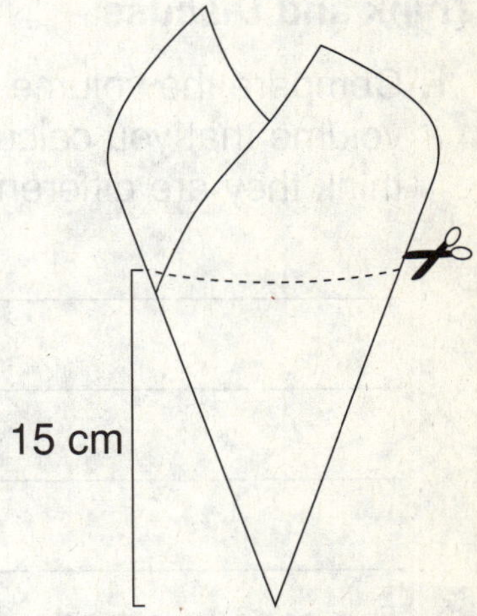

Next, fill the cone with rice. Then pour the rice into a measuring cup to see how many cups of rice filled the cone. There are 236.59 milliliters in one cup of rice. Convert the number of cups of rice it took to fill the cone to milliliters. One milliliter is equal to one cubic centimeter.

If a cone has a diameter of 15 cm and a height of 20 cm, find its volume using the following steps. Use 3.14 for pi.

Step 1 If a cone has a base with a radius of *r* units and a height of *h* units, its volume is given by the formula $V = \frac{1}{3}\pi r^2$. Since πr^2 is the area of the base, or *B*, the formula is also written $V = \frac{1}{3}Bh$.

Step 2 $V = \frac{1}{3}\pi(7.5)^2(20)$

Step 3 $V = \frac{1}{3}\pi(56.25)(20)$

Step 4 $V \approx = \frac{1}{3}(3.14)(56.25)(20)$

Step 5 $V \approx 1{,}177.5 \text{ cm}^3 \approx 1{,}177.5 \text{ ml}$

Name ______________________ Date __________ Class __________

LESSON 8E

Hands-On Lab 1

Volume of Pyramids and Cones, continued

Think and Discuss

1. Compare the volume of rice that filled the cone you made to the volume that you calculated. How are they different? Why do you think they are different?

____

____

____

____

2. The cone you made does not have a base. How does this affect its volume?

____

____

____

Try This

1. Write in words the formula for finding the volume of a cone.

____

____

____

2. Find the volume of a cone with a height of 22.5 cm and a diameter of 8 cm.

____

3. An ice cream shop has waffle cones that are 6.5 in. high and 4.5 in. across the top. What volume of ice cream will fit exactly inside the cone?

____

Name ______________________ Date ____________ Class ____________

LESSON 8E

Hands-On Lab 2

Volume of Pyramids and Cones

Activity

1. Build a pyramid using cubes that are 1 cm by 1 cm. Begin by building a square base that is 4 cm by 4 cm. Build the next layer by overlapping half of each of the cubes on the base layer. Continue adding layers this way.

2. Complete the chart. Count the number of cubes in each layer and multiply this number by 1 cm^3 to find the volume of each layer.

Layer	Number of Cubes	Volume of Cubes
Base		
First		
Second		
Third		
Total:	30 cubes	30 cm^3

3. What is the height of the pyramid?

____

4. What is the area of the base of the pyramid?

____

5. Use the formula $V = \frac{1}{3}Bh$ to calculate the volume of the pyramid you built to the nearest tenth of a cubic centimeter. Show your work.

$V = \frac{1}{3}(16)(4)$

$V = \frac{1}{3}(\_\_)$

$V = \_\_\ cm^3$

6. How does the actual volume of a square pyramid with a side of 4 cm and a height of 4 cm compare to the volume of the pyramid you built?

____

Name ________________________________ Date ____________ Class ____________

LESSON 8E

Hands-On Lab 2

Volume of Pyramids and Cones, continued

Think and Discuss

1. Compare the volumes of the two figures.

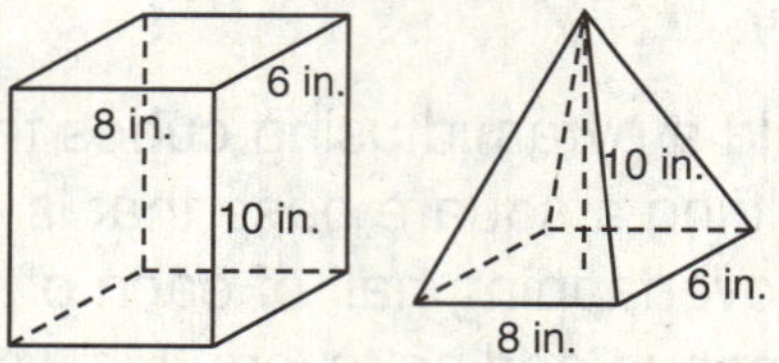

____

2. A pyramid has the same base and height as a rectangular prism. How can you calculate the volume of the pyramid by using the volume of the rectangular prism?

____

____

Try This

1. Calculate the volume of a pyramid that has a triangular base with a base height of 8 ft and a base length of 5 ft if the height of the pyramid is 17 ft. Write your answer to the nearest tenth of a foot. Show your work.

____

____

____

2. A pyramid-shaped object is being packed into a rectangular box for shipping. The box is 22 in. long, 20 in. wide and 18 in. high. The pyramid has the same base as the larger box and is the same height. Subtract the volume of the pyramid from the volume of the box to find the amount of packing material needed to go around the pyramid.

____

____

____

____

____

Name ______________________________ Date ____________ Class ____________

LESSON **8E**

Review for Mastery

Volume of Pyramids and Cones

Pyramid: 3-dimensional figure named forthe shape of its base, which is a polygon; all other faces are triangles

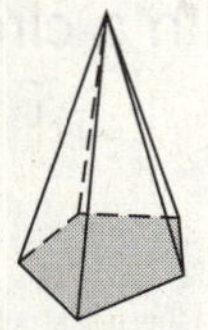

Pentagonal Pyramid

This rectangular pyramid and rectangular prism have congruent bases and congruent heights.

Volume of Pyramid $= \frac{1}{3}$ Volume of Prism

$$\boldsymbol{V = \frac{1}{3} Bh}$$

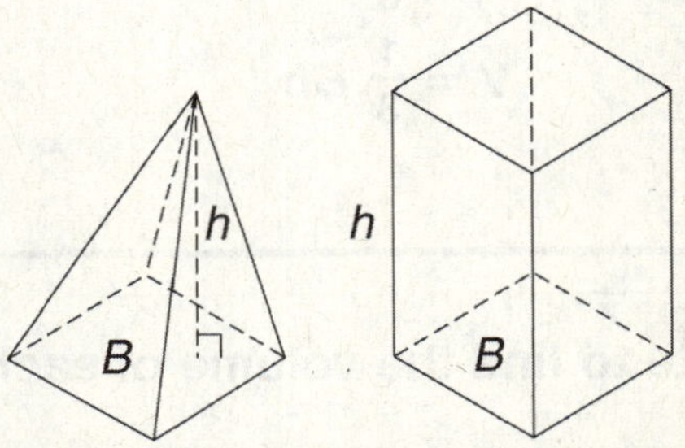

Complete to find the volume of each pyramid.

1. square pyramid

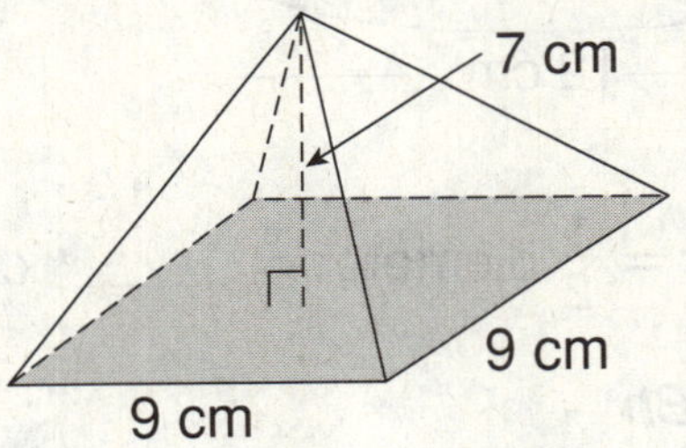

base is a ____________________

$V = \frac{1}{3} Bh$

$V = \frac{1}{3}$ (area of square) × h

$V = \frac{1}{3}$ (____ × ____) × ____

$V = \frac{1}{3}$ (____) × ____

$V =$ ________ cm^3

2. rectangular pyramid

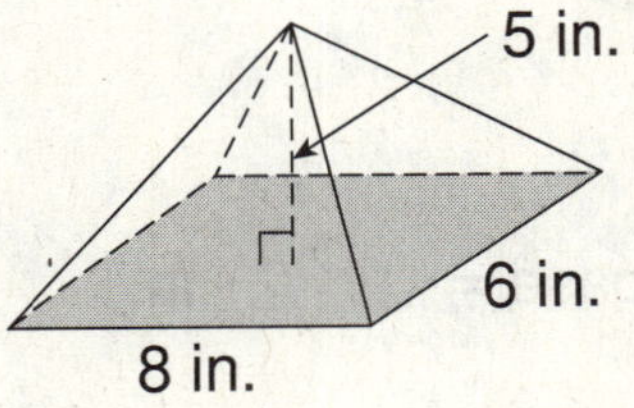

base is a ____________________

$V = \frac{1}{3} Bh$

$V = \frac{1}{3}$ (area of rectangle) × h

$V = \frac{1}{3}$ (____ × ____) × ____

$V = \frac{1}{3}$ (____) × ____

$V =$ ________ in^3

Name ______________________ Date __________ Class __________

LESSON 8E

Review for Mastery

Volume of Pyramids and Cones, continued

Cone: 3-dimensional figure with a circular base

This cone and cylinder have congruent bases and congruent heights.

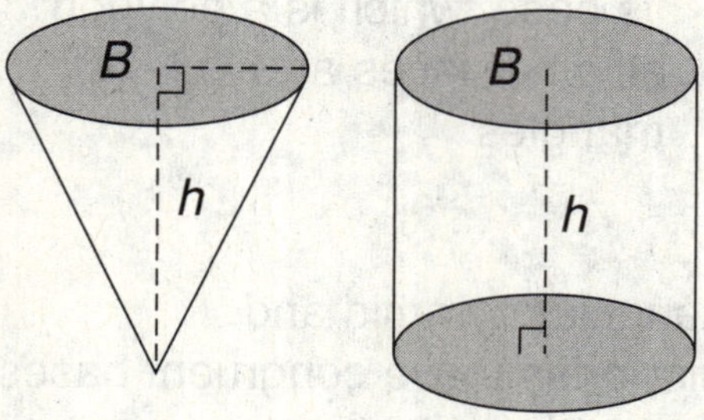

Volume of Cone = $\frac{1}{3}$ Volume of Cylinder

$$V = \frac{1}{3}Bh$$

Complete to find the volume of each cone.

3.

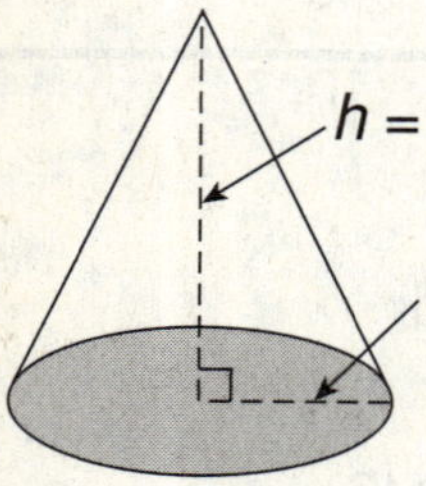

radius r of base = ______ in.

$V = \frac{1}{3}Bh$

$V = \frac{1}{3}(\pi r^2)h$

$V = \frac{1}{3}(\pi \times$ ______ $) \times$ ______

$V = \frac{1}{3}($ ______ $) \times$ ______

$V =$ ______ $\times$ ______

$V =$ ______

$V \approx$ ______ $\times 3.14$

$V \approx$ __________ in^3

4.

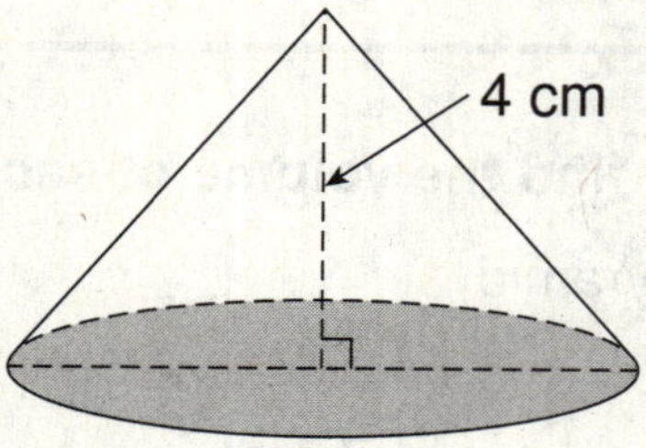

radius $r = \frac{1}{2}$ diameter = ______ cm

$V = \frac{1}{3}Bh$

$V = \frac{1}{3}(\pi r^2)h$

$V = \frac{1}{3}(\pi \times$ ______ $) \times$ ______

$V = \frac{1}{3}($ ______ $) \times$ ______

$V =$ ______ $\times$ ______

$V =$ ______

$V \approx$ ______ $\times 3.14$

$V \approx$ __________ cm^3

Name ______________________ Date __________ Class __________

LESSON **8E**

Homework and Practice

Volume of Pyramids and Cones

Find the volume of each figure to the nearest tenth of a unit. Use 3.14 for π.

1.

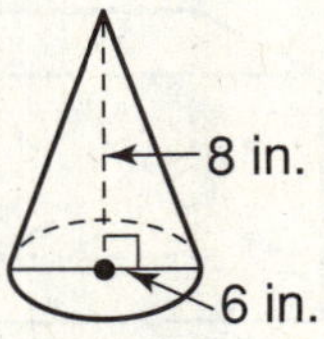

____

2.

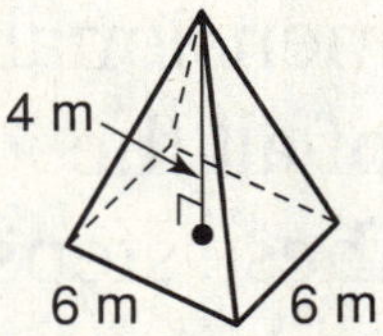

____

3.

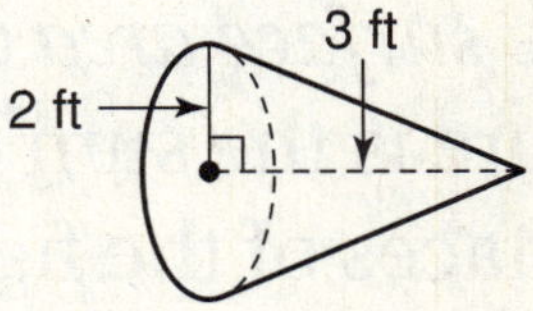

____

4.

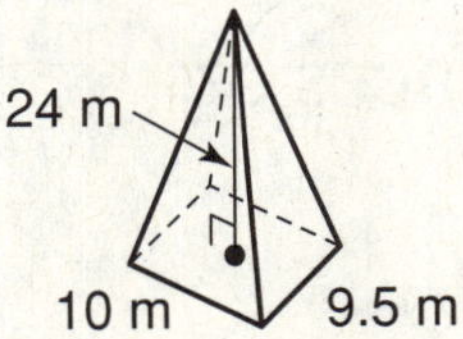

____

5.

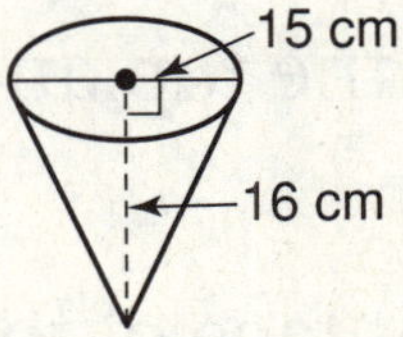

____

6.

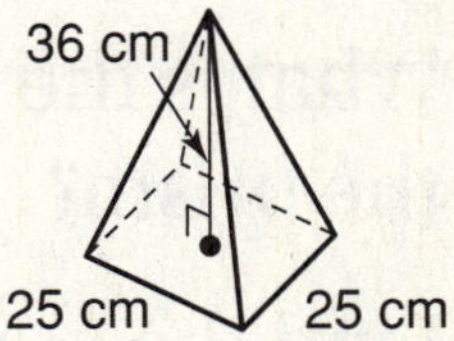

____

7. A funnel has a diameter of 8 in. and is 21 in. deep. What is the volume of the funnel to the nearest tenth of a unit? Use 3.14 for π

____

8. The radius of a cone is 13 ft and its height is 27 ft. Find the volume of the cone to the nearest tenth. Use 3.14 for π.

____

9. Find the volume of a rectangular pyramid if the height is 35 cm and the base sides are 21 cm and 28 cm.

____

10. The base of a regular pyramid has an area of 135 in.^{2}. The height of the pyramid is 8.5 in. Find the volume.

____

EXPLORATION

8F Surface Area of Prisms and Cylinders

Program of Studies and Core Content

MA-8-AT-S-EI2, MA-8-AT-S-VE02, MA-8-M-S-MPA7, MA-08-2.1.5, MA-08-3.1.3

The *surface area* of a three-dimensional figure is the sum of the areas of all the surfaces of the figure. Follow these steps to find the surface area of the prism.

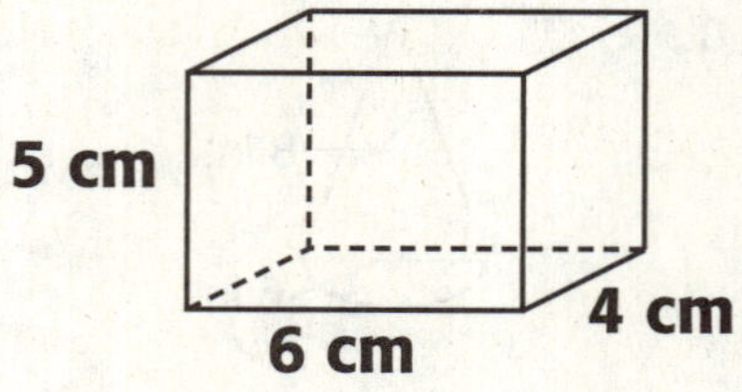

1. Find the area of the bottom face of the prism.
2. What is the total area of the top and bottom faces of the prism?
3. Find the area of the front face of the prism.
4. What is the total area of the front and back faces of the prism?
5. Find the area of one of the side faces of the prism.
6. What is the total area of both side faces of the prism?
7. Add the areas in Steps 2, 4, and 6 to find the surface area.

Think and Discuss

8. **Explain** how you can find the surface area of a rectangular prism.

9. **Describe** a shortcut you can use to find the surface area of a cube.

Name ______________________ Date ____________ Class ____________

LESSON 8F

Hands-On Lab 1

Surface Area of Prisms and Cylinders

Activity

A prism is a three-dimensional figure that has two parallel, congruent sides, called bases. The surface area is the sum of the areas of all surfaces of a figure.

For the purposes of this activity, each square will equal 1 cm^2. Find the surface area of the figure below.

Step 1 Draw the figure shown on a piece of graph paper.

Step 2 Cut out the figure and fold along the lines and tabs shown into a rectangular prism.

Step 3 Determine the surface area.

Once the figure has been folded into a rectangular prism, you will see that the length is 6 cm, the width is 5 cm and the height is 4 cm. To calculate the surface area, the formula is:

2 × (length × width + length × height + width × height)

Step 1 2 × (length × width + length × height + width × height)

Step 2 2 × (6 × 5 + 6 × 4 + 5 × 4)

Step 3 2 × (30 + 24 + 20)

Step 4 2 × (74)

Step 5 148

The surface area is 148 cm^2.

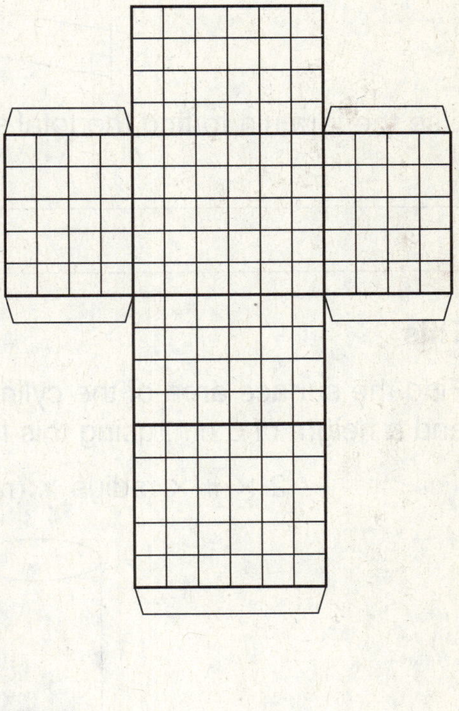

Name ______________________________ Date ____________ Class ____________

LESSON 8F

Hands-On Lab 1

Surface Area of Prisms and Cylinders, continued

Think and Discuss

1. Area is the number of square units needed to cover a given surface of a two-dimensional figure. There are special formulas for prisms, but they deal only with right prisms. A right prism is one with two special characteristics: all lateral edges are perpendicular (a line at right angles to a line or plane) to the bases, and lateral faces are rectangular. The figure below shows a right prism with a length of 12 cm, a width of 4 cm and a height of 4 cm. Discuss and create a formula for finding the total area of the prism.

Prism Area Theorem:
The lateral area L (area of the vertical sides only) of any right prism is equal to the perimeter of the base times the height of the prism ($L = Ph$). The total surface area T of a right prism is equal to two times the area of the base, plus the lateral area.

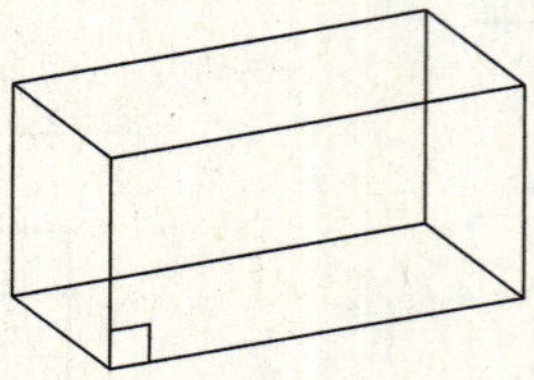

Use the formula to find the total surface area of the prism.

____

____

Try This

1. Find the surface area of the cylinder below with a radius of 2 cm and a height of 6 cm, using this formula.

$$2 \times \pi \times \text{radius} \times (\text{radius} + \text{height})$$

____

Name ______________________ Date ____________ Class ____________

LESSON 8F

Hands-On Lab 2

Surface Area of Prisms and Cylinders

Activity

The net of a cylinder shows what a cylinder would look like if it were unfolded and laid flat. Use a metric ruler and compass to draw the net of a cylinder like the one below. Make the cylinder 6 cm high with a radius of 2 cm. Carefully cut out the net so the bases remain attached to the lateral surface. Then fold the net into a cylinder.

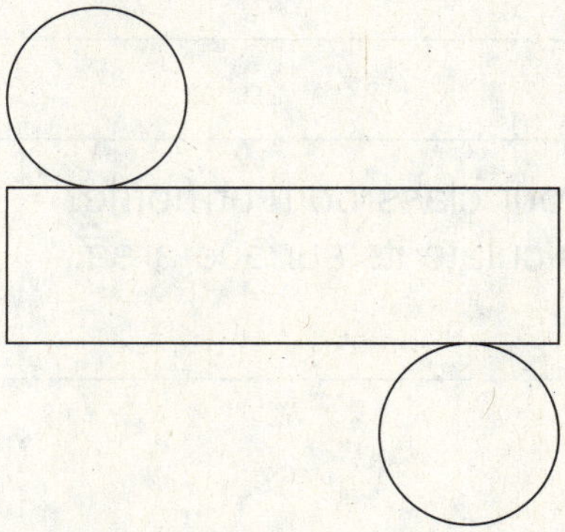

Think and Discuss

1. What is the formula for the circumference of a circle?

____

2. Write out the formula for the surface area of a cylinder. Circle the formula for the circumference of a circle in this formula.

____

3. How does the formula for the lateral area of a cylinder, $L = 2\pi rh$, relate to the formula for the area of a rectangle, $A = lw$?

____

4. What is the length of the lateral surface of the cylinder you created?

____

Name ______________________ Date __________ Class __________

LESSON 8F

Hands-On Lab 2

Surface Area of Prisms and Cylinders, continued

Try This

1. A label on a can of soup covers the lateral surface of the can. What is the area of the label if the can is 4 in. tall and has a diameter of 2 in.? Show your work.

____

____

____

2. Find a cylindrical-shaped object in your classroom or home. Measure its height and diameter. Calculate its surface area.

____

Name ______________________ Date ___________ Class ___________

LESSON 8F

Hands-On Lab 3

Surface Area of Prisms and Cylinders

Activity 1

Measure the length, width, and height of a cardboard box. Make a sketch of the box and its dimensions below. Cut the box apart so that it can lie flat. This is a net of the box. Calculate the surface area of the box by finding the surface areas of parts of the net and adding them together.

What is the surface area of the box?

Remember: Units for area are always square, such as cm^2 or in^2.

Activity 2

You can find the surface area of a prism by drawing its net on graph paper and counting the squares to find the area. Draw the net of the prism below on a piece of graph paper. Each square on the graph paper will represent 1 cm^2.

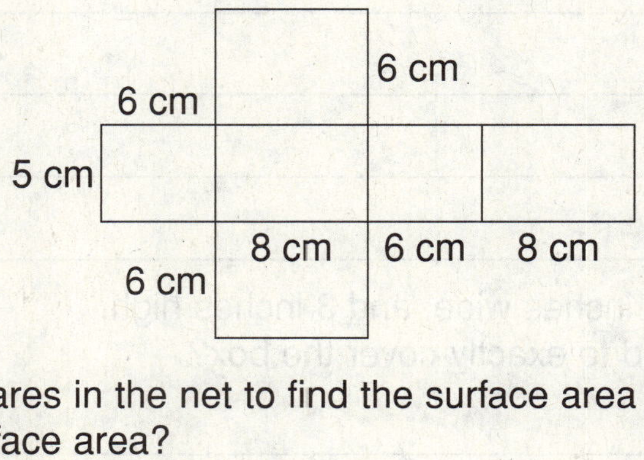

Count the squares in the net to find the surface area of the prism. What is its surface area?

Name ______________________________ Date ______________ Class ______________

LESSON **8F**

Hands-On Lab 3

Surface Area of Prisms and Cylinders, continued

Think and Discuss

1. The bases of a prism can be any polygon. How would you find the surface area of a triangular prism using the formula below given the following dimensions?

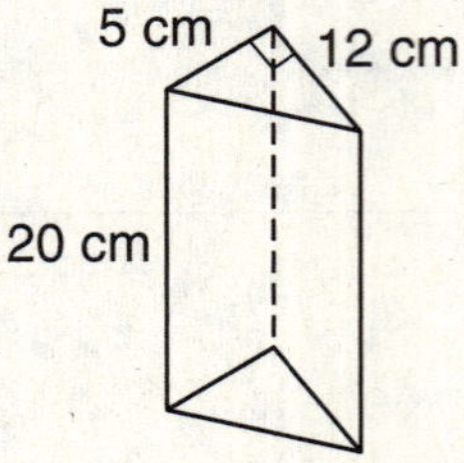

Formula: $S = 2B + Ph$

$$S = 2(\frac{1}{2} \cdot 12 \cdot 5) + (5 + 12 + 13)(20)$$
$$= 2(30) + 30(20)$$
$$= 660$$
$$S = 660 \text{ cm}^2$$

Try This

1. Write in words the formula for finding the surface area of a prism. Then follow these steps to find the surface area of a prism that has a square base with sides of 12 cm and a height of 22 cm.

____

____

____

____

____

2. A gift box is 16 inches long, 24 inches wide, and 3 inches high. How much paper will be needed to exactly cover the box?

____

3. What is the surface area of a triangular prism with a base area of 12 cm^2, a base perimeter of 16 cm and a height of 8 cm?

____

Name ______________________ Date ____________ Class ____________

LESSON 8F

Technology Lab

Volume of Prisms and Cylinders

Activity

A cereal company has asked the students in Mr. Garcia's math class to design a container. The container needs to hold at least 500 cubic centimeters of the new cereal while using the least amount of cardboard. Make a spreadsheet with columns for values, volume, and surface area that will allow you to manipulate sizes of the box to find the best one.

Step 1 Begin a new spreadsheet in Excel. In cell A1, type "Length (cm)"; in cell B1, type "Width (cm)"; in cell C1, type "Height (cm)", in cell D1, type "Surface Area", and in cell E1, type "Volume".

Step 2 Holding down the "shift" key, select columns A, B, and C. Click on "Format" on the toolbar, then click on "column", then click on "auto-fit selection".

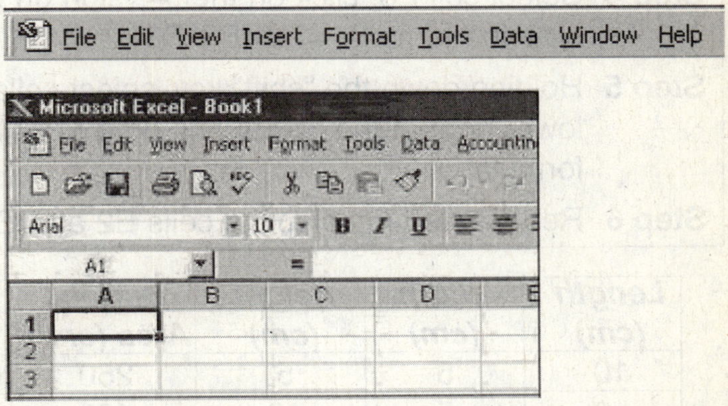

Record the numbers from the table on the right into the corresponding cells on the spreadsheet. These are the values to use for finding container size. Column A contains numbers for length; column B contains numbers for width; and column C contains numbers for height.

Since you will need to find the surface area of each of the containers (rectangular prisms) in order to compare and find which uses the least amount of cardboard, enter formulas into cells D2 through D12. You also need to find the volume, since the company wants a container that holds at least 500 cubic centimeters of cereal. Enter formulas into cells E2 through E12.

Row	A	B	C
2	10	5	5
3	10	5	12
4	5	5	10
5	11	3	12
6	7	6	12
7	6	4	12
8	8	3	12
9	10	3	8
10	12	4	9
11	12	6	9
12	10	5	10

Name ______________________ Date __________ Class __________

LESSON 8F

Technology Lab

Volume of Prisms and Cylinders, continued

To enter formulas:

Step 1 Select cell D2, click on the "=" sign on the formula bar, and in the text box type the formula =2*(A2*C2+B2*C2+A2*B2)

Step 2 Select cell E2, click on the "=" sign on the formula bar, and in the text box type the formula =A2*B2*C2

Step 3 Select cell D3, click on the "=" sign on the formula bar, and in the text box type the formula =2*(A3*C3+B3*C3+A3*B3)

Step 4 Select cell E3, click on the "=" sign on the formula bar, and in the text box type the formula = A3*B3*C3

Step 5 Holding down the "shift" key, select cells D2 and D3; using the handle on the lower right of the box at D3, drag to cell D12. This will automatically copy the formula.

Step 6 Repeat step 5, selecting cells E2 and E3.

Length (cm)	*Width (cm)*	*Height (cm)*	*Surface Area (cm^2)*	*Volume (cm^3)*
10	5	5	250	250
10	5	12	460	600
5	5	10	250	250
11	3	12	402	396
7	6	12	396	504
6	4	12	288	288
8	3	12	312	288
10	3	8	268	240
12	4	9	384	432
12	6	9	468	648
10	5	10	400	500

Think and Discuss

Which container would work best for the cereal company and why?

____

____

Try This

1. Calculate the surface area and volume of a cylinder with a radius of 7 cm and a height of 13 cm. Use, volume = $\pi \times \text{radius}^2 \times$ height and surface area = $2 \times \pi \times$ radius(height + radius).

____

Name ________________________ Date ____________ Class ____________

LESSON 8F

Review for Mastery

Surface Area of Prisms and Cylinders

Find the number of tiles needed to cover the faces of the prism. Unfold the prism to get a better look at its six faces.

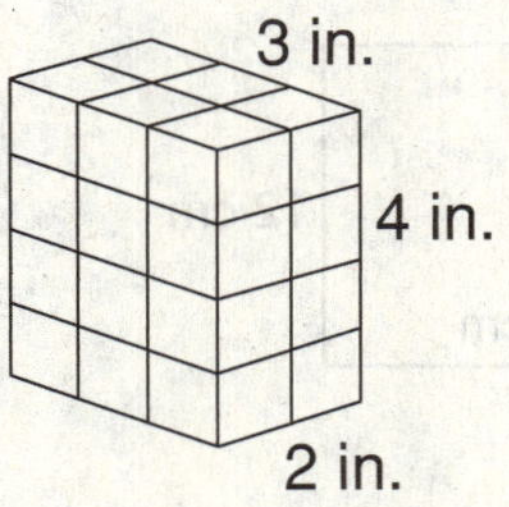

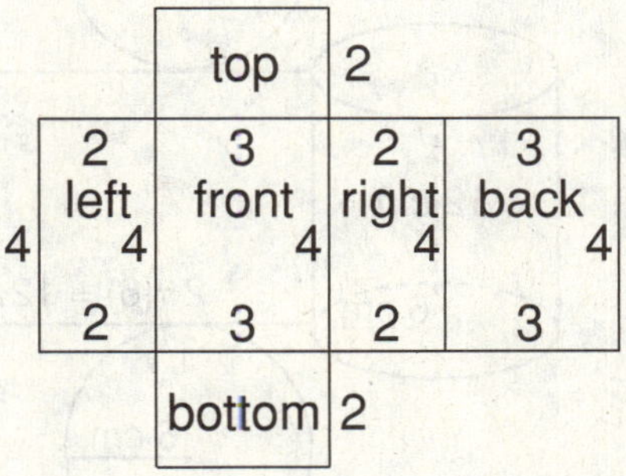

Face	Area	in.^{2}
top	3 × 2	6
bottom	3 × 2	6
front	3 × 4	12
back	3 × 4	12
left	2 × 4	8
right	2 × 4	8
	total =	52 in^2

Surface Area S = the sum of the areas of the faces of the prism

= top + bottom + front + back + left + right

= area of bases + area of lateral faces

= $2B$ + perimeter of the base × height of prism

$\boldsymbol{S = 2B + Ph}$

$S = 2(3 \times 2) + (3 + 2 + 3 + 2) \times 4$

$S = 12 + (10) \times 4$

$S = 12 + 40$

$S = 52 \text{ in.}^2$

1. Complete to find the number of square units needed to cover all the faces of the rectangular prism.

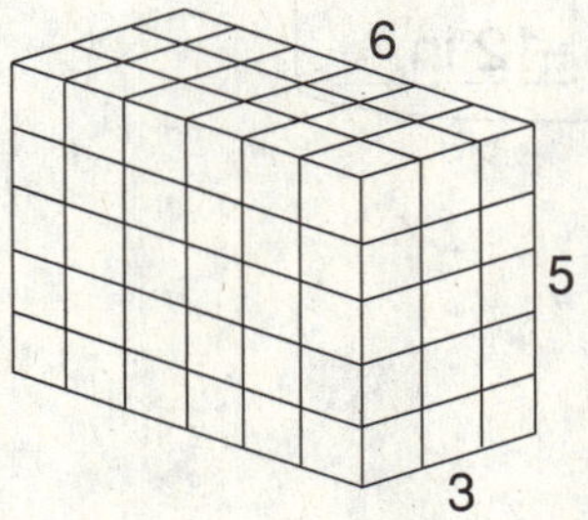

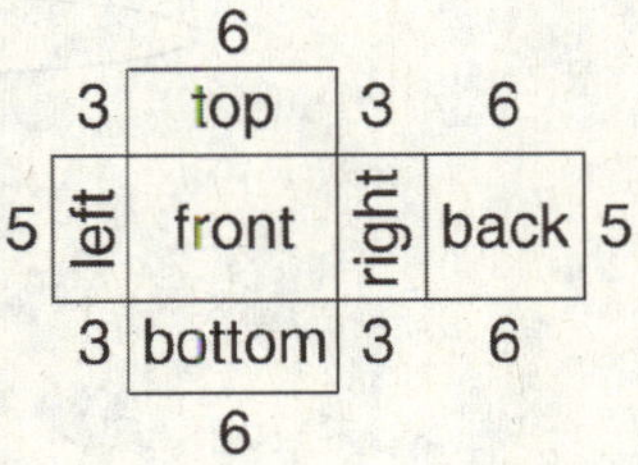

Face	Area	unit^2
top	______	______
bottom	______	______
front	______	______
back	______	______
left	______	______
right	______	______
	total =	______

2. Complete to find the surface area of the prism.

$S = 2B + Ph$

S = 2(________) + (________________________) × ____

S = 2(________) + (________________________) × ____

S = ________ + ________________________ = ________ cm^2

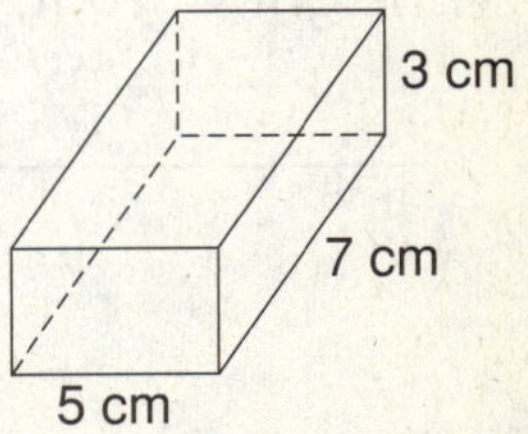

Name ______________________ Date ____________ Class ____________

LESSON **8F**

Review for Mastery

Surface Area of Prisms and Cylinders, continued

An unfolded cylinder results in two circles and a lateral surface drawn as a rectangle.

The base of the rectangle equals the circumference of the circular base.

The height of the rectangle equals the height of the cylinder.

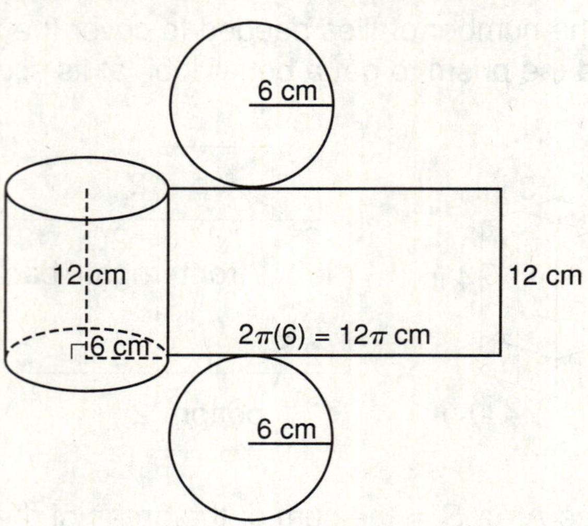

Surface Area S = area of 2 circular bases + area of lateral surface (rectangle)

$= 2(\pi r^2) +$ circumference × height

$= 2\pi r^2 + 2\pi r \times h$

$\boldsymbol{S = 2\pi r^2 + 2\pi rh}$

$S = 2\pi(6^2) + 2\pi(6)(12)$

$S = 72\pi + 144\pi = 216\pi \text{ cm}^2$

$S \approx 216(3.14) \approx 678.24 \text{ cm}^2$

3. Complete to find the surface area of the cylinder.

$\boldsymbol{S = 2\pi r^2 + 2\pi rh}$

$S = 2\pi($______$) + 2\pi \times$ ______ $\times$ ______

$S =$ ______ $\pi +$ ______ π

$S =$ ______ π

$S \approx$ ______ (3.14)

$S \approx$ ______ in.^2

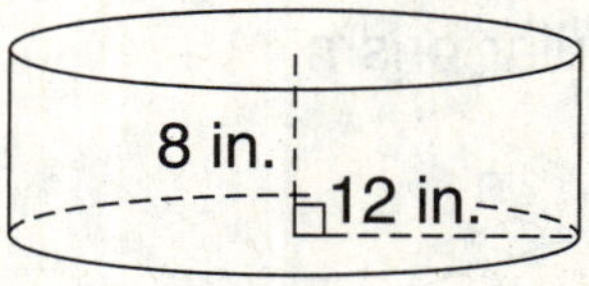

Find the surface area of each cylinder. Round to the nearest whole number.

4. height = 10 ft, radius = 5 ft

____

5. height = 2.5 cm, diameter = 8 cm

____

Name ______________________ Date ______________ Class ______________

LESSON 8F

Homework and Practice
Surface Area of Prisms and Cylinders

Find the surface area of each figure to the nearest tenth of a unit. Use 3.14 for π.

1.

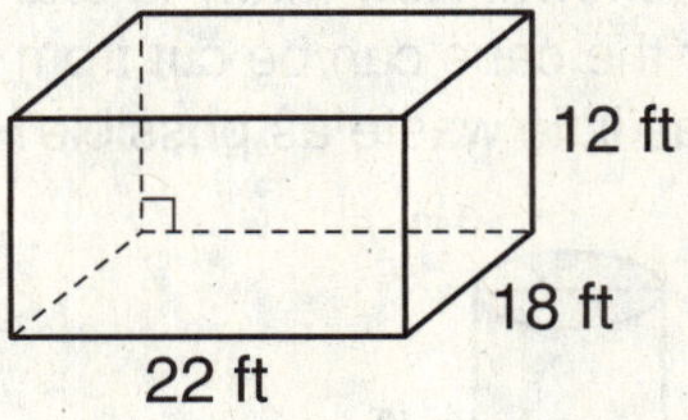

____

2.

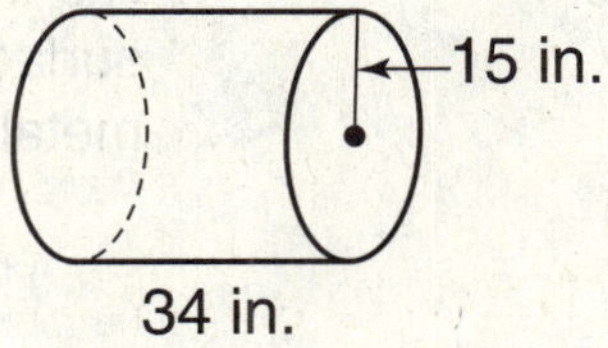

____

3.

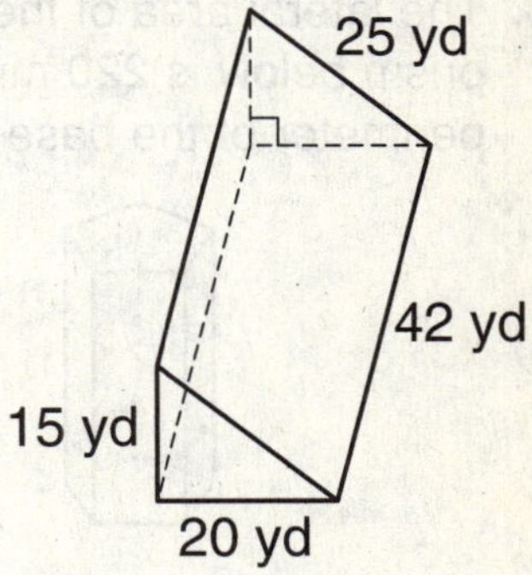

____

4.

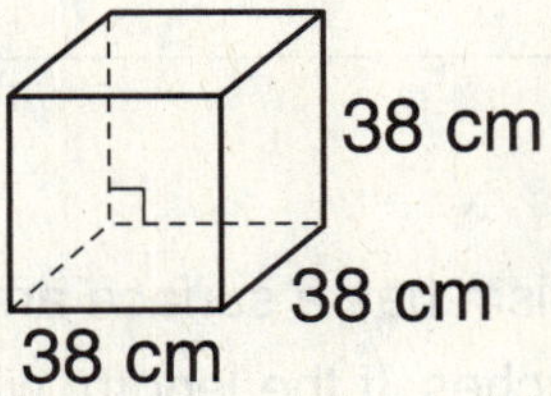

____

5.

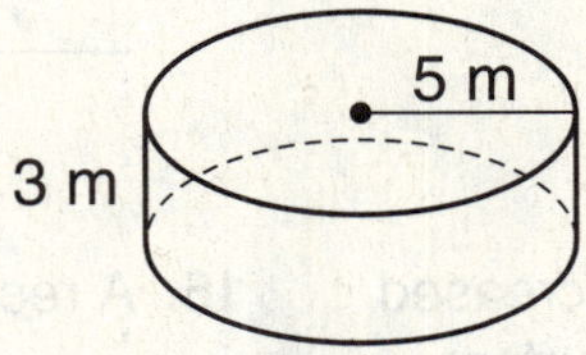

____

6.

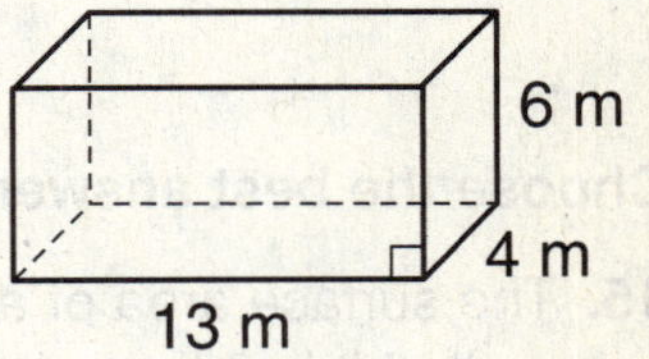

____

7.

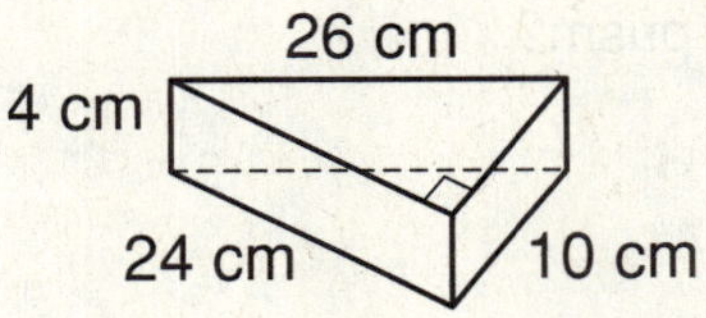

____

8.

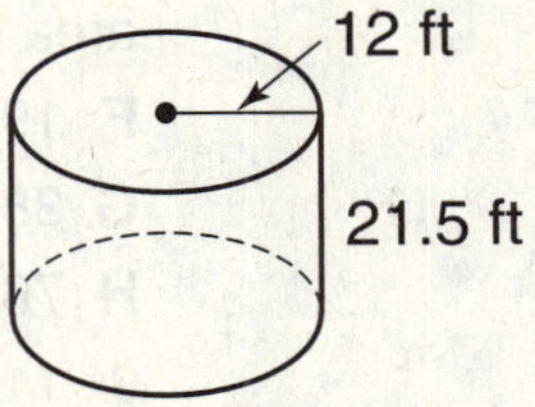

____

9.

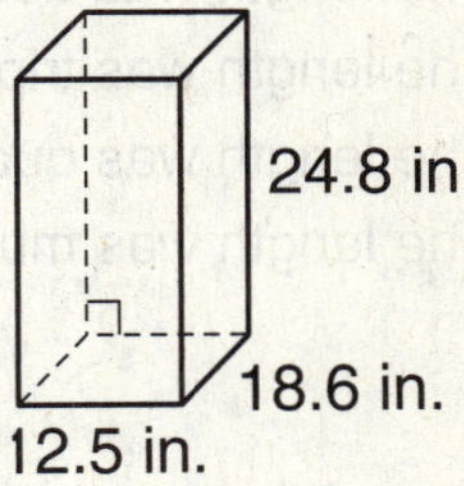

____

10. Find the surface area to the nearest tenth of a unit of a cylinder 84.5 m tall that has a diameter of 50 m. Use 3.14 for π.

____

11. Find the surface area to the nearest tenth of a unit of a rectangular prism with height 24 cm and sides 18 cm and 14 cm.

____

12. A recipe calls for a 9 × 13 × 2 in. baking dish to have the inside of the baking dish coated before adding the recipe contents. What is the surface area of the baking dish that will be coated?

____

Name ______________________ Date __________ Class __________

LESSON 8F

Homework and Practice

Surface Area of Prisms and Cylinders, continued

13. The lateral area of the regular pentagonal prism below is 220 mm^2. What is the perimeter of the base of the prism?

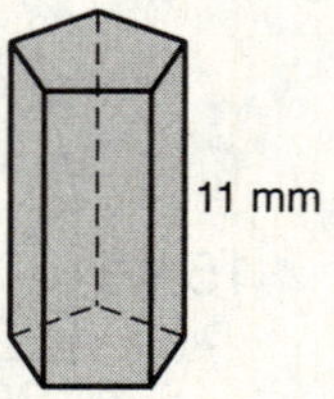

____

14. A sheet of metal 8 feet long and 6 feet wide is to be cut into cylindrical cans like the one shown. How many lateral surfaces for the cans can be cut from the metal with as little waste as possible?

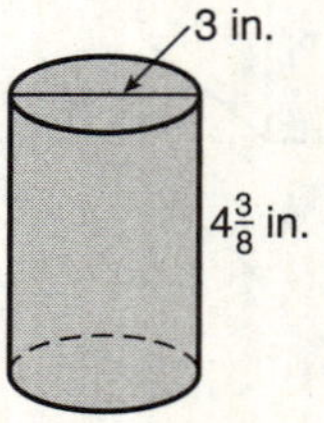

____

Choose the best answer.

15. The surface area of a cube is increased so that it is 9 times its original surface area. How did the length of the cube change?

A The length was doubled.

B The length was tripled.

C The length was quadrupled.

D The length was multiplied by 9.

16. A rectangular prism has a surface area of 152 square inches. If the length, width, and height are all changed to $\frac{1}{2}$ their original size, what will be the new surface area of the prism?

F 19 in^2

G 38 in^2

H 76 in^2

J 114 in^2

17. Determine the surface area of the figure shown. Round to the nearest tenth.

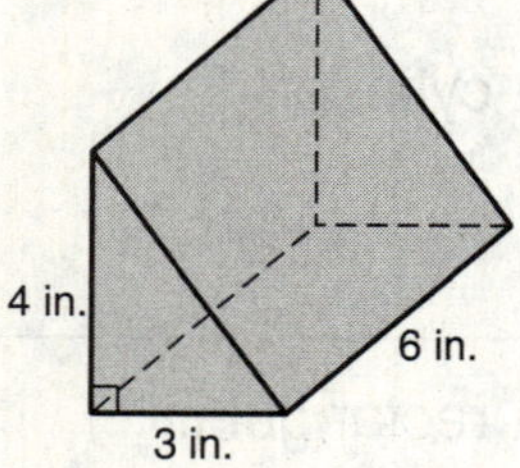

A 60 in^2

B 72 in^2

C 84 in^2

D 96 in^2

18. Which of the two cylindrical cans has a greater surface area?

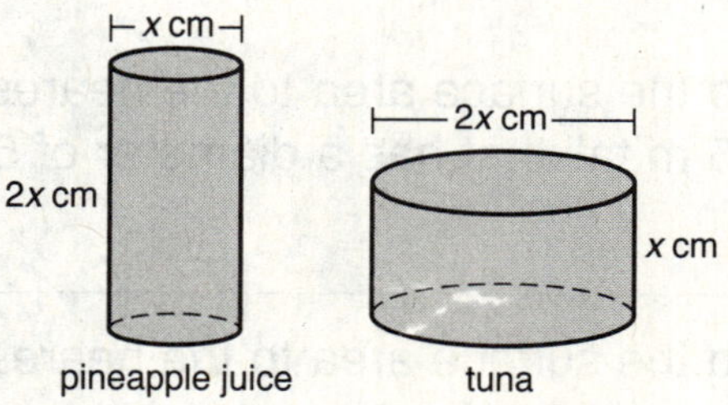

F pineapple juice can

G tuna can

H The two cans have the same surface area.

J It is impossible to determine which can has a greater surface area.

EXPLORATION

Program of Studies and Core Content

MA-8-AT-S-EI2, MA-8-AT-S-VEO2, MA-8-G-S-SR3, MA-8-M-S-MPA4, MA-8-M-S-MPA7, MA-08-3.1.3

8G Surface Area of Pyramids and Cones

Look at how the figures are built to answer each question.

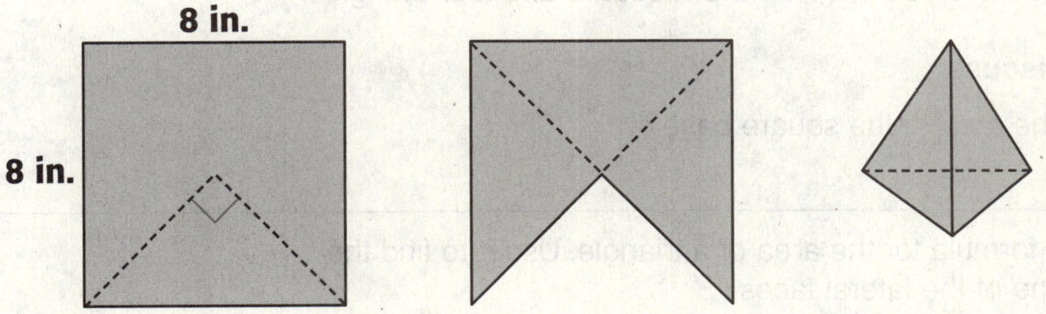

1. What is the surface area of the pyramid without a base?

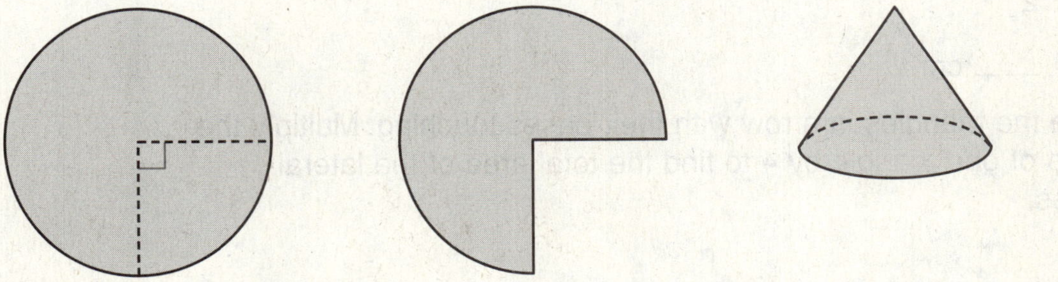

Area of circle = 16π in.^{2}

2. What is the surface area of the cone without a base?

Think and Discuss

3. Explain what you would need to do to find the surface area of the pyramid and cone including the base.

Name ______________________ Date __________ Class __________

LESSON **8G**

Hands-On Lab 1

Surface Area of Pyramids and Cones

Activity

Draw and cut out the net of a square pyramid that has a base with sides of 6 cm and a slant height of 5 cm. Next, cut the lateral faces apart from the base. You should have a square and four triangles.

Think and Discuss

1. What is the area of the square base?

____

2. Write the formula for the area of a triangle. Use it to find the area of one of the lateral faces.

$A = \frac{1}{2}bh$

$A = \frac{1}{2}(\_\_\_\_)(\_\_\_\_)$

$A = \_\_\_\_$ cm^2

3. Line the triangles in a row with their bases touching. Multiply the area of one triangle by 4 to find the total area of the lateral faces.

____

4. Explain how $\frac{1}{2}P\ell$ is equal to $4(\frac{1}{2}bh)$.

____

____

Name ______________________ Date ____________ Class ____________

LESSON 8G

Hands-On Lab 1

Surface Area of Pyramids and Cones, continued

Try This

1. What is the surface area of a square pyramid that has a square base with sides of 15 in. and a slant height of 9 in.?

____

____

____

2. Find the surface area of a square pyramid that has a square base with sides of 11 cm and a slant height of 14 cm.

____

____

____

Name ______________________________ Date ____________ Class ____________

LESSON 8G

Hands-On Lab 2

Surface Area of Pyramids and Cones

Activity

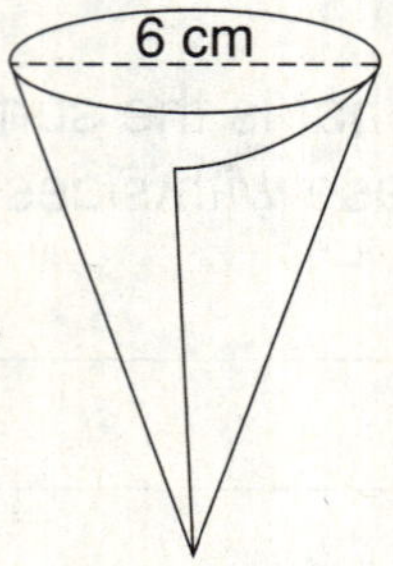

1. Roll a piece of paper to make a right cone that has a slant height of 12 cm and a diameter of 6 cm. Imagine it is an ice cream cone. Many ice cream shops place paper wrappers around their cones. If an ice cream cone has the same diameter and slant height as your paper cone, find the area of the paper wrapper to the nearest square centimeter.

Step 1 Decide which part of the surface area you need to find. You need only to find the lateral area because the wrapper will not cover the top of the cone.

Step 2 $L = \pi r\ell$

Step 3 Divide the diameter by 2 to find the radius. $r = 3$ cm

Step 3 $L \approx (3.14)(3)(12)$

Step 4 $L \approx 113.04$

The area of the wrapper is about 113 cm^2.

Activity 2

Make a wrapper for your "ice cream cone" by drawing a net for your wrapper.

Step 1 Use a compass to draw a circle with a radius of 12 cm.

Step 2 Find the circumference of the base. This is the length of the arc.
$2\pi r = 2\pi \times 3 = 6\pi$

Step 3 Find what fraction of the circle makes up the lateral surface. First find the circumference of the circle.
$C = 2\pi r = 2\pi \times 12 = 24\pi$

Step 4 Find what fraction of the circle is needed for the net by writing the measurement of the arc as a fraction of the circumference. Multiply that fraction by 360° to find the angle necessary to create an arc length of 6π cm.
$\frac{6\pi}{24\pi}(360^\circ) = 90^\circ$.

Step 5 Measure and mark an angle of 90° at the center of your circle.

Step 6 Erase all but the 90° portion of the circle.

Name ______________________________ Date ____________ Class ____________

LESSON 8G

Hands-On Lab 2

Surface Area of Pyramids and Cones, continued

Think and Discuss

1. Explain why you know the arc length of the lateral surface of a right cone is equal to the circumference of the base of the cone.

____

____

____

Try This

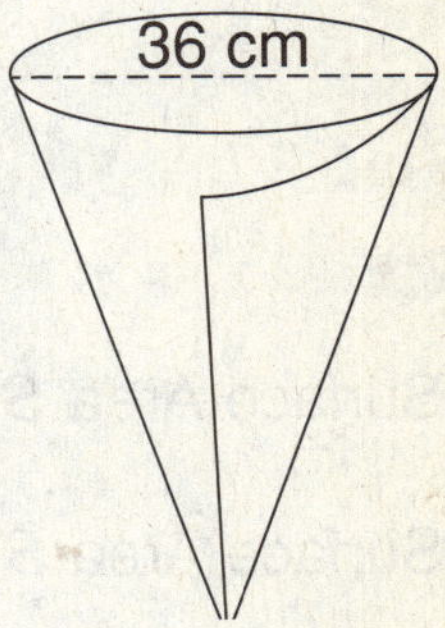

1. Write in words the formula for finding the surface area of a right cone. Imagine you are explaining how to find the surface area of a right cone to a younger student. Then follow your instructions to find the surface area of a cone that has a radius of 18 cm and a slant height of 35 cm. Find the surface area to the nearest square centimeter.

____

____

____

____

____

2. Find an object in your classroom or at home that is cone-shaped. Use a metric ruler to measure its diameter and slant height. Find its radius and its surface area. Pay close attention to whether the area of its base should be included in your calculations.

____

Name ______________________ Date ____________ Class ____________

LESSON 8G

Review for Mastery

Surface Area of Pyramids and Cones

Regular Pyramid: base is a regular polygon; lateral faces are congruent triangles; named after the shape of its base

When a square pyramid is unfolded, there are 5 faces: a square and 4 congruent triangles.

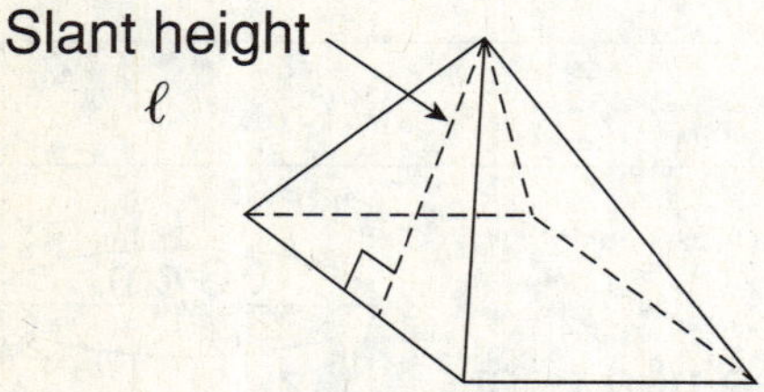

Square Pyramid

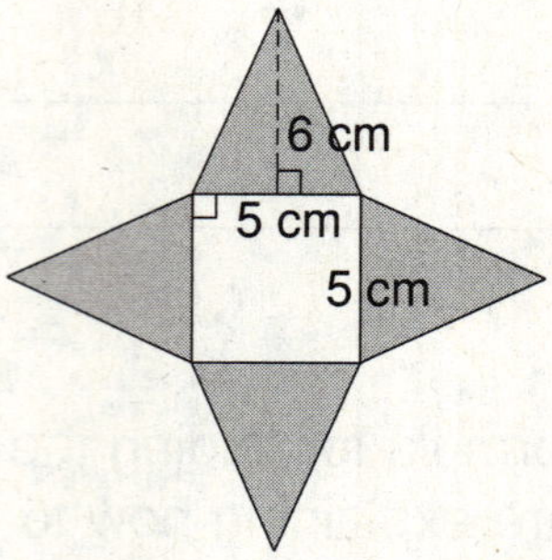

Surface Area S = the sum of the areas of the faces of the pyramid
= area of base + area of lateral faces

Surface Area S = area of square + 4(area of triangle)

= B + $\frac{1}{2}$ perimeter P of base × slant height ℓ

$\boldsymbol{S = B + \frac{1}{2}P\ell}$

$S = (5 \times 5) + \frac{1}{2}(5 \times 4)(6)$

$S = 25 \quad + \quad 60$

$S = 85 \text{ cm}^2$

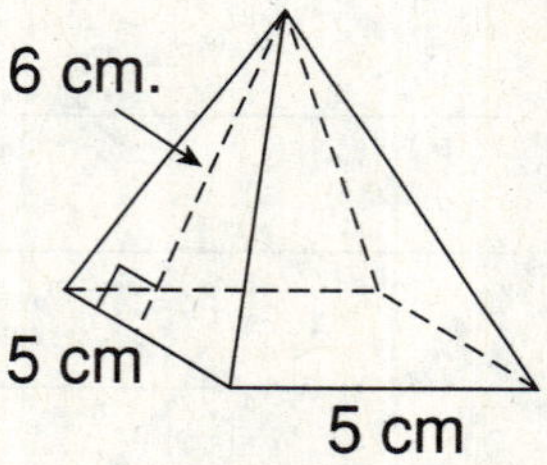

1. Find the surface area of the square pyramid.

S = area of square + 4(area of triangle)

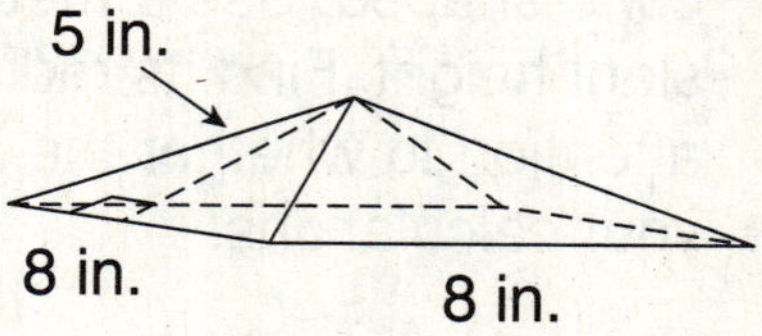

S = __________ + ______________

S = __________ + ______________

S = ________________________ in^2

2. Complete to find the surface area of the square pyramid.

$S = B + \frac{1}{2}P\ell$

S = __________ + $\frac{1}{2}$ __________ × __________

S = __________ + ______________

S = ________________________ ft^2

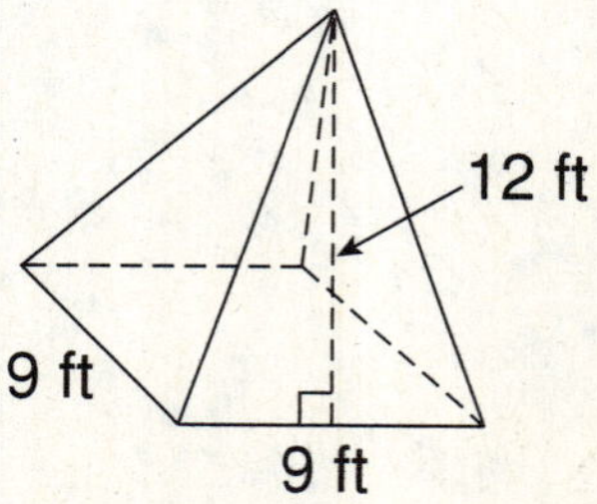

Name ______________________ Date __________ Class __________

LESSON 8G

Review for Mastery

Surface Area of Pyramids and Cones, continued

An unfolded cone results in a circle and a lateral surface drawn as a sector of a circle.

The slant height of the cone is the radius of the circle sector.

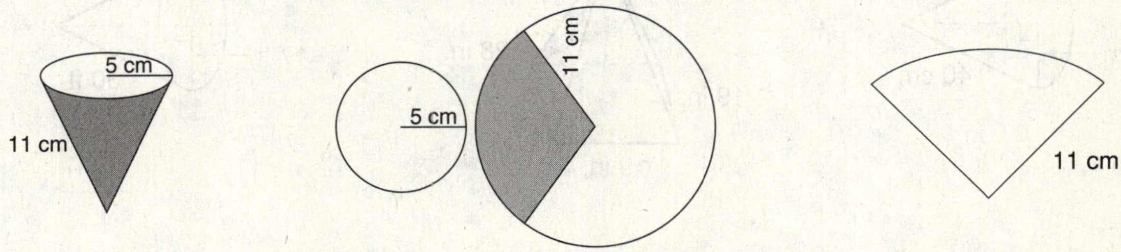

Surface Area S = area of circular base + area of lateral surface (circle sector)

$= \pi r^2 + \frac{1}{2}$ circumference of base × slant height

$= \pi r^2 + \frac{1}{2}(2\pi r) \times \ell$

$\boldsymbol{S = \pi r^2 + \pi r\ell}$

$S = \pi(5^2) + \pi(5)(11)$

$S = 25\pi + 55\pi = 80\pi \text{ cm}^2$

$S \approx 80(3.14) \approx 251.2 \text{ cm}^2$

Complete to find the surface area of the cone.

3. $\boldsymbol{S = \pi r^2 + \pi r\ell}$

$S = \pi($ ________ $) + \pi \times$ ______ $\times$ ______

$S =$ ________ $\pi +$ ________ $\pi =$ ________ π

$S \approx$ ________ (3.14)

$S \approx$ ___________ in^2

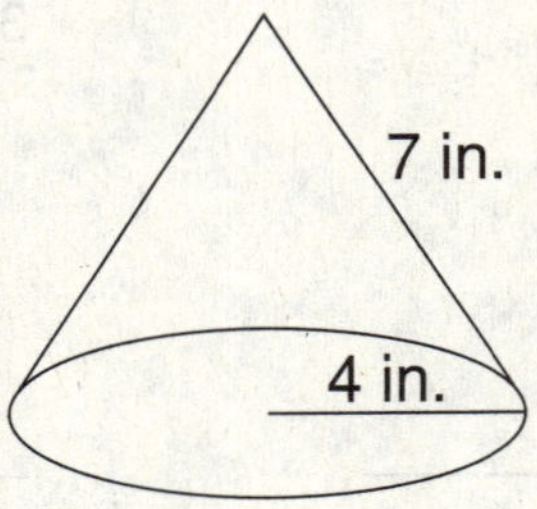

Find the surface area of each cone. Round to the nearest whole number.

4. radius = 3 ft, slant height = 5 ft

____

5. diameter = 8.6 cm, slant height = 10 cm

____

Name ______________________ Date ____________ Class ____________

LESSON **8G**

Homework and Practice

Surface Area of Pyramids and Cones

Find the surface area of each figure to the nearest of a tenth unit. Use 3.14 for π.

1.

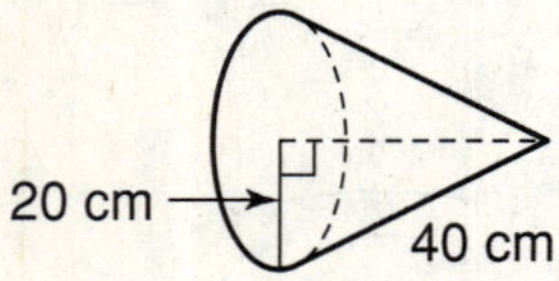

____

2.

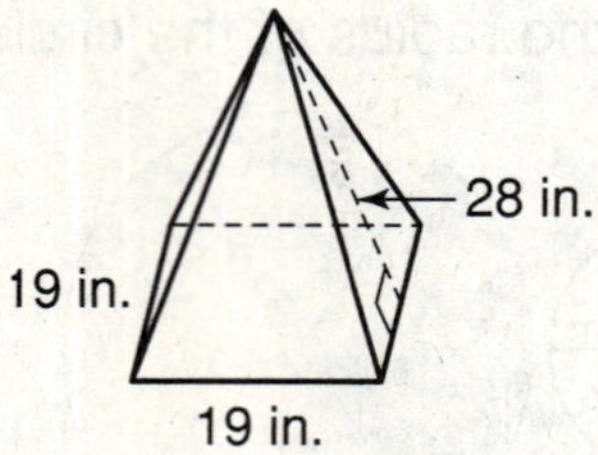

____

3.

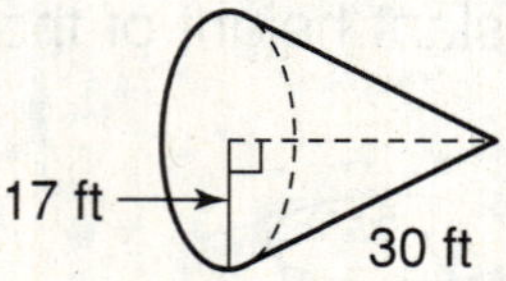

____

4.

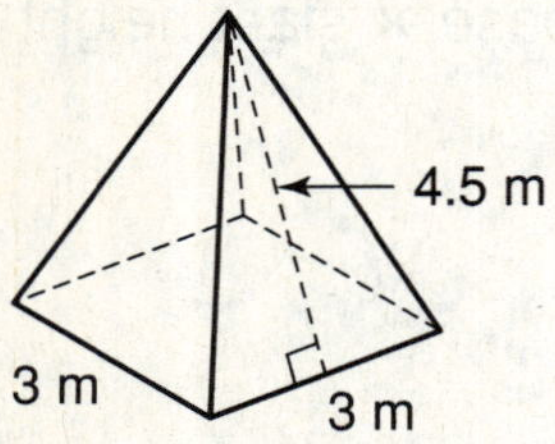

____

5.

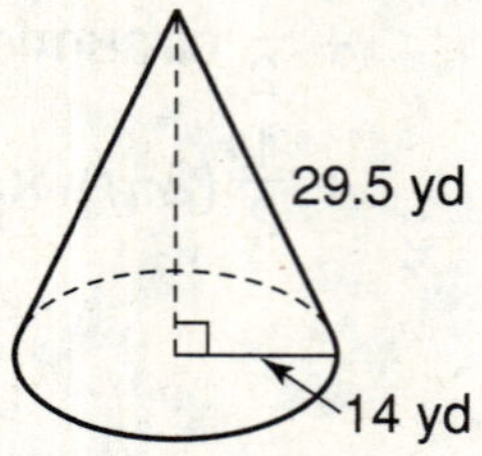

____

6.

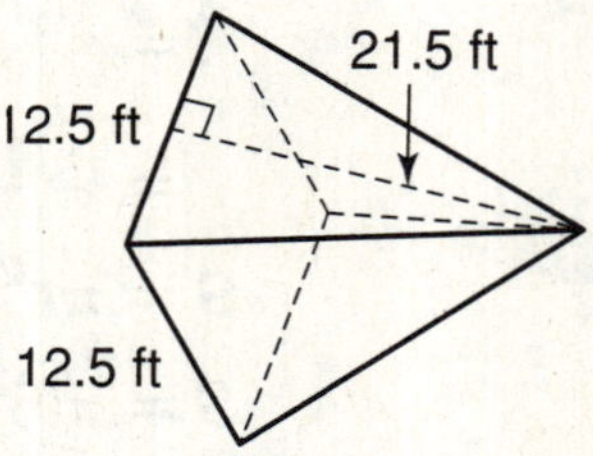

____

7.

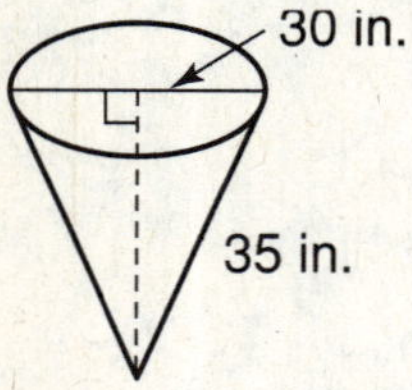

____

8.

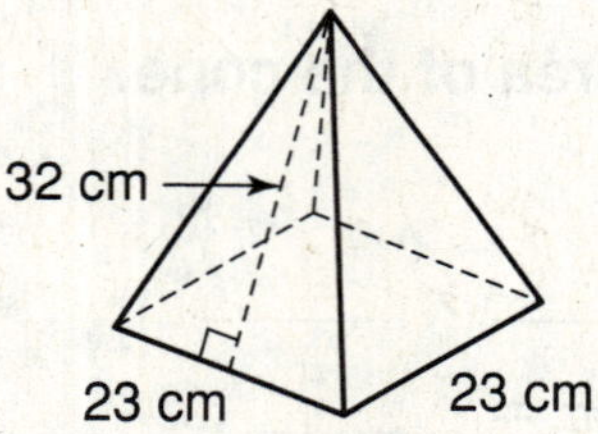

____

9.

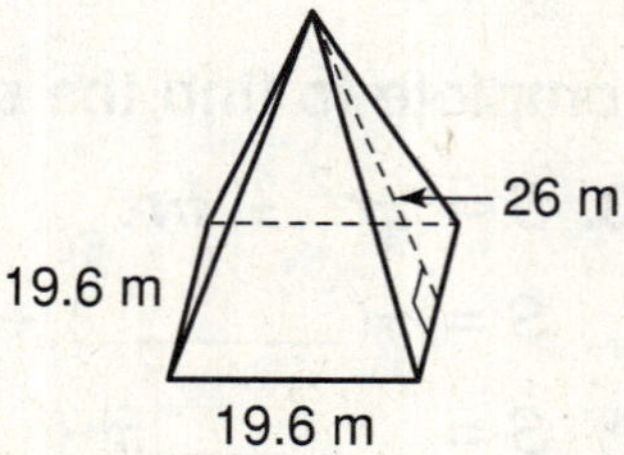

____

10. Find the length of the slant height of a square pyramid if the one side of the base is 26 cm and the surface area is 2,236 cm^2.

____

11. Find the surface area of a regular square pyramid with a slant height of 27 in. and a base perimeter of 96 in.

____

12. Find the length of the slant height of a cone with a radius of 40 ft and a surface area of 9,420 ft^2. Use 3.14 for π.

____

EXPLORATION

8H Spheres

Program of Studies and Core Content
MA-8-AT-S-EI2, MA-8-AT-S-VEO2, MA-8-G-S-SR3, MA-08-3.1.3

A basketball is an example of a sphere. The radius of a basketball is about 4.5 inches. A ***hemisphere*** is half a sphere.

To find the volume of a basketball, imagine the ball is sliced into two halves. Then find the volume of one half and multiply times 2.

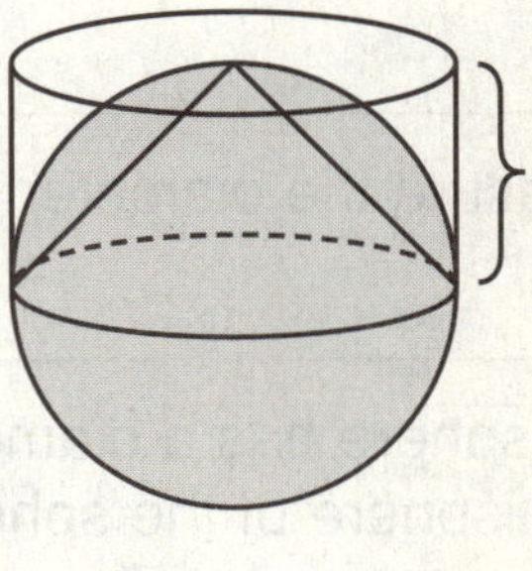

The volume of a hemisphere is exactly halfway between the volume of a cone and the volume of a cylinder that both have the same radius *r* as the hemisphere and a height equal to *r*.

Use a calculator to find the volume of the following spheres.

	Volume of Cylinder Sitting on Top Half of Sphere	Volume of Cone = $\frac{1}{3}$ · Volume of Cylinder	Volume of Hemisphere = Halfway Between Vol. of Cylinder and Vol. of Cone	Volume of Sphere = Volume of Hemisphere · 2
1.	63.62 in^3	$\frac{1}{3} \cdot 63.62 = 21.21\ in^3$	$\frac{63.62 + 21.21}{2} = 42.42\ in^3$	$42.42 \cdot 2 = 84.84\ in^3$
2.	70 in^3			
3.	27 in^3			
4.	123 in^3			

Think and Discuss

5. Describe how you could find the volume of the solid at right. The height of the dome is the same as the height of the cylinder, which is equal to the radius of the base.

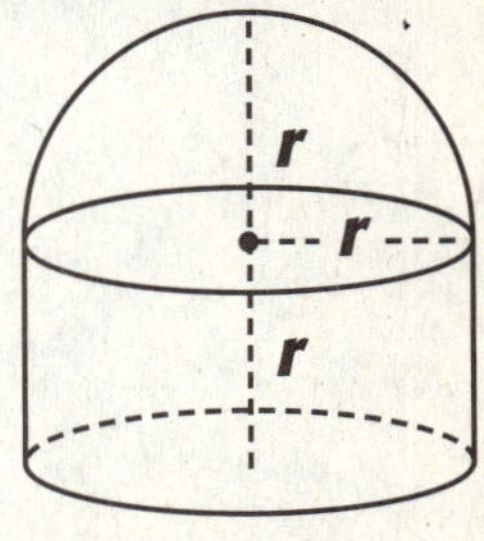

Name ______________________ Date __________ Class __________

LESSON 8H

Hands-On Lab

Spheres

Activity

1. Measure the volume of a bowl that is shaped like a hemisphere by filling it full of water. Carefully pour the water into a graduated cylinder. Measure the number of milliliters of water that the bowl holds.

____

2. Measure the diameter of the bowl.

____

3. If a sphere has a diameter of 13 in., find the volume of a hemisphere of the sphere to the nearest tenth of a cubic inch.

Step 1 Use the formula for volume of a sphere.

$V = \frac{4}{3}\pi r^3$

Step 2 $V \approx \frac{4}{3}(3.14)(6.5)^3$

Step 3 $V \approx \frac{4}{3}(3.14)(274.625)$

Step 4 $V \approx 1{,}149.8 \text{ in}^3$.

Step 5 Divide the volume of the sphere by 2 to find the volume of a hemisphere of the sphere.

$1{,}149.8 \div 2 = 574.9 \text{ in}^3$

The volume of the hemisphere is about 574.9 in^3.

4. Find the volume of the bowl you measured to the nearest tenth of a cubic inch by using the formula for the volume of a sphere and dividing by 2. Show your work.

____

____

Name ______________________________ Date __________ Class __________

LESSON 8H

Hands-On Lab

Spheres, continued

Think and Discuss

1. The volume of a cone with a radius of 5 in. and height of 5 in. is about 130.8 in^3. The volume of a cylinder with the same radius and height is about 392.5 in^3. How can you calculate the volume of a hemisphere with a radius of 5 inches without using the formula for the volume of a sphere? Calculate the volume of the hemisphere this way. Show your work.

____

____

____

____

2. How would you find the surface area of a sphere with a radius of 24 cm using the formula below? Write your answer in terms of π.

Formula: $S = 4\pi r^2$

Step 1

Step 2

Step 3

Try This

1. The radius of the earth at the equator is 6,378.1 km. Find the volume of the earth using this radius, both approximated and in terms of π. Then find the surface area of the earth both approximated and in terms of π. Show your steps.

Name ______________________________ Date ____________ Class ____________

LESSON **8H**

Technology Lab

Spheres

Activity

A sphere is a 3-dimensional figure with all of its points the same distance from its center. The distance from the center to the surface of the sphere is called its radius. Any cross-section of a sphere is a circle. The surface area S of the sphere is given by the formula:

$$S = 4\pi r^2.$$

Use your calculator to determine the surface area of a sphere having a radius of 10 inches.

$S \approx 4 \times (3.14) \times 10^2$

Using your calculator, enter the following:

[4] [×] [3] [.] [1] [4] [×] [1] [0] [x^2] [=]

The surface area of the sphere is about 1256 in^2.

Think and Discuss

1. If you were a sculptor who was planning to make a gigantic sphere and cover it in aluminum foil, why would you need to know the surface area of the sphere?

____

____

____

2. The diameter of this gigantic sphere is 50 feet, but the aluminum foil you buy comes in packages measured in square inches. Why would it be useful to have a calculator for computing the surface area of this giant sphere?

____

____

____

3. Calculate the surface area of the sphere sculpture in inches, using your calculator.

____

Name ____________________________ Date ____________ Class ____________

LESSON 8H

Technology Lab

Spheres, continued

Try This

1. Using drawing software, draw a sphere with a radius of 8 centimeters. Then calculate the surface area of your sphere.

Name ______________________ Date __________ Class __________

LESSON **8H**

Review for Mastery

Spheres

Sphere: the set of points in space at a fixed distance (*radius r*) from a fixed point (*center*)

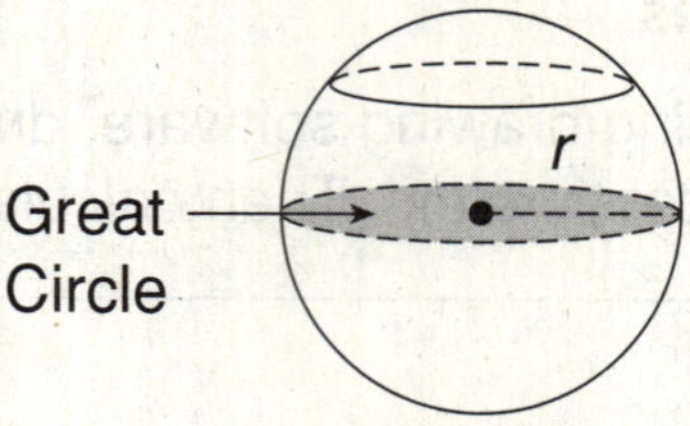

Sphere

Volume V of Sphere $= \frac{4}{3}\pi r^3$

For a sphere of radius = 9 cm,

$V = \frac{4}{3}\pi r^3$

$V = \frac{4}{3}\pi \times 9^3 = \frac{4}{3}\pi \times 729 = \frac{4 \times \cancel{729}^{243}}{\cancel{3}}\pi = 972\pi \text{ cm}^3$

$V \approx 972(3.14) = 3052.08 \text{ cm}^3$

Complete to find the volume of the sphere.

1. $V = \frac{4}{3}\pi r^3$

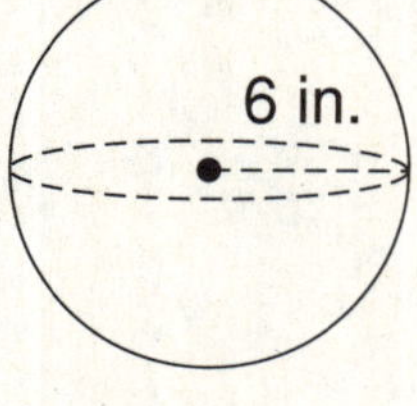

$V = \frac{4}{3}\pi \times$ ______ $= \frac{4}{3}\pi \times$ ______

$V =$ ______ $\pi \text{ in}^3$

$V \approx$ ______ $(3.14) =$ ____________ in^3

2. $V = \frac{4}{3}\pi r^3$

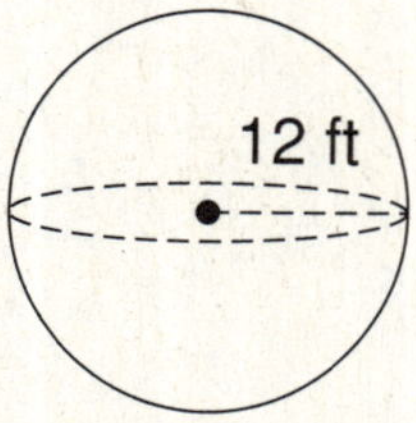

$V = \frac{4}{3}\pi \times$ ______ $= \frac{4}{3}\pi \times$ ______

$V =$ ______ $\pi \text{ ft}^3$

$V \approx$ ______ $(3.14) =$ ____________ ft^3

3. A beach ball has a volume of $4.5\pi \text{ ft}^3$. Complete to find its radius.

$\frac{4}{3}\pi r^3 = \frac{9}{2}\pi$ Rewrite 4.5 as a fraction.

$\frac{4}{3}r^3 = \frac{9}{2}$ Cancel π on each side.

$r^3 = \frac{9}{2} \times$ ______ $=$ ______, so $r =$ ______ ft Multiply by the reciprocal. Then take the cube root.

Name ______________________ Date ____________ Class ____________

LESSON 8H

Homework and Practice

Spheres

Find the volume of each sphere, both in terms of π and to the nearest tenth of a unit. Use 3.14 for π.

1. $d = 48$ ft

2. $r = 4.5$ m

3. $r = 7.5$ cm

4. $r = 33$ cm

5. $d = 32.02$ ft

6. $d = 60$ m

7.

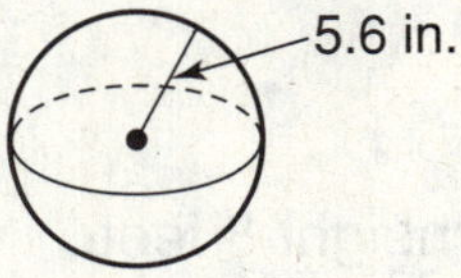

8.

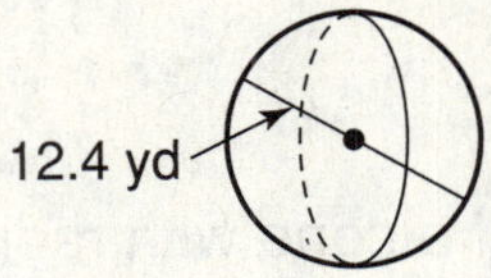

9.

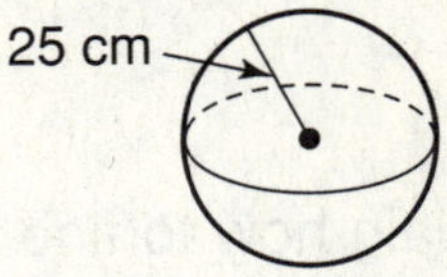

10.

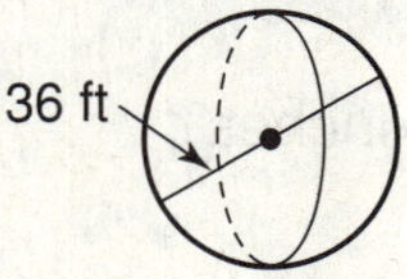

11.

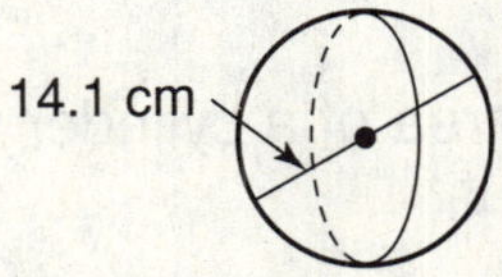

12.

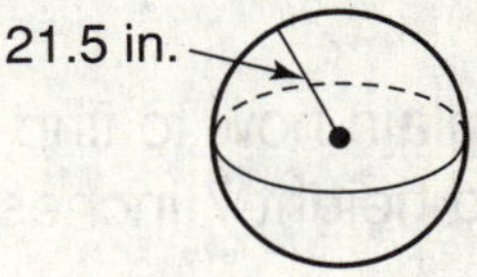

13. Estimate to the nearest tenth the volume of a baseball with a 3 in. diameter. Use 3.14 for π.

14. Eight bocce balls are in a box 18 inches long, 9 inches wide, and 4.5 inches deep. If each ball has a diameter of 4.5 inches, what is the volume of the space around the balls? Round to the nearest tenth.

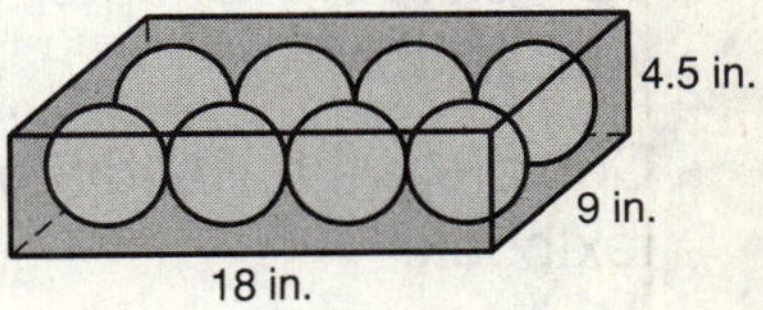

CHAPTER 8

Chapter Summary

Answer these questions to summarize the important concepts from Chapter 8 in your own words.

1. Explain how to find the area of a trapezoid with bases 3 inches and 7 inches, and height 6 inches.

2. Explain how to find the circumference of a circle with radius 6 meters.

3. Explain how to find the volume of a cone with radius 3 feet and height 8 feet.

4. Explain how to find the surface area of a cylinder with diameter 6 inches and height 7 inches.

For more review of Chapter 8:

- Complete the Chapter 8 Study Guide and Review on pages 450–452 of your textbook.
- Complete the Ready to Go On quizzes on pages 404 and 444 of your textbook.

Name ______________________ Date ____________ Class ____________

CHAPTER 8

KCCT Prep

Multiple Choice

1. Which is the best estimate of the area of the triangle?

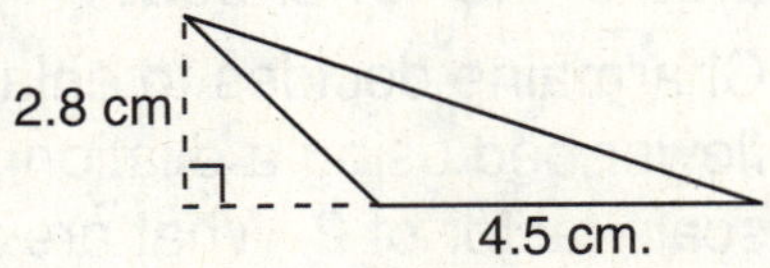

(A) 6 cm^2 (C) 12 cm^2
(B) 8 cm^2 (D) 15 cm^2

2. A triangle has a 38° angle and a 52° angle. Which is the best way to classify the triangle?

(A) Acute (C) Isosceles
(B) Equilateral (D) Right

3. A three-dimensional figure has exactly one vertex. Which of the following could the figure be?

(A) Cone (C) Prism
(B) Cylinder (D) Pyramid

4. A circle has an area of 64π ft^2. What is the circle's circumference?

(A) 8π ft (C) 64π ft
(B) 16π ft (D) 256π ft

5. $\angle M$ and $\angle N$ are supplementary. $\angle N$ and $\angle P$ are complementary. Given that $m\angle M = 121°$, what is $m\angle P$?

(A) 31° (C) 61°
(B) 59° (D) 149°

6. There are children and adults at an art exhibit. There are 15 children, and this is 12% of the total number of people at the exhibit. How many adults are at the exhibit?

(A) 80 (C) 125
(B) 110 (D) 180

7. What is the area of the figure with vertices (1, 1), (7, 1), (6, 4), and (2, 4)?

(A) 12 $units^2$ (C) 18 $units^2$
(B) 15 $units^2$ (D) 30 $units^2$

8. A right triangle has legs that are 4 inches long and 9 inches long. Which is closest to the length of the triangle's hypotenuse?

(A) 3.6 in. (C) 9.8 in.
(B) 8.1 in. (D) 10.8 in.

9. The circumference of a circle is 12π m^2. What is the area of the circle?

(A) 6π m^2 (C) 36π m^2
(B) 18π m^2 (D) 144π m^2

10. A box of crackers is shown below. The box is one-fourth full of crackers. What is the volume of crackers in the box?

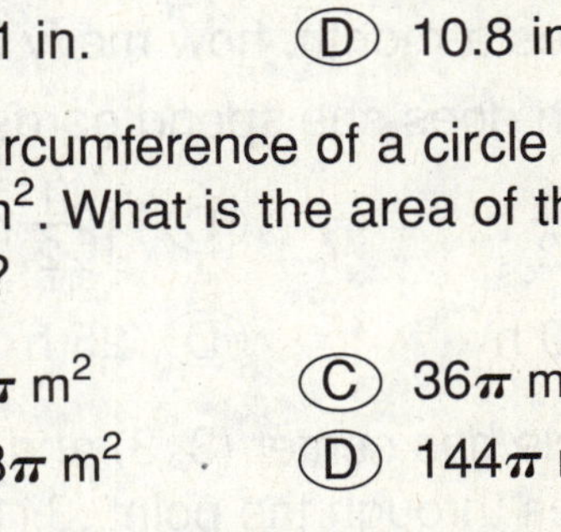

(A) 4 $in.^3$ (C) 32 $in.^3$
(B) 16 $in.^3$ (D) 40 $in.^3$

11. Enrique makes a scale drawing of a building that is 35 meters long. In the drawing, the length of the building is 5 centimeters. What is the scale of the drawing?

(A) 1 cm:5 m (C) 5 cm:1 m
(B) 1 cm:7 m (D) 7 cm:1 m

12. For which of these values of n is the fraction $\frac{n}{10}$ closest to $\sqrt{2}$?

(A) 4 (C) 14
(B) 7 (D) 20

13. A trapezoid has a height of 6.15 cm and bases that are 4.92 cm and 7.11 cm long. Which is the best estimate of the area of the trapezoid?

(A) 30 cm^2 (C) 60 cm^2
(B) 36 cm^2 (D) 72 cm^2

14. Laura does $\frac{2}{5}$ of the gardening in $1\frac{1}{4}$ hours. If she does the entire job 4 times a month, how many hours per month does she spend gardening?

(A) $8\frac{1}{2}$ h (C) $12\frac{1}{2}$ h
(B) 10 h (D) 25 h

15. A circle has center (3, 3) and the circle passes through the point (3, 1). What is the circumference of the circle?

(A) 2π units (C) 4π units
(B) 3π units (D) 8π units

16. Which is the best estimate of $9\frac{1}{10} \div \frac{1}{3}$?

(A) 3 (C) 27
(B) 9 (D) 30

Open Response

17. Charmaine is designing a flower bed for her garden. She starts by drawing a trapezoid with vertices (–3, 3), (3, 3), (1, –3), and (–1, –3).

a. Each unit of the coordinate plane represents one foot. What is the area of the flower bed?

b. Charmaine decides to enlarge the flower bed using a dilation with a scale factor of 2. What are the coordinates of the enlarged flower bed?

c. Charmaine claims that the dilation doubled the area of the flower bed. Do you agree? Why or why not?

d. Charmaine wants to divide the enlarged flower bed into three regions, all having the same area. Show how she can do this and explain how you know the areas of the regions are equal.

18. The figure shows the dimensions of a rectangular shaped aquarium.

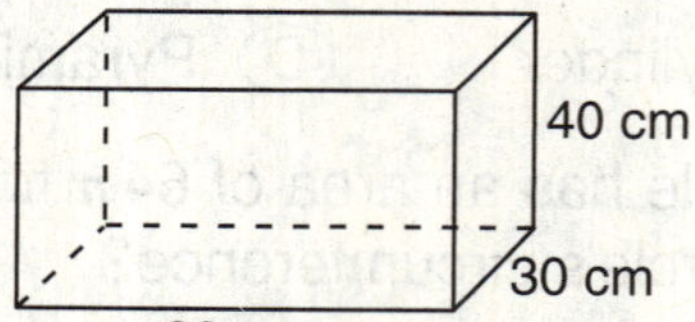

a. What is the volume of the aquarium?

b. One liter is equal to 1000 cm^3. What is the capacity of the aquarium in liters?

c. The aquarium starts out with 15 L of water in it. Then it fills at the rate of 3 L/min. Explain how you can figure out the amount of time it will take to fill the aquarium completely.

CHAPTER 9

Project
Data and Statistics: Poor Pizza

Career: Quality Assurance Specialist

How do manufacturers know that their products are well made? It is the job of the quality assurance specialist. QA specialists design tests and procedures that allow the companies to determine how good their products are. Because checking every product or procedure may not be possible, QA specialists use sampling to predict the margin of error.

Before you start, work through the Making Quality Projects worksheet on the next page.

In this chapter project . . .

You simulate running a pizza parlor.

What to do:

- Divide your class into two teams. One team will be the Pizza Makers and the other team will be the Quality Assurance Team.
- The Pizza Makers will decide how many pizzas out of 100 will be "ruined," but they should not tell the Quality Assurance Team that number. The Pizza Makers will write "good" or "ruined" on 100 slips of paper and place them in a bag.
- The Quality Assurance Team will pick 10 slips from the bag, record the number of "ruined" pizzas, replace the slips, and pick 10 slips again. Repeat this 10 times and find the mean number of ruined pizzas.
- Both teams should discuss strategies to determine the total number of ruined pizzas out of 100 based on the mean of the random samples.

What to present:

- A table listing the actual number of ruined pizzas determined by the Pizza Makers, the results of each sample, and the guess from the Quality Assurance Team.
- A paragraph explaining the methods used to guess the number of pizzas based on the samples.

Where's the math?

- Calculate the mean number of ruined pizzas.
- Use random samples to make predictions.

go.hrw.com
Chapter Project Online
KEYWORD: MT7 Ch9

Name ______________________ Date __________ Class __________

CHAPTER 9

Project Recording Sheet

Data and Statistics

Making Quality Products

The type of product being made often determines how many errors occur during the production process. To judge efficiency, you need to compare companies that produce the same kind of product.

Frozen Pizza Companies	**Errors per Sample** (100 pizzas)												
	1	**2**	**3**	**4**	**5**	**6**	**7**	**8**	**9**	**10**	**11**	**12**	**13**
Taste of New York	2	6	1	7	4	2	3	3	4	9	1	8	4
Momma Mildred's	7	6	12	7	15	14	13	5	18	15	7	16	11
Speedworks	4	8	8	6	4	9	7	3	5	8	7	9	3
Pizza Pizzazz	3	2	4	5	2	1	0	7	5	4	6	8	2

Calculate the mean, median, and mode for each company.

Company	**Mean**	**Median**	**Mode**
Taste of New York			
Momma Mildred's			
Speedworks			
Pizza Pizzazz			

Make a bar graph showing the average number of bad pizzas per hundred produced by each company.

____

Rank the companies from best to worst.

____

If each waste pizza costs the companies $1.50 and each company produces 10,000 pizzas a year, how much does each lose in that time? How does that loss affect the price they must charge for a pizza?

____

____

____

Program of Studies and Core Content
MA-8-DAP-S-ES1, MA-8-DAP-S-CD5, MA-08-4.3.1

9A Samples and Surveys

A survey is being conducted to determine what items should be offered at the school snack bar. Instead of including every student in the survey, a method of selecting a fair representation of all students must be used.

1. Determine which method of selecting students might provide a sample that fairly represents all of the students in the school.

Method 1: selecting students whose last names begin with *A* or *B*

Method 2: selecting students who purchased snack bar items in the past week

Method 3: selecting all eighth-grade students

Method 4: selecting one-fourth of all students at random

Method 5: selecting the first 100 students to come to the office to take the survey

2. Explain one way to make the method you chose in Problem 1 even more likely to give a representative sample.

Think and Discuss

3. Explain your reasons for selecting the method you chose in Problem 1.

Name ______________________ Date __________ Class __________

LESSON **9A**

Review for Mastery

Samples and Surveys

A *survey* uses a *small sample* to represent a *large population.*

Sampling Method	How Members of the Sample Are Chosen
Random	By chance; members have an equal chance of selection
Systematic	According to a rule or pattern
Stratified	At random from randomly-chosen subgroups

A senator's office sends workers to ask constituents at a local mall how they feel about floating a bond to acquire land around a reservoir to be preserved as open space.

The population in this survey are all the eligible voters of the state. The shoppers at the mall are the sample of that population.

Identify the sampling method used.

1. A survey calls every 10th name listed in a local phone book.

____

Identify the population and the sample.

2. Kennedy HS seniors planning to attend the prom were asked if senior dues should include a photo taken at the prom.

Population ______________________________________

Sample ______________________________________

A *biased sample* is not a good representative of the population.

A school principal asks parents attending an art workshop if funding for a theater arts program should be included in the school budget. This is a biased sample since the parents attending an art workshop are likely to be in favor of additional art programs.

Identify the population and the sample. Give a reason why the sample could be biased.

3. Homeowners within a 10-mile radius of a nuclear power plant were asked if they think the plant should be closed.

Population ______________________________________

Sample ______________________________________

Possible Bias: ______________________________________

____

Name ______________________ Date ____________ Class ____________

LESSON 9A

Homework and Practice

Samples and Surveys

Identify the population and sample. Give a reason why the sample could be biased.

1. The customers at a diner are surveyed to determine the number of times that local people eat out in a week.

   population ______________________

   sample ______________________

   possible bias ______________________

2. At a school with school colors of purple and white, the male students are asked to name their favorite color.

   population ______________________

   sample ______________________

   possible bias ______________________

3. The first one thousand adults at a professional baseball game are surveyed and asked to name their favorite sport.

   population ______________________

   sample ______________________

   possible bias ______________________

Identify the sampling method used.

4. The supervisor of the recreational committee visits 5 of the city's pools and collects water samples from the pools at various times of the day.

   ____

5. Commuters on a subway are selected at random and asked to name their favorite flavor of ice cream.

   ____

6. The register receipt of every 10th shopper at a grocery store is automatically compiled into a data base of the total amount spend and the items purchased.

   ____

EXPLORATION

Program of Studies and Core Content

MA-8-DAP-S-CD4, MA-8-DAP-S-DR1, MA-08-4.1.1, MA-08-4.1.2, MA-08-4.1.4

9B Organizing Data

1. Students in a class completed a survey in which they were asked how many hours they watched TV during one month. Their responses are shown below.

15	33	10	25	8
72	21	2	30	21
10	20	36	6	16
27	14	47	25	14
33	14	30	15	25

a. Write the numbers in order from least to greatest.

b. Divide the data into two groups: students who watched 30 hours of TV or less, and students who watched more than 30 hours. Which group is larger?

c. One student from the class was absent on the day of the survey. If the student were asked the same question, what number of hours would you guess might be close to the student's answer?

Think and Discuss

2. Explain how writing the numbers in order from least to greatest helps to organize data.

3. Describe another way to organize the data using a table.

Name ______________________ Date __________ Class __________

LESSON 9B

Technology Lab 1

Organizing Data

Activity

Show a line plot by using the drawing functions of Microsoft Word. For example, make a line plot of how many students watched television for a certain number of hours per day during the last week.

Celeste surveyed students in her class, asking how many hours of television they watched per day. Responses ranged from zero to six hours. For each number of hours reported, Celeste made a table of how many students watched for that length of time. She recorded her data as shown below.

Number of hours per day	0	1	2	3	4	5	6
Number of students	2	5	10	16	8	4	5

To make a line plot:

1. Open up a new Word document.
2. From the main menu, choose **View**, and then **Toolbars**. From the Toolbars dropdown menu, select **Drawing** to make the Drawing toolbar visible.
3. Click the **Text Box** icon on the Drawing toolbar.
4. Click and drag with the mouse to create a text-box area on the page several inches wide and deep.
5. Inside the text box, type the figure "0" followed by a vertical line " | " and follow it with two small *x*'s to represent the two students who reported zero hours of watching television. Press **Enter** to go to the next line.
6. Type the figure "1" followed by a vertical line " | " and follow it with five small *x*'s to represent the five students who reported one hour of watching television. Press **Enter** to go to the next line.

Name ______________________ Date ____________ Class ____________

LESSON **9B**

Technology Lab 1

Organizing Data, continued

7. Continue with the remaining data items in the same fashion.

8. A line plot is usually oriented with a line of numbers at the bottom and small *x*'s above each number. To display your line plot in that orientation, change the text in your text box to a vertical orientation. From the main menu, choose **Format**, and then **Text Direction…** In the **Text Direction** dialog box, click the icon at the left representing **vertical text running upwards**. Click **OK**.

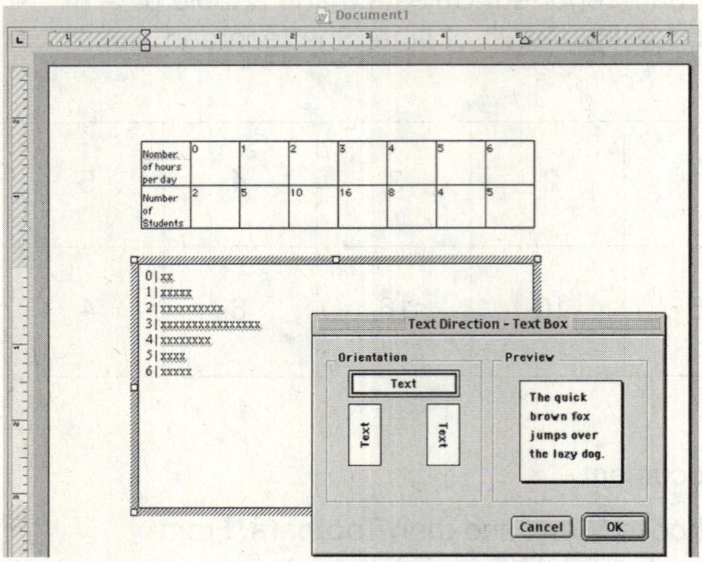

Think and Discuss

1. Could you also display the data above as a frequency tally? Would that be an appropriate display for the purpose? Discuss any possible advantages or disadvantages of using a frequency tally or a line plot.

____

____

2. Would a stem-and-leaf plot be helpful to display the data above? ____________

Try This

1. Make a line plot from the data in the table below.

Number of pets in household	Number of Students
0	2
1	9
2	6
3	4
4	1

Name ______________________ Date __________ Class __________

LESSON 9B

Technology Lab 2

Organizing Data

Activity

Bradley's soccer team spends an hour each week at team practice. Each team member also practices on his own. The team members record the total number of minutes each one spends practicing. Bradley saves the data in a spreadsheet table. He wants to organize it in a stem-and-leaf plot.

Begin by making two column headings, one for the student and one for the time spent practicing.

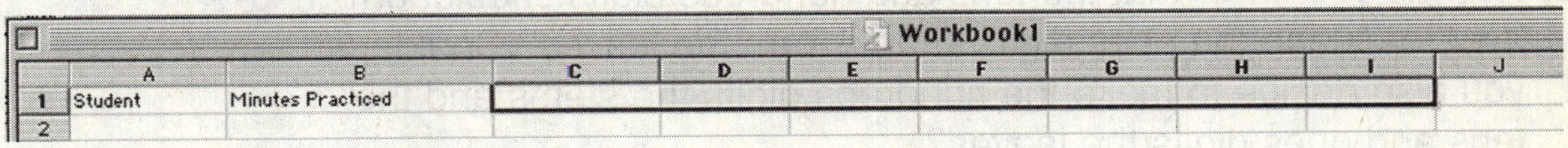

Input the data into the column using the table below.

Student	Minutes Practiced
1	97
2	118
3	148
4	121
5	91
6	124
7	116
8	122
9	128
10	111
11	105

Use the sort function of the spreadsheet to order the Minutes Practiced column in order from least to greatest. The tens and hundreds digits will become the stems and the ones digits will be the leaves.

Make a separate worksheet for the stem-and-leaf plot. Label the first column "Stems" and the second column "Leaves". Input each of the stems in the first column. Then add the leaves in the second column.

Stems	Leaves
9	1, 7
10	5
11	1, 6, 8
12	1, 2, 4, 8
13	
14	8

Key: 10 | 5 = 105

Stems	Leaves
9	1, 7
10	5
11	1, 6, 8
12	1, 2, 4, 8
13	
14	8

Name ______________________________ Date ____________ Class ____________

LESSON 9B

Technology Lab 2

Organizing Data, continued

Think and Discuss

1. How does using a spreadsheet make it easier to identify all of the possible stems?

____

____

2. How do you know that the tens and hundreds digits should form the stems and the ones digits should form the leaves? Could you also decide to make the hundreds digits the stems and the tens and ones digits the leaves?

____

____

____

Try This

1. Use a spreadsheet to help you put the data in order and create a stem-and-leaf plot for the test scores on a math exam: 83, 73, 78, 95, 92, 99, 85, 60, 85, 68.

2. Use a spreadsheet to help you put the data in order and create a stem-and-leaf plot for the number of points scored by the girls' intramural basketball team this season: 107, 88, 75, 59, 66, 64, 52, 56, 71, 60, 84.

Name ______________________ Date ___________ Class ___________

LESSON **9B**

Technology Lab 3

Organizing Data

Activity

A spreadsheet can also be used to organize back-to-back stem-and-leaf plots. For example, the Cougars baseball team has been playing in Fayetteville for 10 years. Their record for those ten years is in the table.

	Wins	Losses
1995	9	15
1996	8	16
1997	12	13
1998	11	15
1999	16	9
2000	16	10
2001	18	9
2002	19	8
2003	17	11
2004	20	6

Start by entering the data in a worksheet. Then use the **Sort** function to order the wins from least to greatest number of winning games per season.

Create a new worksheet that has three columns, one for the losses, one for the stems, and one for the wins.

	A	B	C
1	Losses	Stems	Wins
2			
3			
4			

Enter the stems and leaves for the Wins column on the right.

	A	B	C
1	Losses	Stems	Wins
2		|0|	8,9
3		|1|	1,2,6,6,7,8,9
4		|2|	0

Name ______________________________ Date ____________ Class ____________

LESSON 9B

Technology Lab 3

Organizing Data, continued

Now return to your first worksheet of season records. Sort the worksheet by the Losses this time. Enter the leaves for the Losses column on the left of your second worksheet of the stem-and-leaf plot.

	A	B	C
1	Losses	Stems	Wins
2	9,9,8,6	|0|	8,9
3	6,5,5,3,1,0	|1|	1,2,6,6,7,8,9
4		|2|	0

Think and Discuss

1. Is a spreadsheet useful for sorting the leaves of a stem-and-leaf plot? Why or why not?

____

____

____

2. When preparing a back-to-back stem-and-leaf plot, it is important to remember to include a key. Why?

____

____

Try This

Use a spreadsheet to help you organize data in a back-to-back stem-and-leaf plot for the season record of the Fighting Aardvarks basketball team.

	1995	1996	1997	1998	1999	2000	2001	2002	2003	2004
Wins	25	26	30	38	40	37	30	24	20	27
Losses	15	14	10	2	0	3	10	16	20	13

Name ______________________ Date __________ Class __________

LESSON 9B

Technology Lab 4

Organizing Data

Activity

Use your word processor to show the relationship between data using a Venn diagram.

Open a new document and write a description of the data. In this case, Sheila is doing a report on how many students in her class wear sweaters, sweatshirts or both on cold days.

Her survey shows the following data:

	Sweaters	Sweatshirts	Both
Number of Students	13	17	5

She writes in her word processor document:

This survey shows the number of students who wear sweaters, sweatshirts, or both with their uniforms on cool days.

Then she uses the Drawing toolbar to create a Venn diagram. From the toolbar, she selects Basic Shapes and then the circle (oval) shape from the choices offered. She draws two overlapping circles and then adds text boxes to label the circles and to add the data to the circles. (Hint: double-click on each circle after drawing it to change its fill to "No Fill." That will make the overlap of the circles visible.)

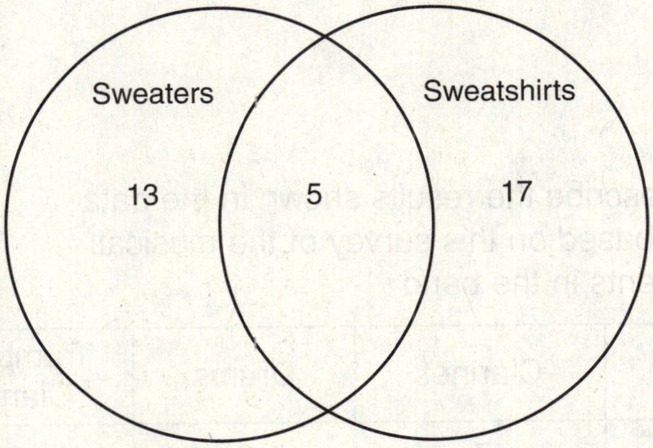

Name ______________________ Date ____________ Class ____________

LESSON 9B

Technology Lab 4

Organizing Data, continued

Think and Discuss

1. When would a Venn diagram in a report be more useful than a table listing the data values?

____

____

____

2. If there were no students who wore both a sweater and a sweatshirt, how would the Venn diagram show that fact?

____

Try This

1. Use a word processor to describe the results shown in the data and make a Venn diagram based on this survey of the pets owned by students in Mr. Reid's class.

	Cats	Dogs	Both
Students	25	15	4

2. Use a word processor to describe the results shown in the data and make a Venn diagram based on this survey of the musical instruments played by students in the band.

	Trumpet	Clarinet	Drums	Trumpet & Clarinet
Students	9	7	3	2

Name ______________________ Date __________ Class __________

LESSON 9B

Review for Mastery

Organizing Data

In a stem-and-leaf plot, horizontal displays of numbers can be written in a compact form by eliminating repetition.

60 61 63 66 66 67 can be written 6 | 0 1 3 6 6 7

All the numbers start with 6, the **stem.**

Just write the second digit for each number, the **leaves.**

Display each set of numbers in compact form, using a stem and leaves.

1. 72 75 75 76 76 76 79

____

2. 120 123 124 125 125 127

____

Here are the scores on the last test in Ms. Kahn's math class.

76 84 88 93 97 65 100 86 91 97
93 79 81 99 92 78 78 79 87 100

To display these scores in a **stem-and-leaf plot:**

Use the given order to record scores.

6	5
7	6 9 8 8 9
8	4 8 6 1 7
9	3 7 1 7 3 9 2
10	0 0

Key: 7 | 6 represents 76.

Order each set of leaves.

6	5
7	6 8 8 9 9
8	1 4 6 7 8
9	1 2 3 3 7 7 9
10	0 0

Key: 7 | 6 represents 76.

Complete a stem-and-leaf plot for the data.

3. Daily High Temperatures

46 52 48 47 56 59 61 50 37 35 34 37 44 49 43
43 44 50 50 52 53 50 48 46 44 39 37 32 32 32

3	7 5 4 7 9 7 2 2 2
4	
5	
6	

Key: represents 52.

3	2 2 2 4 5 7 7 7 9
4	
5	
6	

Key: represents 52.

4. Create a stem-and-leaf plot for heights in inches of students in Mrs. Gray's class.

40 48 49 53 60 62 48 62 55 53 54 60 65 63 49 55

Name ______________________ Date ____________ Class ____________

LESSON 9B

Review for Mastery

Organizing Data, continued

Two sets of data can be compared by using a **back-to-back stem-and-leaf plot**.

This plot compares the number of games won to the number of games lost by each of the pennant winners in the American League East for the years 1995–2000.

The stems are in the center. The left leaves are read in reverse.

The greatest number of games won was 114.

The greatest number of games lost was 74.

American League East Pennant Winners 1994–2000

Games Lost		Games Won
8	4	
8	5	
4 4	6	
4 0	7	
	8	6 7
	9	2 8 8
	10	
	11	4

Key: | 8 | 6 represents 86 games.

Key: 8 | 4 | represents 48 games.

Refer to the American League East back-to-back stem-and-leaf plot shown above.

5. The least number of games won was: ____________

6. The least number of games lost was: ____________

Using the given data set, complete to make a back-to-back stem-and-leaf plot.

7. **American League West Pennant Winners**

Year	Winner	Won	Lost
1994	Texas	52	62
1995	Seattle	79	66
1996	Texas	90	72
1997	Seattle	90	72
1998	Texas	88	74
1999	Texas	95	67
2000	Oakland	91	70

Games Lost		Games Won
	5	
	6	
	7	
	8	
	9	

Key: | 8 | 8 represents 88 games.

Key: 2 | 6 | represents 62 games.

Refer to the American League West back-to-back stem-and-leaf plot.

8. The difference between the greatest number of games won and the least number of games won was: ____________

9. The difference between the greatest number of games lost and the least number of games lost was: ____________

Name ______________________ Date __________ Class __________

LESSON 9B

Homework and Practice

Organizing Data

1. Use the given data to make a table.

Two hundred people were asked which was their favorite type of movie—comedy, drama, or action. The men's choices were 10 for comedy, 35 for drama and 55 for action. The women's choices were 45 for comedy, 50 for drama and 5 for action.

List the data values in the stem-and-leaf plot.

2.

1	2 5 7 8
2	0 0 6
3	1 1 2 2
4	5 6

key: 4 | 5 = 45

____

____

3.

4	4 6 8
5	0 2 4 6
6	7
8	1 3 5 7

key: 8 | 1 = 81

____

____

4. Use the given data to make a stem-and-leaf plot.

Michael's Test Scores					
86	85	92	75	72	94
81	86	94	77	67	80

key:

5. Use the given data to make a back-to-back stem-and-leaf plot.

MLB American League—Central Division 2000–2001 Final Standings		
MLB Team	**Wins**	**Losses**
Cleveland	91	71
Minnesota	85	77
Chicago	83	79
Detroit	66	96
Kansas City	65	97

Wins		Losses

key:

EXPLORATION

Program of Studies and Core Content
MA-8-DAP-S-CD3, MA-08-4.2.1

9C Measures of Central Tendency

1. Students in a class completed a survey in which they were asked what their heights in inches are. Their responses are shown below.

49	53	60	55	48
72	65	66	58	68
75	65	64	57	59
61	67	64	58	62
63	59	61	55	65

a. Write the numbers in order from least to greatest.

b. What number appears most often?

c. What number is in the middle of the data set?

d. Add the numbers and divide the total by 25 to find the mean.

Think and Discuss

2. **Discuss** which number from Problems **1b, 1c,** and **1d** best represents the entire set of numbers. Why?

3. **Describe** a set of numbers arranged from least to greatest in which the middle number is not close to the value of the mean.

Name ______________________ Date __________ Class __________

LESSON 9C

Technology Lab

Mean, Median, and Mode

The National Collegiate Athletic Association (NCAA) tournaments determine the champion of women's and men's college basketball. The victory margins for the championship games from 1995 through 2001 are shown below.

Margin of Victory , NCAA Championship Games							
Year	1995	1996	1997	1998	1999	2000	2001
Men's Game (points)	11	9	5	9	3	13	10
Women's Game (points)	6	18	9	18	17	19	2

Activity

1 Use a spreadsheet to find the mean, median, and mode of the men's championship-game victory margins from the table. Fill in rows 1 and 2 with the data and labels shown in the spreadsheet below.

The **AVERAGE**, **MEDIAN**, and **MODE** functions find the mean, median, and mode of the data in a given range of spreadsheet cells.

- Enter = **AVERAGE(B2:H2)** into cell H3 to find the mean of the data in cells B2 through H2.
- Enter = **MEDIAN(B2:H2)** into cell H4 to find the median of the data.
- Enter = **MODE(B2:H2)** into cell H5 to find the mode of the data.

	A	B	C	D	E	F	G	H
1	Year	1995	1996	1997	1998	1999	2000	2001
2	Margin (points)	11	9	5	9	3	13	10
3							Mean	8.571429
4							Median	9
5							Mode	9

Name ______________________ Date ____________ Class ____________

LESSON 9C

Technology Lab

Mean, Median, and Mode, continued

Think and Discuss

1. If an eighth game with a victory margin of 30 points were added, what would happen to these three calculated values?

____

____

Try This

1. Use a spreadsheet to find the mean, median, and mode for the women's championship games (shown in the table on the previous page).

____

2. The *range* of a set of data is the difference between the greatest and least data values. Compare the ranges for the women's and men's championship games.

____

3. Compare the mean of the data sets. Which group do you think has more close championship games?

____

4. Compare the medians for the men's and women's margin of victory data. Which is a better indicator for that group's typical margin of victory? Explain.

____

____

Name ______________________________ Date ____________ Class ____________

LESSON 9C

Review for Mastery

Measures of Central Tendency

The data show the mileages (in miles per gallon) for the top-selling SUVs of two car companies.

Company A	12	8	11	10	14	16
Company B	15	11	17	11	15	29

You can calculate the range and measures of central tendency for the SUVs from Company A:

Range = greatest value − least value = 16 − 8 = 8 mi/gal

$$\textbf{Mean} = \text{average value} = \frac{12 + 8 + 11 + 10 + 14 + 16}{6} = \frac{71}{6} \approx 11.8 \text{ mi/gal}$$

The **mode** is the most frequently occurring data value. There is no mode.

The **median** is the middle value when the data are listed in order:

8 10 11 12 14 16

When there is an even number of data values, the median is equal to the mean of the two middle values:

$$\text{Median} = \frac{11 + 12}{2} = 11.5 \text{ mi/gal}$$

1. Find the range, mean, mode, and median for Company B.

____

____

2. Company B sells a hybrid SUV that's great on gas. What is its mileage?

____

An **outlier** is a value that is much greater or much less than the others in the data set. The value 29 is an outlier for the Company B SUV data above.

3. Calculate the mean without the outlier for the Company B data. Remember that there is one less data value.

____

Name ______________________ Date __________ Class __________

LESSON 9C

Homework and Practice

Measures of Central Tendency

Find the mean, median, mode, and range of each student's set of quiz scores.

1. Tom: 77 91 84 84 79

mean: ______________________

median: ______________________

mode: ______________________

range: ______________________

2. Anvi: 53 97 100 66 99

mean: ______________________

median: ______________________

mode: ______________________

range: ______________________

Use your answers from exercises 1 and 2 for exercises 3–4.

3. What do the mean and range of each set of quiz scores say about how Tom and Anvi did in class? Explain.

____

____

4. Is the median a fair representation for a typical quiz score for Tom? For Anvi? Explain.

____

____

Use the data to find each answer.

Oceans	Area in sq. mi.	Greatest Known Depth in ft.
Pacific Ocean	64,000,000	36,198
Atlantic Ocean	31,815,000	30,246
Indian Ocean	25,300,000	24,460
Arctic Ocean	5,440,200	18,456

5. Find the mean of the areas in square miles.

____

6. Find the median of the area in square miles.

____

7. Find the mean and median of the greatest known depth in feet.

____

EXPLORATION

Program of Studies and Core Content

MA-8-DAP-S-CD1, MA-8-DAP-S-DR1, MA-08-4.1.1, MA-08-4.2.1

9D Outliers, Gaps, and Clusters

Vicky and Miguel measure the weights of 20 fish caught at a local lake. They display their results in the line plot below.

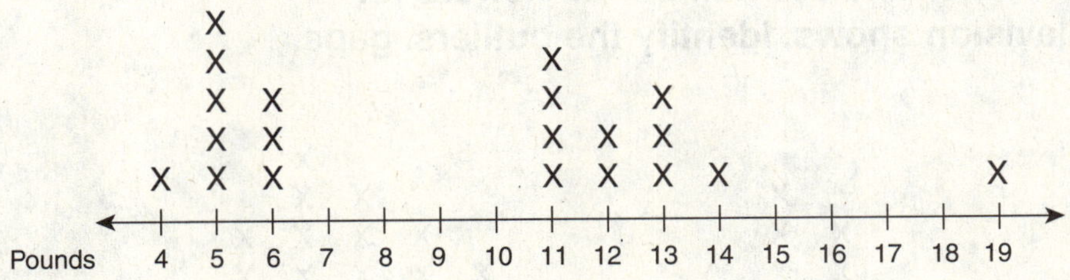

1. Is there an outlier in this data set? If so, identify it.

2. What is the mean of the data?

3. Vicky and Miguel are going to add another fish to this plot. Would the mean weight be a good guess for the new fish? Why or why not?

Think and Discuss

4. Explain what might have caused the data to fall into two separate clusters.

Name ______________________ Date __________ Class __________

LESSON 9D

Review for Mastery

Outliers, Gaps, and Clusters

In a previous lesson, you learned about outliers. Outliers are data points that are much larger or much smaller than the rest of the data in a set.

Data sets can also have gaps and clusters. A gap is any range of values that has no data points, and a cluster is a range of values that has many data points.

The line plot below shows the number of viewers for 18 different television shows. Identify the outliers, gaps, and clusters.

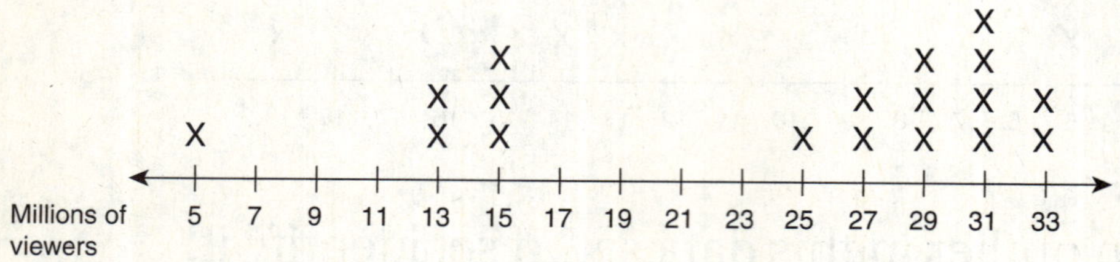

There is an outlier at 5 million viewers.

There are two gaps in the data, between 5 and 13 million viewers and between 15 and 25 million viewers.

There are two clusters of data. There are 5 shows clustered at 13 and 15 million viewers and 12 shows clustered at 25 to 33 million viewers.

Use the line plot above to answer Exercises 1–2.

1. What is the mode of the data set? Is the mode a good representation of the data set? Why or why not?

____

____

2. One of the clusters represents shows that were repeats, and the other data cluster represents shows that were new. Which cluster do you think represents new shows? Why?

____

Name ______________________ Date ____________ Class ____________

LESSON 9D

Homework and Practice

Outliers, Gaps, and Clusters

1. A nutritionist approaches 15 students during lunch and asks to examine their drinks. She records the grams of sugar per ounce (g/oz) of each student's drink. The results are given below. Use her data to make a line plot.

   g/oz: 1.5, 4.0, 3.5, 1.5, 4.0, 0, 1.5, 0, 4.0, 3.5, 5.5, 1.5, 0, 4.0, 1.5

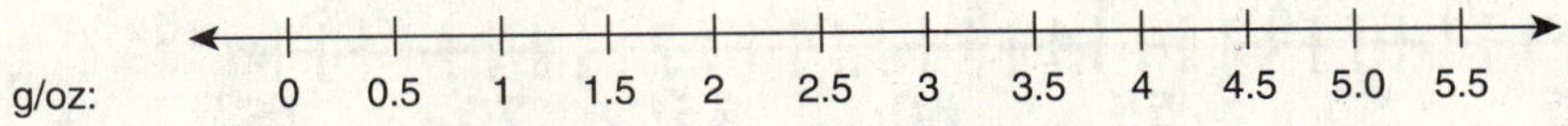

2. Is there an outlier in the data? If yes, identify it. ______________

3. Describe the three clusters in the data.

   ____

4. Describe the three gaps in the data.

   ____

5. Find the mean of the data. Is the mean a good descriptor of the data set? Explain why or why not.

   ____

   ____

6. What beverage might correspond to each of the three clusters? Research in books or on the Internet to support your theories.

   ____

   ____

EXPLORATION

Program of Studies and Core Content

MA-8-DAP-S-CD1, MA-8-DAP-S-CD3, MA-8-DAP-S-CD4, MA-8-DAP-S-DR1, MA-08-4.1.1, MA-08-4.1.2, MA-08-4.1.4, MA-08-4.1.5, MA-08-4.2.1

9E Variation

When placed on a number line, values in a data set can be spread out or clustered together.

1. Order the data sets from the ones you think are the least spread out to the most spread out.

Set 1:

Set 2:

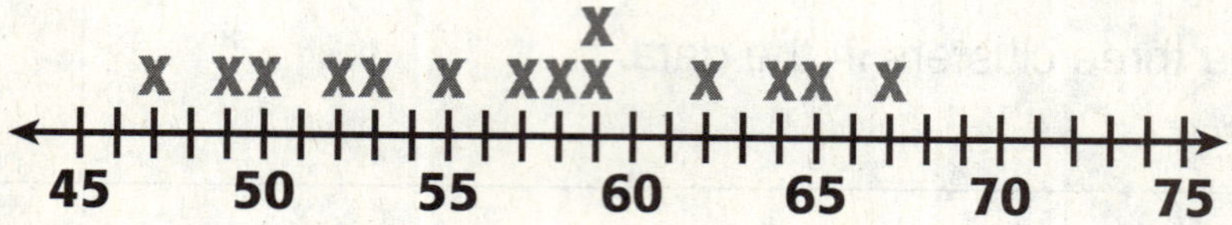

Set 3:

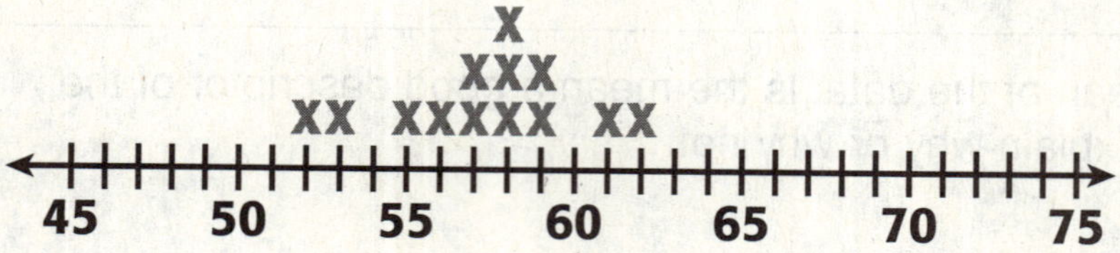

Set 4:

Think and Discuss

2. Discuss the strategies you used to order the sets according to the spread of the data.

3. Explain how the spread of set 1 would change if the value 72 were added.

Name ______________________ Date __________ Class __________

LESSON 9E

Review for Mastery

Variation

Quartiles divide a data set into four equal parts.

[1 1 3 3 3 4 4 6 6 6 6] 6 [7 7 7 7 7 8 9 9 9 9 9]

The **1st quartile** is the median of the lower half.

The **median,** or **2nd quartile,** divides the data into two halves.

The **3rd quartile** is the median of the upper half.

For each data set, circle and label the median *M*. Circle and label the first quartile Q_1. Circle and label the third quartile Q_3.

1. 3 3 3 4 5 6 7

2. 3 3 3 4 4 5 6 7 9 10 10

_______________________ _______________________

A **box-and-whisker** plot displays the quartile values as well as the lowest and highest numbers of a data set. The sides of the box are the first and third quartiles; the median is inside the box. The whiskers connect the box to the lowest and highest values.

This plot displays the values for the data set at the top of this page.

lowest value = 1
1st quartile = 4
median = 6
3rd quartile = 8
highest value = 9

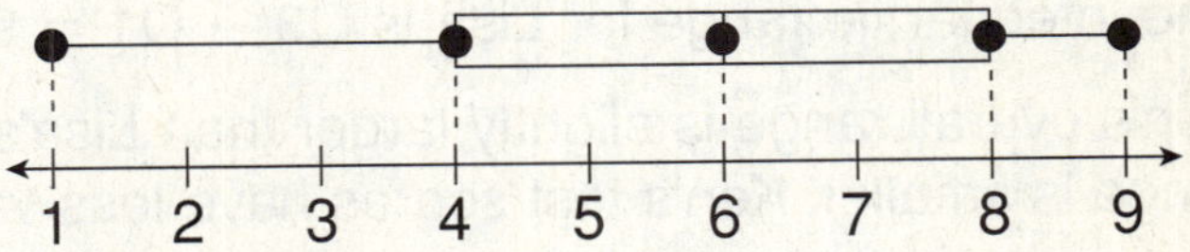

Complete to make a box-and-whisker plot for the data set 45, 47, 47, 48, 48, 49, 53.

3. First, calculate three significant values for the data set.

median = ______ 1st quartile = ______ 3rd quartile = ______

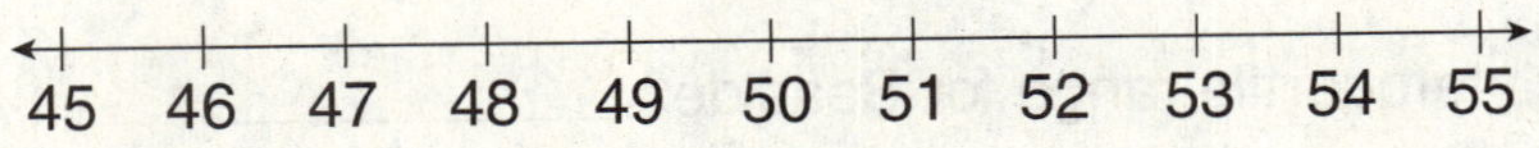

Name ______________________ Date __________ Class __________

LESSON 9E

Review for Mastery

Variation, continued

The *variation* of a data set is a way of describing how spread-out the data is. Two measures of variation are the range and the interquartile range.

The *interquartile range* is the third quartile minus the first quartile. Half of the data points are in between the first and third quartiles, so if the interquartile range is small, then 50% of the data values are clustered close to the median. This means there is less variation in the data.

The two box-and-whisker plots show the test scores of two students. Find the ranges and interquartile ranges of the two data sets. Whose test scores have less variation?

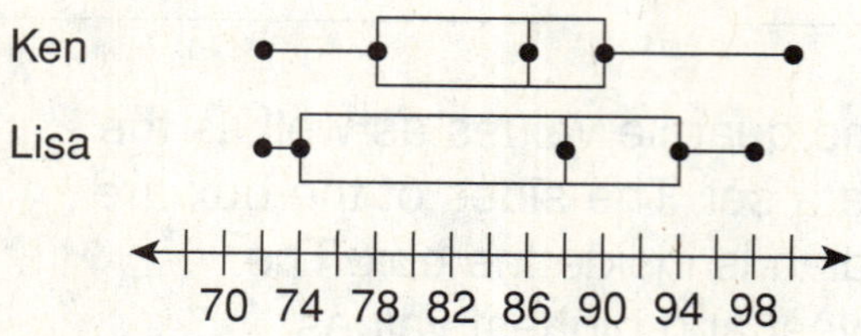

The range for Ken is 100 − 72 = 28.
The interquartile range for Ken is Q3 − Q1 = 90 − 78 = 12.
The range for Lisa is 98 − 72 = 26.
The intequartile range for Lisa is Q3 − Q1 = 94 − 74 = 20.

Ken's overall range is slightly larger than Lisa's, but his interquartile range is smaller. Ken's test scores have less variation.

The box-and-whisker plots show the high temperatures for two towns. Use the plots to answer Exercises 4–6.

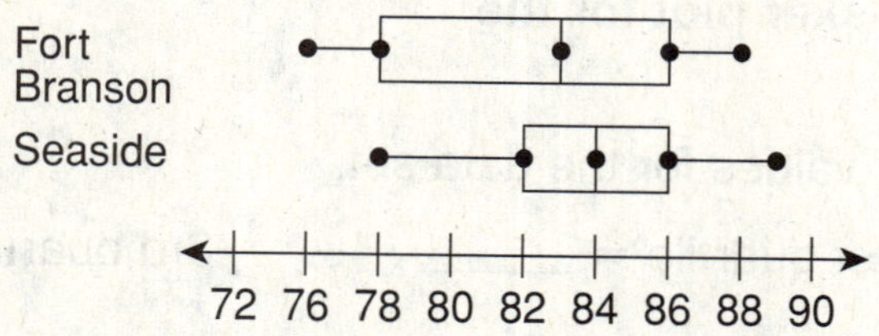

4. What is the interquartile range for Fort Branson? ______________

5. What is the interquartile range for Seaside? ______________

6. Which town has more variation in high temperatures? ______________

Name ________________ Date ________ Class ________

LESSON **9E**

Homework and Practice

Variation

Find the range and the first and third quartiles for each data set.

1. 52, 36, 41, 48, 55, 31, 57

range: ________________

first quartile: ________________

third quartile: ________________

2. 26, 28, 42, 43, 22, 56, 51, 36

range: ________________

first quartile: ________________

third quartile: ________________

Use the given data to make a box-and-whisker plot.

3. 44, 56, 30, 73, 34, 55, 62, 51

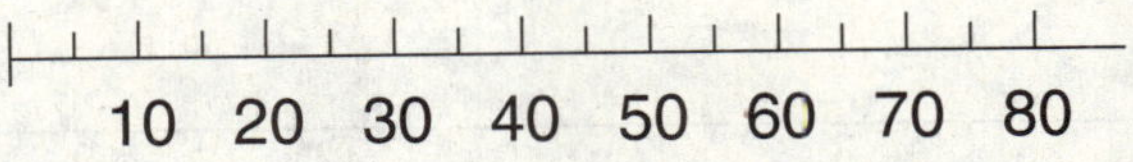

4. 26, 37, 45, 62, 55, 11, 27, 72, 51, 48, 43, 38, 47

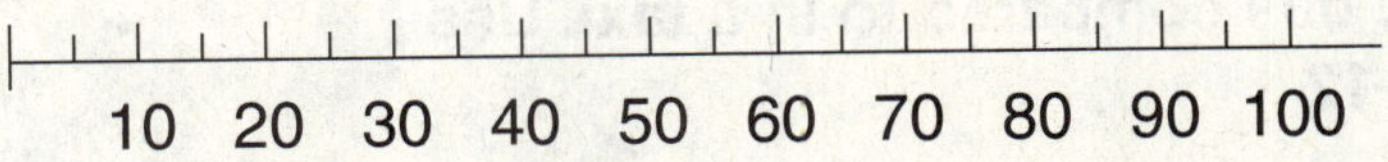

Use the box-and-whisker plots to compare the data sets.

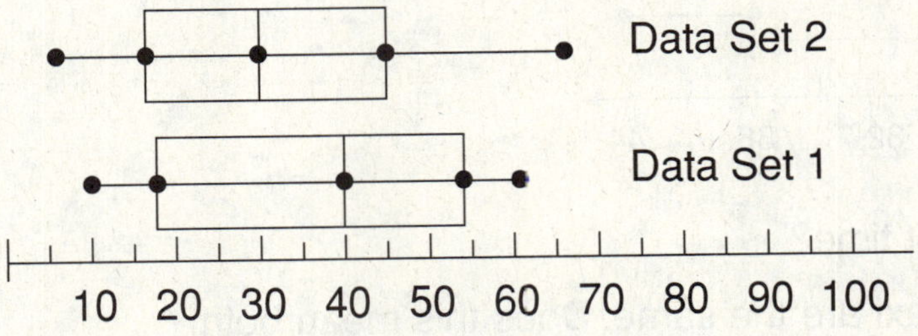

5. Compare the ranges and medians.

____

____

6. Compare the interquartile range for each data set.

____

____

Name ____________________ Date __________ Class __________

LESSON 9E

Homework and Practice

Variation, continued

The box-and-whisker plot shows the daily sales for two juice bars. Use the plot to answer Exercises 7–10.

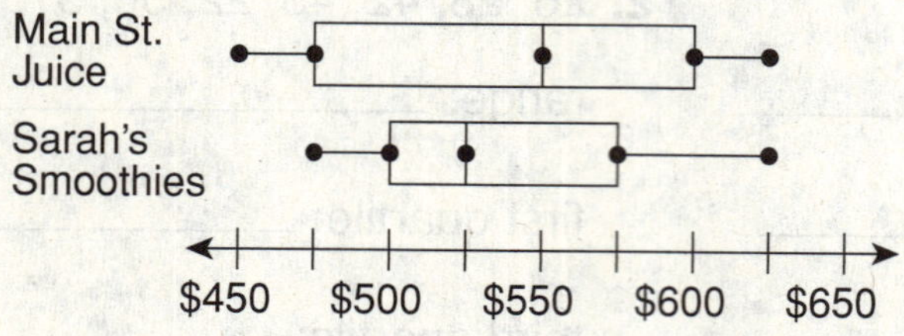

7. Which shop has the greater median sales? ____________

8. Which shop has the smaller range? ____________

9. Which shop has the smaller interquartile range? ____________

10. Which shop appears to have more predictable sales? Explain.

____

____

The box-and-whisker plot shows the time in minutes that it takes to get to the airport on the bus compared to in a taxi. Use the plot to answer Exercises 11–12.

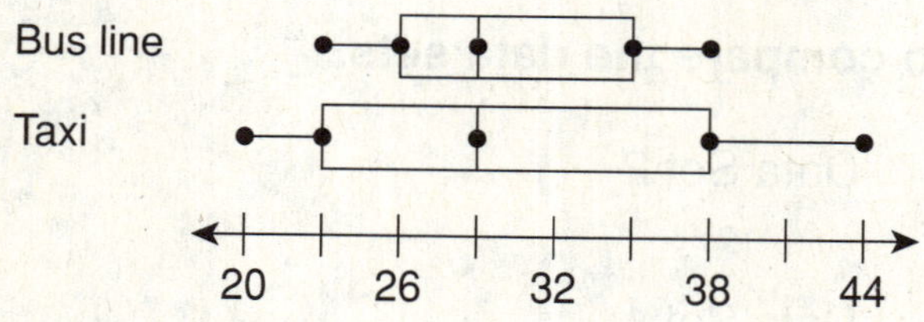

11. Which method has the shortest time? ____________

12. The medians for the bus and taxi are the same. Does this mean both methods are equally reliable? Explain.

____

____

EXPLORATION

Program of Studies and Core Content

MA-8-DAP-S-CD4, MA-8-DAP-S-DR1, MA-8-NPO-S-E2, MA-08-4.1.1, MA-08-4.1.2, MA-08-4.1.4

9F Displaying Data

The bar graph shows the results of a survey about students' lunch beverage preferences.

1. Do at least half the students prefer sports drink? Why?

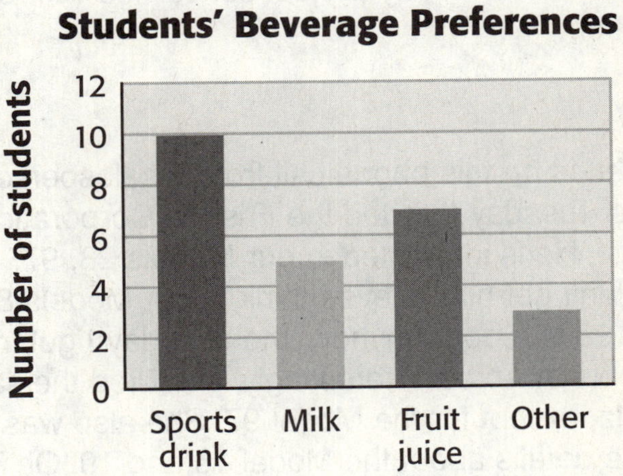

The histogram shows the results of a survey about the amount of time in hours per week that adults spend on the Internet.

2. Did at least half say they spend 20 hours or less? How do you know?

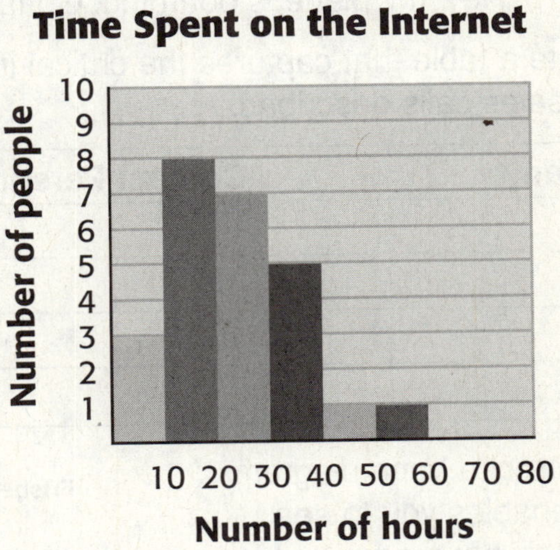

Think and Discuss

3. Compare the types of data displayed by the bar graph with the types of data displayed by the histogram.

Name ______________________________ Date ____________ Class ____________

LESSON **9F**

Hands-On Lab 1

Venn Diagrams

Very often, information from the real world doesn't present itself in neat tables and charts. It can be buried in collections of news reports or it can be contained in the thousands of words that make up a single discussion or report. Digging out information from various sources and putting it into some kind of order is an important activity in any project concerned with organizing information.

Activity

1. Listen in on this phone call from a salesperson to her manager: "Last Tuesday I visited the Frisbee Corporation and spoke to Hal Hertz. He is interested in our Models 23, 97, 12, and 44. I couldn't get him to even think about Models 89 and 99. I'm afraid that 99 is a dog. Anyhow, the next day I got in to see Marge Dubois at Acme. Interestingly, she liked the same things as Hertz, except for the Model 97. She also was very interested in more details about the Model 15 and 19. On Thursday I had lunch with Tony Romero over at Benson Corporation. He's really something else! The only two models he would even think about are 13 and 52. And just this morning I got to see Carmen Chang at Holloway. She really wants to know more about Models 25 and 15. Hey, my plane is boarding. Gotta run!"

2. Create a table that captures the critical information about the four sales calls described.

Company	Contact Person	Models Interested In

3. Construct a Venn diagram that enables you to see clearly what models each company is interested in and patterns of how the choices of the different companies overlap.

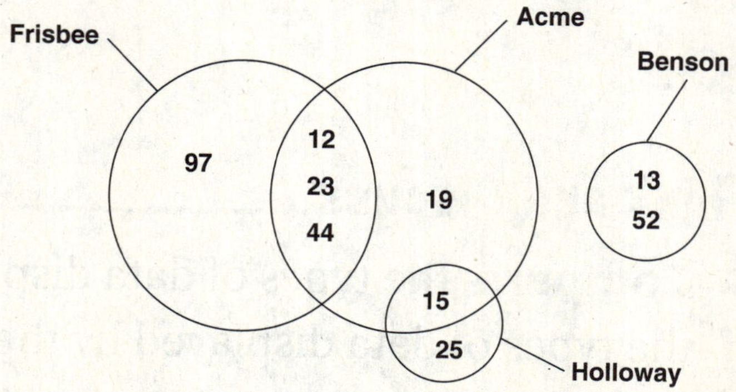

Name ______________________ Date __________ Class __________

LESSON 9F

Hands-On Lab 1

Venn Diagrams, continued

Think and Discuss

1. Name some other situations where the flow of information would be oral and somewhat rushed.

____

____

2. How would an analysis of the information from the table differ from an analysis using the Venn diagram?

____

3. Would your answer to Question 2 be different if the salesperson had described 15 sales calls instead of 4? Explain.

____

Try This

1. Use the Internet to look up 3 or 4 recipes for the same dish, such as beef stew. Draw a Venn diagram comparing the ingredients used in the recipes. What observations can you make from studying these Venn diagrams?

____

____

____

____

Name ______________________________ Date ____________ Class ____________

LESSON **9F**

Hands-On Lab 2

Displaying Data

Activity

1. Ask at least 10 classmates how many minutes they spend getting ready for school in the morning. Write the data on the line below.

____

2. Use the data to make a frequency table. Add more rows if necessary. Choose appropriate intervals of minutes. For example, if most students took between 30 and 45 minutes to get ready, 5-minute intervals would be appropriate.

Minutes	Frequency

3. Use the data in the table to make a histogram on the section of graph paper below.

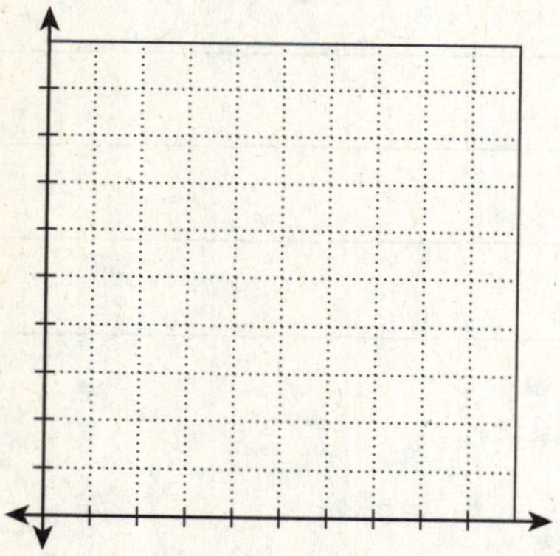

Remember: Give your graph a title. Be sure each axis has a title to describe the data.

Name ______________________ Date ____________ Class ____________

LESSON 9F

Hands-On Lab 2

Displaying Data, continued

Think and Discuss

1. Suppose you asked 10 elementary school students and 10 middle school students how many minutes they spend getting ready for school. What type of graph could you use to compare the data for the elementary students to the data for the middle school students? Explain.

____

____

____

2. If you knew how many minutes you and a friend spent getting ready for school when you were in first, third, fifth, and seventh grades, what type of graph would you use to study the changes over time? Explain.

____

____

____

Try This

1. Use the Internet or newspaper to find the high temperatures in your community for the last week. Make a frequency table of the data.

Temperature	Frequency

2. On a separate sheet of paper, make a bar graph and a line graph of the data. Explain how the two graphs display the data differently.

____

____

____

____

Name ______________________ Date ____________ Class ____________

LESSON 9F

Review for Mastery
Displaying Data

Depending on the data, one kind of graph is usually more appropriate than another.

Bar Graph

- Used for data that can be organized into *categories*.
- Makes it easy to make *comparisons*.
- A **double bar graph** can be used to compare two sets of data.

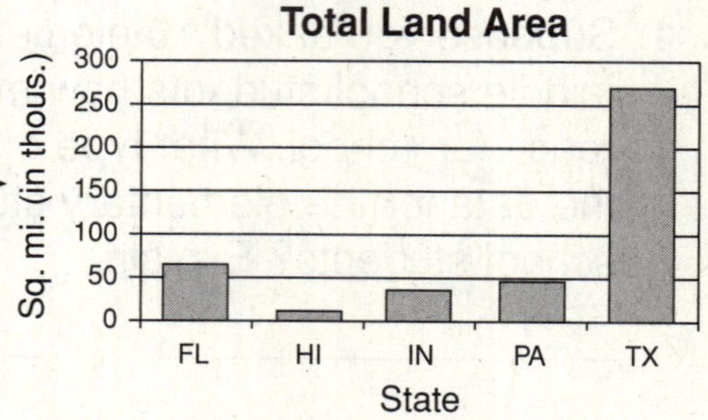

Line Graph

- Used for displaying data that changes over a period of *time*.
- Makes it easy to see *change*.
- A **double line graph** is used to compare two related sets of data that change over time.

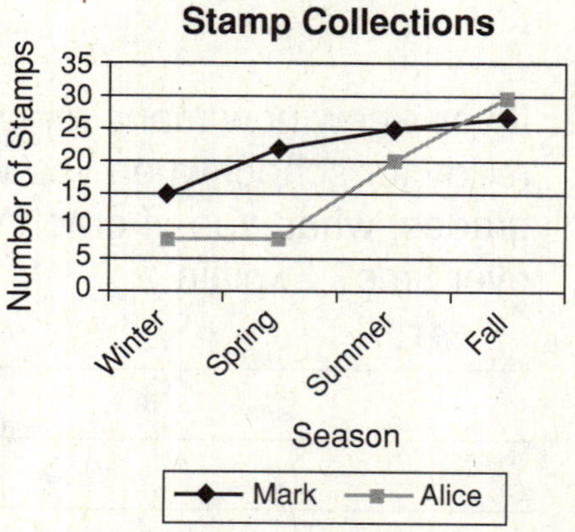

Circle Graph

- Used for data that represents *part of a whole*.
- Makes it easy to compare each category to the whole set.
- The whole circle represents 100% and each sector represents a *percent of the total*.

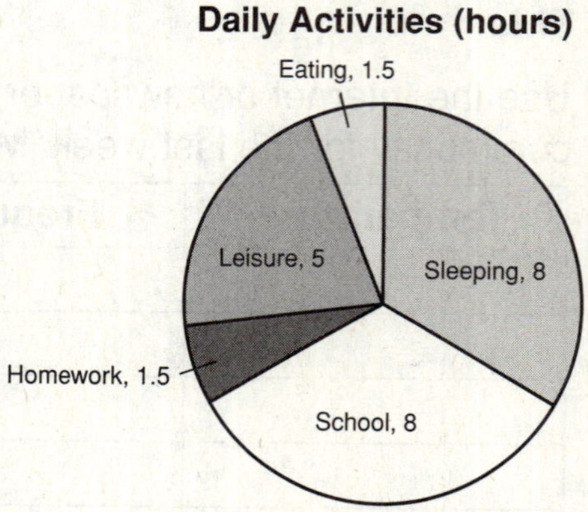

Complete the following.

1. There are five categories in the bar graph. List them. ____________
2. Look at the line graph. When does Alice have more stamps than Mark? ____________
3. Add the hours in each sector of the circle graph. ____________
   Divide 1.5 by 24 to find the percent of the day spent on homework. ____________

Name the type of graph that would be best to display the data described.

4. the cost of two brands of cereal over time ____________
5. the percent of a person's budget spent on various things ____________
6. the life expectancy of several species of reptile ____________
7. the change in school enrollment from 2000 to 2005 ____________

Name ______________________ Date ____________ Class ____________

LESSON 9F

Review for Mastery

Displaying Data, continued

If data:	Then use:
is organized into categories	bar graph/double bar graph
changes over a period of time	line graph/double line graph
categories can be compared to whole set	circle graph

Use the data at right to make a graph. Explain why you chose that type of graph.

Recipe for Bubbles	
Ingredient	**Cups**
bubble soap	8
dishwashing liquid	1
glycerin	1
water	10

Because the categories (the ingredients) can be compared to a whole set (the recipe), a circle graph is best.

Step 1: Total the number of cups.

$8 + 1 + 1 + 10 = 20$

Step 2: Calculate the percent of each ingredient.

- buble soap $\frac{8}{20} = 40\%$
- glycerin: $\frac{1}{20} = 5\%$
- dishwashing liquid: $\frac{1}{20} = 5\%$
- water: $\frac{10}{20} = 50\%$

Step 3: Find the angle measure for each sector of the graph.

- bubble soap: $40\%(360°) = 144°$
- glycerin: $5\%(360°) = 18°$
- dishwashing liquid: $5\%(360°) = 18°$
- water: $50\%(360°) = 180°$

Use a compass and protractor to draw the graph.

Recipe for Bubbles

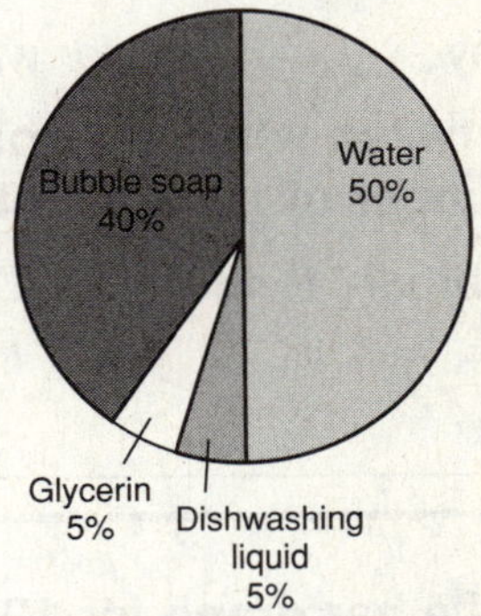

Write bar, double-bar, line, double-line, or circle to indicate the type of graph that would best display the data described.

8. math scores of one student over the school year ____________

9. attendance at an exercise class by age group, as it relates to total attendance ____________

10. numbers of different types of animals seen at a farm ____________

11. A store owner uses an entire wall to display toys as shown in the table. Use the data to make a graph. Then explain why you chose that type of graph.

Toys for Sale	
Toys	**Shelves**
games	10
puzzles	5
dolls	3
trains	2

Name ____________________ Date __________ Class __________

LESSON 9F

Review for Mastery

Displaying Data, continued

Comparisons between two groups can be made easily on a double-bar graph.

Which class shows the greatest difference between residents and non-residents?

Dance — *Compare bars for each class. Choose the bars with the greatest difference.*

How many more people were enrolled in soccer than dance?

300 — *Find total for soccer: 350 + 300 = 650.*
Find total for dance: 100 + 250 = 350.
Subtract: 650 − 350 = 300.

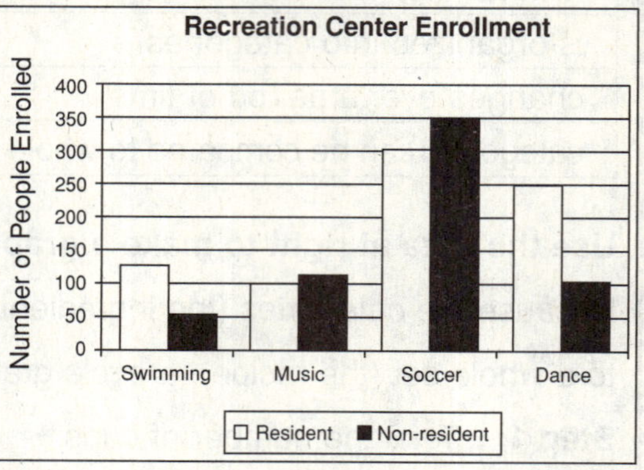

With a double-line graph, you can easily see how groups change over time.

During which month(s) were the number of male instructors equal to the number of female instructors?

May — *Look for where the data points overlap.*

Between what two consecutive months did the number of female instructors increase the most?

February to March — *Look at female data points only. Find the steepest positive slope.*

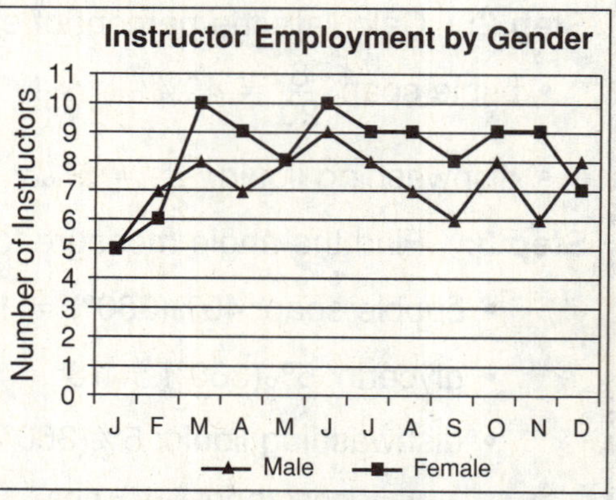

Use the bar graph for 12–14.

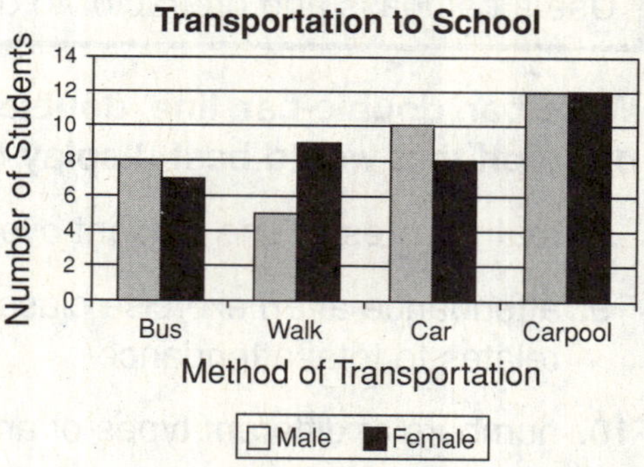

12. Which method of transportation is used by most students? ____________

13. How many more girls than boys walk to school? ____________

14. How many boys go by car or ride in a carpool? ____________

Use the circle graph for 15–17.

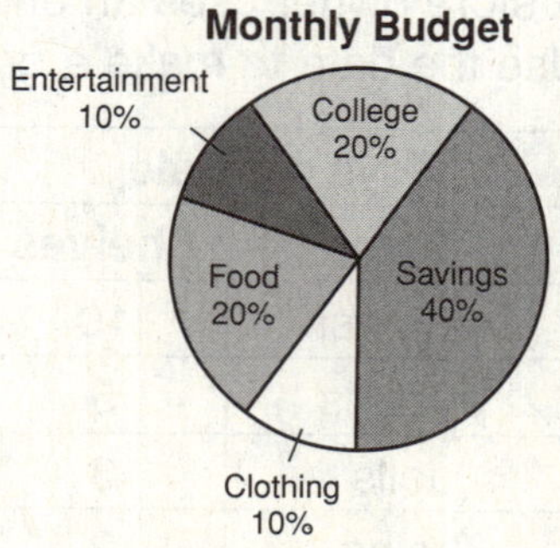

15. Which category accounts for the highest percentage of the monthly budget? ____________

16. Which categories account for the smallest percentages of the monthly budget? ____________

17. If the budget is $500, how much is spent on food?

____

Name ______________________ Date ______________ Class ______________

LESSON 9F

Review for Mastery

Displaying Data, continued

A **stem-and-leaf plot** arranges data by dividing each data value into two parts: a stem (the digit or digits other than the last digit), and a leaf (the last digit).

The amount of money collected by each student for the drama club is shown below. Use the data to make a stem-and-leaf plot.

55, 85, 90, 113, 100, 90, 93, 68, 66, 108, 116, 56, 85, 89, 102, 103

Step 1: List the stems.

The least value is 55; the greatest value is 116. List stems from 5 to 11. Do not omit any stems.

Step 2: List the leaves.

For each stem, write the ones digits from least to greatest.

Step 3: Write a key explaining one value.

Stem	Leaves
5	5 6
6	6 8
7	
8	5 5 9
9	0 0 3
10	0 2 3 8
11	3 6

Key: 8|2 means 82

The test scores from two different math classes are shown below. Use the data to make a back-to-back stem-and-leaf plot.

Class A: 50, 68, 95, 80, 92, 100, 98, 85, 82, 81
Class B: 75, 81, 100, 63, 52, 94, 100, 100, 87, 99

Step 1: List the stems.

The lowest value is 50; the highest value is 100. List stems from 5 to 10. Do not omit any stems.

Step 2: List the leaves.

For each stem, write the ones digits from least to greatest.

Step 3: Write a key explaining one value from each side.

Leaves Class A	Stems	Leaves Class B
0	5	2
8	6	3
	7	5
5 2 1 0	8	1 7
8 5 2	9	4 9
0	10	0 0 0

Key: |8|1 means 81
5|9| means 95

18. The daily low and high temperatures in degrees Fahrenheit in a town in the Northeast are given below. Use the data to make a back-to-back stem-and-leaf plot.

Daily Low Temperatures (°F)					
40	56	50	60	62	63
49	48	49	40	36	59
57	52	53	42	44	39

Daily High Temperatures (°F)					
70	84	71	73	71	70
73	78	76	65	65	67
66	76	69	70	70	58

Low Temp		High Temp

Name ______________________ Date ____________ Class ____________

LESSON 9F

Review for Mastery

Displaying Data, continued

The ages of people visiting a water park during a certain time period are given below. Use the data to make a frequency table with intervals. Then make a histogram.

5, 12, 22, 15, 17, 13, 25, 34, 7, 9, 12, 32, 12, 15, 18

Step 1: Find the difference between the greatest and least values.

Least: 5 Greatest: 34

$34 - 5 = 29$

Step 2: Use the difference to decide on intervals ⇨

Finding the Interval		
If width of interval is:	Then divide	The number of intervals is:
10	$\frac{29}{10} = 2.9$	3 (too few)
3	$\frac{29}{3} \approx 9.7$	10 (too many)
5	$\frac{29}{5} = 5.8$	6 (good)

Try different widths for your intervals to determine the number of bars in the histogram.

Step 3: Create a frequency table.

Ages of Visitors	
Age	Frequency
5 - 9	3
10 - 14	4
15 - 19	4
20 - 24	1
25 - 29	1
30 - 34	2

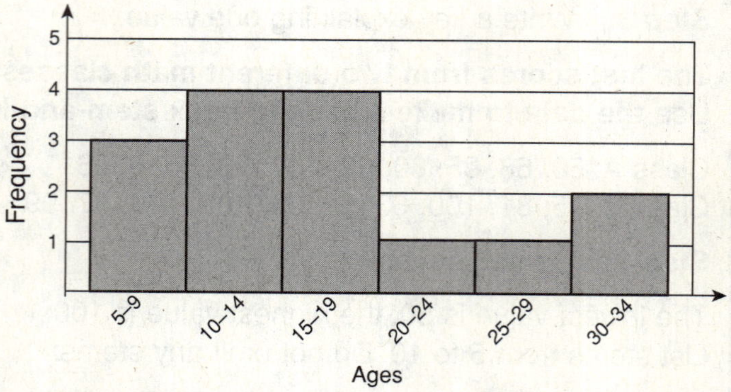

Step 4: Use the frequency table to create the histogram. Draw each bar to the corresponding frequency.

19. The estimated miles per gallon for selected cars are shown in the table. Use the data to make a frequency table with intervals. Then make a histogram.

26	28	32	33	26	15	21
35	17	18	25	29	30	26
27	30	24	25	24	32	25
19	22	32	25	31	28	23
27	23	24	20	38	44	18

Car Gas Mileage	
mi/gal	Frequency

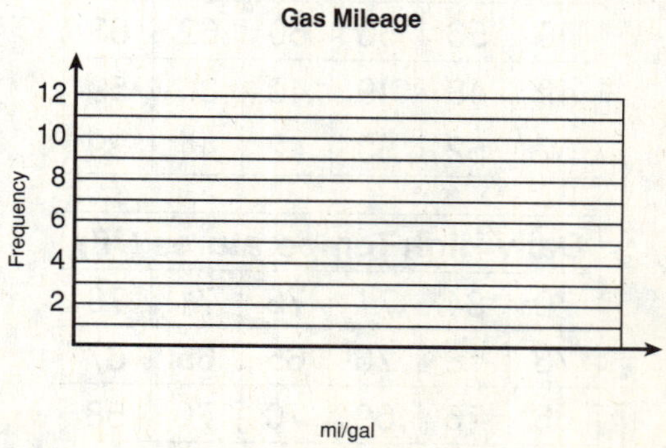

Name ______________________ Date ____________ Class ____________

LESSON 9F

Homework and Practice

Displaying Data

The table below shows the top five ringtone genres reported during a survey of cell phone subscribers in August, 2005.

Ringtone genre	Males (millions)	Females (millions)
Hip Hop or Rap	2.232	2.472
Rock or Alternative	2.219	2.082
Pop	1.044	2.069
TV/Movie/Game Themes	1.161	1.371
Classic Rock	0.949	0.918

1. Use the data to make a graph below. Explain why you chose that type of graph.

____

____

____

2. From the graph (*not* the table), explain how you can tell that the Pop genre is nearly twice as popular with females as males.

____

____

____

In 2000, there were approximately 27 million people 18–24 years old in the United States. The circle graph below shows the highest levels of education for that population. Use the graph to select the best answers for questions 3–5.

Educational Attainment of the Population 18–24 Years Old, 2000

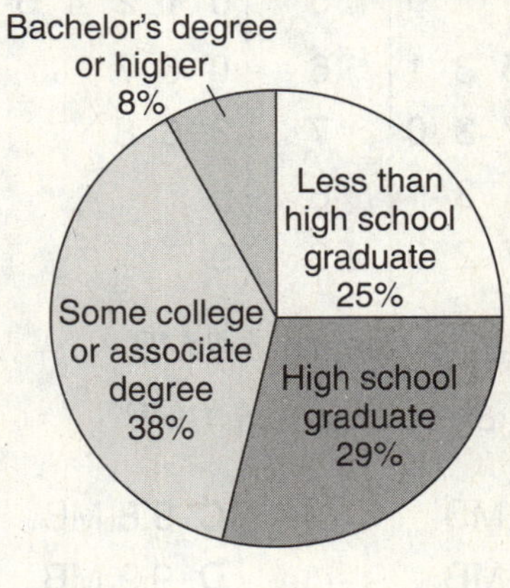

3. Which level of education was least represented?

A Less than high school graduate
B High school graduate
C Some college or associate degree
D Bachelor's degree or higher

4. Approximately how many 18–24-year-olds had attained a bachelor's degrees or higher?

F 2.16 million
G 8 million
H 10.26 million
J 12.42 million

5. Approximately how many 18–24-year-olds never attended college?

A 6.75 million
B 7.83 million
C 12.42 million
D 14.58 million

Name ______________________________ Date ____________ Class ____________

LESSON **9F**

Homework and Practice

Displaying Data, continued

The heights in inches of the 2005 NBA All-Star Game players are given below.

Player's Heights (in.)											
75	78	80	87	72	80	81	83	85	78	76	81
77	78	83	83	78	82	79	80	75	84	82	90

6. Use the data to make a frequency table with intervals. Use an interval of 5.

Player's Heights	
Heights (in.)	**Frequency**

7. Use your frequency table to make a histogram for the data.

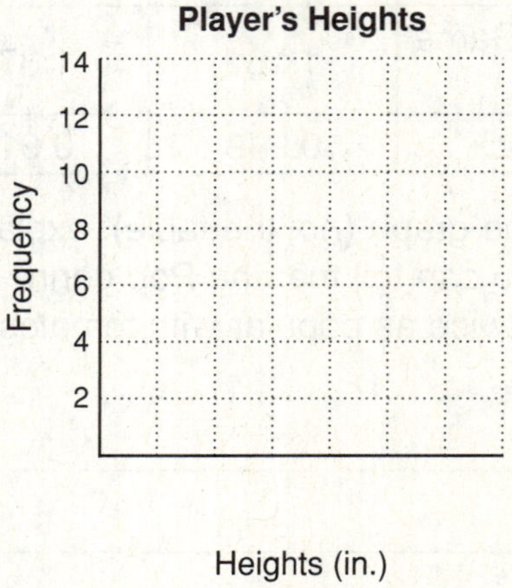

Select the best answer.

8. The stem-and-leaf plot below gives the file sizes of the songs on two albums after being converted to mp3 files. What is the largest file size for Album 1?

MP3 File Sizes (megabytes)

Album 1		Album 2
0	5	0 1 2 4 5 5 8
6 6 5 3 1	6	0 6 7
7 3 0	7	3 6 8
3 1	8	6
8 7 2 0	9	9

Key: | 8 | 6 *means 8.6 MB*

1 | 8 | *means 8.1 MB*

A 6.6 MB

B 8.9 MB

C 9.8 MB

D 9.9 MB

9. The cumulative frequency table below gives the scores of 100 students on a standardized mathematics test. How many students scored between 400 and 499?

Standardized Test Scores	
Scores	**Cumulative Frequency**
200–299	1
300–399	3
400–499	19
500–599	50
600–699	85
700–799	100

F 15

G 16

H 19

J 31

EXPLORATION

9G Misleading Graphs and Statistics

KY Program of Studies and Core Content

MA-8-DAP-S-DCD5, MA-8-DAP-S-DR1, MA-8-DAP-S-DR4, MA-08-4.1.1, MA-08-4.1.2, MA-08-4.1.4

The bar graph shows the results of a survey asking people who drink flavored water which brand they prefer.

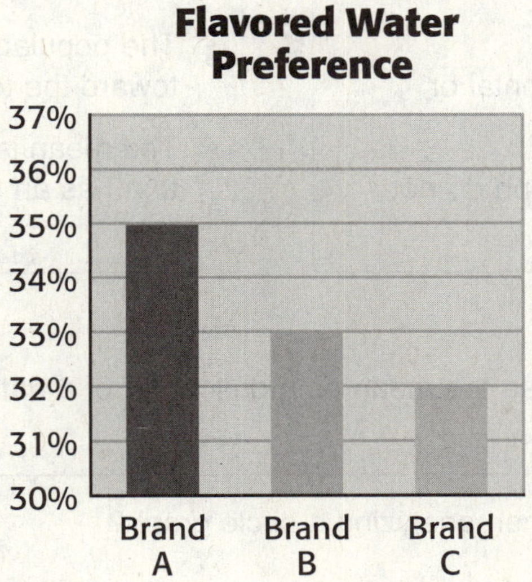

1. According to the survey, which brand is preferred the most?

2. How would the graph look if the percent scale started at 0% instead of 30%?

3. Why might someone want to display a graph like this?

Think and Discuss

4. Explain how the graph's appearance can be misleading.

5. Discuss situations that might be likely to involve misleading graphs or statistics.

Name ______________________ Date __________ Class __________

LESSON 9G

Review for Mastery

Misleading Graphs and Statistics

There are several ways for graphs and statistics to be misleading.

Graphs	Statistics
• The scale on the vertical axis does not start at 0, or extends beyond what is needed. • The intervals on the horizontal or vertical axis are not equal. • The sectors of a circle graph do not total 100%.	• The sample size is too small to be representative. • The population surveyed has a bias toward the topic. • The mean is used to represent data that has an outlier.

Answer the following.

1. When is it misleading to use the mean as a descriptor of a data set?

____

2. What should you check when analyzing a circle graph?

____

3. Which would be more trustworthy: a survey of 10 people or a survey of 100 people?

____

Look at the line graphs below.

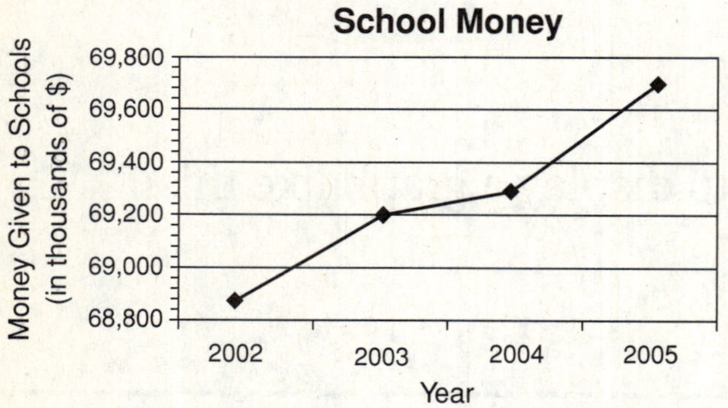

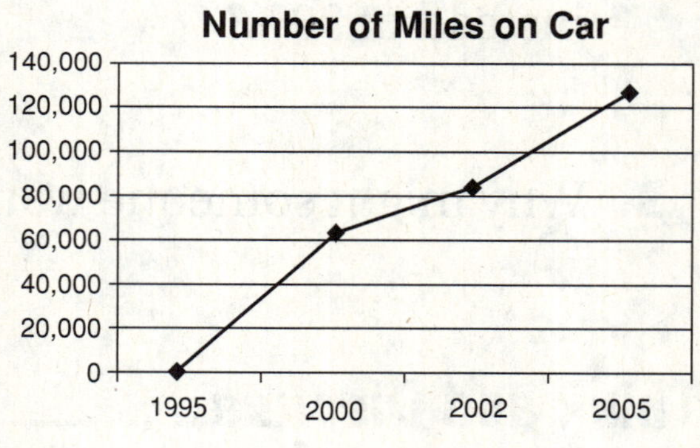

4. How is the first graph misleading? What might someone believe because of it?

____

____

5. How is the second graph misleading? What might someone believe because of it?

____

____

Name ______________________________ Date ____________ Class ____________

LESSON 9G

Review for Mastery

Misleading Graphs and Statistics, continued

Graphs can be used to mislead people.

Bar Graphs and Line Graphs: If the vertical scale does not start at 0, the difference between categories or time intervals can look larger than it is. If the intervals on the horizontal scale are not equal, the rate of change can look steeper or less steep than it is.

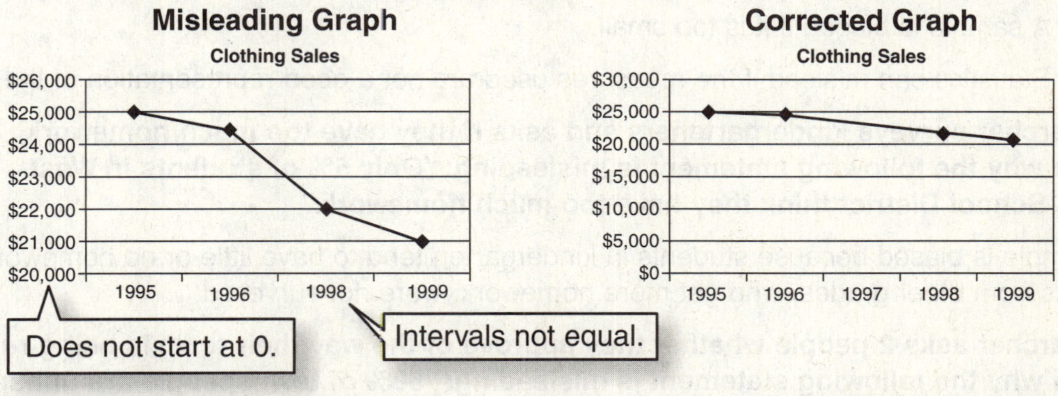

Circle Graphs: If the sections in a circle graph do not sum to 100%, sections will appear larger than they actually are.

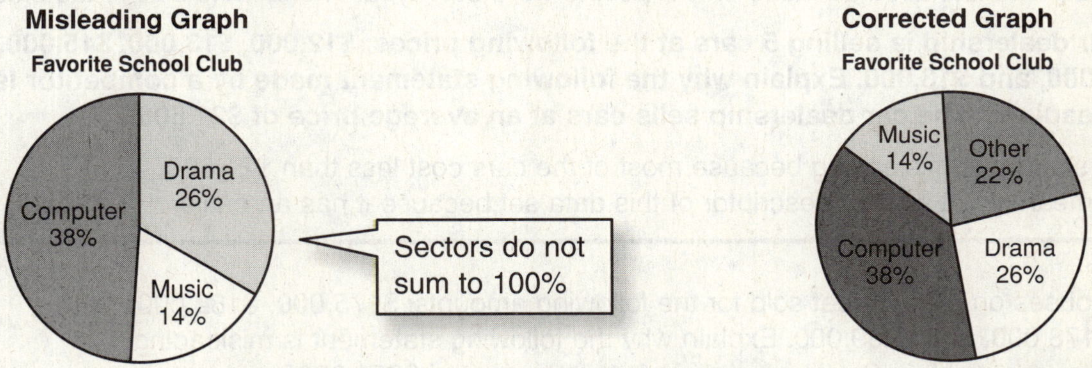

6. The graph shows the number of men and women who have enrolled in a school.

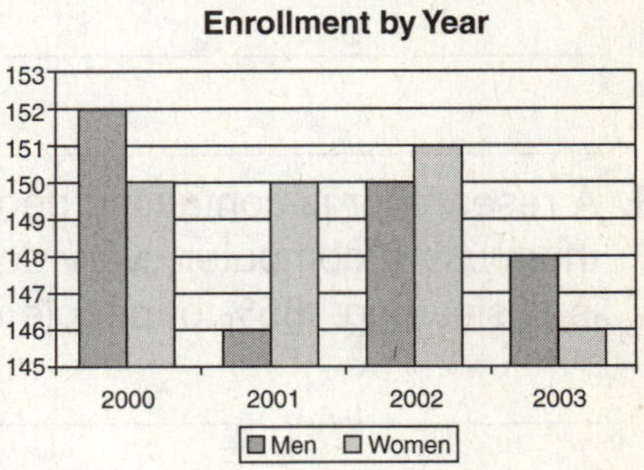

a. Explain why the graph is misleading.

____

____

b. What might someone believe because of the graph?

____

____

c. What might want to use this graph? ______________________________

____

Name ______________________ Date ____________ Class ____________

LESSON **9G**

Review for Mastery

Misleading Graphs and Statisitics, continued

Statistics can be misleading because of the way the data is collected or because of the way the results are reported.

- A sample is biased if it only surveys a certain group of people who may be biased in some way.
- A sample is biased if it is too small.
- Statistics can mislead if the measures used are not a good representation of the data.

A researcher surveys kindergarteners and asks if they have too much homework. Explain why the following statement is misleading: "Only 5% of students in West Branch School District think they have too much homework."

The sample is biased because students in kindergarten tend to have little or no homework. Students from other grades, who get more homework, were not surveyed.

A researcher asks 2 people whether they approve of the way their town is being run. Explain why the following statement is misleading: "50% of townspeople are unhappy with local government."

This sample is biased because only 2 people were surveyed. The sample size is too small.

A car dealership is selling 5 cars at the following prices: $12,000, $13,000, $15,000, $55,000, and $13,000. Explain why the following statement made by a competitor is misleading: "The car dealership sells cars at an average price of $21,600."

This statistic is misleading because most of the cars cost less than $21,600. The mean is not a good descriptor of this data set because it has an outlier.

7. Houses on Main Street sold for the following amounts: $175,000, $182,000, $178,000, and $389,000. Explain why the following statement is misleading: "Houses on Main Street are selling for an average of $231,000."

____

____

8. A researcher is contacting people by e-mail to see what proportion of them use a computer every day. Explain why the following statement is misleading: "85% of people use a computer every day."

____

____

9. A researcher asks 5 students if they think the cost of school lunches is too high. Explain why the following statement is misleading: "Four-fifths of all students think school lunches are too expensive."

____

Name ______________________ Date ____________ Class ____________

LESSON **9G**

Homework and Practice

Misleading Graphs and Statistics

Explain why each graph is misleading.

1.

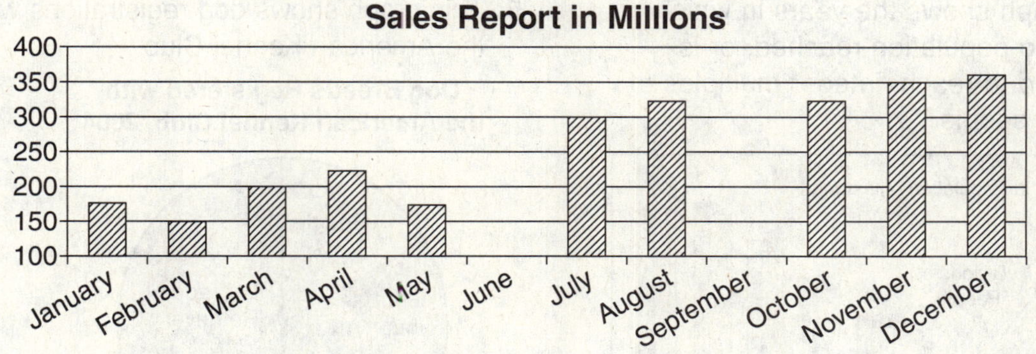

____

____

____

2.

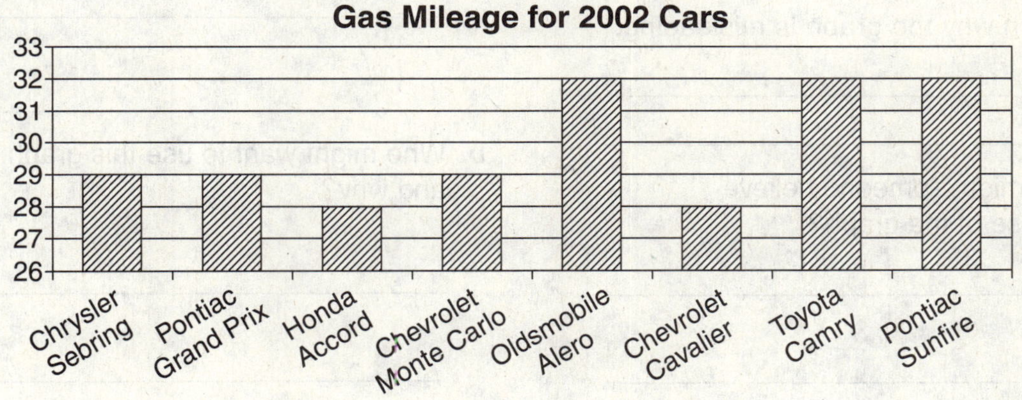

____

____

Explain why the statistic is misleading.

3. A toothpaste company advertises that 9 out of 10 dentists surveyed recommend their brand of toothpaste to their patients.

____

____

____

____

Name ______________________ Date __________ Class __________

LESSON 9G

Homework and Practice

Misleading Graphs and Statistics, continued

Analyze these misleading graphs.

4. This graph shows the years in which the world population reached, or is projected to reach, integer multiples of 1 billion people.

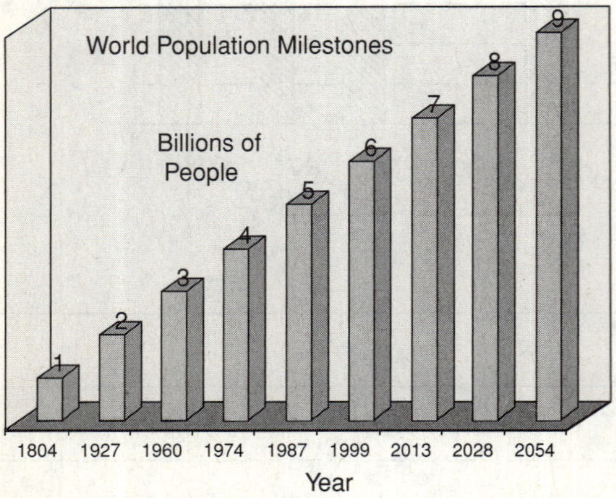

a. Explain why the graph is misleading.

____

____

b. What might someone believe because of the graph?

____

____

5. This graph shows dog registrations with the American Kennel Club.

Dog Breeds Registered with the American Kennel Club, 2004

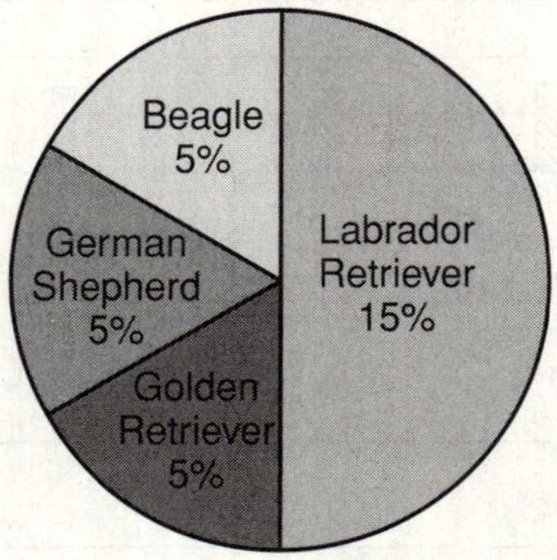

a. Explain why the graph is misleading.

____

____

b. Who might want to use this graph, and why?

____

____

Use the graph to select the best answer.

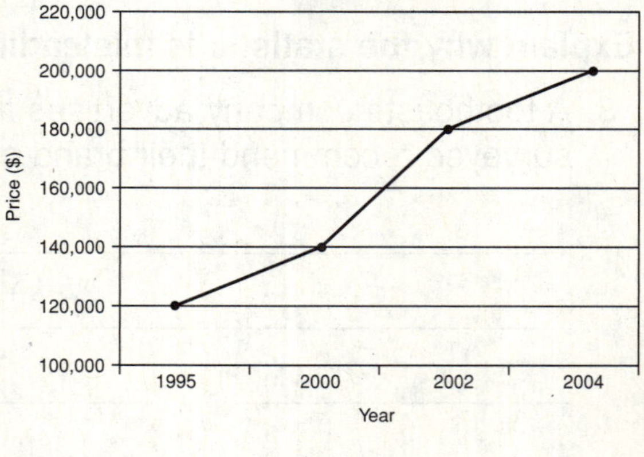

6. The graph is misleading because of which scale(s)?

A horizontal only   **C** both

B vertical only   **D** neither

7. Which statement is certainly true?

F The average house price increased every year between 1995 and 2004.

G The average house prices have increased $80,000 between 1995 and 2004.

H The average house price in 2001 was about $160,000.

J None of the above.

EXPLORATION

Program of Studies and Core Content

MA-8-DAP-S-CD4, MA-8-DAP-S-DR1, MA-8-DAP-S-P1, MA-08-4.1.1, MA-08-4.1.2, MA-08-4.1.4

9H Scatter Plots

1. The table shows the numbers of pages in some paperback books and the books' prices. Plot the data on the graph provided.

Pages	Price ($)
300	2.25
200	1.75
130	1.65
450	3.00
180	1.75
75	1.25
250	2.50

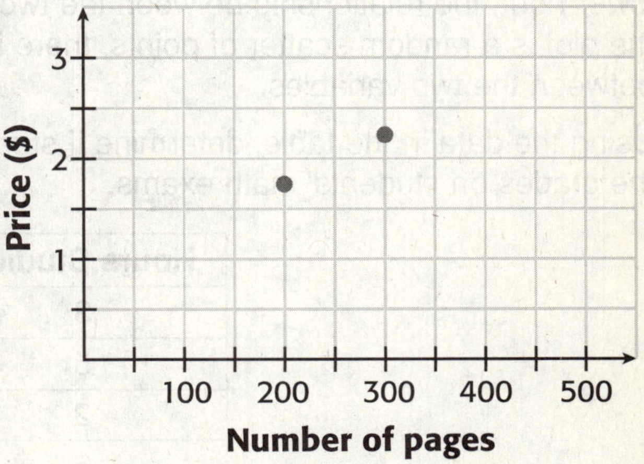

2. The table shows the number of hours some people spent outside and spent watching TV in one day. Plot the data on the graph provided.

TV hours	Outside hours
2.0	1.5
0.5	3.5
1.5	1.5
1.0	2.5
2.5	0.5
1.5	2.0
2.0	0.5

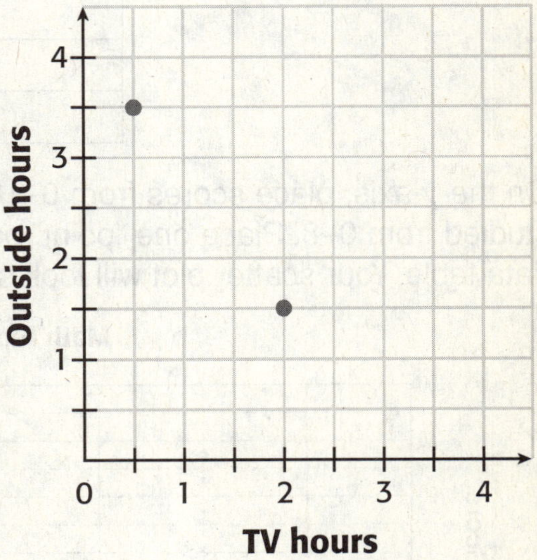

Think and Discuss

3. Explain what the shape of a graph tells you about the data being displayed.

Name ______________________ Date ____________ Class ____________

LESSON 9H

Hands-On Lab

Scatter Plots

Activity

Scatter plots are used to show relationships between two variables. The data displayed on the graph may resemble a line. If the line around which the points tend to cluster runs from lower left to upper right, the relationship between the two variables is positive. If the line around which the points tend to cluster runs from upper left to lower right, the relationship between the two variables is negative. If the plot is a random scatter of points, there is no correlation between the two variables.

Using the data in the table, determine if studying longer will affect the grades on students' math exams.

Hours Studied	Score
3	80
5	90
2	75
6	80
7	90
1	50
2	65
7	85
1	40
7	100

On the *y*-axis, place scores from 0–100. On the *x*-axis, place hours studied from 0–8. Place one "point" on the graph for each row on the data table. Your scatter plot will look similar to this.

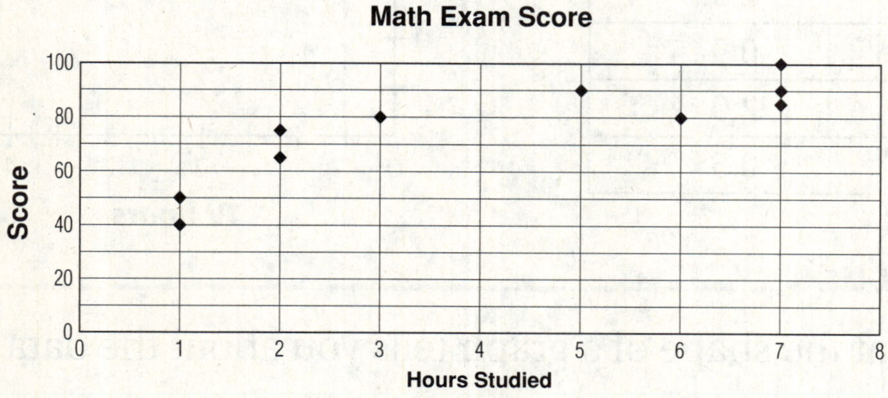

Name ______________________ Date ____________ Class ____________

LESSON 9H

Hands-On Lab

Scatter Plots, continued

Think and Discuss

The data displayed on the graph resembles a line rising from left to right. What type of correlation does this show between the two sets of data? What does this mean in terms of the data compared?

____

____

____

____

____

____

Name ______________________ Date ____________ Class ____________

LESSON 9H

Hands-On Lab

Scatter Plots, continued

Try This

Plot the following points on the graph below:

(0.5, 1.5), (1, 0.5), (1, 2), (1, 4), (2, 1), (2.5, 2), (2.5, 3),
(3, 4), (4, 1.5), (4, 3), (4, 3.5), (5, 2), (5, 3.5), (5, 5)

What type of correlation between data is represented on the scatter plot below?

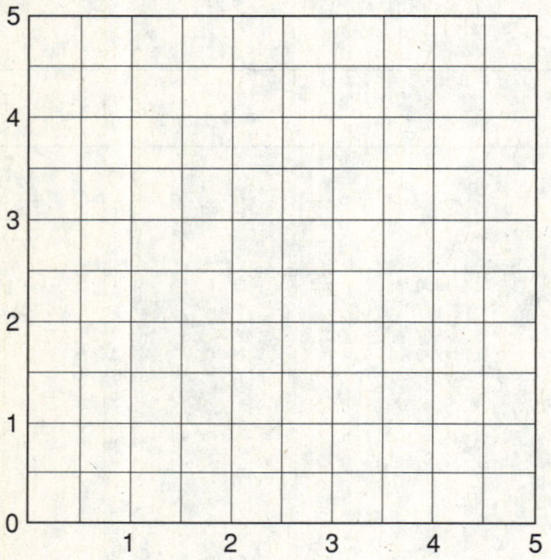

____

____

Name ______________________ Date ____________ Class ____________

LESSON 9H

Technology Lab 1
Scatter Plots

Activity

The 4th grade class has collected information from Internet research with which to make a scatter plot. The class tracked the age of gym students and the numbers of pushups they were able to do. The data they collected is in the table below and they will use it with a spreadsheet program to make a scatter plot.

Age	10	19	15	20	13	17	16	12	14	15	18
Pushups	3	36	20	21	10	28	22	13	17	25	32

First they open a worksheet in the spreadsheet program and enter the data in the top left corner of a new worksheet. Then they highlight the data and select the Chart menu on the toolbar at the top of the document.

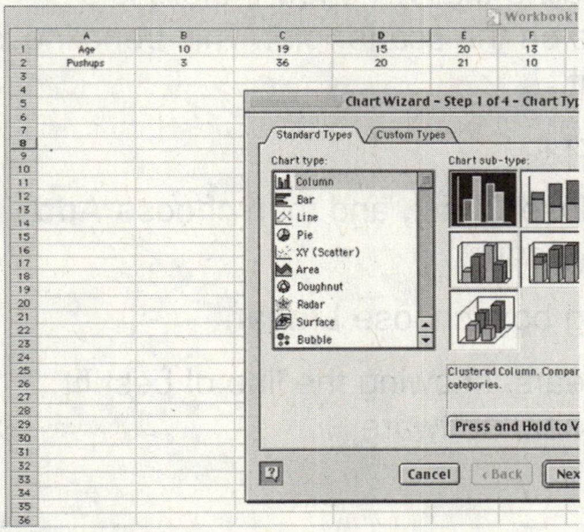

Next they select the Scatter Plot option and follow the instructions to select data and labels for the chart. The class makes the ages the *x* axis of the plot and the number of pushups the *y* axis on the plot.

Name ______________________ Date ____________ Class ____________

LESSON **9H**

Technology Lab 1

Scatter Plots, continued

Think and Discuss

1. Do the data show a general correlation? If so, what is it?

____

____

2. If you extended the scatter plot data to cover the pushups done by people in their 20s, 30s, 40s, 50s, and 60s, would you expect the same correlation to hold and the same trend to continue?

____

Try This

You can choose to add a trendline that will show if there is a correlation between the points on the scatter plot. This trendline is the same as the line of best fit.

1. Click on the chart to select it.
2. From the main menu, choose **Chart,** and then choose **Add Trendline** from the Chart menu.
3. In the Add Trendline dialog box, choose **Linear.**
4. Click **OK.** A trendline appears, showing the line of best fit calculated by the spreadsheet software.

____

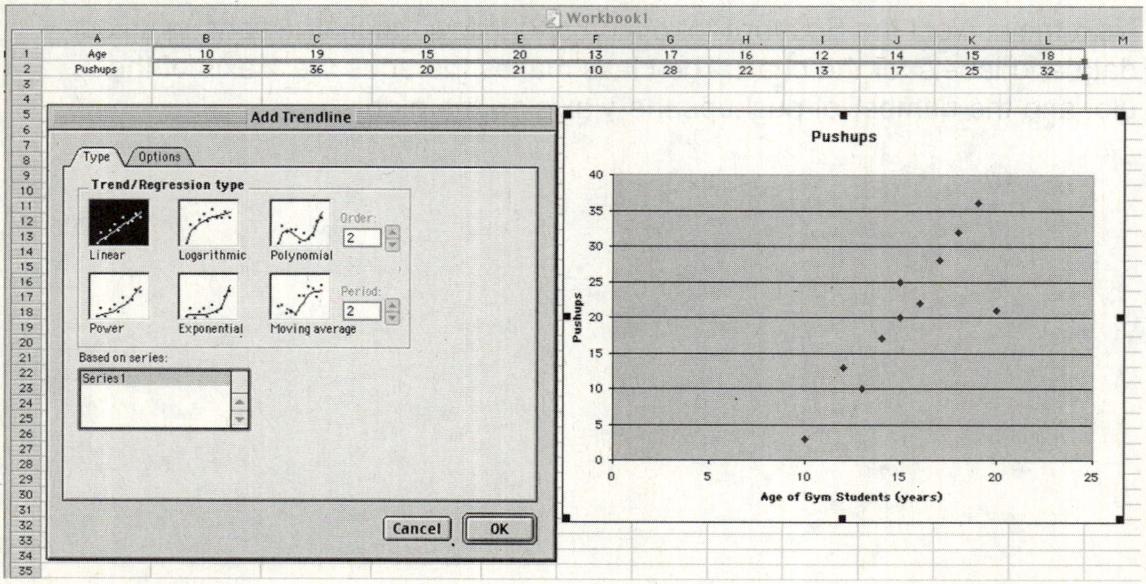

Name ______________________________ Date ____________ Class ____________

LESSON 9H

Technology Lab 2

Collect Data and Match to a Linear Function

You have learned that the sides of similar triangles are in proportion. You can verify that for yourself by measuring several similar triangles and seeing how accurately your measurements agree with the proportions the triangles should show. If your measurements are accurate, the heights of the triangles should be a linear function of the bases.

Activity

On a piece of 8.5 in. by 11 in. paper, use a centimeter ruler to draw four baselines with the following lengths: 4 cm, 5 cm, 6 cm, and 7 cm. Using a protractor, draw a ray at a 30° angle starting from the left end of the 4-cm line segment. From the right side of the 4-cm segment, draw a vertical segment intersecting the ray you just drew. Do the same with the other line segments to create four similar triangles.

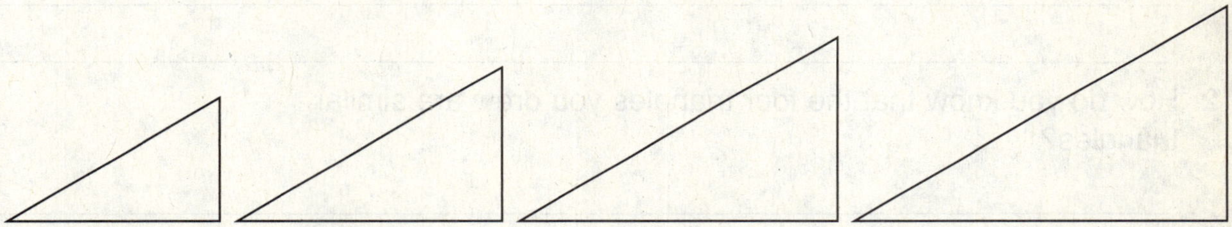

Measure the vertical line segments of the four triangles and complete the table below.

Base (cm)	4	5	6	7
Height (cm)				

Use a graphing calculator to plot the data as a scatter plot.

1. Press STAT 1 to enter statistical editing mode.
2. Under the heading L1, enter the bases of the four triangles.
3. Under the heading L2, enter the four heights of the triangles in the same order.
4. Press 2nd Y= to summon up the Stat Plots window. Press 1 to modify Plot 1.
5. Under "Plot 1," use the arrow keys to move the cursor to On and press ENTER. After Type, select the Scatter Plot icon (the first entry), and choose L1 as the XList and L2 as the YList. For Mark, select the + sign (second choice).

Name ______________________ Date __________ Class __________

LESSON 9H

Technology Lab 2

Collect Data and Match to a Linear Function, continued

6. Press WINDOW. Set Xmin as 0, Xmax as 7.5, Ymin as 0, and Ymax as 7.5.
7. Press GRAPH. Your measurements appear as a scatter plot.
8. Press Y= and press X,T,O,n ÷ 2nd x^2 3 to enter the equation $y = x/\sqrt{3}$.
9. Press GRAPH again to see how well the linear function you just entered agrees with the data you obtained by measuring.

Think and Discuss

1. This activity calls for a centimeter ruler. Would an inch ruler be just as good? Explain.

____

____

____

2. How do you know that the four triangles you drew are similar triangles?

____

____

____

Try This

1. If the scatter plot and the function line are not showing, press GRAPH to display them. Press TRACE and use the forward and back arrow keys to check the *y*-values of the four points you plotted on the scatter plot. Then press the Down arrow to shift the trace onto the function line closest to each of your four plotted points. Read the *y*-value of the point as plotted on the function line and compare it with your measurement for that point. Were you close?

Name ______________________ Date __________ Class __________

LESSON 9H

Review for Mastery

Scatter Plots

Two sets of data can be graphed as points in a **scatter plot.** If there is a relationship between the data sets, a **line of best fit** can be drawn. To draw a line of best fit, draw a line with about half the data points above and about half below.

Positive correlation: both sets of data increase together.

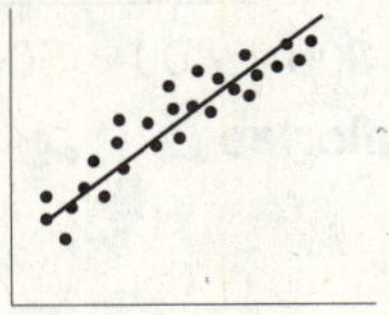

Negative correlation: values of one set increase while values of the other set decrease.

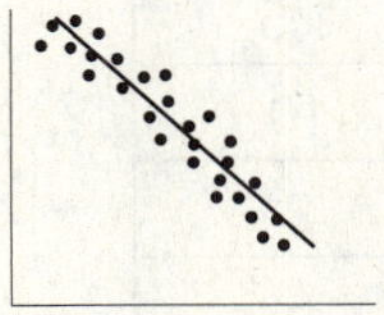

No correlation: points neither increase nor decrease together.

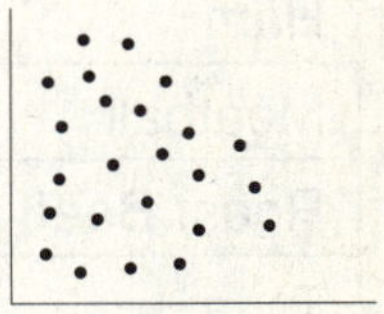

Make a scatter plot. Include a line of best fit if there is a correlation. Describe the correlation.

1.

Time (hours)	1	2	2.5	6
Distance (miles)	50	150	175	270

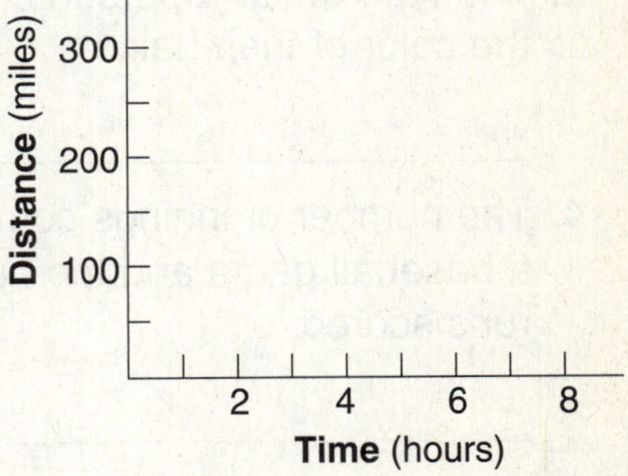

As the values for time increase, the distances: ______________.

So, there is a ______________ correlation.

A prediction for a time of 4 hours is

______________ miles.

2.

Number of Workers	6	4	2	1
Number of Days	1	2	5	7

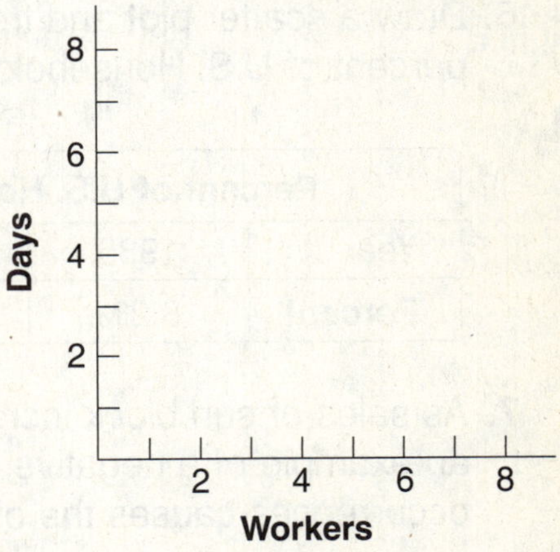

As the values for the number of workers ______________, the number of days ______________.

So, there is a ______________ correlation.

A prediction for 3 workers is

______________ days.

Name ______________________ Date ____________ Class ____________

LESSON 9H

Homework and Practice

Scatter Plots

1. Use the given data to make a scatter plot.

Sandwich	Calories	Total Fat Grams
Chicken Breast	318	2
Chicken Teriyaki	374	1.5
Club	323	2
Cold Cut	441	7
Ham	288	1.5
Meatball	527	10
Roast Beef	293	2
Tuna	445	6
Turkey Breast	281	1.5
Veggie	226	1

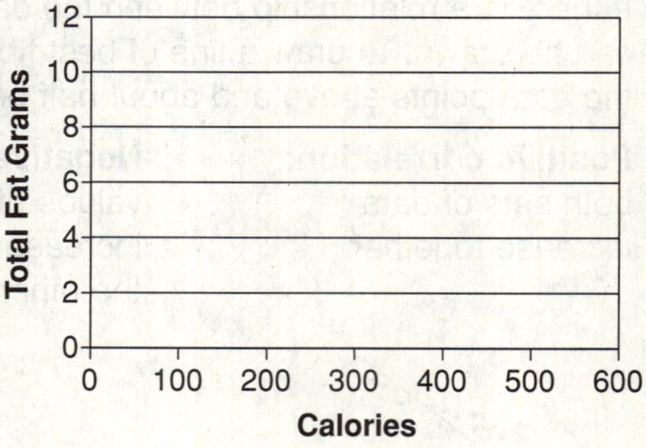

Do the data sets have a positive, a negative, or no correlation?

2. The type of car a person drives and the color of their hair.

____

3. The number of people working on a job and the hours to completion.

____

4. The number of innings completed in a baseball game and the number of runs scored.

____

5. The number of shares of stock owned and the amount received in dividends.

____

6. Draw a scatter plot and trend line for the data. Predict the percent of U.S. Households with a computer in 1998.

Percent of U.S. Households with a Computer				
Year	1985	1990	1995	2000
Percent	8.2%	15%	24.1%	51%

7. As sales of sun block increase, sales of mittens decrease. Is this an example of a negative correlation? Do you think one of these occurrences causes the other? Explain.

____

EXPLORATION

Program of Studies and Core Content

MA-8-DAP-S-DR1, MA-8-DAP-S-CD4, MA-8-DAP-S-P1, MA-08-4.1.1, MA-08-4.1.2, MA-08-4.1.4

9I Choosing the Best Representation of Data

Some students sell water bottles at football games to raise money for their club. They want to know how many cases of water bottles to order, so they look at graphs showing the sales for last year.

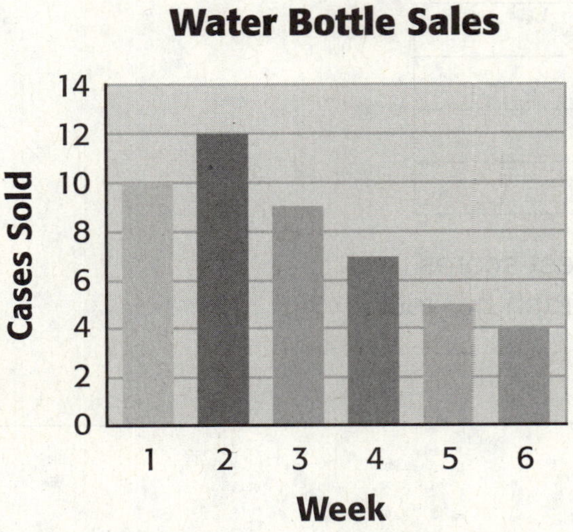

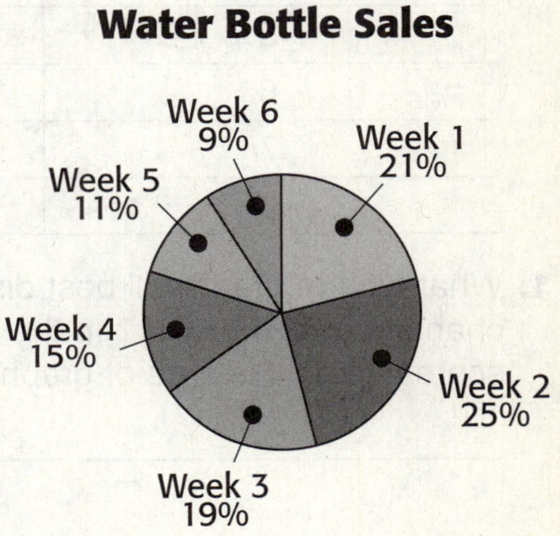

1. According to the bar graph, how many cases of water bottles were sold over the six-week period?

2. According to the circle graph, what percentage of the water bottles were sold in the first two weeks?

Think and Discuss

3. Describe the advantages of displaying this data in a bar graph.

4. Describe the advantages of displaying this data in a circle graph.

Name ______________________ Date __________ Class __________

LESSON 9-1

Hands-On Lab
Choosing the Best Representation of Data

Activity

Anika's test scores in math for the school year are given in the table below.

Test	Score (out of 100)
1	84
2	86
3	84
4	91
5	92
6	89
7	95
8	95

1. What type of graph will best display how her test scores changed over the year? In the space below, graph her test scores using this type of graph.

____

____

2. What type of graph will best show the distribution and spread of her test scores? In the space below, graph her test scores using this type of graph.

____

____

Name ______________________ Date __________ Class __________

LESSON 9I

Hands-On Lab

Choosing the Best Representation of Data, continued

Think and Discuss

1. Data answers questions. How does the question that is asked affect the type of graph used to display data? Give an example of a question and the best type of graph to answer the question.

Try This

1. In one week Sasha recorded the amount of time she spent studying for math, science, history, and English. What type of graph would best display her data if she would like to compare the amount of time she spent studying for each subject to the entire amount of time she spent studying? Explain.

2. Kirk went on a five-day bicycling tour. He recorded the number of miles he rode each day in the table below. What type of graph will show him how the number of miles he rode each day changed over the five days? Explain.

Day	1	2	3	4	5
Miles	75	102	57	72	63

Name ______________________ Date __________ Class __________

LESSON 9I

Technology Lab 1
Choosing the Best Representation of Data

Activity

On New Year's Eve, Samantha and Sean are watching fireworks at their community center with their parents. They keep track of how many of each color of fireworks go off in the display. The table below shows their data.

Fireworks	Gold	Silver	Green	Red	Blue
Number	73	65	30	22	28

Using a spreadsheet, they decide to use a chart to represent their data. First they have to choose the most appropriate representation for their data.

They start by putting the data into a spreadsheet. Then they highlight the data that will be represented in the chart.

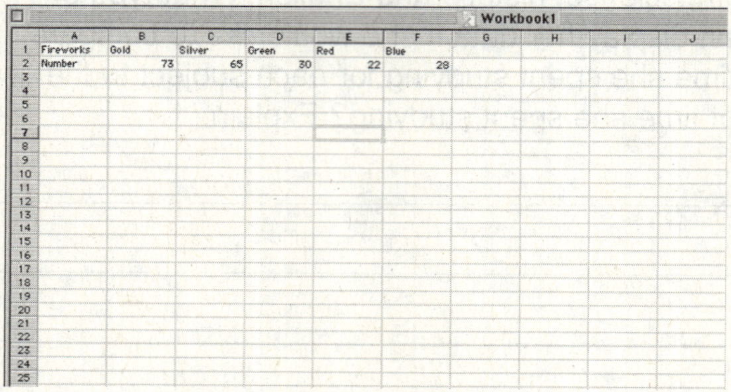

To see the difference in numbers of each color of fireworks, they choose to show the data in a bar graph. The spreadsheet Chart menu includes a choice for Bar Graph. They choose to use the colors on the *x*-axis and numbers in increments of 5 on the *y*-axis.

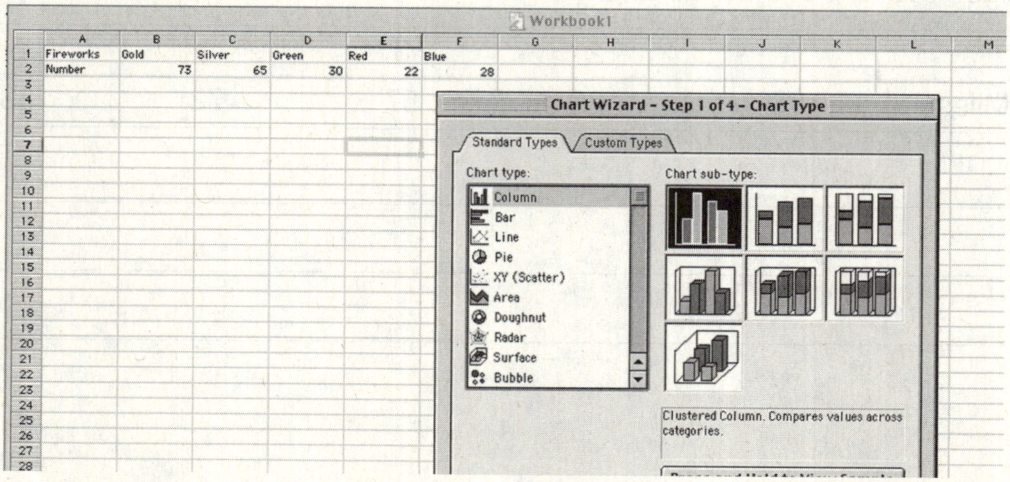

Name ______________________ Date __________ Class __________

LESSON 9I

Technology Lab 1

Choosing the Best Representation of Data, continued

Think and Discuss

1. Would the relationship have been easier or harder to see if Samantha and Sean had represented their data using a circle graph? Explain your answer.

____

____

2. Would a line graph be an appropriate representation for Samantha and Sean's data? Why or why not?

____

____

Try This

1. Use a spreadsheet to create an appropriate representation of the data gathered by the community center to find out how many town residents have been using the center in the past five years.

Year	Residents using the center
2000	12,800
2001	13,500
2002	14,900
2003	14,756
2004	15,012

Name ______________________ Date ____________ Class ____________

LESSON 9I

Technology Lab 2

Choosing the Best Representation of Data

Activity

Angela asked all of her classmates how many music CDs they owned. She collected her data in a table but now she wants to use a spreadsheet to save the data and make a graphic representation of the data as a graph or chart.

CDs	0-5	6-10	11-15	16-20
Number of students	4	8	15	13

She started by putting the data into a spreadsheet. Then she highlighted the data that would be represented in the chart.

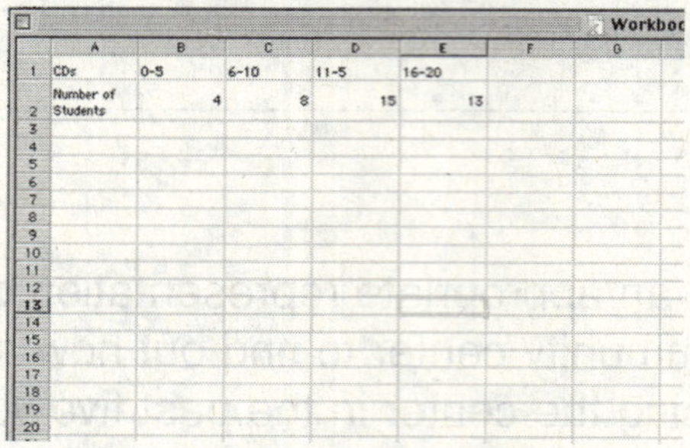

To see the difference in the frequency of students owning CDs, Angela made sure there were even increments in the range of CDs. She thought this would make it a good data set to show as a histogram. The spreadsheet Chart menu includes a choice for Column, which she selected. She input the CD ranges for the *x*-axis and increments of 5 students for the *y*-axis.

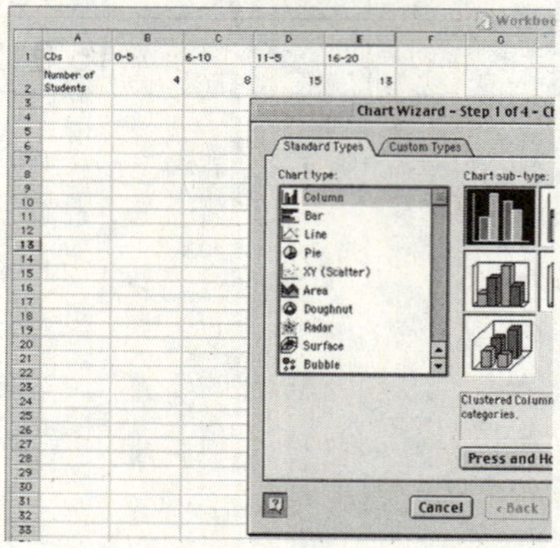

When the chart was generated from the data, she could see that the most students own 11–15 CDs. She could also see how the number of students changed in the ranges to either side of the most popular range.

Name ______________________ Date ____________ Class ____________

LESSON 9I

Technology Lab 2

Choosing the Best Representation of Data, continued

Think and Discuss

1. Would Angela be able to show her data as a pie chart? Why or why not?

____

____

2. Could Angela show her data as a scatter plot? Why or why not?

____

____

____

Try This

1. Use a spreadsheet to create an appropriate representation of the data compiled from students in the 4th grade who reported how long they spent on their math homework each week.

Minutes spent on math homework	Frequency (students)
0-9	1
10-19	2
20-29	7
30-39	13
40-49	10

Name ______________________________ Date ____________ Class ____________

LESSON **9I**

Technology Lab 3

Choosing the Best Representation of Data

Activity

The students in Ms. Kinsley's class made a list of their favorite types of stores to spend their allowances in. Then they surveyed all of the students in their school to see how many students liked each type of store. Below is the table they made with the data. Now they are going to try to find a way to represent it graphically.

Store	Candy	Video	Games	Movies	Books
Students	320	91	187	12	65

Start by putting the data into a spreadsheet. Use the "Sum" function to add up the total number of students surveyed. Then highlight the data that will be represented in the chart.

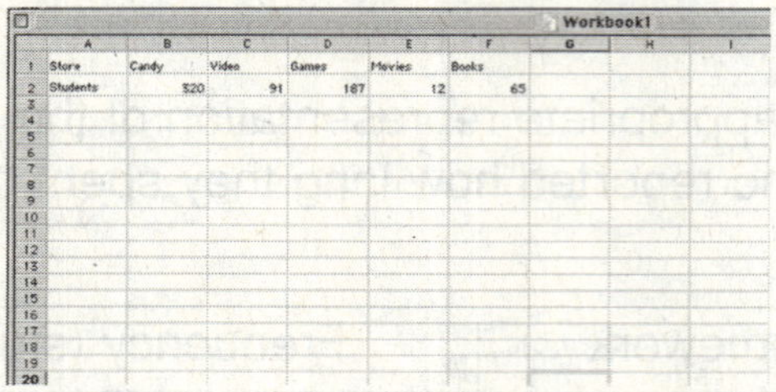

The students want to show which stores were selected by the most students, as a percentage of the total number of students. To do this, they decide to show the data in a circle graph. On the spreadsheet menu, they choose to make a Pie Chart. Then they input labels for each of the data categories included in the circle graph.

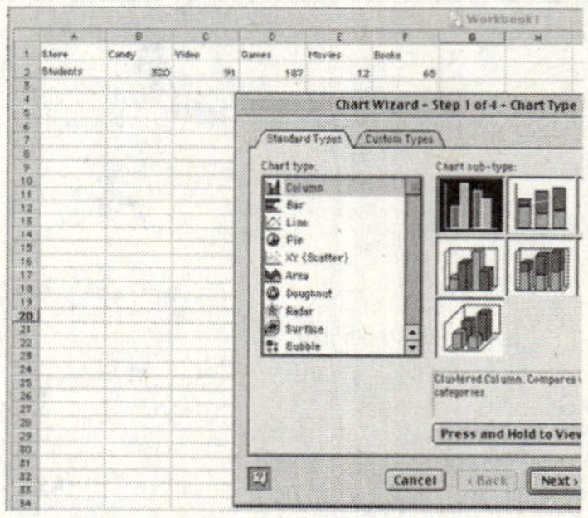

The results show what part of the total each type of store represents.

Favorite Stores for Spending

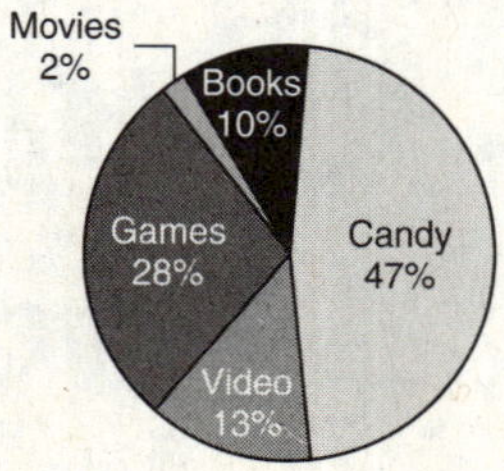

Name ______________________________ Date ______________ Class ______________

LESSON 9I

Technology Lab 3

Choosing the Best Representation of Data, continued

Think and Discuss

Would the students be able to use a circle graph if some of the students responded to the survey more than once? Why or why not?

____

____

Try This

1. A group of students surveyed their school to find out how each student gets to school in the morning. Use the data below and a spreadsheet program to make an accurate representation of the transportation used relative to the total number of students in the school.

Transportation	Number of Students
Car	514
Bus	120
Subway	65
Bike	4
Walk	8

Name ______________________ Date ____________ Class ____________

LESSON 9I

Review for Mastery

Choosing the Best Representation of Data

A line graph is a good way to show changes over time. The amount of money in Tasha's account changes from month to month.

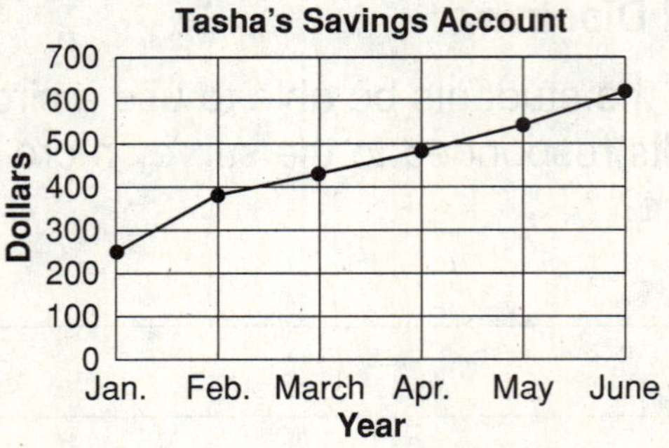

A bar graph is a good way to compare different quantities. The amounts of money in the students' bank accounts are being compared.

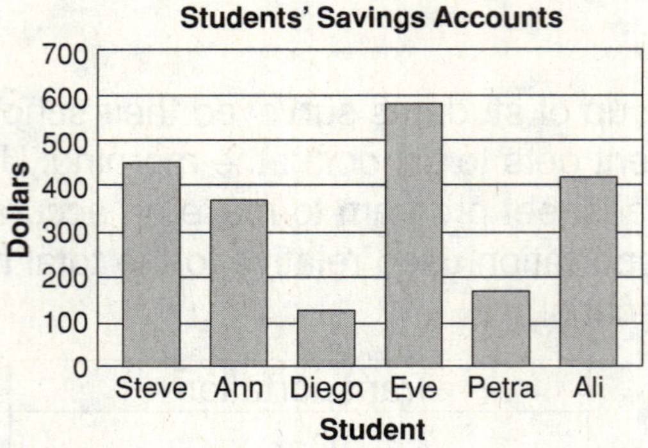

A circle graph is a good way to compare parts of a whole. Lionel's assets are the whole; the parts are his savings account, checking account, and mutual funds.

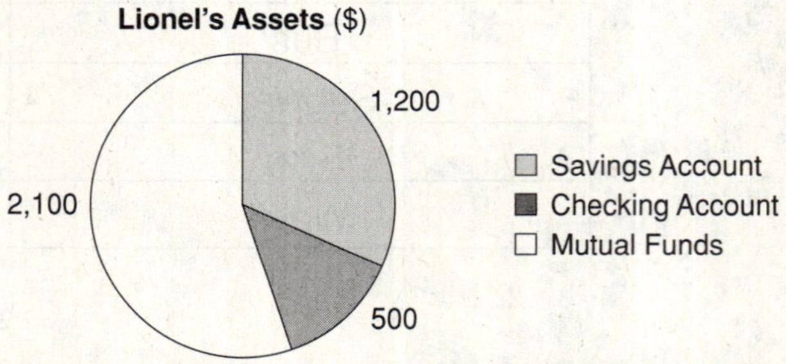

Determine which kind of graph would be best to display the described data. Write *line graph, bar graph,* or *circle graph.*

1. Total number of hits by each member of a softball team

____

2. Number of miles that a boy rides his bike each week

____

3. Daily profits of a hot dog stand at a county fair

____

4. Part of a hot dog stand's profits that come from popcorn sales

____

Name _______________ Date _______ Class _______

LESSON **9I**

Homework and Practice

Choosing the Best Representation of Data

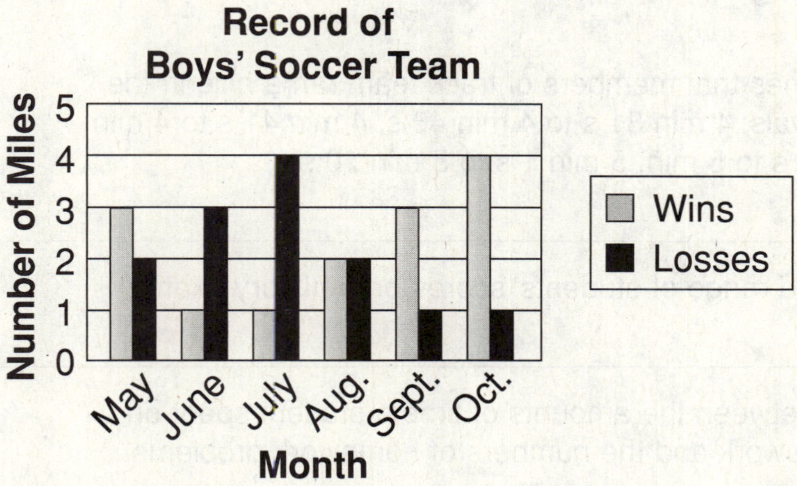

Refer to the graph above.

1. What does the title of the graph tell about the graph?

____

____

2. What does the legend tell about the graph?

____

____

3. What does the scale of the graph show?

____

____

4. What is the range of the scale?

____

5. Which month did the team have the most losses?

____

6. Which month did the team have the most wins?

____

Name ______________________ Date ____________ Class ____________

LESSON 9I

Homework and Practice

Choosing the Best Representation of Data, continued

Write what kind of graph would be best to display the described data.

7. Numbers of times that members of track team ran a mile in the following intervals: 4 min 31 s to 4 min 40 s, 4 min 41 s to 4 min 50 s, 4 min 51 s to 5 min, 5 min 1 s to 5 min 10 s

____

8. Distribution and range of students' scores on a history exam

____

9. Relationship between the amounts of time a student spent on her math homework and the numbers of homework problems she solved

____

10. Total numbers of victories of eight teams in an intramural volleyball league

____

11. Part of calories in a meal that come from protein

____

12. Numbers of books that a student reads each month over a year

____

Choose the letter for the best answer.

13. A bar graph is a good way to display

A data that changes over time.

B parts of a whole.

C distribution of data.

D comparison of different groups of data.

14. A circle graph is a good way to display

F range and distribution of data.

G distribution of data.

H parts of a whole.

J changes in data over time.

15. A scatter plot is a good way to display

A comparison of different groups of data.

B distribution and range of data.

C the relationship between two sets of data.

D parts of a whole.

16. A box-and-whisker plot is a good way to display

F range and distribution of data.

G the relationship between two sets of data.

H data that changes over time.

J parts of a whole.

CHAPTER 9

Chapter Summary

Answer these questions to summarize the important concepts from Chapter 9 in your own words.

1. Explain how to find the mean, median, mode, and range of the data set.

45, 53, 78, 29, 84, 25, 74, 57, 29, 36

2. Explain how to use a matrix to organize data.

3. Explain when data in a scatter plot shows a positive correlation.

For more review of Chapter 9:

- Complete the Chapter 9 Study Guide and Review on pages 510–512 of your textbook.
- Complete the Ready to Go On quizzes on pages 482 and 506 of your textbook.

Name ______________________ Date ____________ Class ____________

CHAPTER 9

KCCT Prep

Multiple Choice

1. The line plot shows Derrick's quiz scores. One of his scores is not shown. Given that the median of all his scores is 8, which of these could NOT be the missing score?

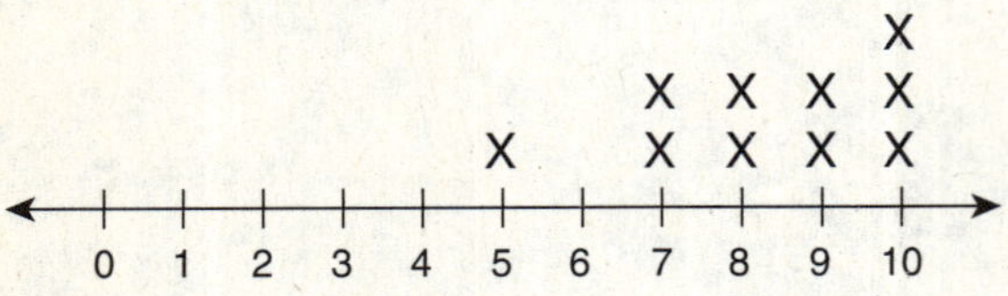

Ⓐ 5　Ⓑ 7　Ⓒ 8　Ⓓ 9

2. The measure of ∠*R* is 64°. ∠*R* and ∠*S* are vertical angles. ∠*T* is supplementary to ∠*S*. What is m∠*T*?

Ⓐ 26°　Ⓑ 64°　Ⓒ 116°　Ⓓ 154°

3. Marissa asks 20 friends and relatives for their heights and their shoe sizes. Then she makes a scatter plot of the data. Which of these is she most likely to see in her scatter plot?

Ⓐ A positive correlation
Ⓑ A negative correlation
Ⓒ No correlation
Ⓓ Both a positive and negative correlation

4. The area of a circular park is $25{,}600\pi$ m^2. Mai walks across the park along its diameter at a rate of 80 m/min. How long does it take her to cross the park?

Ⓐ 2 min　Ⓑ 4 min　Ⓒ 6 min　Ⓓ 8 min

5. The table shows the number of miles that two friends ran on four days last week. On which day was the mean number of miles that the friends ran the greatest?

	Mon.	Tues.	Wed.	Thurs.
Ellie	3.6	2.5	0	3.2
Wes	1.2	2.5	4.1	2.1

Ⓐ Monday　Ⓑ Tuesday　Ⓒ Wednesday　Ⓓ Thursday

6. Alex adds one data value to this stem-and-leaf plot. After he does so, the range of the data is 33. Which of these could be the value that he adds?

Stems	Leaves
2	4 4 6 7 8
3	1 3 5
4	5
5	1 5 7

Key: 2|4 means 24

Ⓐ 23　Ⓑ 37　Ⓒ 58　Ⓓ 60

7. A company packages popcorn in a cardboard box that is a cube. Each edge of the box is 19.8 cm long. Which is the best estimate of the volume of the box?

Ⓐ 60 cm^3　Ⓑ 400 cm^3　Ⓒ 2400 cm^3　Ⓓ 8000 cm^3

8. Which of these is the best estimate of $\sqrt{2} + \sqrt{3}$?

Ⓐ 2　Ⓑ 3　Ⓒ 5　Ⓓ 13

9. A survey of 15 students asked the number of times they had visited a miniature golf course. The frequency table shows the results. Which of these statements is true?

Number of Visits	Frequency
0	7
1	2
2	5
3	1

Ⓐ Mode > Mean
Ⓑ Range < Median
Ⓒ Mean = Median
Ⓓ Mode = Mean

10. Which type of graph would be the best choice for displaying the annual number of visitors to Mammoth Cave National Park since 1995?

Ⓐ Box-and-whisker plot
Ⓑ Line graph
Ⓒ Scatter plot
Ⓓ Stem-and-leaf plot

11. Imelda builds a model of a ship using the scale 2 cm:5 m. Her model is 29.5 cm long. Which is the best estimate of the actual length of the ship?

Ⓐ 12 m Ⓒ 75 m
Ⓑ 30 m Ⓓ 150 m

12. Which equation has the same solution as $\frac{3}{4}n - \frac{1}{4} = \frac{1}{8}$?

Ⓐ $3n - 1 = 1$ Ⓒ $6n - 1 = 1$
Ⓑ $3n - 1 = 2$ Ⓓ $6n - 2 = 1$

13. A circle has center (–1, –2) and area 9π units^2. Which of these points does NOT lie on the circle?

Ⓐ (–3, –2) Ⓒ (2, –2)
Ⓑ (–1, 1) Ⓓ (–1, –5)

Open Response

14. The table shows data about the distance of a train ride and the price of a ticket.

Distance (mi)	20	40	50	10
Price ($)	15	10	20	5

Distance (mi)	50	35	30	20
Price ($)	25	20	15	8

a. Make a scatter plot of the data.
b. Describe the correlation of the data in the scatter plot.
c. Draw a line of best fit on your scatter plot.
d. Use the line of best fit to predict the price of a ticket for a train ride that is 75 miles long.

15. The box-and-whisker plot shows the number of runs a baseball team scored in 20 games.

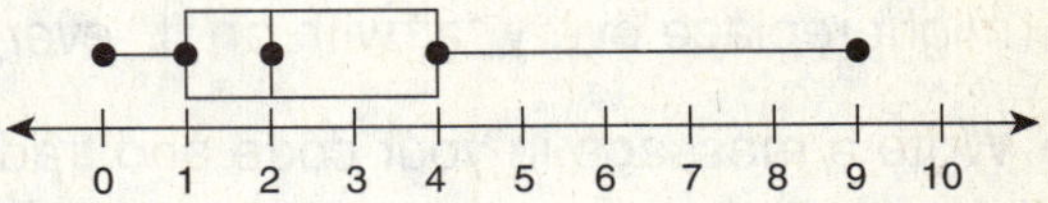

a. What is the range of the data?
b. What is the median of the data?
c. Is it possible to find the mean of the data? If so, what is the mean? If not, explain why it is not possible to find the mean.
d. Construct a data set that could show the number of runs the team scored in its 20 games. Demonstrate how the data set you created fits the information in the box-and-whisker plot.

CHAPTER 10

Project

Probability: Breaking the Code

Career: Cryptographer

1001001101011110000011010011100101110010011
10001001000100100100110011101000111

Is this pattern of zeros and ones some kind of message or secret code? A cryptographer could find out. Cryptographers create and break codes by assigning number values to letters of the alphabet. Almost all text sent over the Internet is encrypted to ensure security for the sender. Codes made up of zeros and ones, or *binary codes,* are frequently used in computer applications.

Before you start, work through the Code Breaking worksheet on the next page.

In this chapter project . . .

You create and break codes.

What to do:

- Create what's called a *substitution cipher.* In this type of code, each letter of the alphabet is changed to a different letter. For example, you might replace every "a" with an "f," every "b" with a "q," and so on.
- Write a message in your code and trade messages with a classmate.
- Use probability to help crack your classmate's code. For example, "e" is the most commonly used letter in English, so finding the most commonly used letter in your classmate's code is likely to be the coded version of "e." Research the frequency of letters of the alphabet to help you crack the rest of the code.

What to present:

- Your classmate's coded message, the key to breaking the code, and the unencrypted message.
- A table listing the frequency of letters in the alphabet, and the corresponding frequency of each letter in your classmate's code.

Where's the math?

- Use probability to determine which letters are substituted for which in a code.

go.hrw.com
Chapter Project Online
KEYWORD: MT7 Ch10

Name ______________________ Date __________ Class __________

CHAPTER 10

Project Recording Sheet

Probability

Code Breaking

You might think that only spies worry about codes but in this modern world that just isn't true any more. Today, anyone who sends electronic information wants it to be safe. Encrypting a code is the way to do it. Cryptographers are the people who create codes. If you like discovering and creating patterns then cryptography might be for you.

Use the chart to figure out the message below.

10010011010111100000110100111001011100100110001001000100100100110011101000111

Letter	Code	Letter	Code
A	1000001	N	1001110
B	1000010	O	1001111
C	1000011	P	1010000
D	1000100	Q	1010001
E	1000101	R	1010010
F	1000110	S	1010011
G	1000111	T	1010100
H	1001000	U	1010101
I	1001001	V	1010110
J	1001010	W	1010111
K	1001011	X	1011000
L	1001100	Y	1011001
M	1001101	Z	1011010

What is the message? ______________________

Create a message of your own using the code above. Exchange messages with your classmates.

Look at the code. What patterns do you see? ______________________
What other patterns might be used to create a code using 1's and 0's? ______________________

Develop a code of your own. Write a message. See if a classmate can decipher the message using your code.

Research: The code above uses the binary system of 1's and 0's. What other uses for the binary system can you discover?

EXPLORATION

Program of Studies and Core Content
MA-8-DAP-S-P2, MA-8-DAP-S-P7, MA-8-DAP-S-CD2

10A Probability

A plastic container contains 10 marbles: 4 red, 3 yellow, 2 green, and 1 blue. An experiment consists of shaking the container and then drawing a number of marbles.

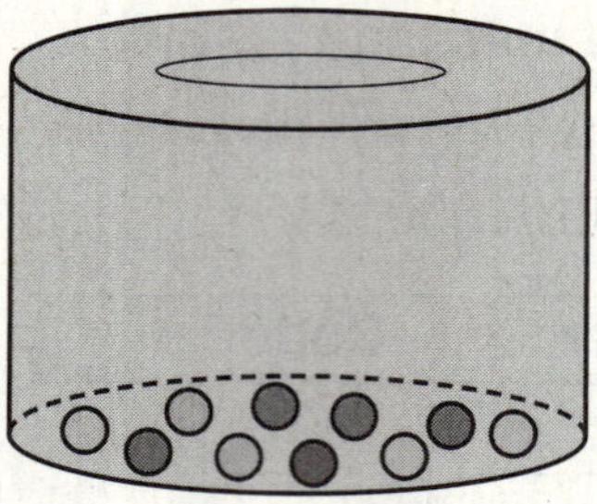

1. If you draw one marble, which color is it most likely to be?
2. If you draw one marble, which color is it least likely to be?
3. If you draw two marbles at the same time, which colors might you draw two of?
4. If you repeatedly draw one marble and return it to the container before the next draw, which color do you think you would draw most often?

Think and Discuss

5. **Discuss** real-world situations in which probability is used.

6. **Explain,** using the marble experiment, why the probability of drawing a blue marble is $\frac{1}{10}$. (*Hint:* Compare the number of blue marbles to the total number of marbles.)

Name ______________________ Date __________ Class __________

LESSON 10A

Review for Mastery

Probability

The **probability** that something will happen is how often you can expect that **event** to occur. This depends upon how many outcomes are possible, the **sample space.**

In the spinner shown, the circle is divided into four equal parts. There are 4 possible outcomes.

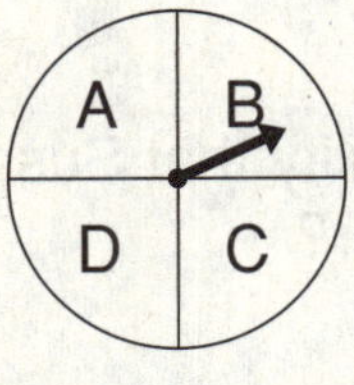

So, in a single spin:

$$P(A) = P(B) = P(C) = P(D) = 25\% = \frac{1}{4}$$

Complete to give the probability for each event.

	1. A fair coin is tossed.	**2.** A number cube is rolled.
List all the possible outcomes.	________ ________	________ ________
How many outcomes in sample space?	________	________
Find the probability of the event shown.	*P*(heads) = ______	*P*(5) = ______

- A probability of 0 means the event is **impossible,** or can never happen. On the spinner above, *P(F)* = 0.
- A probability of 1 means the event is **certain,** or has to happen. In one roll of a number cube, *P*(a whole number from 1 through 6) = 1.

Give the probability for each event.

3. selecting a rectangle from the set of squares

P(rectangle) = ______

4. selecting a negative number from the set of whole numbers

P(negative number) = ______

- The sum of the probabilities of all the possible outcomes in a sample space is 1. If the probability of *snow* is 30%, then the probability of *no snow* is 70%.

$$P(\text{snow}) + P(\text{no snow}) = 1$$

5. If the probability of selecting a senior for a committee is 60%, then the probability of not selecting a senior is:

____

6. If the probability of choosing a red ball from a certain box is 0.35, then the probability of not choosing a red ball is:

____

Name ______________________ Date __________ Class __________

LESSON 10A

Review for Mastery

Probability, continued

To find the probability that an event will occur, add the probabilities of all the outcomes included in the event.

This bar graph shows the midterm grades of the 30 students in Ms. Lin's class.

What is the probability that Susan has a grade of C or higher?

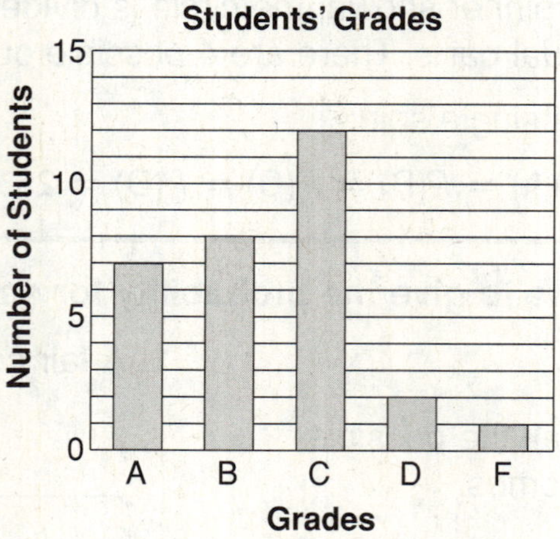

The event "a grade of C or higher" consists of the outcomes C, B, A.

$P(\text{C or higher}) = P(\text{C}) + P(\text{B}) + P(\text{A})$

$$= \frac{12}{30} + \frac{8}{30} + \frac{7}{30} = \frac{12 + 8 + 7}{30} = \frac{27}{30} = \frac{9}{10} = 90\%$$

So, the probability that Susan's midterm grade is C or higher is $\frac{9}{10}$ or 90%.

Use the bar graph above to find each probability.

7. B or higher is honor roll. What is the probability that Ken made the honor roll?

$P(\text{B or higher}) = P(\text{B}) + \_\_\_\_\_$

$= \frac{\_\_}{30} + \frac{\_\_}{30}$

$= \frac{\_\_}{30} = \_\_\_\_\_$

So, the probability that Ken made the honor roll is: ______ or ______%.

8. In this class, D is a failing grade. What is the probability that Tom failed?

$P(\text{D or lower}) = P(\text{D}) + \_\_\_\_\_$

$= \frac{\_\_}{30} + \frac{\_\_}{30}$

$= \frac{\_\_}{30} = \_\_\_\_\_$

So, the probability that Tom failed is: ______ or ______%.

Name ______________________ Date ____________ Class ____________

LESSON **10A**

Homework and Practice

Probability

The following table indicates the inventory totals for a shoe store. Find the probability for choosing the following types of shoes.

	Two-Tone	Black	White
High-Top	25	54	21
Low Cut	15	24	11

1. P(two-tone high-top) ____________

2. P(black low cut) ____________

3. P(white high top) ____________

4. P(black shoe) ____________

5. P(two-tone) ____________

6. P(two-tone low cut) ____________

7. P(blue shoe) ____________

8. P(black or white) ____________

9. P(low cut or high top) ____________

Find the probability of drawing the following colored candies from a jar containing 25 red, 15 blue, 20 green, 30 yellow, and 10 orange pieces of candy.

10. P(yellow) ____________

11. P(red) ____________

12. P(orange) ____________

13. P(blue) ____________

14. P(purple) ____________

15. P(blue or green) ____________

16. P(candy) ____________

17. P(red or yellow) ____________

18. There are 25 red, 16 green, 30 purple, 14 white, 20 black, and 15 orange marbles in a bag. What is the probability of picking a marble that is not purple?

____

19. What is the probability of tossing a 4 on a regular numerical cube?

____

EXPLORATION

Program of Studies and Core Content

MA-8-DAP-S-CD2, MA-8-DAP-S-P1, MA-DAP-S-P2, MA-08-4.4.2, MA-08-4.4.3

10B Experimental Probability

Use a plastic lid to make a spinner such as the one below. Color $\frac{1}{2}$ of the lid red, $\frac{1}{4}$ yellow, and the other $\frac{1}{4}$ green. Use a paper clip to secure the spinner to the center.

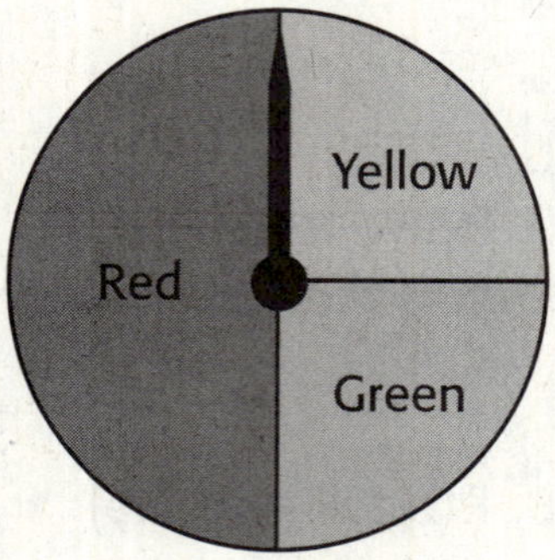

1. Spin the spinner 20 times, and record the results in the table.

Color	Number of Spins
Red	
Yellow	
Green	

a. Write a ratio to compare the number of spins for each color to 20, the total number of spins.

b. Add the three ratios in **a.** What is this sum equal to?

Think and Discuss

2. Discuss which color had the largest ratio in Problem **1a.**

3. Explain whether the sum in Problem **1b** is what you expected.

Name ______________________ Date __________ Class __________

LESSON 10B

Review for Mastery

Experimental Probability

A machine is filling boxes of apples by choosing 50 apples at random from a selection of six types of apples. An inspector records the results for one filled box in the table below.

Type	Pink Lady	Red Delicious	Granny Smith	Golden Delicious	Fuji	McIntosh
Number	8	12	6	4	15	5

The inspector then expands the table to find the experimental probability.

$$\text{probability} = \frac{\text{number of type of apple}}{\text{total number of apples}}$$

Type	Pink Lady	Red Delicious	Granny Smith	Golden Delicious	Fuji	McIntosh
Experimental Probability (ratio)	$\frac{8}{50}$, or $\frac{4}{25}$	$\frac{12}{50}$, or $\frac{6}{25}$	$\frac{6}{50}$, or $\frac{3}{25}$	$\frac{4}{50}$, or $\frac{2}{25}$	$\frac{15}{50}$, or $\frac{3}{10}$	$\frac{5}{50}$, or $\frac{1}{10}$
Experimental Probability (percent)	16%	24%	12%	8%	30%	10%

Find each sum for the apple experiment.

1. The sum of the experimental probability ratios.

   $\text{probability} = \frac{8}{50} + \frac{12}{50} + \frac{6}{50} + \frac{4}{50} + \frac{15}{50} + \frac{5}{50} = \frac{\_\_\_}{50}$ or ____

2. The sum of the experimental probability percents.

   probability = 16% + 24% + 12% + 8% + 30% + 10% = ____% or ____

Complete the table to find the experimental probability.

3. Five types of seed are inserted at random in a pre-seeded strip ready for planting.

Type	Marigold	Impatiens	Snapdragon	Daisy	Petunia
Number	40	100	80	60	120
Experimental Probability (ratio)	$\frac{\_\_\_}{400}$, or ____	$\frac{\_\_\_}{400}$, or ____	$\frac{\_\_\_}{400}$, or ____	$\frac{\_\_\_}{400}$, or ____	$\frac{\_\_\_}{400}$, or ____
Experimental Probability (percent)					

Name ______________________________ Date ____________ Class ____________

LESSON **10B**

Homework and Practice

Experimental Probability

1. A number cube was thrown 125 times. The results are shown in the table below. Complete the table with the experimental probability for each outcome.

Outcome	**1**	**2**	**3**	**4**	**5**	**6**
Frequency	15	25	20	26	18	21
Probability	12%					

The school spirit wear shop sells special sweatshirts with the school logo imprinted on them in sizes small, medium, large, x-large and xx-large. In the first hour the store is open, the first 50 customers buy 2 small, 4 medium, 5 large, 15 x-large, and the rest buy xx-large. Find the probability of the purchase of each of the different size sweatshirts from the store.

2. P(x-large sweatshirt)

____

3. P(medium sweatshirt)

____

4. P(xx-large sweatshirt)

____

5. P(small or large sweatshirt)

____

If the store has 225 customers in the second hour they are open, predict how many sweatshirts of each size will be sold based on the purchases during the first hour.

6. number of small sweatshirts sold

____

7. number of medium sweatshirts sold

____

8. number of xx-large sweatshirts sold

____

9. number of large and x-large sweatshirts sold

____

10. Jay practiced his foul shots for the upcoming basketball game. He attempted 45 shots and made 36 of them. What is the probability that Jay would make the foul shot? ______________

11. If Jay took 20 foul shots in the basketball game, predict how many shots he would make based on his practicing. ______________

EXPLORATION

Program of Studies and Core Content
MA-8-DAP-S-CD2, MA-08-4.4.2, MA-08-4.4.3

10C Use a Simulation

You can use a calculator to simulate flipping a coin 30 times.

- Press the MATH key.
- Select **PRB.**
- Select **5:randInt(.**
- Key in 0 , 1 , 6) .
- Press the ENTER key five times.

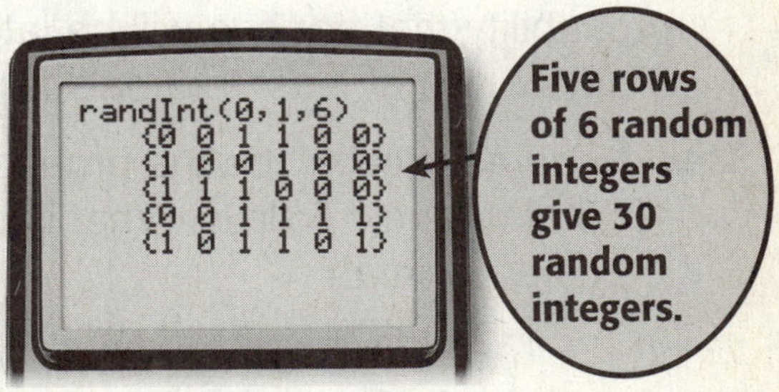

Assume that 1 represents heads and 0 represents tails to answer each question.

1. How many heads did you get in the 30 simulated flips?
2. How many tails did you get in the 30 simulated flips?
3. What is the experimental probability of flipping heads? (*Hint:* Write a fraction that compares the number of heads to 30 flips.)
4. What is the experimental probability of flipping tails? (*Hint:* Write a fraction that compares the number of tails to 30 flips.)
5. Use a calculator to simulate 100 flips, and compute the experimental probability of flipping heads.

Think and Discuss

6. **Explain** how to perform simulations on a calculator.
7. **Discuss** what you think will happen to the experimental probability as you increase the number of simulated flips.

Name ______________________ Date ____________ Class ____________

LESSON 10C

Hands-On Lab

Use a Simulation

Activity

1. Use a computer or calculator to create a table of random numbers. Use the table to simulate the following probabilities.

Lisa's bus for school is late about 25% of the time. What is the probability that her bus will be late at least 3 days in the next 5 days? Use at least ten trials for the simulation.

Step 1 The answer will be the probability that Lisa's bus will be late 3 of the next 5 days. What is this outcome, 3 out of 5, written as a decimal?

____

Step 2 Use 2-digit numbers for your simulation. Out of all the possible 2-digit numbers, how many make up 25%? What is one way to divide the numbers into groups representing outcomes (late and not late)?

____

Step 3 Determine how many numbers will represent one trial. Since you want to find the probabilities that Lisa's bus will be late for each of the next 5 days, 5 numbers will make up a trial.

Step 4 For every 5 numbers, count the numbers that represent Lisa's bus being late. For example, given the random numbers:
12 85 65 24 95

The probability that Lisa's bus will be late is $\frac{2}{5}$ or 0.4.

Step 5 Use at least 10 trials for the simulation. Of the 10 trials, how many represented a probability of Lisa's bus being late at least 3 of the next 5 days?

Based on the fact that Lisa's bus is late only about 25% of the time, is your answer reasonable? Explain.

____

____

____

Name ____________________ Date __________ Class __________

LESSON 10C

Hands-On Lab

Use a Simulation, continued

Think and Discuss

1. Redo the simulation of Lisa's bus being late using 75% as the probability that it will be late. What is the probability that it will be late at least 3 days out of 5 according to the simulation? Does this result support the usefulness of random-number simulations?

____

____

____

____

2. Describe what *random* means in your own words.

____

____

____

Try This

Create a table of random numbers to simulate the two situations. Use at least 10 trials for each simulation.

1. 18 of the 20 students in a math class passed a quiz. What is the experimental probability that one student randomly chosen from the class did not pass the quiz?

____

____

2. The probability that you will roll a 4 when you roll a number cube is $\frac{1}{6}$. What is the experimental probability you will roll a 4 in 3 out of the next 10 rolls?

____

____

Name ______________________ Date ____________ Class ____________

LESSON **10C**

Review for Mastery

Use a Simulation

Situation: Strout's Market is having a contest. They give a puzzle piece to each customer at the checkout. A customer who collects all 10 different puzzle pieces gets $100 in store credit.

Using a table of random numbers, you can model the situation to estimate how many times a customer would have to shop to collect all 10 puzzle pieces.

3	1	9	4	1	1	8	8
5	7	4	5	7	7	9	0
7	0	3	0	1	3	5	0
0	4	3	8	9	5	3	8
2	6	1	7	6	7	6	9
0	8	2	6	5	5	9	2

- Start anywhere in the table. Count the numbers you pass as you "collect" the digits 0–9.

Suppose you start at the top of Column 3 and move to the right. List each number until you have collected all the numbers 0–9.

9 4 1 1 8 8 5 7 4 5 7 7 9 0 7 0 3 0 1 3 5 0 0 4 3 8 9 5 3 8 2 6

You had to go through 32 numbers to get each number at least once (underscored).

- Do the experiment again.

Suppose you start at the bottom of Column 4 and move to the right. When you reach the end of the row, go to the beginning of the table.

6 5 5 9 2 3 1 9 4 1 1 8 8 5 7 4 5 7 7 9 0

You had to go through 21 numbers to get each number at least once (underscored).

- Find the average of your results. $\frac{32 + 21}{2} = \frac{53}{2} = 26.5$

So, on average, you need to shop 27 times to get all 10 pieces to win $100 credit.

Model each situation. Use the list of random numbers shown above. Do two trials. Tell where you start for each trial.

1. A box of Whammos contains a toy dinosaur. If there are 10 different model dinosaurs in the collection, estimate how many boxes of Whammos you would have to buy to get all 10 dinosaurs.

____

____

2. For this spinner, estimate how many times you would have to spin the pointer to get the numbers 1–10.

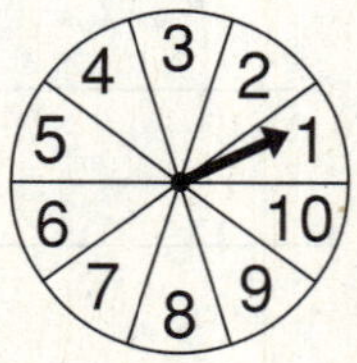

____

____

Name ______________________ Date ____________ Class ____________

LESSON **10C**

Homework and Practice

Use a Simulation

Use the table of random numbers for the problems below.

2596	3623	1053	2161	8550	9672	1044	9852	8935	8727
2393	1231	2795	2036	5138	7488	5814	2086	5397	8073
5086	9057	2050	7424	1706	1391	3585	3124	4164	9498
3274	5099	1789	7021	5636	2404	7180	3395	4734	0395
1387	7460	9196	0463	1388	5104	6408	7463	3289	9321
5891	8081	6222	9656	5606	8739	8698	3982	4974	9961
1544	2817	5296	1340	9750	3943	5697	1585	1097	8365
0442	3881	9080	7964	9357	8202	9987	5681	1288	7894
7170	4188	7099	7590	2158	0242	5299	4095	6986	3935
6039	5890	3996	0217	4045	6545	3744	8368	2034	8670

Ms. Patrick gave the same math test to three of her math classes. Estimate the probability that at least 13 of 20 students will receive a grade above 70 in the third class.

1. Using the first row as the first trial, count the successful outcomes and name the unsuccessful outcomes.

____

2. Using the second row, count the successful outcomes and name the unsuccessful outcomes.

____

Determine the number of successful outcomes in the remaining rows of the random number table.

3. third row ____________

4. fourth row ____________

5. fifth row ____________

6. sixth row ____________

7. seventh row ____________

8. eighth row ____________

9. ninth row ____________

10. tenth row ____________

11. Based on the simulation, estimate the probability that at least 13 of 20 students will receive a grade above 70.

____

EXPLORATION

Program of Studies and Core Content
MA-8-DAP-S-P7, MA-8-DAP-S-P8, MA-08-4.4.2, MA-08-4.4.3, MA-08-4.4.4

10D Theoretical Probability

The *theoretical probability* of an event tells you the probability of the event without your having to conduct an experiment.

For example, the experiment of rolling two standard number cubes and adding the two numbers that each cube shows does not have to be conducted to know the possible sums of numbers.

2	3	4	5	6	7
3	4	5	6	7	8
4	5	6	7	8	9
5	6	7	8	9	10
6	7	8	9	10	11
7	8	9	10	11	12

1. Use the number of times each sum occurs to complete the table.

Sum	2	3	4	5	6	7	8	9	10	11	12
Outcomes	1	2									
Theoretical Probability	$\frac{1}{36}$	$\frac{2}{36}$									

Think and Discuss

2. Explain which sum is most likely to occur.

Name ______________________ Date ____________ Class ____________

LESSON **10D**

Hands-On Lab

Comparing Experimental and Theoretical Probability

Materials: cardboard to make a spinner, spinner

Activity

There are two types of probability: theoretical and experimental. You will explore both of these in this lab. Theoretical probability is calculated as the ratio of the number of ways an event can occur to the number of total possibilities. For example the probability of rolling a four on one roll of one die is 1 to 6. Experimental probability is based on carrying out an experiment, keeping track of the results and using those results to calculate the probabilities as a ratio of the actual times the event occurred to the number of times you carried out the experiment.

You need to make a "spinner" with six sections. Use a sturdy piece of cardboard or plastic that can withstand you spinning. The spinner should be a circle sectioned off into equally-spaced sections. Color each section red, blue, green, yellow, black, and white. Attach the spinner to the center of the circle or use a pencil and paperclip. Make sure the spinner spins freely all the way around the circle.

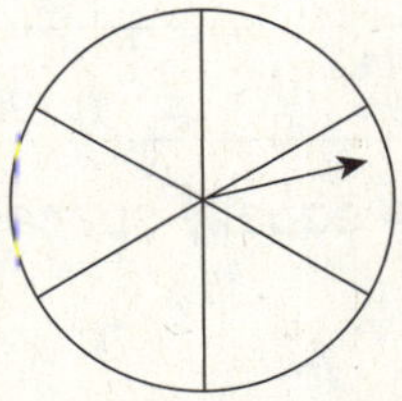

Carry out the experiment by spinning the spinner at least 50 times. More is better! Use the table below and keep track of the number of times that the spinner lands on each color.

Color	**Red**	**Blue**	**Green**	**Yellow**	**Black**	**White**	**TOTAL**
Tick marks for each color →							Number of times you spin the Spinner:
Total for each color							
Experimental Probability							
Theoretical Probability							

Name ______________________ Date __________ Class __________

LESSON 10D

Hands-On Lab

Comparing Experimental and Theoretical Probability, cont.

Try This

1. Based on your results, calculate the experimental probability of landing on each color. (The total for each color divided by the total number of times you spun the spinner.) Record your calculation in the table.

2. Fill in the table with the theoretical probability for each color (this is the same value for each color).

3. Why do you think the experimental and theoretical probabilities are not equal?

____

____

4. Critical Thinking If you spin the spinner 200 times, how do you think this will affect the difference between the theoretical and the experimental probability of your data? Explain.

____

____

5. Create a different spinner, this time with 4 equally-spaced sections. Spin the spinner at least 100 times.

a. Complete the table.

Color	Red	Blue	Green	Yellow	TOTAL
Tick marks for each color →					Number of times you spin the spinner:
Total for each color					
Experimental Probability					
Theoretical Probability					

b. How does the theoretical probability compare to that of the experimental probability for each color?

____

____

c. Predict how many times you will spin yellow if you spin the spinner 250 times. Explain your prediction.

____

____

Name ______________________ Date ____________ Class ____________

LESSON **10D**

Review for Mastery
Theoretical Probability

The sample space for a fair coin has 2 possible outcomes: heads or tails. Both possibilities have the same chance of occurring; they are **equally likely.**

The probability of each outcome is $\frac{1}{2}$.

$P(\text{heads}) = P(\text{tails}) = \frac{1}{2}$

Complete to find each probability.

1. 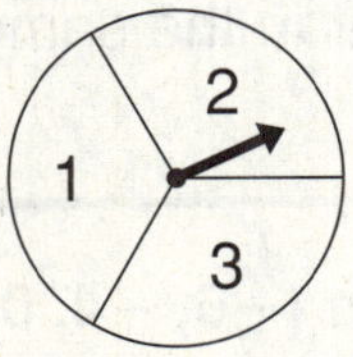

2. 

How many outcomes in the sample space? ________ ________

Are the outcomes equally likely? ________ ________

What is the probability for each outcome? ________ ________

For this spinner, there are 10 possible outcomes in the sample space. The outcomes are equally likely.

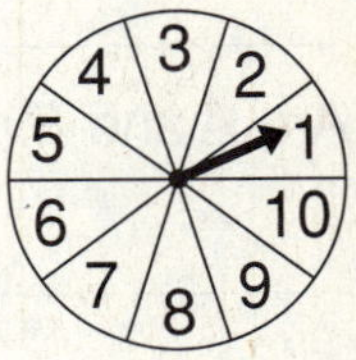

$P(7) = \frac{1}{10}$ $\quad P(\text{even number}) = \frac{5}{10}$, or $\frac{1}{2}$

$P(\text{a number greater than 4}) = \frac{6}{10}$, or $\frac{3}{5}$

When the possible outcomes are equally likely, you calculate the probability that an event E will occur by using a ratio.

$$P(E) = \frac{\textbf{number of favorable outcomes}}{\textbf{total number of possible outcomes}}$$

Find each probability.

3. 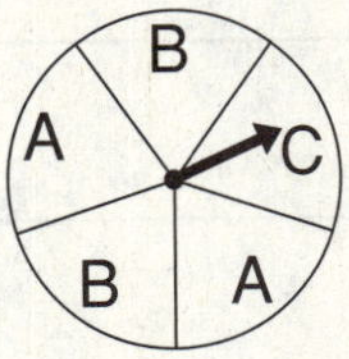

4. 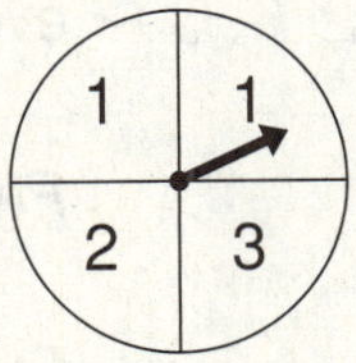

5. 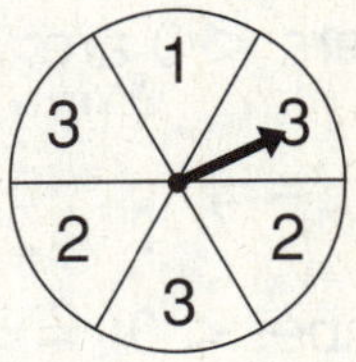

$P(C) =$ ____ $\qquad P(1) =$ ____, or ____ $\qquad P(2) =$ ____, or ____

$P(A) =$ ____ $\qquad P(\text{even}) =$ ____ $\qquad P(\text{odd}) =$ ____, or ____

Name ______________________ Date __________ Class __________

LESSON **10D**

Review for Mastery

Theoretical Probability, continued

For this spinner:

$P(\text{odd}) = \frac{3}{6}$, or $\frac{1}{2}$ $P(\text{even}) = \frac{1}{6}$

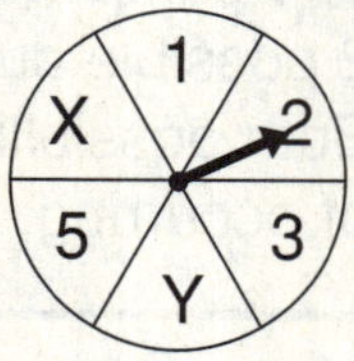

You cannot get an odd number and an even number in the same spin. $P(\text{odd } \textit{and} \text{ even}) = 0$

Events that cannot occur in the same trial are called **mutually exclusive.**

A number is drawn from {−6, −4, 0, 2, 4, 7, 9}. List the possible favorable results for each event. Tell if the events are mutually exclusive.

6. *Event A*: get an odd number

____

Event B: get a negative number

____

Are *A* and *B* mutually exclusive? Explain.

____

7. *Event C*: get a multiple of 3

____

Event D: get an even number

____

Are *C* and *D* mutually exclusive? Explain.

____

For the spinner at the top of this page:

$P(\text{odd}) = \frac{3}{6}$, or $\frac{1}{2}$ $P(\text{even}) = \frac{1}{6}$ $P(\text{odd } \textit{or} \text{ even}) = \frac{3}{6} + \frac{1}{6} = \frac{4}{6}$, or $\frac{2}{3}$

A number is drawn from {−6, −4, 2, 5, 6, 7, 9}. Find the indicated probabilities.

8. odd numbers are: 5, 7, 9

numbers < 0 are: ______________

$P(\text{odd}) = \frac{\quad}{7}$

$P(\text{number} < 0) = \frac{\quad}{7}$

$P(\text{odd number or number} < 0) =$

$\frac{\quad}{7} + \frac{\quad}{7} = \frac{\quad}{7}$

9. numbers > 6 are: ______________

even numbers: ______________

$P(\text{number} > 6) =$ ______________

$P(\text{even number}) =$ ______________

$P(\text{number} > 6 \text{ or even number}) =$

______ + ______ = ______

Name ______________________ Date __________ Class __________

LESSON 10D

Homework and Practice

Theoretical Probability

An experiment consists of rolling two fair dice. Find each probability.

1. P(total shown = 4) ________

2. P(total shown = 6) ________

3. P(total shown = 10) ________

4. P(total shown = 11) ________

5. P(total shown = 1) ________

6. P(total shown = 5) ________

7. P(total shown ≥ 2) ________

8. P(total shown ≤ 7) ________

9. P(total shown > 8) ________

Find the theoretical probability of having a thrown dart land in the indicated area.

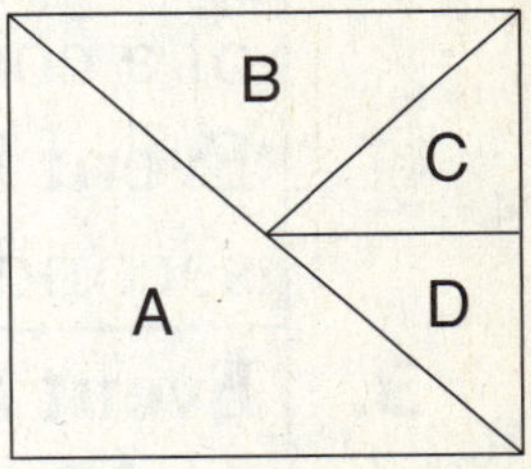

10. P(landing in area A) ________

11. P(landing in area B) ________

12. P(landing in area C) ________

13. P(landing in area D) ________

14. P(landing in areas A or C) ________

15. What is the theoretical probability of winning a raffle if 375 tickets were sold and you bought 25?

____

16. The second place winner of the raffle could choose between 12 envelopes each with money in them. Two of the envelopes contained a $20 bill, 4 contained $10 bills, and 6 contained $5 bills. What is the theoretical probability that the winner will choose an envelope with a $20 bill in it?

____

EXPLORATION

Program of Studies and Core Content

MA-8-DAP-S-CD2, MA-8-DAP-S-P2, MA-8-DAP-S-P6, MA-08-4.4.2, MA-08-4.4.4

10E Independent and Dependent Events

When the occurrence of one event affects the probability of a second event, the events are ***dependent.*** Otherwise, they are ***independent.***

Classify each pair of events as dependent or independent.

	Events	Independent	Dependent
1.	**Event A:** Drawing a 3 from a deck without replacing it **Event B:** Drawing another 3 from the same deck		
2.	**Event A:** Tossing heads on a flip of a coin **Event B:** Tossing heads on a second flip of the same coin		
3.	**Event A:** Running between 5:00 P.M. and 5:30 P.M. **Event B:** Drinking water between 5:30 P.M. and 6:00 P.M.		
4.	**Event A:** Drawing a blue marble and putting it back in a bag that contains 6 blue marbles and 4 red marbles **Event B:** Drawing another blue marble from the same bag		

Think and Discuss

5. Discuss real-world examples of dependent events and independent events.

Name ______________________ Date ____________ Class ____________

LESSON 10E

Hands-On Lab

Independent and Dependent Events

Activity 1

1. Find the probability of tossing two 5's when you toss two six-sided number cubes.

Step 1 Decide whether the two events are dependent or independent.
Does the probability of tossing a 5 on one number cube depend on the probability of tossing a 5 on the other number cube? ____________

Are the events dependent or independent? ____________

Step 2 Determine the probability of tossing a 5 on one number cube.There are six possible numbers on a six-sided number cube. 5 is one of the numbers. What is the probability of tossing a 5 on the first number cube? ____________

What is the probability of tossing a 5 on the second number cube? ____________

Step 3 Multiply the probability of tossing a 5 on the first toss by the probability of tossing a 5 on the second toss.

____

Activity 2

1. A bag has 3 blue socks, 2 red socks, and 5 white socks. Calculate the probability of pulling out a pair of blue socks. Make 10 slips of paper to represent the 10 socks. Place them in a bag.

Step 1 What is the probability of pulling out a blue sock on the first draw?

____

Step 2 Pull out one blue sock from the bag. How many blue socks are left in the bag? ____________

How many socks are left in the bag? ____________

Step 3 What is the probability of pulling a blue sock out of 9 socks if 2 of the socks are blue?

____

Step 4 Multiply the probability of pulling out the first blue sock by the probability of pulling out the second blue sock to find the probability of pulling out a pair of blue socks.

____

Name ______________________ Date __________ Class __________

LESSON 10E

Hands-On Lab

Independent and Dependent Events, continued

Think and Discuss

1. Look back at the first activity where you calculated the probability of tossing double 5's on two number cubes. Find the experimental probability of this event happening in 36 trials. Toss two number cubes 36 times and record the number of times you toss double 5's. The experimental probability is the number of times you toss double 5's over the number of times you toss the number cubes. What is the experimental probability of tossing double 5's?

____

____

2. How is the experimental probability different or similar to the theoretical probability you calculated in the first activity?

____

____

____

Try This

1. Calculate the probability of tossing heads each time when tossing a quarter 5 times. Show your work.

____

2. Charlotte has 10 coins in her pocket. 3 coins are nickels, 3 are dimes, and 4 are quarters. What is the probability that when she pulls 3 coins out of her pocket, they will all be nickels? Show your calculations.

____

Name ______________________________ Date ____________ Class ____________

LESSON 10E

Review for Mastery

Independent and Dependent Events

You can use an area model to help determine the probability of independent and dependent events. The notation $P(B \mid A)$ represents the probability of event B, given that event A has occurred.

For independent events A and B, $P(A \text{ and } B) = P(A) \cdot P(B)$.

Ralph tosses a number cube and flips a coin. What is the probability that he tosses a 2 and gets tails?

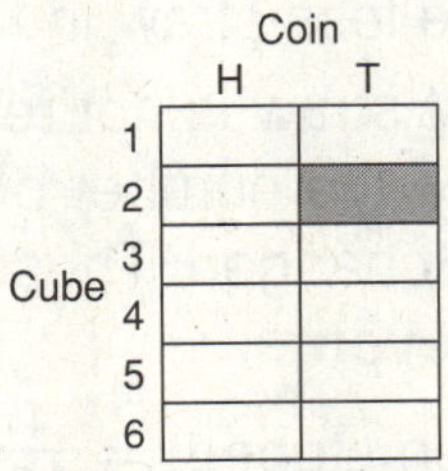

The events are independent. Using the area model, $P(2 \text{ and tails}) = \frac{1}{12}$.

For dependent events A and B, $P(A \text{ and } B) = P(A) \cdot P(B \mid A)$.

A bag contains 2 green marbles and 2 red marbles. You draw one marble, do not replace it, and then draw another. What is the probability that you draw a red marble followed by a green marble?

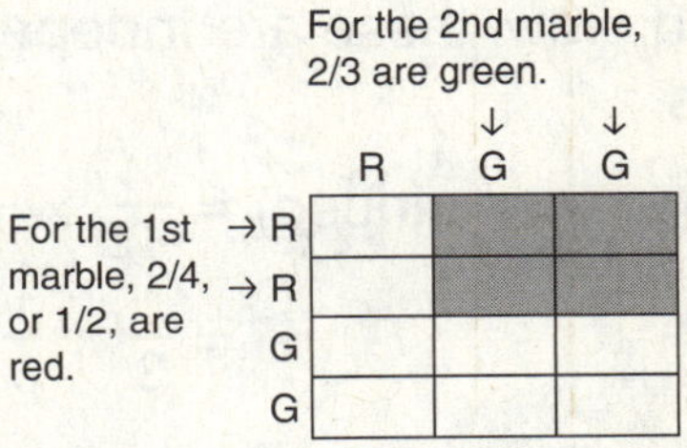

The events are dependent. Using the area model, $P(1^{st}$ marble red and 2nd marble green$) = \frac{4}{12} = \frac{1}{3}$.

Complete to check the answers from the area models above.

1. Let A = toss a 2; $P(A)$ = _____ Let B = get tails; $P(B)$ = _____

$P(A \text{ and } B) = P(A) \cdot P(B) =$ _____ • _____ =

2. Let $A = 1^{st}$ marble red; $P(A)$ = _____ Let $B = 2^{nd}$ marble green.

$P(B \mid A) = P(2^{nd}$ marble green, given that the first marble is _______________)

= _____

$P(A \text{ and } B) = P(A) \cdot P(B \mid A) =$ _____ • _____ = _____

3. Use the area model in the example at right above. What is the likelihood of drawing a second marble of the same color as the first marble? Explain.

____

____

Name ______________________ Date __________ Class __________

LESSON **10E**

Review for Mastery

Independent and Dependent Events, continued

Carlos is to draw 2 straws at random from a box of straws that contains 4 red, 4 white, and 4 striped straws.

$P(\text{1st straw is striped}) = \frac{4}{12}$ ← number of striped straws / ← total number of straws

If Carlos *returns* the 1st straw to the box before drawing the 2nd straw, the probability that the 2nd straw is striped remains the same.

$P(\text{2nd straw is striped})$

$= \frac{4}{12}$ ← same number of striped straws / ← same total number of straws

Since the 1st straw is returned before the 2nd draw, these are **independent events.**

$P(\text{striped then striped}) = \frac{4}{12} \times \frac{4}{12}$

$= \frac{1}{3} \times \frac{1}{3} = \frac{1}{9}$

If Carlos *does not return* the first straw to the box before drawing the second straw, the probability that the second straw is striped changes.

$P(\text{2nd straw is striped})$

$= \frac{3}{11}$ ← one striped straw has been taken / ← one less straw in total number

Since the 1st straw is not returned before the 2nd draw, the number of straws remaining is changed. These are **dependent events.**

$P(\text{striped then striped}) = \frac{4}{12} \times \frac{3}{11}$

$= \frac{1}{3} \times \frac{3}{11} = \frac{1}{11}$

Describe the events as independent or dependent.

4. Josh tosses a coin and spins a spinner. ______________

5. Ana chooses one chocolate from a box of chocolates. Then Lucy chooses one. ______________

6. Sue draws a card from a deck of cards and replaces it. Then she draws a second card from the deck. ______________

Each situation begins with a box of marbles that contains 2 red, 3 blue, 4 green, and 3 yellow marbles. Complete to find each probability.

7. A 1st marble is drawn and replaced. Then a 2nd marble is drawn.

$P(\text{red and blue}) = \frac{}{12} \times \frac{}{12} =$ ______

8. A 1st marble is drawn and not replaced. A 2nd marble is drawn.

$P(\text{red and blue}) = \frac{}{12} \times \frac{}{11} =$ ______

9. A 1st marble is drawn and replaced. Then a 2nd marble is drawn.

$P(\text{red and red}) = \frac{}{12} \times \frac{}{} =$ ______

10. A 1st marble is drawn and not replaced. A 2nd marble is drawn.

$P(\text{red and red}) = \frac{}{12} \times \frac{}{} =$ ______

Name ______________________ Date __________ Class __________

LESSON 10E

Homework and Practice

Independent and Dependent Events

Determine if the events are dependent or independent.

1. spinning a spinner; rolling a number cube

____

2. flipping a coin and getting tails; flipping the same coin again and getting tails again

____

3. selecting a piece of cake; selecting a drink

____

A jar contains 4 white chips, 5 purple chips, and 1 black chip. Chips are selected randomly one at a time, and are not replaced. Find the probability of the following.

4. P(purple then black) ________________

5. P(black then white) ________________

6. P(white then purple) ________________

7. P(purple then white) ________________

8. P(2 whites) ________________

9. P(2 purples) ________________

The table shows how the students in Mr. Wong's science class voted to take a field trip. Use the information in the table to answer exercises 10–15.

Field Trip Voting Results

	Yes	No
Boys	9	3
Girls	5	6

10. How many students are in Mr. Wong's science class? ________

11. How many are boys? ________ How many are girls? ________

12. Find *P*(student is a boy and voted no). ________

13. Find *P*(student is a girl and voted yes). ________

14. Find *P*(student is a boy | student voted yes). ________

15. Find *P*(student is a girl | student voted no). ________

EXPLORATION

Program of Studies and Core Content

MA-8-DAP-S-P1, MA-8-DAP-S-P2, MA-8-DAP-S-P5, MA-08-4.4.2

10F Making Decisions and Predictions

Kate sells shirts and caps that have a school logo on them. Both the shirts and the caps are available in red or black. The table shows the number of shirts and caps that Kate sold during her first week of business.

	Red	Black
Shirts	125	75
Caps	60	40

1. What was the total number of shirts sold during the first week?
2. What fraction of the shirts that were sold were red shirts?
3. Kate assumes that future sales will be similar to the first week's sales. She plans to order 2000 more shirts. How many of these should be red shirts?
4. In general, when Kate sells a shirt, what is the probability that it is a red shirt?
5. What was the total number of caps sold during the first week?
6. What fraction of the caps that were sold were black caps?
7. Kate places an order for 5000 more caps. How many of these should be black caps?

Think and Discuss

8. **Explain** how you determined the number of red shirts that Kate should order in Problem 3.
9. **Explain** how you determined the number of black caps that Kate should order in Problem 6.

Name ______________________ Date __________ Class __________

LESSON **10F**

Review for Mastery

Making Decisions and Predictions

Probability can be used to make predictions.

The spinner has 5 equal sections. Predict how many times you will land on the number 1 in 30 spins.

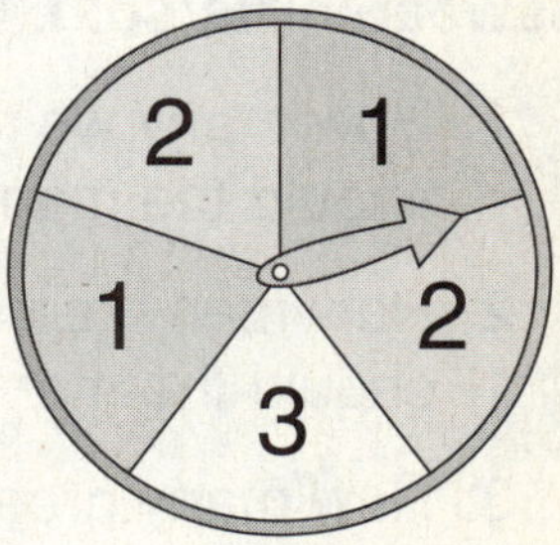

Step 1 Find the probability of landing on the number 1.

$$\frac{\text{number of 1's}}{\text{number of equal sections}} = \frac{2}{5}$$

Step 2 Set up a proportion.

Let n = the number of times you land on the number 1.

$$\frac{2}{5} = \frac{n}{30}$$

Step 3 Solve the proportion.

$$\frac{2}{5} = \frac{n}{30}$$

$2(30) = 5(n)$ *Find the cross products.*

$60 = 5n$ *Simplify.*

$\frac{60}{5} = \frac{5n}{5}$ *Divide both sides by 5.*

$12 = n$ *Simplify.*

In 30 spins, you will land on the number 1 approximately 12 times.

Use the spinner shown at right to answer the questions.

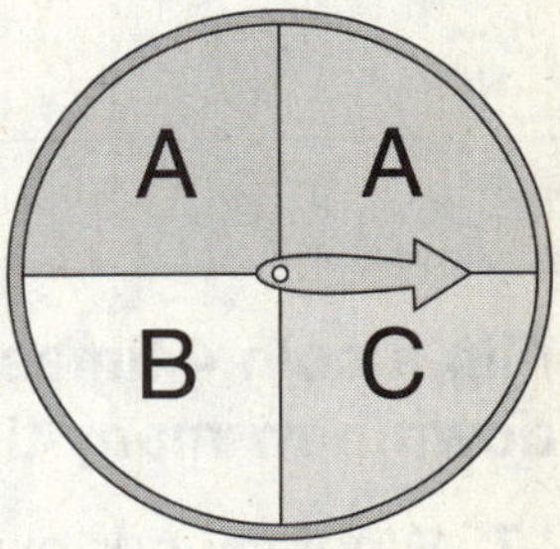

1. Use a proportion to predict how many times you will land on C in 32 spins.

____

2. Use a proportion to predict how many times you will land on B in 48 spins.

____

3. Use a proportion to predict how many times you will land on A in 50 spins.

____

Name ______________________ Date __________ Class __________

LESSON **10F**

Homework and Practice

Making Decisions and Predictions

An athletic wear store sells hooded sweatshirts in five colors. The table shows the colors of the last 60 sweatshirts sold. The manager plans to order 200 more.

Color	Number Sold
White	6
Red	12
Blue	13
Green	9
Black	20

1. How many white sweatshirts should the manager order? ____________

2. How many red sweatshirts should the manager order? ____________

3. How many green sweatshirts should the manager order? ____________

Use the diagram of the spinner for Exercises 4–6.

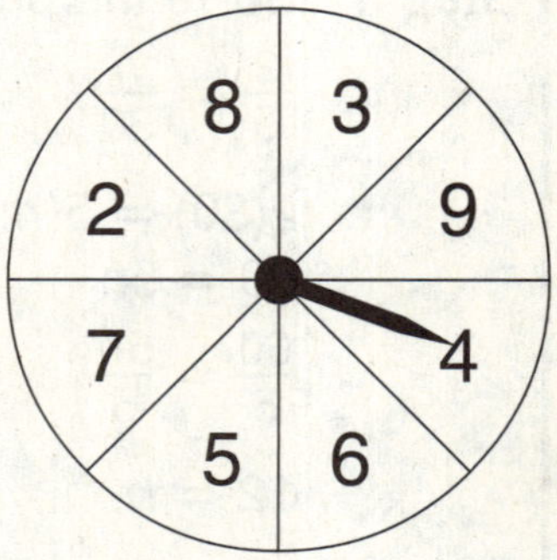

4. Predict the number of times the spinner will land on 5 in 24 spins.

____

5. Predict the number of times the spinner will land on an odd number in 40 spins.

____

6. Irene spins the spinner 64 times. The spinner lands on 7 five times. Is this more or less than the number of times Irene expected a 7? Explain.

____

____

Flip a coin 4 times. Perform 40 trials of this experiment. Write down how many times heads came up exactly twice.

7. What is your experimental probability for getting exactly 2 heads? Use the results from your experiment.

____

8. The theoretical probability of getting exactly two heads when flipping a coin 4 times is $\frac{3}{8}$. Use a proportion to determine how many times you would have expected exactly 2 heads in the experiment.

____

EXPLORATION

Program of Studies and Core Content

MA-8-DAP-S-P3, MA-8-DAP-S-P4, MA-8-DAP-S-P5, MA-08-4.4.1, MA-08-4.4.2

10G Counting Principles

A movie theater sells popcorn in small, medium, or large containers. Each size is also available for regular or lightly buttered popcorn.

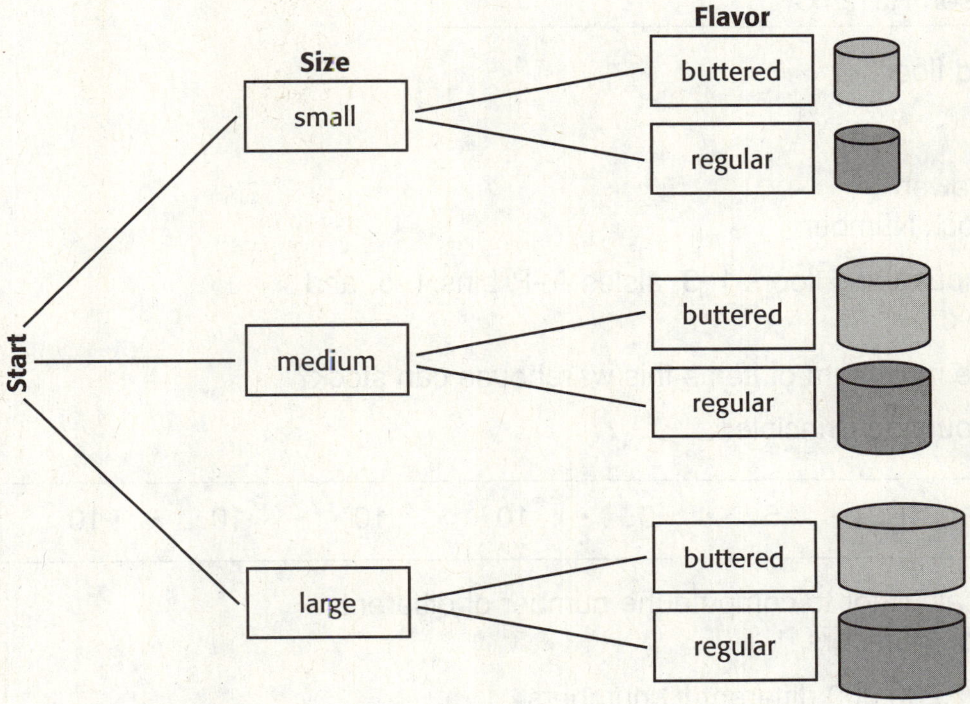

1. How many possible options are there for buying popcorn at the movie theater?
2. How many options are there if the theater adds two new flavors—cheddar cheese and caramel?

Think and Discuss

3. **Explain** how you counted the number of options in Problem 2.
4. **Discuss** whether the number of options in Problem 1 would be different if you were to start with flavor followed by size.

Name ______________________ Date __________ Class __________

LESSON **10G**

Technology Lab

The Fundamental Counting Principle

Activity

Warehouses use a combination of letters and numbers to identify their stock.

Description of Item	ID Number
Champion Spark Plug Tecumseh RJ12YC	2B3a5735

2 = second floor
B = aisle
3 = Bin
a = Bin Drawer
5735 = Stock Number

The warehouse has floors 1–3, aisles A–P, bins 1–5, and drawers a–c.

What is the maximum of items this warehouse can stock?

Use the Counting Principle:

$$3 \cdot 16 \cdot 5 \cdot 3 \cdot 10 \cdot 10 \cdot 10 \cdot 10$$

Use your calculator to compute the number of different identification numbers.

There are 7,200,000 different ID numbers.

Think and Discuss

1. Why did you compute 10^4 for the stock number when applying the counting principle?

____

____

____

____

Name ______________________ Date __________ Class __________

LESSON 10G

Technology Lab

The Fundamental Counting Principle, continued

2. Which would give more possible outcomes, adding another floor or adding another bin drawer?

Try This

The Library of Congress call numbers are a combination of letters and numbers. The first letter indicates the subject area. The second letter indicates a category of the subject area. For example, Q stands for General Science and QA stands for mathematics.

Suppose a call number for a math book is QA1589.

1. Use your calculator to compute the number of different call numbers like the one above, with 4 digits after the QA.

2. If each possible second letter stands for a different category of general science, compute the number of different call numbers with 4 digits for General Science books.

Name ______________________ Date ____________ Class ____________

LESSON 10G

Review for Mastery

Counting Principles

The Fundamental Counting Principle can help you solve some problems about situations that involve more than one activity.

the number of ways in which one activity can be performed × **the number of ways in which a second activity can be performed** = **the total number of ways in which both activities can be performed**

Apply the Fundamental Counting Principle to find the total number of possibilities in each situation.

1. Kelly has 6 shirts and 4 coordinating pants.
The number of possible shirt-pants outfits is: ______________, or ________

2. The menu for dinner lists 2 soups, 4 meats, and 3 desserts. How many different meals that have one soup, one meat, and one dessert are possible? ______________, or ________

A **tree diagram** helps you see all the possibilities in a sample space.

If three coins are tossed at the same time, list all the possible outcomes.

List, in a column, the 2 possibilities for the 1st coin.

For each possibility for the 1st coin, list the 2 possibilities for the 2nd coin.

For each possibility for the 2nd coin, list the 2 possibilities for the 3rd coin.

Read the diagram across to write the list of all possible outcomes.

In this situation, there are $2 \times 2 \times 2 = 8$ possible outcomes.

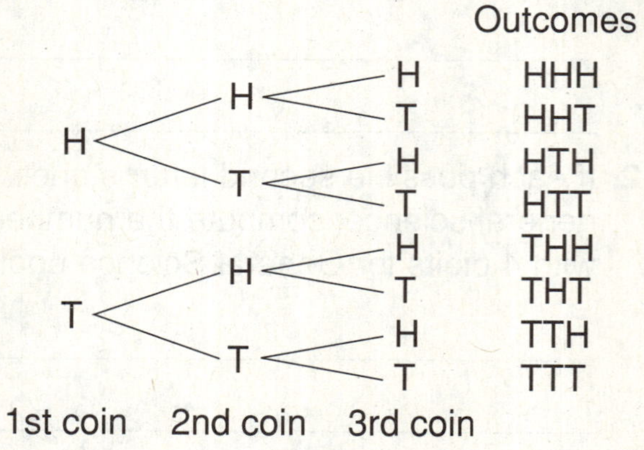

Draw a tree diagram and list the outcomes.

3. A vendor is selling cups of ice cream. There are 2 different sizes of cups: small (S), or large (L). There are 2 different flavors of ice cream: vanilla (V) or chocolate (C). There are 2 different toppings: fudge (F) or pineapple (P).

Name ______________________ Date __________ Class __________

LESSON 10G

Review for Mastery

Counting Principles, continued

How many different 5-letter "words" are possible using the letters of TRIANGLE? Letters can be used only once in each "word."

There are 8 choices for the 1st letter, 7 choices for the 2nd letter, 6 choices for the 3rd letter, 5 choices for the 4th letter, and 4 choices for the 5th.

Apply the Fundamental Counting Principle.

$$\underset{\text{1st letter}}{\underline{8}} \times \underset{\text{2nd letter}}{\underline{7}} \times \underset{\text{3rd letter}}{\underline{6}} \times \underset{\text{4th letter}}{\underline{5}} \times \underset{\text{5th letter}}{\underline{4}} = 6720 \text{ possibilities}$$

If a "word" is selected at random from the 6720 possibilities, what is the probability that it will be the "word" ANGLE?

There is only one outcome ANGLE. $P(\text{ANGLE}) = \frac{1}{6720}$

If a "word" is selected at random from the 6720 possibilities, what is the probability that it will not contain the letter G?

Find the number of favorable outcomes.

Eliminate the letter G from the choices. So, there are 7 choices to begin.

$$\underset{\text{1st letter}}{\underline{7}} \times \underset{\text{2nd letter}}{\underline{6}} \times \underset{\text{3rd letter}}{\underline{5}} \times \underset{\text{4th letter}}{\underline{4}} \times \underset{\text{5th letter}}{\underline{3}} = 2520 \text{ possibilities}$$

$$P(\text{5-letter "word" with no G}) = \frac{\text{number of favorable outcomes}}{\text{total number of possible outcomes}} = \frac{2520}{6720}, \text{ or } \frac{3}{8}$$

Apply the Fundamental Counting Principle.

4. Consider the letters of the word MEDIAN.

a. How many different 4-letter "words" are possible? Letters can be used only once.

$$\underset{\text{1st letter}}{\underline{\quad\quad}} \times \underset{\text{2nd letter}}{\underline{\quad\quad}} \times \underset{\text{3rd letter}}{\underline{\quad\quad}} \times \underset{\text{4th letter}}{\underline{\quad\quad}} = \underline{\quad\quad} \text{ possibilities}$$

b. If a 4-letter "word" is selected at random from all the possibilities, then: $P(\text{DEAN}) = \underline{\quad\quad}$

c. If a 4-letter "word" is selected at random from all the possibilities, what is the probability that it will not contain the letter D?

favorable outcomes: $\underset{\text{1st letter}}{\underline{\quad\quad}} \times \underset{\text{2nd letter}}{\underline{\quad\quad}} \times \underset{\text{3rd letter}}{\underline{\quad\quad}} \times \underset{\text{4th letter}}{\underline{\quad\quad}} = \underline{\quad\quad}$

$$P(\text{4-letter "word" with no D}) = \frac{\text{number of favorable outcomes}}{\text{total number of possible outcomes}} = \underline{\quad\quad}, \text{ or } \underline{\quad\quad}$$

Name ______________________ Date ____________ Class ____________

LESSON **10G**

Homework and Practice

The Fundamental Counting Principle

School identification codes at Truman School contain 1 letter followed by 3 numbers. All codes are equally likely.

1. Find the number of possible identification codes.

____

2. Find the probability that an ID code of the school does not contain the number 0.

____

3. The Dairy King Shop serves 3 flavors of ice cream: vanilla, chocolate, and vanilla/chocolate swirl. The ice cream can be served in a dish, sugar cone, or waffle cone and may be a small, medium, large, or jumbo size. Make a tree diagram indicating the possible outcomes for each ice cream served.

4. How many different ice cream selections does the Dairy King Shop have?

____

Find the probability for each of the following. Assume all selections are equally likely.

5. P(dish of ice cream)

____

6. P(vanilla and small ice cream)

____

7. P(medium size)

____

8. P(not a dish)

____

EXPLORATION

10H Permutations and Combinations

Program of Studies and Core Content

MA-8-DAP-S-P3,
MA-08-4.4.1

Alfonso and Barb race against each other in a 5K race. There are two ways they can finish.

First Place	Second Place
Alfonso	Barb
Barb	Alfonso

1. Carl joins Alfonso and Barb. Complete the table to show all of the different ways they can finish.

	First Place	Second Place	Third Place
	Alfonso	Barb	Carl
a.			
b.			
c.			
d.			
e.			

2. Complete the table to show all of the different ways 2 of the 3 runners can finish in a tie for first place.

	First Place (tie)	Second Place
	Alfonso and Barb	Carl
a.		
b.		

Think and Discuss

3. Explain how you could find the number of ways 5 runners could finish.

Name ______________________ Date ____________ Class ____________

LESSON 10H

Hands-On Lab

Odds

Activity

1. You roll a single die. What is the probability that you roll a four?

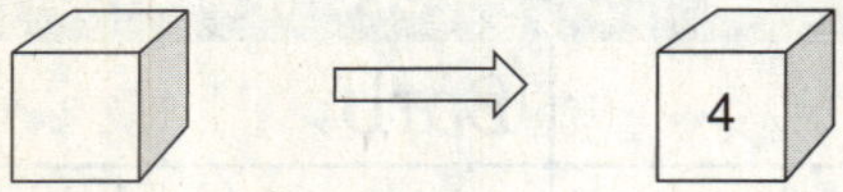

Probability $= \frac{\text{number of favorable outcomes}}{\text{number of total outcomes}}$

Probability $= \frac{1}{6} \approx 16.7\%$

The idea of odds is a little different from the idea of probability. What are the odds in favor of rolling a four?

Odds in favor $= \frac{\text{number of favorable outcomes}}{\text{number of unfavorable outcomes}}$

On a single die, there is one 4, so the number of favorable outcomes is 1. There are 5 other numbers on the die, so the number of unfavorable outcomes is 5.

Odds in favor $= \frac{1}{5} = 1{:}5$

2. You roll a fair die. What are the odds that you will roll an even number? First, figure the probability that you will roll an even number.

Probability $= \frac{3}{6} = \frac{1}{2}$

Odds $= 1{:}(2 - 1) = 1{:}1$

Name ______________________ Date ____________ Class ____________

LESSON 10H

Hands-On Lab

Odds, continued

Think and Discuss

1. Explain the difference between odds and probability.

____

____

____

____

Try This

1. Roll a die 10 times and record your outcomes in a table.

1	2	3	4	5	6	7	8	9	10

2. How many times did you roll a number less than 3? What is the probability of rolling a number less than 3?

____

3. What are the odds of doing so?

____

____

Name ______________________ Date __________ Class __________

LESSON 10H

Hands-On Lab

Odds, continued

4. Roll three dice 50 times and record your outcomes in a table. What was the experimental probability of rolling three numbers less than 4?

____

5. What was the experimental probability of rolling three numbers greater than 4? Is it the same as above? Why or why not?

____

____

____

____

Name ______________________ Date ____________ Class ____________

LESSON **10H**

Technology Lab 1

Permutations and Combinations

Activity

1. Washington state is famous for its apples. At an apple festival prizes are awarded for the best apple pies. Fifteen pies have made it to the last round. Prizes are awarded for first, second, and third best-tasting pie. How many permutations of three winning pies are possible?

$${}_{15}P_3 = 15 \cdot 14 \cdot 13 = 2730$$

There are 2730 permutations.

If a prize were awarded for fourth place as well, how many different permutations would there be?

$${}_{15}P_4 = 15 \cdot 14 \cdot 13 \cdot 12 = 32{,}760$$

2. The judges cannot decide which one pie tastes best. So they are giving first prizes to the three best-tasting pies. How many choices do the judges have for awarding 1st prize to the three best-tasting pies?

$${}_{15}C_3 = \frac{{}_{15}P_3}{3!} = \frac{15 \cdot 14 \cdot 13}{3 \cdot 2} = 455$$

The judges have 455 choices.

Name ______________________ Date ____________ Class ____________

LESSON 10H

Technology Lab 1

Permutations and Combinations, continued

Think and Discuss

1. Why is the number of combinations of a set generally smaller than the number of permutations of the same set?

____

____

____

____

____

____

Try This

1. Find the value of each expression. Use your calculator.

${}_{20}P_4$	
${}_{12}P_5$	
${}_{8}P_4$	
${}_{8}C_4$	
${}_{12}C_8$	
${}_{9}C_4$	
${}_{23}C_3$	

Name ________________________________ Date ____________ Class ____________

LESSON **10H**

Technology Lab 2

Permutations and Combinations

Graphing calculators have features to help with computing factorials, permutations, and combinations.

Activity

1 In a stock-car race, 11 cars finish the race. The number of different orders in which they can finish is 11! A calculator can help you do the computation. Both ways are shown–the direct way, using the definition of *factorial*, and the calculator factorial command.

To compute 11! on a graphing calculator, enter 11 MATH, press > to go to the **PRB** menu, and select **4:!** ENTER.

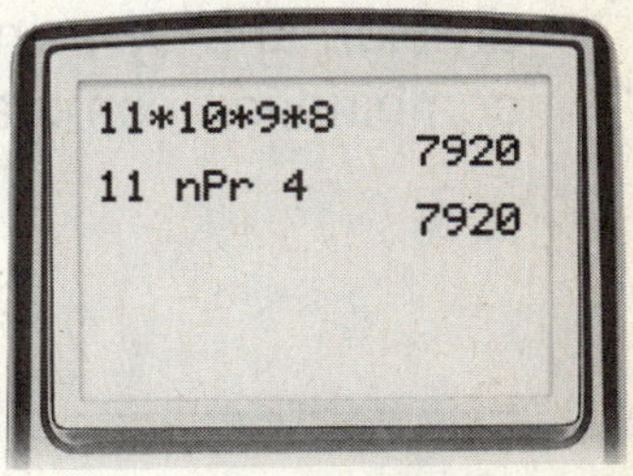

The number of ways the 11 cars can finish first, second, third, and fourth is given by 11·10·9·8, or in *permutation* notation, ${}_{11}P_4$, 11 things taken 4 at a time. Both the direct and calculator *nPr* command methods are shown. The *nPr* command is also found in the **PRB** menu.

To compute ${}_{11}P_4$, enter 11, press MATH, press > to go to the **PRB** menu, select **2:nPr**, type 4, and press ENTER.

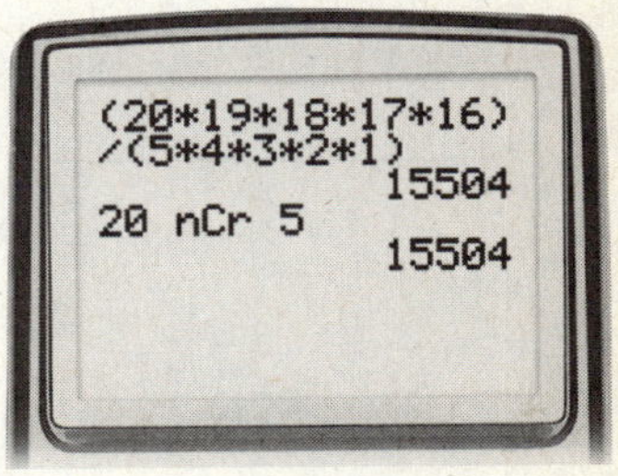

2 Twenty girls try out for 5 open places on a hockey team. Since order is not considered, the number of different *combinations* of these girls that can be chosen is given by ${}_{20}C_5$, the number of combinations of 20 things taken 5 at a time. Both the direct and calculator *nCr* command computations are shown.

To compute ${}_{20}C_5$, press 20 MATH, press > to go to the PRB menu, select **3:nCr**, and press 5 ENTER.

Name ______________________ Date __________ Class __________

LESSON 10H

Technology Lab 2

Permutations and Combinations, continued

Think and Discuss

1. Explain why *nPr* is usually greater than *nCr* for the same values of *n* and *r*.

____

____

2. Can *nPr* ever equal *nCr*?

____

____

Try This

Compute each value by direct calculator multiplication and division and by using the calculator permutation and combination commands.

1. ${}_{14}P_6$ ________

2. ${}_{25}P_{17}$ ________

3. ${}_{8}P_{3}$ ________

4. ${}_{8}C_{3}$ ________

5. ${}_{16}C_{4}$ ________

6. ${}_{40}C_{6}$ ________

Name ______________________ Date __________ Class __________

LESSON 10H

Review for Mastery

Permutations and Combinations

Factorial: a string of factors that counts down to 1

$$6! = 6 \cdot 5 \cdot 4 \cdot 3 \cdot 2 \cdot 1$$

To evaluate an expression with factorials, cancel common factors.

$$\frac{5!}{3!} = \frac{5 \cdot 4 \cdot \cancel{3} \cdot \cancel{2} \cdot \cancel{1}}{\cancel{3} \cdot \cancel{2} \cdot \cancel{1}} = 5 \cdot 4 = 20$$

Complete to evaluate each expression.

1. $\frac{7!}{4!} = \frac{7 \cdot 6 \cdot 5 \cdot \_ \cdot \_ \cdot \_ \cdot \_}{\_ \cdot \_ \cdot \_}$

   $= 7 \cdot$ ______ $=$ ______

2. $\frac{6!}{(5-2)!} = \frac{6!}{\_!} = \frac{\_ \cdot \_ \cdot \_ \cdot \_ \cdot \_ \cdot \_}{\_ \cdot \_ \cdot \_}$

   $=$ ______ $=$ ______

Permutation: an arrangement in which order is important

wxyz is not the same as *yxzw*

Apply the Fundamental Counting Principle to find how many permutations are possible using all 4 letters *w*, *x*, *y*, *z* with no repetition.

$\frac{4}{\text{1st letter}} \times \frac{3}{\text{2nd letter}} \times \frac{2}{\text{3rd letter}} \times \frac{1}{\text{4th letter}} = 4! = 24$ possible arrangements

When you arrange *n* things, *n*! permutations are possible.

Complete to find the number of permutations.

3. In how many ways can 6 people be seated on a bench that seats 6?

   $6! = \_ \cdot \_ \cdot \_ \cdot \_ \cdot \_ \cdot \_$

   $=$ ______ possibilities

4. How many 5-digit numbers can be made using the digits 7, 4, 2, 1, 8 without repetitions?

   ___! $= \_ \cdot \_ \cdot \_ \cdot \_ \cdot \_$

   $=$ ______ possibilities

Apply the Fundamental Counting Principle to find how many permutations are possible using 4 letters 2 at a time, with no repetitions.

$\frac{4}{\text{1st letter}} \times \frac{3}{\text{2nd letter}} = 12$ possible 2-letter arrangements

Apply the Foundamental Counting Principle.

5. In how many ways can 6 people be seated on a bench that seats 4?

   $\frac{\_\_\_}{\text{1st seat}} \times \frac{\_\_\_}{\text{2nd seat}} \times \frac{\_\_\_}{\text{3rd seat}} \times \frac{\_\_\_}{\text{4th seat}} =$ ______ possibilities

6. How many 3-digit numbers can be made using the digits 7, 4, 2, 1, 8 without repetitions?

   $\frac{\_\_\_}{\text{1st digit}} \times \frac{\_\_\_}{\text{2nd digit}} \times \frac{\_\_\_}{\text{3rd digit}} =$ ______ possibilities

Name ______________________ Date __________ Class __________

LESSON **10H**

Review for Mastery

Permutations and Combinations, continued

When using fewer than the available number of items in an arrangement, instead of the Fundamental Counting Principle, you can use a formula to find the number of possible permutations.

To arrange *n* things *r* at a time, the number of possible permutations *P* is: $_nP_r = \frac{n!}{(n-r)!}$

Find how many permutations are possible using 4 letters 2 at a time, with no repetitions.

$$_4P_2 = \frac{4!}{(4-2)!} = \frac{4!}{2!} = \frac{4 \cdot 3 \cdot \cancel{2} \cdot \cancel{1}}{\cancel{2} \cdot \cancel{1}} = 12 \text{ possible 2-letter arrangements}$$

Complete to apply the permutations formula.

7. In how many ways can 6 people be seated on a bench that seats 4?

$$_6P_4 = \frac{6!}{(6-\ \ )!} = \frac{\ \ !}{\ \ !} = \underline{\qquad\qquad} = \text{ possible seating arrangements}$$

8. How many 3-digit numbers can be made using the digits 7, 4, 2, 1, 8 without repetitions?

$$_5P_{\_\_\_} = \frac{5!}{(5-\ \ )!} = \frac{\ \ !}{\ \ !} = \underline{\qquad\qquad} = \underline{\qquad} \text{ possible 3-digit numbers}$$

Combination: an arrangement in which order is not important

How many 2-letter combinations can be made from the 4 letters *w*, *x*, *y*, *z* without repetition?

The combinations *w x* and *x w* are the same. After all the same combinations are removed, there are 6 different combinations possible.

wx	~~*xw*~~	~~*yw*~~	~~*zw*~~
wy	*xy*	~~*yx*~~	~~*zx*~~
wz	*xz*	*yz*	~~*zy*~~

There are fewer combinations than permutations.

$$_nC_r = \frac{_nP_r}{r!} = \frac{n!}{(n-r)!\,r!}$$

The number of combinations *C* of *n* things taken *r* at a time is:

$$_4C_2 = \frac{_4P_2}{2!} = \frac{4!}{(4-2)!\,2!} = \frac{4!}{2!\,2!} = \frac{4^2 \cdot 3 \cdot \cancel{2} \cdot \cancel{1}}{\cancel{2} \cdot 1 \cdot \cancel{2} \cdot \cancel{1}} = 2 \cdot 3 = 6$$

Complete to apply the combinations formula.

9. How many different 4-person committees can be formed from a group of 6 people?

$$_6C_4 = \frac{_6P_4}{4!} = \frac{6!}{(6-\ \ )!\ \ !} = \frac{6!}{\ \ !\ \ !} = \underline{\qquad\qquad} = \underline{\qquad} \text{ possible 4-person committees}$$

Name ______________________ Date ____________ Class ____________

LESSON 10H

Homework and Practice

Permutations and Combinations

Evaluate each expression.

1. 12! ______

2. 9! ______

3. 13! − 10! ______

4. 11! − 8! ______

5. $\frac{22!}{19!}$ ______

6. $\frac{20!}{15!}$ ______

7. $\frac{18!}{(24 - 12)!}$ ______

8. $\frac{21!}{(21 - 5)!}$ ______

9. $\frac{17!}{(23 - 14)!}$ ______

10. How many different 3 people committees can be formed from a group of 25 people?

______

11. From a class of 28 students, how many different ways can 4 students be selected to serve on the student council as president, vice president, secretary and treasurer?

______

12. The boys' volleyball team has 24 players. If the coach chooses 9 boys to play at a time, how many different teams can be formed?

______

13. The golf tournament has 120 players signed up to play. How many different 4-person sets can be formed?

______

14. Mr. Cruz bought new tires for his car. The dealer advised Mr. Cruz to have the tires rotated every 5,000 miles. If Mr. Cruz takes the advice of the dealer and has the tires rotated a different way every 5,000 miles, how many possible miles would he drive before using all the possible rotations?

______

CHAPTER 10

Chapter Summary

Answer these questions to summarize the important concepts from Chapter 8 in your own words.

1. Explain the difference between experimental probability and theoretical probability.

2. Describe how independent and dependent events are different.

3. Explain how to find the number of possible arrangements of several objects.

For more review of Chapter 10:

- Complete the Chapter 10 Study Guide and Review on pages 572–574 of your textbook.
- Complete the Ready to Go On quizzes on pages 538 and 568 of your textbook.

Name ______________________ Date ____________ Class ____________

CHAPTER 10

KCCT Prep

Multiple Choice

1. The table gives information on the last 160 vehicles to pass through the toll plaza at a bridge. Which is the best estimate of the experimental probability that a randomly-chosen vehicle on the bridge is a car?

Bridge Traffic	
Vehicle	**Number**
Car	92
Truck	42
Bus	8
Other	18

(A) 33% (C) 60%
(B) 40% (D) 75%

2. Kaitlyn rolls a number cube. What is the probability that she rolls a prime number?

(A) $\frac{1}{6}$ (C) $\frac{1}{2}$
(B) $\frac{1}{3}$ (D) $\frac{2}{3}$

3. Jake spins a spinner with equal sections numbered 1 through *n*. He also flips a coin at the same time. There are 30 outcomes in the sample space. What is the value of *n*?

(A) 2 (C) 28
(B) 15 (D) 30

4. Which of these is an irrational number?

(A) $\sqrt{4}$ (C) $\frac{4}{5}$
(B) $\sqrt{5}$ (D) $4.\overline{5}$

5. The figure shows a shortcut that has been created by people walking across the corner of a rectangular lawn. Which is the best estimate of the length of the shortcut?

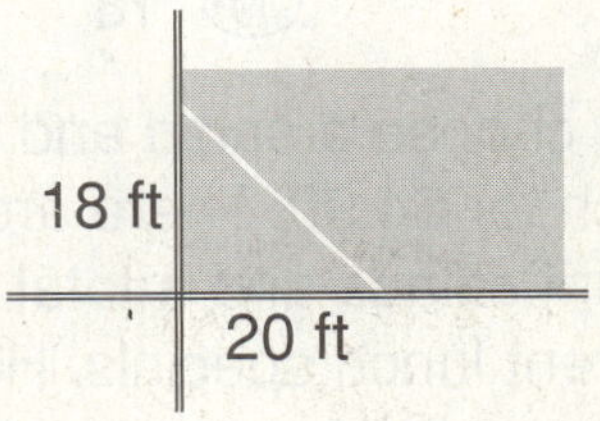

(A) 22 ft (C) 35 ft
(B) 27 ft (D) 38 ft

6. A bag contains 12 green cubes, 8 red cubes, and 4 blue cubes. You choose a cube without looking. What is the probability that you choose a green or red cube?

(A) $\frac{1}{2}$ (C) $\frac{3}{4}$
(B) $\frac{2}{3}$ (D) $\frac{5}{6}$

7. Which is the best estimate of $\sqrt{5} + \sqrt{11}$?

(A) 2.5 (C) 5.5
(B) 4 (D) 8

8. The vertices of a rectangle are (–1, 2), (4, 2), (4, –1), and (–1, –1). The rectangle is enlarged by a dilation with a scale factor of 3. Which of these is NOT a vertex of the enlarged rectangle?

(A) (12, 6) (C) (–3, 12)
(B) (–3, 6) (D) (12, –3)

9. Which of these is closest to 65%?

Ⓐ 0.07 Ⓒ $\frac{2}{3}$

Ⓑ 5.5 Ⓓ $\frac{3}{5}$

10. What is the least possible value of the expression $18 \div 2 + n^2 + 1$ when n is an integer?

Ⓐ 6 Ⓒ 10

Ⓑ 9 Ⓓ 18

11. You can choose a salad and a sandwich for $4.99. There are 5 different salads and a total of 20 different lunch specials. How many different sandwiches are available?

Ⓐ 2 Ⓒ 15

Ⓑ 4 Ⓓ 100

12. Ernesto has a set of eight cards numbered 1 through 8. He chooses a card at random. What is the probability that he chooses a card with a number greater than 6?

Ⓐ $\frac{1}{8}$ Ⓒ $\frac{3}{8}$

Ⓑ $\frac{1}{4}$ Ⓓ $\frac{3}{4}$

13. Two out of every 20 boxes of Crunchy-O's cereal contains a free gel pen. What is the probability that you will NOT find a gel pen in your box of Crunchy-O's?

Ⓐ $\frac{1}{10}$ Ⓒ $\frac{9}{10}$

Ⓑ $\frac{1}{2}$ Ⓓ $\frac{19}{20}$

14. The experimental probability that Daria makes a free throw is $\frac{3}{5}$. If she continues to make free throws at this rate, how many free throws will she make in her next 100 attempts?

Ⓐ 32 Ⓒ 75

Ⓑ 60 Ⓓ 92

Response

15. Mei rolls a number cube 30 times. The table shows her results.

Outcome	Frequency
1	𝍸 I
2	𝍸 I
3	𝍸 II
4	𝍸 III
5	II
6	I

a. What is the theoretical probability of rolling a 5 or a 6?

b. Based on Mei's results, explain clearly why the experimental probability of rolling a 5 or a 6 is the same as or different from your answer to part **a**.

c. Suppose Mei rolls the number cube a total of 150 times. How do you think the frequency of rolling a 5 and the frequency of rolling a 6 will be affected?

16. At Ashton's school, students are assigned a locker with a 3-digit number. Each digit in the number can be 1, 2, 3, 4, 5, 6, 7, 8, or 9. All locker numbers are equally likely.

a. How many different locker numbers are there?

b. Ashton says he has a pretty good chance of getting a locker where all three digits are the same, such as 111 or 777. Assess Ashton's statement and explain your response.

CHAPTER 11

Project

Multi-Step Equations and Inequalities: Riding the Rapids

Career: Hydrologist

Hydrologists measure water flow between rivers, streams, lakes, and oceans. They map their results to record locations and movement of water above and below the earth's surface. Hydrologists are involved in projects such as water-resource studies, field irrigation, flood management, soil-erosion prevention, and the study of water discharge from creeks, streams, and rivers.

Before you start, work through the Water Flows Downhill worksheet on the next page.

In this chapter project . . .

You make a scale drawing of a man-made river that would be good for white-water rafting.

What to do:

- Research the properties of rivers that have white-water rapids. Use this information to design a man-made river that would have two or three sections of rapids.
- Use the equations given in the Water Flows Downhill worksheet to calculate the discharge of your river at the rapids.
- Make a scale drawing of your river and label the dimensions of the cross section. Include the discharge and water velocity in your diagram.

What to present:

- Your scale drawing.
- A table listing the water discharge and velocity for each section of rapids on your river.

Where's the math?

- Use equations to calculate river cross-section and discharge.

go.hrw.com
Chapter Project Online
KEYWORD: MT7 Ch11

Name ______________________ Date __________ Class __________

CHAPTER 11

Project Recording Sheet

Multi-Step Equations and Inequalities

Water Flows Downhill

You are monitoring a river in which the water's velocity has been changing. What is the effect of this velocity change on the amount of water flowing past the measuring station? The river is 30 m across with a mean depth of 2.5 m. Use the relationships below to complete the table.

Discharge (m^3/s) = water velocity (m/s) × stream cross section (m^2)
Stream cross section = width (m) × mean depth (m)
1 $m^3 \approx$ 264.2 gallons
1 hour = 3,600 seconds

My River
Stream Cross Section (75 m^2)

Velocity (m/s)	**Discharge** (m^3/s)	**Hourly Discharge** (m^3)	**Hourly Discharge** (gallons)
0.49	36.75	132,300	34,953,660
0.76			
1.22			
1.86			
2.10			

Answer the following questions:

1. Dixon Creek in Texas has a discharge of 0.11 m^3/s. What happens to the discharge:

a. if the velocity of the stream doubles? ______________

b. if the cross section becomes smaller? ______________

c. if both the velocity and cross section triple? ______________

2. The Mississippi River at Pointe Coupee Parish has a typical discharge of 16,735 m^3/s.

a. How many billions of gallons of water pass in a typical day? ______________

b. If the flow has a velocity of 2.0 m/s what might the cross section be? ______________

3. Create a river of your own. Draw a picture to demonstrate its flow. What would its characteristics be during:

a. normal times? ______________

b. drought? ______________

c. flood season? ______________

EXPLORATION

11A Simplifying Algebraic Expressions

Program of Studies and Core Content

MA-8-AT-S-EI1, MA-8-AT-S-VE01, MA-8-AT-S-VE03, MA-08-5.1.2, MA-08-5.2:1, MA-08-5.2.2, MA-08-5.3.1

Like terms are combined separately from unlike terms. Look at how the like terms are combined in the diagram.

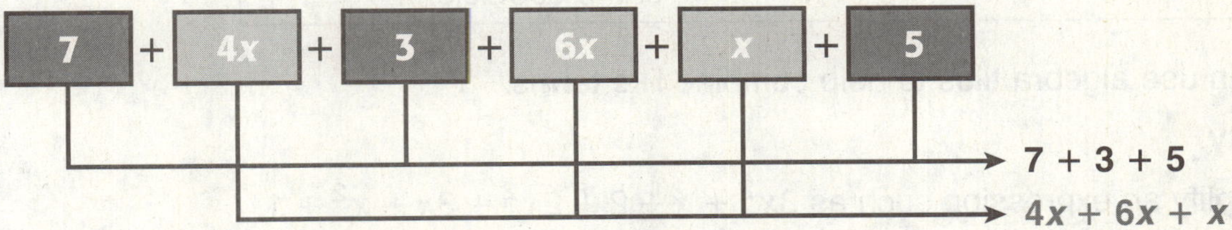

Express the perimeter of each figure with all the like terms combined.

1.

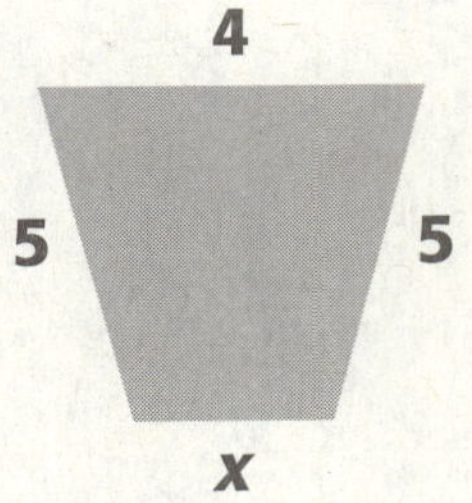

perimeter = ______

2.

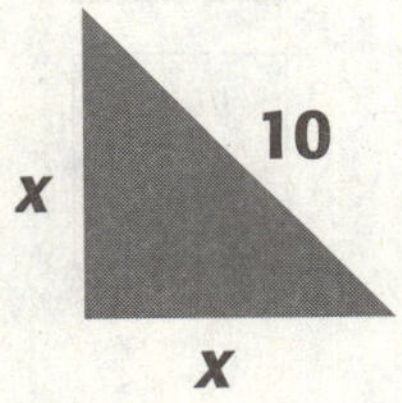

perimeter = ______

3.

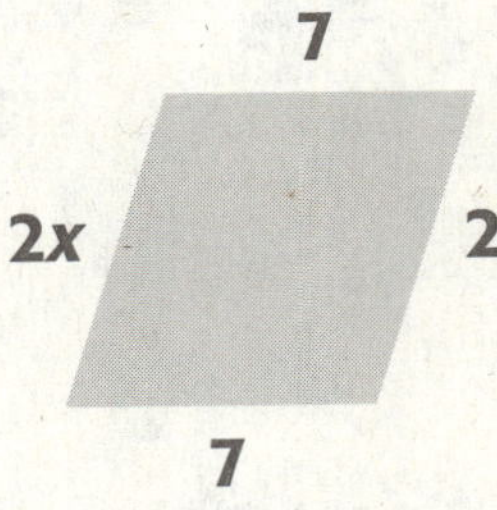

perimeter = ______

4.

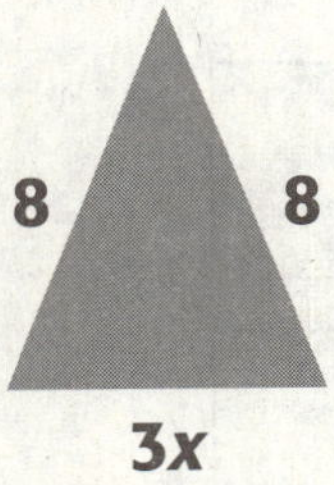

perimeter = ______

Think and Discuss

5. Explain what it means to combine like terms.

6. Describe a real-world situation in which you combine like terms.

Name ______________________ Date ___________ Class ___________

LESSON **11A**

Hands-On Lab

Model Combining Like Terms

KEY	REMEMBER:
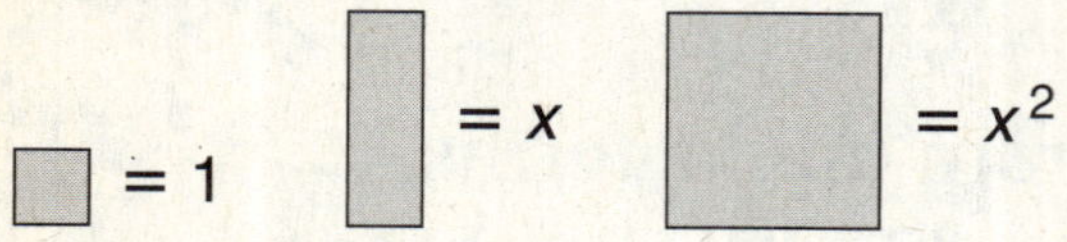	Like terms in an equation have the same variable and exponent. They do not need to have the same coefficient.

You can use algebra tiles to help combine like terms.

Activity

To simplify an expression such as $3x^2 + x + 2 + 2x^2 + 3x + x^2 + 1$, combine the like terms.

$$3x^2 + x + 2 + 2x^2 + 3x + x^2 + 1$$

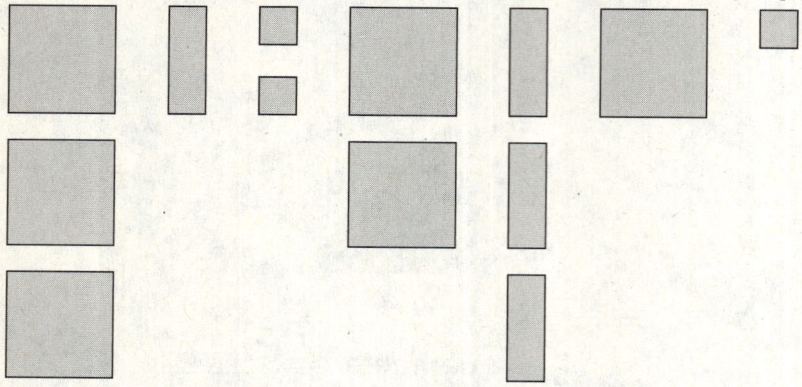

Group the similar tiles together. Simplify.

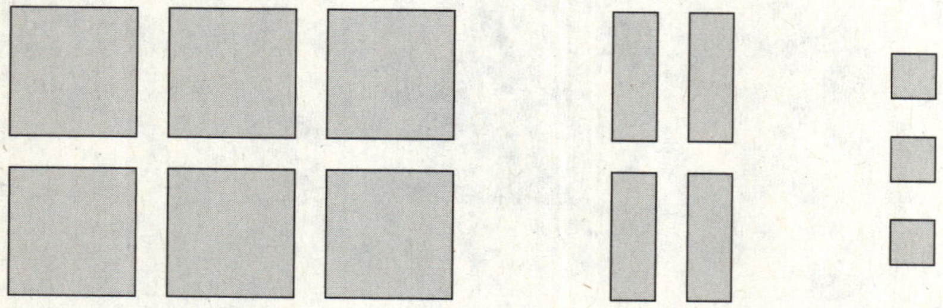

$$6x^2 \quad + \quad 4x \quad + \quad 3$$

Name ______________________ Date ____________ Class ____________

LESSON 11A

Hands-On Lab

Model Combining Like Terms, continued

1. Use algebra tiles to model and simplify each expression.

a. $2 + x^2 + 3x + 2x^2 + 2$

____

b. $2x + 3x^2 + 3x + 2x^2 + 2 + x^2 + x + 1$

____

c. $x + x^2 + 2x + x^2 + 1 + x^2 + x + 1 + x^2 + x^2 + 1$

____

d. $x^2 + 2x + x + 2x^2 + x^2 + 3 + 3x + 1 + x^2 + 3x^2$

____

Black tiles can represent terms that are subtracted rather than added.

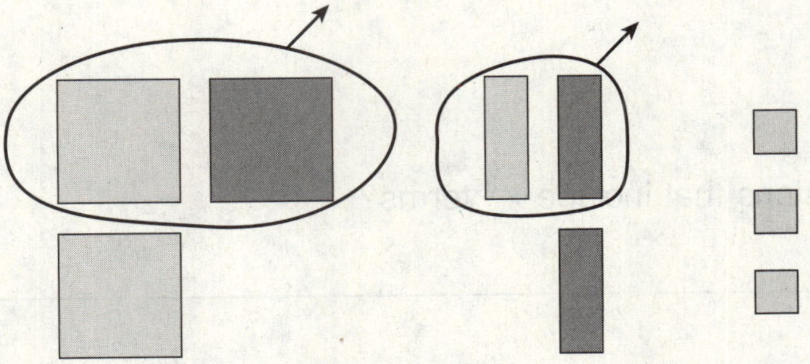

$\mathbf{2x^2 \quad - \quad x^2 \quad + \quad x - 2x \quad + \quad 3}$

Take away any pairs of tiles that are the same size but different colors. Then simplify.

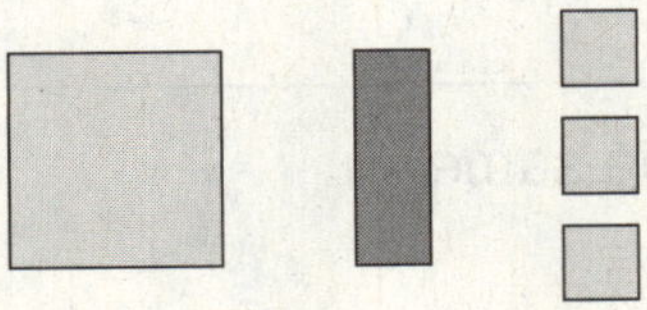

Name ______________________ Date __________ Class __________

LESSON 11A

Hands-On Lab

Model Combining Like Terms, continued

2. Use algebra tiles to model and simplify each expression.

a. $2 - x^2 + 2x + 2x^2 - 2$

____

b. $3x^2 - 3x - 2x^2 + 2 - x^2 + x - 1$

____

c. $-x - x^2 - 2x + 2x^2 - 1 + x^2 - x^2 + 3$

____

d. $x^2 - 2x - 2x^2 + x^2 - 3 + 3x + 1 - x^2 + 3x^2$

____

Think and Discuss

1. How would you model expressions that include x^3 terms?

____

____

2. How you would model the expression $x^2 + 2y + y^2 + 1 + x$ that has two variables, x and y?

____

____

3. Explain why you can take away a pair of tiles that have the same shape but different colors.

____

Name ______________________ Date ____________ Class ____________

LESSON 11A

Hands-On Lab

Model Combining Like Terms, continued

Try This

Use algebra tiles to model and simplify each expression.

1. $3x^2 - x + 2x - x^2 - 2$

____

2. $x^2 - 2x - x^2 + 1 + 2x^2 + x - 1$

____

3. $x + y^2 - 2x + 2x^2 - 1 + 2y^2 - x^2 + 3 + y$

____

Name ____________________ Date ____________ Class ____________

LESSON **11A**

Review for Mastery

Simplifying Algebraic Expressions

The **terms** of an expression are separated by plus or minus signs.

$5a + 7b - 3a + 6a^2$

Like terms have the same variable raised to the same power.

The expression shown above has four terms. You can combine two of these terms to **simplify** the expression.

Simplify $x^3 - 4y^2 + 2x^3 - y$.

$x^3 - 4y^2 + 2x^3 - y$	
$x^3 + 2x^3 - 4y^2 - y$	Group like terms.
$3x^3 - 4y^2 - y$	Combine x^3 and $2x^3$.

Equivalent expressions, such as $x^3 - 4y^2 + 2x^3 - y$ and $3x^3 - 4y^2 - y$, have the same value for all values of the variables.

Simplify.

1. $7m + 3n - m + 2n$ ____________

2. $15r^2 + 4 - 3r^2 - 2$ ____________

3. $6x + 3z - y$ ____________

Algebraic equations can be solved by first combining like terms.

Solve $4w - w = 24$.

$4w - w = 24$	Identify like terms; w is $1w$.
$3w = 24$	Combine $4w$ and w.
$\frac{3w}{3} = \frac{24}{3}$	Divide both sides by 3.
$w = 8$	

Complete to solve.

4. $5h + 2h = 21$

$\_\_h = 21$

$\frac{\_\_h}{\_\_} = \frac{21}{\_\_}$

$h = \_\_$

5. $16w - 5w = 44$

$\_\_w = 44$

$\frac{\_\_w}{\_\_} = \frac{44}{\_\_}$

$w = \_\_$

6. $48 = 13x - x$

$48 = \_\_x$

$\frac{48}{\_\_} = \frac{\_\_x}{\_\_}$

$\_\_ = x$

Name ______________________________ Date ____________ Class ____________

LESSON 11A

Homework and Practice

Simplifying Algebraic Expressions

Simplify.

1. $12b + 14p - 7$

2. $7a + 11a - a$

3. $15 + 3x^2 - 9 + 8x^2$

4. $6x + y + x + 9y$

5. $8m + 14 + 7m - 11n$

6. $13x^2 - 4 - 12x + 9x$

7. $5(x^2 + 4) - 7$

8. $9(6 + y^3) + 3y^4$

9. $6(7 + 2a) - 8a$

Solve.

10. $5r + 2r = 21$

11. $15t - 9t = 54$

12. $6x + 8x = 42$

13. Janelle has d dimes and n nickels. Her sister has 5 times as many dimes and 6 times as many nickels as Janelle has.

a. Write the sum of the number of coins they have, then simplify.

b. If Janelle has 9 dimes and 18 nickels, how many total coins do Janelle and her sister have?

14. On four quizzes, Peter scored a total of 331 points. On the second, Peter improved his performance by 11 points. On the third he scored 14 points higher than on the first quiz. For the fourth quiz, Peter only increased his score by 6 points from his first quiz. What was the score on each quiz?

EXPLORATION

11B Solving Multi-Step Equations

Program of Studies and Core Content

MA-8-AT-S-EI1, MA-8-AT-S-EI4, MA-8-AT-S-VEO1, MA-8-AT-S-VEO3, MA-08-5.1.2, MA-08-5.2.1, MA-08-5.3.1

At the park, José rents inline skates and Rebecca rents a motor scooter.

Inline skates	Motor scooter
\$3.00, plus \$1.50 per hour	**\$10.00, plus \$2.75 per hour**

Individual costs (c): $c = 3 + 1.50x$ and $c = 10 + 2.75x$

Combined cost (C): $C = 3 + 1.50x + 10 + 2.75x$

Use the equation $C = 3 + 1.50x + 10 + 2.75x$ to answer each question.

1. How much do José and Rebecca pay for 2 hours' rental?
2. How much do José and Rebecca pay for 3 hours' rental?
3. If their total rental cost is \$38.50, for how many hours did they use the equipment?

Think and Discuss

4. **Explain** how to combine like terms in the equation $C = 3 + 1.50x + 10 + 2.75x$.
5. **Discuss** whether it makes sense to combine like terms when each item of sports equipment is rented for a different number of hours.

Name ______________________ Date ___________ Class ___________

LESSON **11B**

Hands-On Lab

Solving Multi-Step Equations

Multi-step equations require you to solve them in several steps.The steps could involve addition, subtraction, multiplication, or division. The steps could also involve setting up the equation. In a multi-step equation, you need to define your variable and the relationship of the variable to the rest of the equation.

Activity

You know the perimeter of a triangle is 74 centimeters. You do not know the lengths of any sides. What you do know is that the second side is 10 centimeters less than the first side and the third side is 4 centimeters longer than half of the first side. How can you find the length of each side?

Step 1 Often it helps to visualize a problem before you try to solve it. To visualize the triangle, draw the first side with any length:

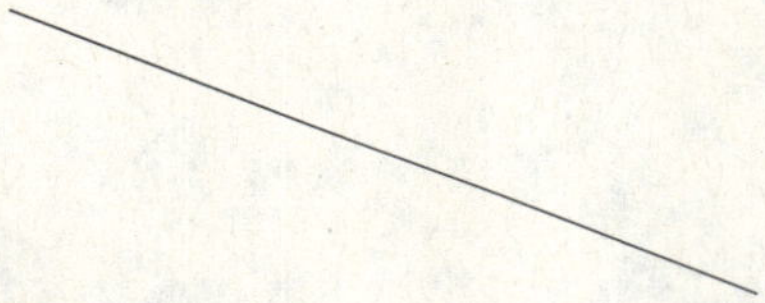

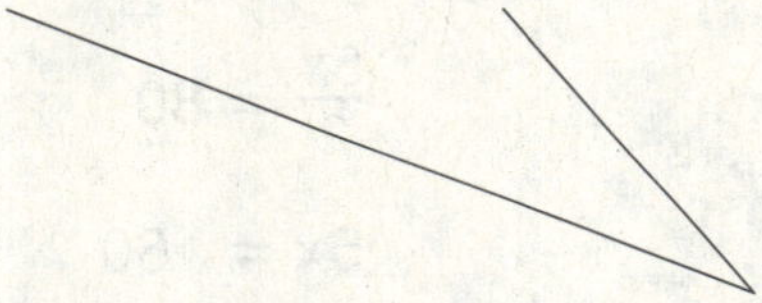

Step 2 You do not know what the lengths of the sides are yet, but you can adjust them later. Draw the second side smaller than the first side. You know the third side is a little more than half the first, so draw it in.

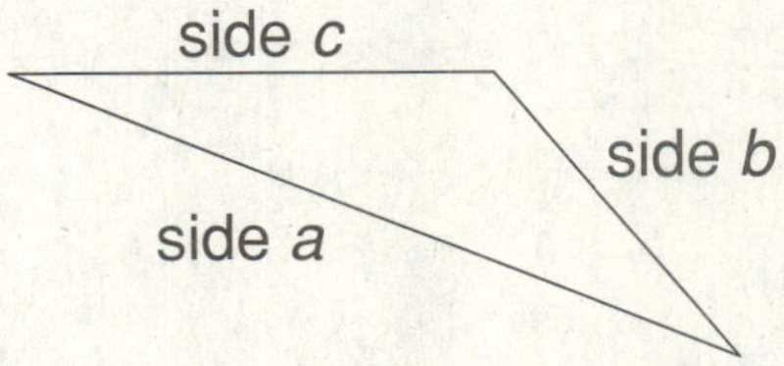

Name ______________________ Date ____________ Class ____________

LESSON **11B**

Hands-On Lab

Solving Multi-Step Equations, continued

Step 3 Now that you can see what the triangle might look like, it is easier to set up the equation:

Perimeter = side a + side b + side c, let side a = the variable x

Side b is 10 centimeters less than side a, so side $b = x - 10$.

Side c is 4 centimeters more than half of side a, so side $c = (x \div 2) + 4$.

To solve the equation, the sum of all three sides of the triangle should equal the perimeter of 74 cm.

$x + (x - 10) + [(x \div 2) + 4] = 74$

$$x + x + \frac{x}{2} - 10 + 4 = 74$$

$$\frac{5x}{2} - 6 = 74$$

$$\frac{5x}{2} = 80$$

$$5x = 160$$

$$x = 32$$

side $b = 32 - 10$, or 22 side $c = (32 \div 2) + 4$, or 20

Check your answer by substituting the side values into the original equation.

Perimeter = side a + side b + side c, or $74 = 32 + 22 + 20$

Name ______________________ Date ____________ Class ____________

LESSON 11B

Hands-On Lab

Solving Multi-Step Equations, continued

Think and Discuss

1. What if you had been given the area of the triangle instead of the perimeter, and the same descriptions of the side lengths? Could you have found the length of each side in the same way as you found the perimeter? Explain.

____

____

____

____

____

____

Try This

1. You know the perimeter of a right triangle is 48 cm. You do not know the lengths of the sides, but you do know that the second side is 4 centimeters shorter than the first side, and you know that the third side is 4 centimeters shorter than 2 times the second side.

Figure out the lengths of the sides and then draw the triangle.

Show your work.

____

Name ____________________ Date ____________ Class ____________

LESSON 11B

Review for Mastery

Solving Multi-Step Equations

To solve an equation, first combine the like terms.

$2m + 3m = 35 - 25$

$5m = 10$ Combine like terms.

$\frac{5m}{5} = \frac{10}{5}$ Divide by 5.

$m = 2$

Check: Substitute into the original.

$2m + 3m = 35 - 25$

$2(2) + 3(2) \stackrel{?}{=} 35 - 25$

$4 + 6 \stackrel{?}{=} 10$

$10 = 10$ ✓

$x + 6 - 3x + 5 = 13$

$-2x + 11 = 13$ Combine like terms.

$\underline{-11 \quad -11}$ Subtract 11.

$-2x = 2$

$\frac{-2x}{-2} = \frac{2}{-2}$ Divide by −2.

$x = -1$

Check: $x + 6 - 3x + 5 = 13$

$-1 + 6 - 3(-1) + 5 \stackrel{?}{=} 13$

$-1 + 6 + 3 + 5 \stackrel{?}{=} 13$

$-1 + 14 \stackrel{?}{=} 13$

$13 = 13$ ✓

Complete to solve and check each equation.

1. $4z - 7z = -20 - 1$

____ z = ____ Combine like terms.

$\frac{z}{\_\_\_\_}$ = $\frac{\_\_\_\_}{\_\_\_\_}$ Divide.

z = ____

Check: $4z - 7z = -20 - 1$

4(____) − 7(____) $\stackrel{?}{=}$ −20 − 1

____ − ____ $\stackrel{?}{=}$ ____

____

2. $t + 1 - 4t + 8 = 21$

____ t + ____ = ____ Combine like terms.

________ ____ Subtract.

____ t = ____

$\frac{t}{\_\_\_\_}$ = $\frac{\_\_\_\_}{\_\_\_\_}$ Divide.

t = ____

Check: $t + 1 - 4t + 8 = 21$

____ + 1 − 4(____) + 8 $\stackrel{?}{=}$ ____

____ + 1 + ____ + 8 $\stackrel{?}{=}$ 21

____ + ____ $\stackrel{?}{=}$ 21

____

Name ______________________ Date __________ Class __________

LESSON 11B

Review for Mastery

Solving Multi-Step Equations, continued

To clear fractions in an equation, multiply every term by the least common denominator (LCD).

$\frac{x}{4} + 3 = \frac{1}{2}$	The LCM of 4 and 2 is 4.	**Check:** $\frac{-10}{4} + 3 \stackrel{?}{=} \frac{1}{2}$
$4 \cdot \frac{x}{4} + 4 \cdot 3 = 4 \cdot \frac{1}{2}$	Multiply *every* term by the LCD.	$\frac{-10}{4} + \frac{12}{4} \stackrel{?}{=} \frac{1}{2}$
$x + 12 = 2$		$\frac{2}{4} \stackrel{?}{=} \frac{1}{2}$
$\underline{-12 \quad -12}$	Subtract 12.	$\frac{1}{2} = \frac{1}{2}$ ✓
$x = -10$		

Complete to solve and check.

3. $\frac{t}{2} + \frac{t}{6} = 2$ — Determine the LCD. — **Check:** $\frac{\_\_}{2} + \frac{\_\_}{6} \stackrel{?}{=} 2$

$\_\_\_ \cdot \frac{t}{2} + \_\_\_ \cdot \frac{t}{6} = \_\_\_ \cdot 2$ — Multiply *every* term by LCD. — $\frac{\_\_}{2} + \frac{\_\_}{2} \stackrel{?}{=} 2$

$\_\_\_t + t = \_\_\_$ — The fractions are cleared. — $\frac{\_\_}{2} \stackrel{?}{=} 2$

$\_\_\_t = \_\_\_$ — Combine like terms. — ______

$\frac{t}{\_\_} = \frac{\_\_}{\_\_}$ — Divide.

$t = \_\_\_$

4. $\frac{m}{3} + \frac{m}{4} - 1 = \frac{5}{2}$ — Determine the LCD.

$\_\_\_ \cdot \frac{m}{3} + \_\_\_ \cdot \frac{m}{4} - \_\_\_ \cdot 1 = \_\_\_ \cdot \frac{5}{2}$ — Multiply *every* term by the LCD.

$\_\_\_\, m + \_\_\_\, m - \_\_\_ = \_\_\_$ — The fractions are cleared.

$\_\_\_\, m - \_\_\_ = \_\_\_$ — Combine like terms.

$\_\_\_ \quad \_\_\_$ — Add.

$\_\_\_\, m = \_\_\_$

$\frac{m}{\_\_} = \frac{\_\_}{\_\_}$ — Divide.

$m = \_\_\_$

Check: $\frac{\_\_}{3} + \frac{\_\_}{4} - 1 \stackrel{?}{=} \frac{5}{2}$

$\frac{\_\_}{12} + \frac{\_\_}{12} - \frac{\_\_}{12} \stackrel{?}{=} \frac{\_\_}{12}$

______

Name ______________________ Date ____________ Class ____________

LESSON **11B**

Homework and Practice

Solving Multi–Step Equations

Solve.

1. $8y - 4 - 7y = 11$

2. $2x + 2 + 5x = 23$

3. $8 = 7a - 8 - 3a$

4. $r + 5r + 5 = -25$

5. $9w - 12w - 2 = 7$

6. $38 = 24 - 7x + 5x$

7. $\frac{2}{7}y - \frac{5}{6} + \frac{5}{7}y = 1\frac{1}{6}$

8. $\frac{5}{8}x + 1 - \frac{7}{8}x = \frac{3}{4}$

9. $6x + 16 - 15x = -11$

10. $\frac{4}{9}w + 6 - \frac{1}{3}w = 4$

11. $0.05d - 0.2d - 17 = 13$

12. $0.6m - 4.4 - 4.8m = 2.4$

13. $8.2s + 4 - 13.7s = -7$

14. $11y - 7.5 - 6y = 2$

15. $13x - 4.6x - 3.9 = 8.7$

16. The measure of an angle is 34° less than its supplement. Find the measure of each angle.

17. The measure of an angle is 9° more than twice its complement. Find the measure of each angle.

18. The perimeter of the triangle is 139. Find the measure of each side.

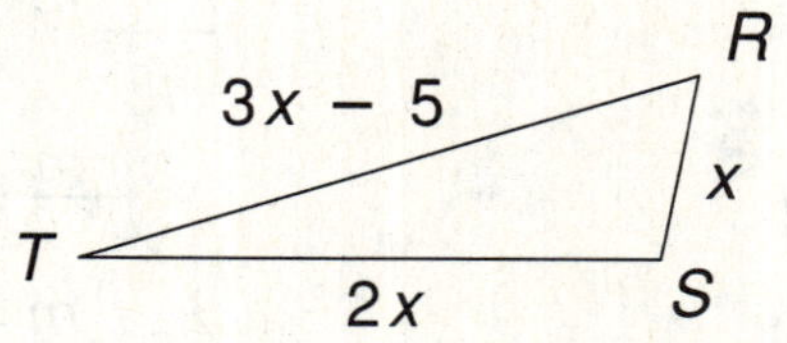

Name ______________________ Date ____________ Class ____________

LESSON 11B

Homework and Practice

Solving Multi-Step Equations, continued

Write the correct answer.

19. Stephen belongs to a movie club in which he pays an annual fee of \$39.95 and then rents DVDs for \$0.99 each. In one year, Stephen spent \$55.79. Write and solve an equation to find how many DVDs *d* he rented.

____

____

20. In 2003, the population of Zimbabwe was about 12.6 million, which was 1 million more than 4 times the population in 1950. Write and solve an equation to find the population *p* of Zimbabwe in 1950.

____

____

21. Maggie's brother is three years younger than twice her age. The sum of their ages is 24. How old is Maggie?

____

22. Kate is saving to take an SAT prep course that costs \$350. So far, she has saved \$180, and she adds \$17 to her savings each week. How many more weeks must she save to be able to afford the course?

____

Use the graph below to answer questions 23–25. Select the best answer. The graph shows the population density (number of people per square mile) of various states given in the 2000 census.

23. One seventeenth of Rhode Island's population density minus 17 equals the population density of Colorado. What is Rhode Island's population density?

A 425 **C** 714

B 697 **D** 1003

24. One more than sixteen times the population density of New Mexico equals the population density of Texas. To the nearest whole number, what is New Mexico's population density?

F 5 **H** 51

G 14 **J** 71

25. Three times the population density of Missouri minus 26 equals the population density of California. What is Missouri's population density?

A 64 **C** 98

B 81 **D** 729

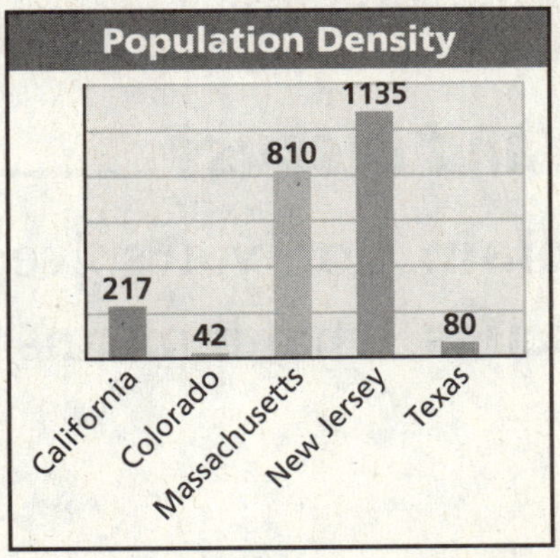

EXPLORATION

11C Solving Equations with Variables on Both Sides

Program of Studies and Core Content

MA-8-AT-S-EI1, MA-8-AT-S-EI4, MA-8-AT-S-VEO1, MA-8-AT-S-VEO3, MA-08-5.1.2, MA-08-5.2.1, MA-08-5.2.2, MA-08-5.3.1

Samara wants to rent a kayak. Two rental places offer the following deals.

Company A
\$8.00, plus \$4.50 per hour

$8 + 4.50h$

Company B
\$10.00, plus \$2.50 per hour

$10 + 2.50h$

Use the expressions above to solve each problem.

1. Find the cost of a 1-hour rental from company A ($h = 1$).
2. Find the cost of a 1-hour rental from company B ($h = 1$).
3. Find the cost of a 2-hour rental from company A ($h = 2$).
4. Find the cost of a 2-hour rental from company B ($h = 2$).
5. Find the number of hours that a kayak would need to be rented from both companies to make the costs equal by solving $8 + 4.50h = 10 + 2.50h$ for h. (*Hint:* Combine like terms first.)

Think and Discuss

6. **Explain** how you solved the equation in Problem 5.
7. **Discuss** what it means to set $8 + 4.50h$ equal to $10 + 2.50h$.

Name ______________________ Date ____________ Class ____________

LESSON **11C**

Technology Lab

Solve Two-Step Equations by Graphing

A graphing calculator is another tool used to solve equations.

Activity

To solve the equation $2x - 3 = 4x + 1$, use the Y= menu. Enter the left side of the equation in **Y1** and the right side in **Y2.**

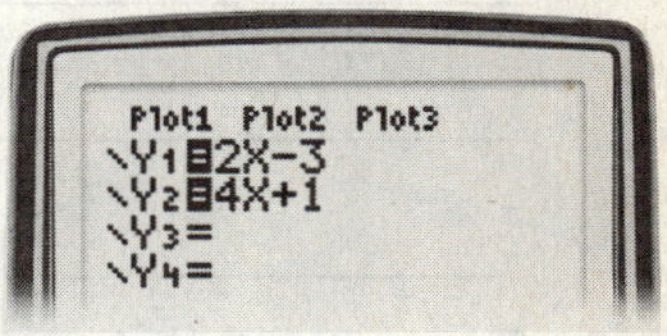

1. Press Y= 2 X,T,θ,n − 3 ENTER 4 X,T,θ,n + 1.

2. To select the standard viewing window and graph the two equations, press ZOOM **6:Zstandard.**

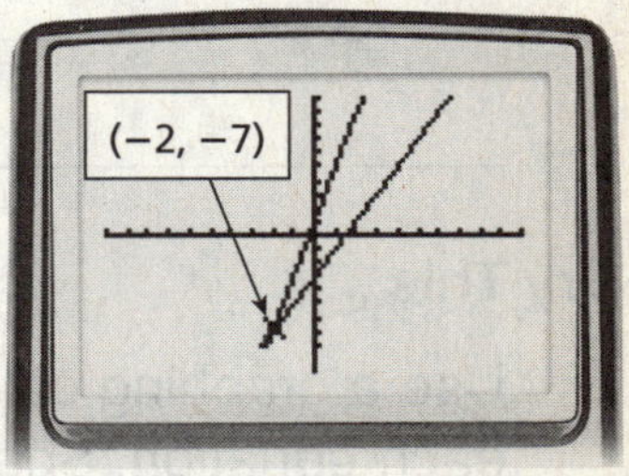

On the figure, the two graphs *appear* to intersect at the point (−2, −7). The coordinates −2 and −7 are *approximations.* The solution of the original equation is the *x*-coordinate of the point of intersection of the two lines. The solution is approximately $x = -2$.

If the equation is solved algebraically by adding $-4x$ and 3 to both sides, the result is $2x - 3 + (-4x) + 3 = 4x + 1 + (-4x) + 3$. This becomes $-2x = 4$, or $x = -2$, thus confirming the estimated graphical solution.

3. Press TRACE and the left arrow key 9 times to get the screen shown. As you press the arrow keys, observe how the coordinates change. This is as close to $x = -2$ as you can get using this window. The *x*-value −1.914...is only an estimate of the exact solution, −2.

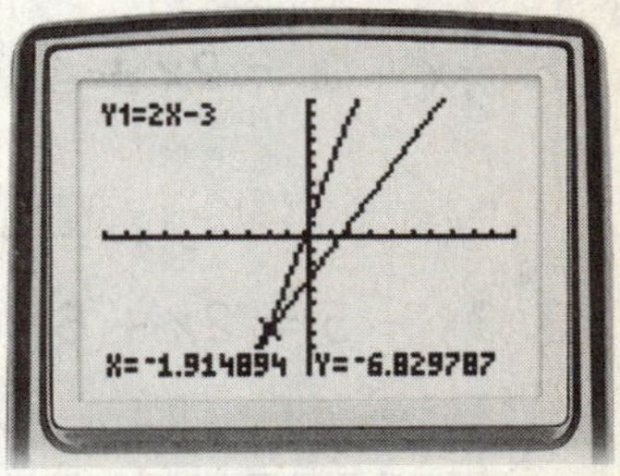

Name ______________________ Date __________ Class __________

LESSON 11C

Technology Lab

Solve Two-Step Equations by Graphing, continued

Think and Discuss

1. Explain why using the TRACE key may show only estimates.

____

____

____

2. Graph both sides of the equation $x + 4 = -2x + 7$ in the standard viewing window. What must be done to solve the equations graphically?

____

____

Try This

Use a graphing calculator to find an approximate solution to each equation. Specify the window used. Confirm your estimate by solving algebraically.

1. $2x + 1 = x - 4$

____

2. $\frac{1}{2}x - 3 = 2x + 4$

____

3. $3x - 5 = 2x + 6$

____

4. $3x + 5 = 4 - 2x$

____

Name ______________________ Date __________ Class __________

LESSON 11C

Review for Mastery

Solving Equations with Variables on Both Sides

You can solve an equation with variables on both sides by combining like terms, then isolating the variable on one side.

$$5x = 2x + 12$$
$$-2x \quad -2x$$
Subtract 2x from both sides.

$$3x = 12$$
$$\frac{3x}{3} = \frac{12}{3}$$
Divide by 3.

$$x = 4$$

Check: Substitute into the original equation.

$$5x = 2x + 12$$
$$5(4) \stackrel{?}{=} 2(4) + 12$$
$$20 \stackrel{?}{=} 8 + 12$$
$$20 = 20 \checkmark$$

$$-6z + 28 = 9z - 2$$
$$+6z \qquad +6z$$
Add 6z to both sides.

$$28 = 15z - 2$$
$$+2 \qquad +2$$
Add 2.

$$30 = 15z$$
$$\frac{30}{15} = \frac{15z}{15}$$
Divide by 15.

$$2 = z$$

Check: $-6z + 28 = 9z - 2$

$$-6(2) + 28 \stackrel{?}{=} 9(2) - 2$$
$$-12 + 28 \stackrel{?}{=} 18 - 2$$
$$16 = 16 \checkmark$$

Complete to solve. Then check each equation.

1. $9m = 4m - 25$

$-$ ______ $-$ ______ Combine the variable terms.

$5m = -25$

$\frac{m}{\quad} = \frac{-25}{\quad}$ Divide.

$m =$ ______

Check: $9m = 4m - 25$

9(______) $\stackrel{?}{=}$ 4(______) $- 25$

______ $\stackrel{?}{=}$ ______ $- 25$

____

2. $3h - 7 = 5h + 1$

$-$ ______ $-$ ______ Combine the variable terms.

$-7 =$ ______ $h + 1$

______ ______ Subtract.

______ $=$ ______ h

$\frac{\quad}{\quad} = \frac{h}{\quad}$ Divide.

______ $= h$

Check: $3h - 7 = 5h + 1$

3(______) $- 7 \stackrel{?}{=}$ 5(______) $+ 1$

______ $- 7 \stackrel{?}{=}$ ______ $+ 1$

____

Name ______________________________ Date ____________ Class ____________

LESSON 11C

Review for Mastery

Solving Equations with Variables on Both Sides, continued

To solve multi-step equations with variables on both sides: 1) *clear* fractions, 2) *combine* like terms, 3) *collect* variable terms on one side, and 4) *isolate* the variable by using properties of equality.

$\frac{t}{3} - \frac{5t}{6} + \frac{1}{2} = t - 1$ — Clear fractions. LCD is 6.

$6 \cdot \frac{t}{3} - 6 \cdot \frac{5t}{6} + 6 \cdot \frac{1}{2} = 6 \cdot t - 6 \cdot 1$ — Multiply *every* term by the LCD.

$2t - 5t + 3 = 6t - 6$

$-3t + 3 = 6t - 6$ — Combine like terms.

$\underline{+3t} \quad \underline{+3t}$ — Add 3*t* to both sides.

$3 = 9t - 6$

$\underline{+6} \quad \underline{+6}$ — Add 6 to both sides.

$9 = 9t$

$\frac{9}{9} = \frac{9t}{9}$ — Divide both sides by 9.

$1 = t$

Complete to solve.

3. $\frac{w}{4} + \frac{w}{2} + \frac{1}{4} = w$ — Clear fractions. LCD is 4.

$\_\_\_\_ \cdot \frac{w}{4} + \_\_\_\_ \cdot \frac{w}{2} + \_\_\_\_ \cdot \frac{1}{4} = \_\_\_\_ \cdot w$ — Multiply *every* term by the LCD.

$\_\_\_\_ w + \_\_\_\_ w + \_\_\_\_ = \_\_\_\_ w$

$\_\_\_\_ w + \_\_\_\_ = \_\_\_\_ w$ — Combine like terms.

$-\_\_\_\_ \quad -\_\_\_\_$ — Subtract.

$\_\_\_\_ = w$

4. $\frac{3x}{5} + 7 = \frac{x}{2} - 1$ — Clear fractions.

$(\_\_\_\_)\frac{3x}{5} + (\_\_\_\_)7 = (\_\_\_\_)\frac{x}{2} - (\_\_\_\_)1$ — Multiply every term by the LCD.

$\_\_\_\_ x + \_\_\_\_ = \_\_\_\_ x - \_\_\_\_$

$-\_\_\_\_ \quad -\_\_\_\_$ — Subtract.

$\_\_\_\_ x + \_\_\_\_ = -\_\_\_\_$ — Combine like terms.

$-\_\_\_\_ \quad -\_\_\_\_$ — Subtract.

$x = \_\_\_\_$

Name ______________________ Date ____________ Class ____________

LESSON **11C**

Homework and Practice

Solving Equations with Variables on Both Sides

Solve.

1. $5x - 1 = 21 + 3x$

2. $9y + 3 = 6y - 15$

3. $7w - 3 = w - 12.6$

4. $12 + 8a = 3a - 3$

5. $\frac{2}{9}n - 5 = 2 + \frac{5}{9}n$

6. $1.8d + 31 = 3.4d + 19$

7. $2b + 30 = 4b + 3$

8. $4x + 2.3 = 2x - 3.7$

9. $3(m - 5) = m - 7$

10. $5(3h - 2) = 2(h + 8)$

11. $5 + \frac{1}{2}y = \frac{3}{8}y + 6$

12. $2\left(x - \frac{1}{2}\right) = 4\left(x + \frac{1}{4}\right)$

13. $4(3r - 2) = 4 - 12r$

14. $\frac{1}{4}x - 7 = \frac{1}{3}x + 8$

15. $\frac{3a - 4}{2} = 5 - 2a$

16. Mandy and Mackenzie spent the same amount of money at the arcade. They both played the same number of games. Mandy played \$0.50 games and spent \$10 on refreshments. Mackenzie played \$0.75 games and spent \$6 on refreshments. How many games did each person play?

17. The square and the equilateral triangle at the right have the same perimeter. Find the lengths of the sides of the square.

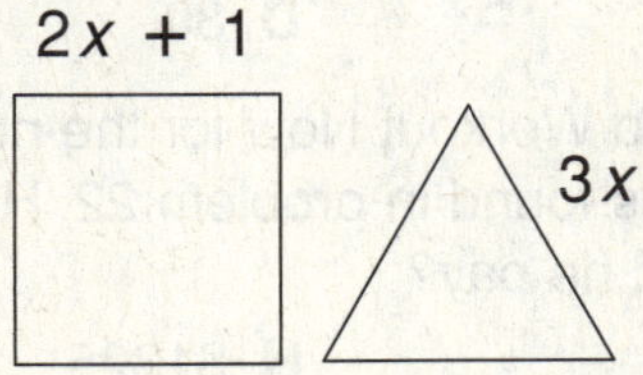

Name ______________________ Date ____________ Class ____________

LESSON **11C**

Homework and Practice

Solving Equations with Variables on Both Sides, continued

Write the correct answer.

18. Claire purchased just enough fencing to border either a rectangular or a triangular garden, as shown, whose perimeters are the same.

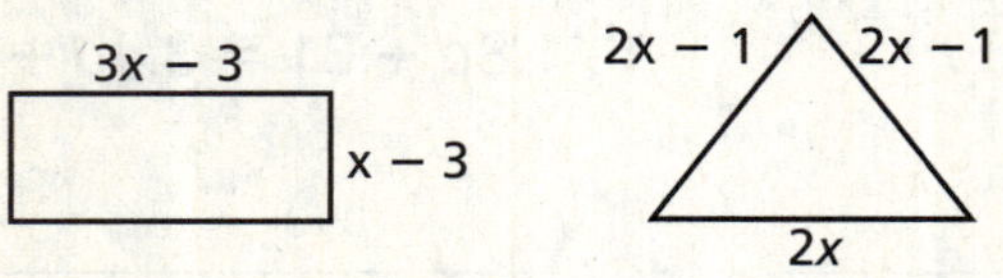

How many feet of fencing did she buy?

____

19. Celia and Ryan are starting a nutrition program. Celia currently consumes 1200 calories a day and will increase that number by 100 calories each day. Ryan currently consumes 3230 calories a day and will decrease that number by 190 each day. They will continue this pattern until they are both consuming the same number of calories per day. In how many days will that be?

____

20. A moving company charges $800 plus $16 per hour. Another moving company charges $720 plus $21 per hour. How long is a job that costs the same no matter which company is used?

____

21. Aaron needs to take out a loan to purchase a motorcycle. At one bank, he would pay $2500 initially and $150 each month for the loan. At another bank, he would pay $3000 initially and $125 each month. After how many months will the two loan payments be the same?

____

Use the table below to answer questions 22–24. Select the best answer. The table shows the membership fees of three different gyms.

Gym	Fees
Workout Now	$200 plus $45 per month
Community Gym	$50 plus $55 per month
Ultra Sports Club	$20 plus $60 per month

22. After how many months will the fees for Workout Now and Community Gym be the same?

A 2.5 **C** 25

B 15 **D** 30

23. Sal joined Workout Now for the number of months found in problem 22. How much did he pay?

F $695 **H** $1325

G $875 **J** $1550

24. After how many months will the fees for Workout Now and Ultra Sports Club be the same?

A 7 **C** 12

B 10 **D** 15

EXPLORATION

Program of Studies and Core Content

MA-8-AT-S-EI3, MA-8-AT-S-EI4, MA-8-AT-S-VEO3, MA-08-5.1.2, MA-08-5.2.2, MA-08-5.3.1

11D Solving Inequalities by Multiplying or Dividing

You can discover an important property of inequalities by looking for patterns.

1. Multiply both sides of each inequality by 2. Write the resulting numbers under the original inequality as shown in **1a.** Then insert the correct inequality symbol, < or >, to make the new inequality true.

a.

3	>	2
6		4

b.

−4	<	−1

c.

2	>	−3

2. Multiply both sides of each inequality by −2. Write the resulting numbers under the original inequality as shown in **2a.** Then insert the correct inequality symbol, < or >, to make the new inequality true.

a.

3	>	2
−6		−4

b.

−4	<	−1

c.

2	>	−3

Think and Discuss

3. Describe any patterns you notice.

4. Explain what happens when you multiply both sides of an inequality by a negative number.

Name ______________________ Date ____________ Class ____________

LESSON **11D**

Hands-On Lab

Solving Inequalities

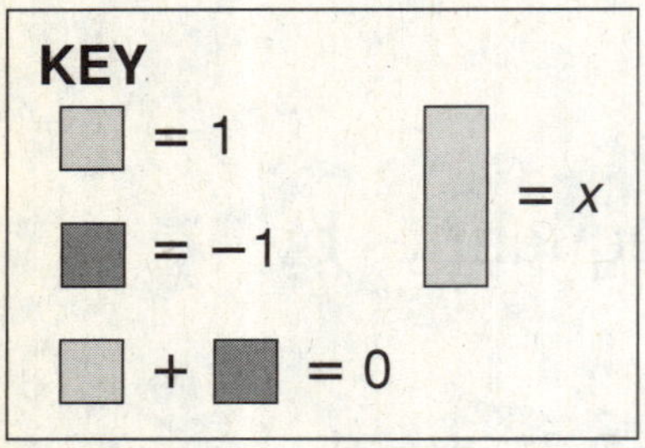

Algebra tiles can help you solve inequalities.

Activity

Solving inequalities is similar to solving equations. You must isolate the variable on one side of the inequality sign. You can add or remove tiles as long as you add the same amount or remove the same amount on both sides.

Solve $x + 4 < 7$.

$\boldsymbol{x + 4}$ $<$ **7** Remove 4 tiles from each side. $\boldsymbol{x}$ $<$ **3**

Use algebra tiles to model and solve each inequality.

1. $x + 1 < 6$ ____________

2. $x + 5 > 3$ ____________

Name ______________________ Date ____________ Class ____________

LESSON **11D**

Hands-On Lab

Solving Inequalities, continued

If the inequality includes negative values, you can add tiles to each side.
Solve $x + (-4) < 5$.

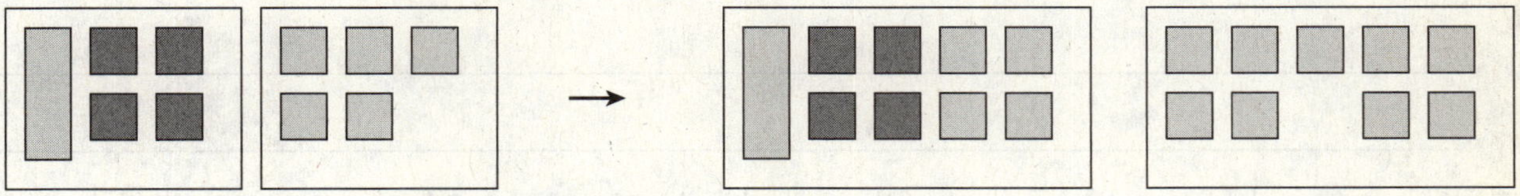

$\boldsymbol{x + (-4)}$ **<** **5** Add 4 tiles to each side. $\boldsymbol{x + (-4) + 4}$ **<** **5 + 4**

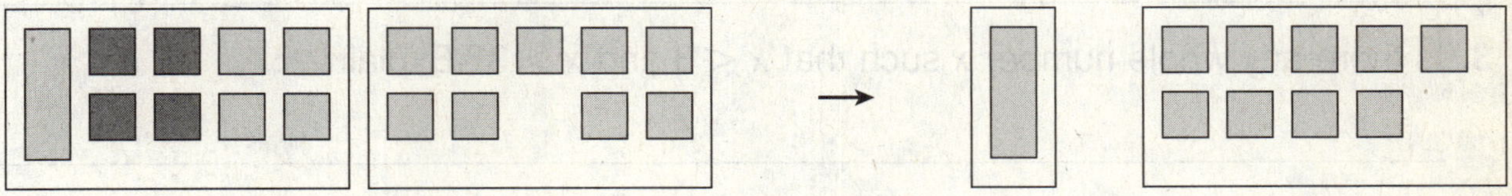

$\boldsymbol{x + (-4) + 4}$ **<** **5 + 4** Combine terms $\boldsymbol{x}$ **<** **9**

Use algebra tiles to model and solve each inequality.

3. $x + (-9) < 3$ ____________

4. $x + (-3) > -1$ ____________

Both sides of an inequality can be multiplied or divided by the same number.
Solve $3x < 9$.

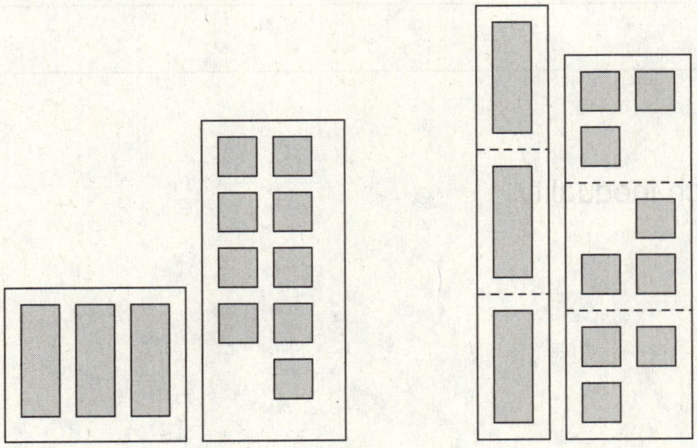

$\boldsymbol{3x}$ **< 9** Regroup $\boldsymbol{x}$ **< 3**

Use algebra tiles to model and solve each inequality.

5. $2x < 2$ ____________

6. $3x > -9$ ____________

Name ______________________ Date ____________ Class ____________

LESSON 11D

Hands-On Lab

Solving Inequalities, continued

Think and Discuss

1. How are inequalities different from equations?

____

____

2. If $a < b$ and $b < c$, then what is the relation between a and c?

____

____

3. Is there any whole number x such that $x < 3$ and $x > 3$? Explain.

____

____

4. What is the first step in solving an inequality such as $3x - 9 > 6$?

____

____

5. Describe the steps required to get x alone on one side of the inequality $4x + 5 < 25$.

____

____

Try This

Use algebra tiles to model and solve each inequality.

1. $x - 4 > 6$ __________

2. $x - 1 < 3$ __________

3. $x + 3 > 6$ __________

Name ______________________ Date __________ Class __________

LESSON 11D

Hands-On Lab

Solving Inequalities, continued

4. $x + (-4) > 1$ __________

5. $x + (-8) > 1$ __________

6. $x - 6 < 2$ __________

7. $2x > 12$ __________

8. $4x < 8$ __________

Name ______________________ Date __________ Class __________

LESSON **11D**

Review for Mastery

Solving Inequalities by Multiplying or Dividing

To solve an inequality, multiply and divide the same way you would solve an equation. The one exception is, if you multiply or divide by a negative number, you must reverse the inequality sign.

Divide by a Positive Number

$2x < 14$

$\frac{2x}{2} < \frac{14}{2}$

$x < 7$

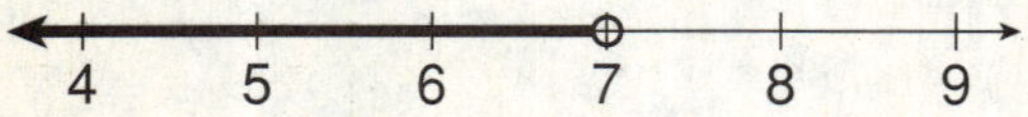

Check
According to the graph, 5 should be a solution, but 10 should not be.

$2x < 14$	$2x < 14$
$2(5) \overset{?}{<} 14$	$2(10) \overset{?}{<} 14$
$10 < 14$ ✓	$20 < 14$ ✗

Divide by a Negative Number

$-2x < 14$

$\frac{-2x}{-2} > \frac{14}{-2}$ Reverse the inequality sign.

$x > -7$

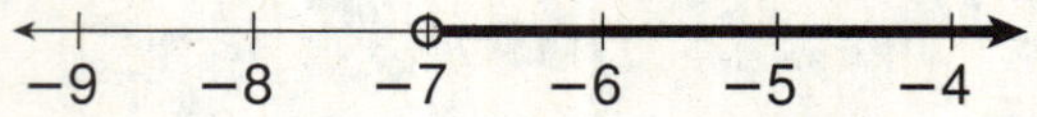

Check
According to the graph, −6 should be a solution, but −8 should not be.

$-2x < 14$	$-2x < 14$
$-2(-6) \overset{?}{<} 14$	$-2(-8) \overset{?}{<} 14$
$12 < 14$ ✓	$16 < 14$ ✗

Complete to solve. Then graph the equation and check.

1. $-3y \geq 24$

$\frac{-3y}{-3}$ ☐ $\frac{24}{\_\_\_}$

y ☐ ____

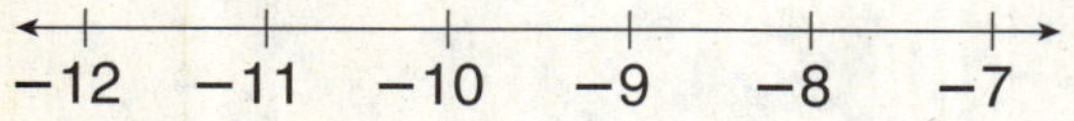

____

____

____

____

2. $\frac{s}{-9} < 4$

____ $\cdot \frac{s}{-9}$ ☐ ____ $\cdot$ 4

s ☐ ____

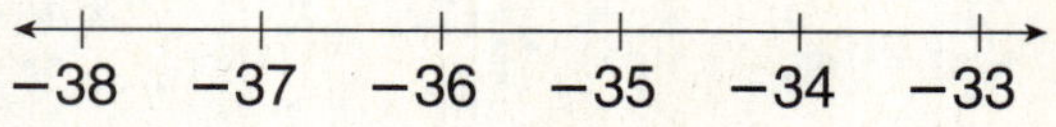

____

____

____

____

Name ______________________ Date __________ Class __________

LESSON 11D

Homework and Practice

Solving Inequalities by Multiplying or Dividing

Solve and graph.

1. $-9c \leq 90$

2. $\frac{m}{-8} < -4$

3. $\frac{z}{4} \leq -7$

4. $-15f \geq 75$

5. $\frac{n}{-6} > -3$

6. $15r \geq -45$

7. $\frac{h}{-1} < 6$

8. $\frac{b}{9} \geq \frac{5}{3}$

9. $-\frac{3}{4}s > 6$

10. It costs Elizabeth $220 to make the shirts that she sells in her store. How many shirts must she sell at $15 each to make a profit?

11. The distance from Martin's house to the sub shop is no more than 5 times the distance from his house to the mall. If Martin's house is 9 miles from the sub shop, what is the least distance his house could be from the mall?

12. Melissa spends no less than $\frac{1}{4}$ the amount of time on her math homework per night that Jane does. If Jane spends $1\frac{1}{2}$ hours per night on her math homework, how much time does Melissa spend working on her math homework per night?

Name ______________________________ Date ____________ Class ____________

LESSON 11D

Homework and Practice

Solving Inequalities by Multiplying or Dividing, continued

Write and solve an inequality for each situation.

13. Karin had \$3 to spend in the arcade. The game she likes costs 50¢ per play. What are the possible numbers of times that she can play?

____

____

14. Tyrone has \$21 and wants to buy juice drinks for his soccer team. There are 15 players on his team. How much can each drink cost so that Tyrone can buy one drink for each person?

____

____

15. A swimming pool is 7 feet deep and is being filled at the rate of 2.5 feet per hour. How long can the pool be left unattended without the water overflowing?

____

____

16. Megan is making quilts that require 11 feet of cloth each. She has 50 feet of cloth. What are the possible numbers of quilts that she can make?

____

____

Alyssa, Reggie, and Cassie are meeting some friends at the movies and have stopped at the refreshment stand. The table below shows some of the items for sale and their prices. Use this information to answer questions 17–19.

Menu Item	Price (\$)
Popcorn	3.50
Drink	3.00
Hot Dog	2.50
Nachos	2.50
Fruit Snack	2.00

17. Alyssa has \$7 and would like to buy fruit snacks for as many of her friends as possible. Which inequality below can be solved to find the number of fruit snacks f she can by?

A $2f \leq 7$ **C** $7f \leq 2$

B $2f < 7$ **D** $7f < 2$

18. Reggie brought \$13 and is going to buy popcorn for the group. Which answer below shows the possible numbers of popcorns p Reggie can buy for his friends?

F 0, 1, or 2 **H** 0, 1, 2, 3, or 4

G 0, 1, 2, or 3 **J** 0, 1, 2, 3, 4, or 5

19. The movie theater donates 12% of its sales to charity. From Cassie's purchases, the theater will donate at least \$2.15. Which inequality below shows the amount of money m that Cassie spent at the refreshment stand?

A $m \geq 17.92$ **C** $m \geq 25.80$

B $m \leq 17.92$ **D** $m \leq 25.80$

EXPLORATION

11E Solving Two-Step Inequalities

Program of Studies and Core Content

MA-8-AT-S-EI3, MA-8-AT-S-EI4, MA-8-AT-S-VEO3, MA-08-5.1.2, MA-08-5.2.2, MA-08-5.3.1

Rosa is offered two telephone service options when she buys her new cell phone.

Cell phone option A

\$25, plus 12.5¢ per minute

Cell phone option B

\$90, plus unlimited minutes

1. Complete the table to compare the costs under each option for the given number of minutes.

Minutes	Cost Under Option A	Cost Under Option B
220	$25 + 0.125 \cdot 220 = \$52.50$	\$90
320		
420		
520		
620		

2. Solve the inequality $25 + 0.125x < 90$. What does the solution tell you about option A and option B?

3. Solve the inequality $25 + 0.125x > 90$. What does the solution tell you about option A and option B?

Think and Discuss

4. **Explain** how you solved the inequalities in Problems 2 and 3.

Name ________________________________ Date __________ Class __________

LESSON **11E**

Review for Mastery

Solving Two-Step Inequalities

Solve an inequality the same way you would an equation. But, when multiplying or dividing by a negative number, reverse the inequality symbol.

$3x + 2 > 11$		
$\underline{\quad -2}$	$\underline{-2}$	Subtract 2 from both sides.
$3x$	> 9	
$\frac{3x}{3}$	$> \frac{9}{3}$	Divide both sides by 3.
x	> 3	

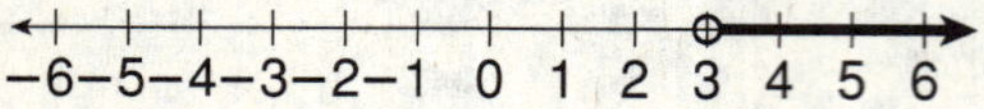

$-3x + 2 > 11$		
$\underline{\quad -2}$	$\underline{-2}$	Subtract 2 from both sides.
$-3x$	> 9	
$\frac{-3x}{-3}$	$> \frac{9}{-3}$	Divide both sides by −3.
x	< -3	Change > to <.

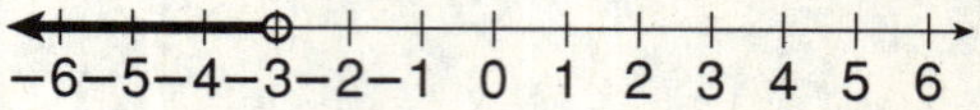

Complete to solve and graph.

1. $2t + 1 \le 9$

______ ___

$2t \quad \le$ ____

$\frac{2t}{\square} \quad \square \quad \frac{\square}{\square}$

$t \quad \square$ ____

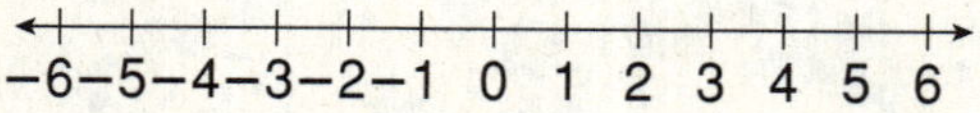

2. $-2t + 1 \le 9$

______ ___

$-2t \quad \le 8$

$\frac{-2t}{\square} \quad \square \quad \frac{\square}{\square}$

$t \quad \square$ ____

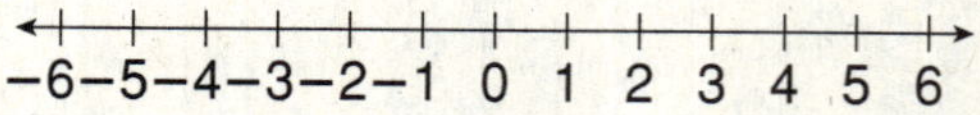

3. $-3z - 2 \;>\; 1$

______ ______

$-3z \quad > \quad \square$

$\frac{-3z}{\square} \quad \square \quad \frac{\square}{\square}$

$z \quad \square$ ____

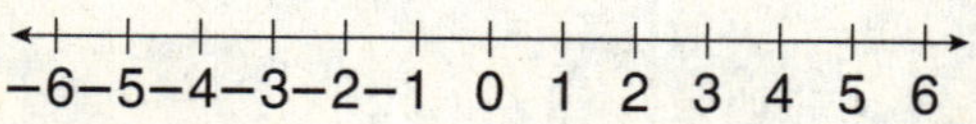

4. $3z - 2 \;>\; 1$

______ ______

$3z \quad > \quad \square$

$\frac{3z}{\square} \quad \square \quad \frac{\square}{\square}$

$z \quad \square$ ____

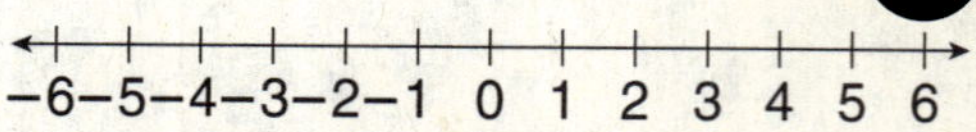

Name ______________________ Date ____________ Class ____________

LESSON **11E**

Review for Mastery

Solving Two-Step Inequalities, continued

To solve multistep inequalities, you may need to first clear fractions by multiplying both sides by the LCD.

$\frac{v}{4} + \frac{1}{4} > -\frac{1}{2}$ The LCD is 4.

$4 \cdot \frac{v}{4} + 4 \cdot \frac{1}{4} > 4 \cdot -\frac{1}{2}$ Multiply by the LCD.

$v + 1 > -2$

$\underline{-1} \quad \underline{-1}$ Subtract from both sides.

$v \quad > -3$

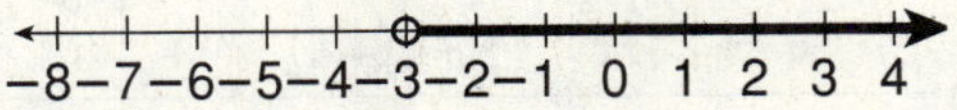

Complete to solve and graph.

5. $-\frac{b}{4} - \frac{7}{12} \le \frac{2}{3}$ Find the LCD.

_____ $\cdot -\frac{b}{4} -$ _____ $\cdot \frac{7}{12} \le$ _____ $\cdot \frac{2}{3}$ Multiply by the LCD.

_____ $b -$ _____ $\le$ _____

_____ _____ Add.

_____ $b \le$ _____

$\frac{b}{\_\_\_}$ ☐ $\frac{\_\_\_}{\_\_\_}$ Divide and change symbol.

b ☐ _____ Check direction.

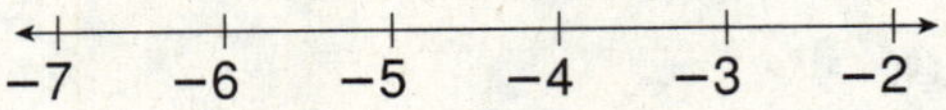

6. $\frac{2}{7} + \frac{y}{14} \ge -\frac{1}{2}$

_____ $\cdot \frac{2}{7} +$ _____ $\cdot \frac{y}{14} \ge$ _____ $\cdot -\frac{1}{2}$

_____ $+$ _____ $\ge$ _____

_____ _____

_____ $\ge$ _____

7. $-\frac{1}{3} > \frac{x}{9} + \frac{2}{3}$

_____ $\cdot -\frac{1}{3} >$ _____ $\cdot \frac{x}{9} +$ _____ $\cdot \frac{2}{3}$

_____ $> x +$ _____

_____ _____

_____ $> x$

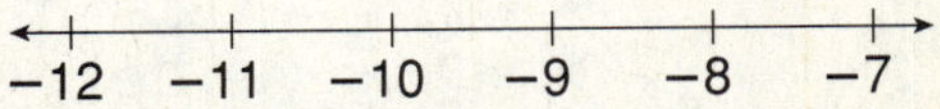

Name ______________________ Date ____________ Class ____________

LESSON 11E

Homework and Practice

Solving Two-Step Inequalities

Solve and graph.

1. $5x + 3 \geq 28$

____

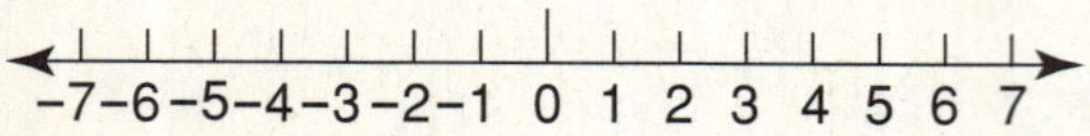

2. $10 - \frac{1}{2}w > 12$

____

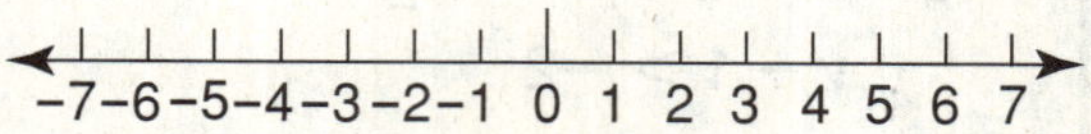

3. $7y - 15 - 4y \leq -18$

____

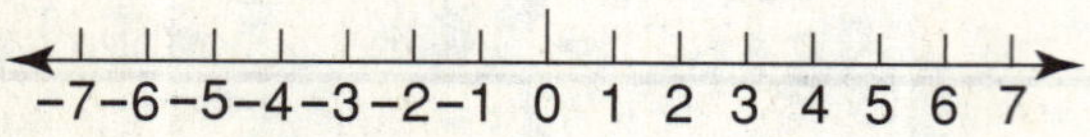

4. $7a < 12a + 10$

____

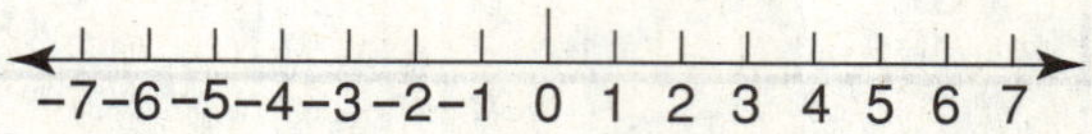

5. $19 - 4m > 4m + 11$

____

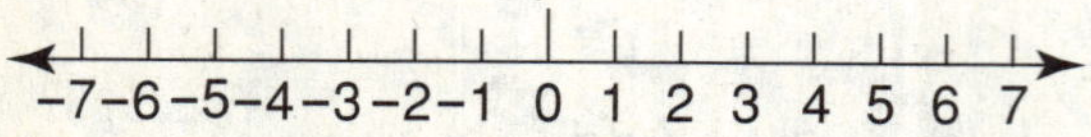

6. $-\frac{5}{6}d + 8 > 13$

____

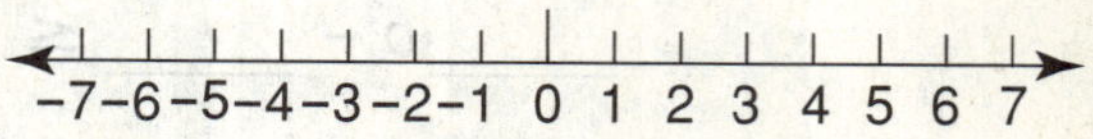

7. $2(3x + 1) \leq 4(2x - 3)$

____

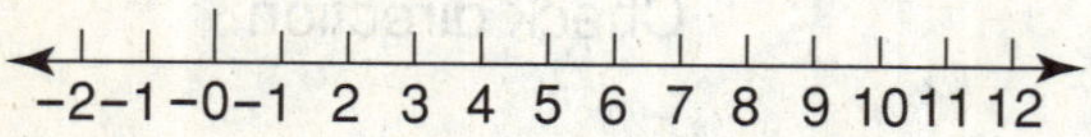

8. $14.1 - \frac{2}{5}h \geq 13.9 - \frac{3}{10}h$

____

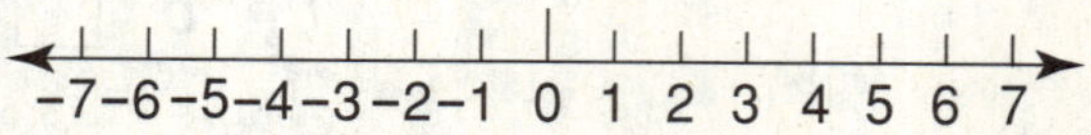

9. Six more than three-fourths a number is greater than or equal to one more than twice the number. Find the number.

____

10. Ms. Garcia wants to carpet her bedroom which requires 24 square yards. Her budget will allow her to spend no more than $600 for the project. If the installation of the carpet will cost $162, what is the most she can pay per square yard for the carpeting?

____

Name ______________________ Date __________ Class __________

LESSON 11E

Homework and Practice

Solving Two-Step Inequalities, continued

Write and solve an inequality for each situation.

11. Jillene is playing in a basketball tournament and scored 24 points in her first game. If she averages over 20 points for both games, she will receive a trophy. How many points can Jillene score in the second game and receive a trophy?

____

____

12. Marcus has accepted a job selling cell phones. He will be paid $1500 plus 15% of his sales each month. He needs to earn at least $2430 to pay his bills. For what amount of sales will Marcus be able to pay his bills?

____

____

13. A 15-foot-tall cedar tree is growing at a rate of 2 feet per year beneath power lines that are 58 feet above the ground. The poser company will have to prune or remove the tree before it reaches the lines. How many years can the power company wait before taking action?

____

____

14. Binh brought $23 with her to the country fair. She purchased a $5 T-shirt and now wants to buy some locally grown plants for $2.50 each. What number of plants can she purchase with her remaining money?

____

____

Benedict, Ricardo, and Charlie are considering opportunities for summer work. The table below shows the jobs open to them and the pay for each. Use this information to answer questions 15–17.

Job	Pay
Mowing Lawns	$15 per lawn
Baby-Sitting	$5.50 per hour
Tutoring	$9 per session

15. Benedict has saved $91 from last year and would like to baby-sit to earn enough to buy a mountain bike. A good quality bike costs at least $300. What numbers of hours h can Benedict baby-sit to reach his goal?

A $h \geq 14$ **C** $h \geq 38$

B $h \geq 23$ **D** $h \geq 71$

16. Ricardo has agreed to tutor for the school. He owes his older brother $59 and would like to end the summer with at least $400 in savings. How many sessions s can Ricardo tutor to meet his goal?

F $s \geq 31$ **H** $s \geq 51$

G $s \geq 38$ **J** $s \geq 83$

17. Charlie has agreed to mow his neighbor's lawn each week and will also baby-sit for some hours. If he makes $100 or more each week, his parents will charge him rent. How many hours h should Charlie agree to baby-sit each week to avoid paying rent?

A $h \leq 15$ **C** $h \leq 21$

B $h \geq 15$ **D** $h \geq 21$

CHAPTER 11

Chapter Summary

Answer these questions to summarize the important concepts from Chapter 9 in your own words.

1. Explain how to simplify the expression $4(7x - 3) + 2x$.

2. Explain the first step in solving the equation $\frac{x}{2} + 3 = \frac{2}{3}$.

3. Explain how to solve the equation $4b + 8 - 6b = -18 + 2b + 10$.

4. Explain why the inequality sign reverses when solving the inequality $12 \leq \frac{c}{-5}$.

5. Explain what an open circle and a closed circle means when graphing inequalities on a number line.

For more review of Chapter 11:

- Complete the Chapter 11 Study Guide and Review on pages 616–618 of your textbook.
- Complete the Ready to Go On quizzes on pages 598 and 612 of your textbook.

Name ______________________ Date ______________ Class ______________

CHAPTER 11

KCCT Prep

Multiple Choice

1. Which equation has the same solution as $3m + 4 - 5m = 12 + m - 2$?

Ⓐ $4 - 2m = 10 + m$

Ⓑ $2m + 4 = 10 + m$

Ⓒ $4 - 2m = 12m - 2$

Ⓓ $2m + 4 = 12m - 2$

2. Laurence cuts a circular hole in a rectangular piece of wood, as shown in the figure. Which of these is closest to the area of the remaining piece of wood?

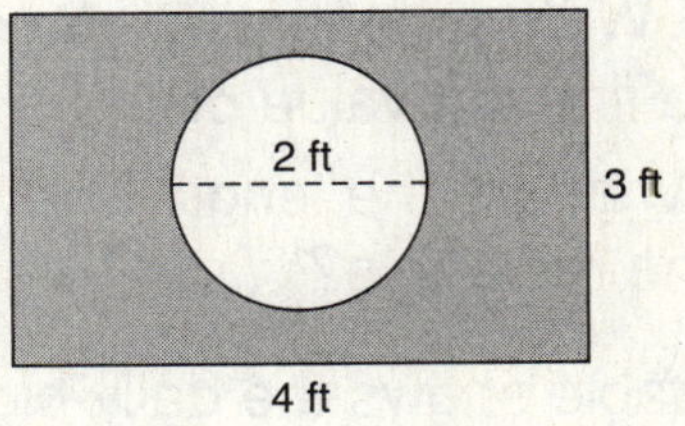

Ⓐ $3.1\ \text{ft}^2$ Ⓒ $6.2\ \text{ft}^2$

Ⓑ $5.7\ \text{ft}^2$ Ⓓ $8.9\ \text{ft}^2$

3. Chandelle buys 5 pairs of jeans that all have the same price. The equation $5(x - y) = 105$ represents the total cost of the jeans, where x is the regular price of one pair and y is the discount on each pair. If Chandelle gets a \$2 discount on each pair, what is the regular price of one pair of jeans?

Ⓐ \$19 Ⓒ \$23

Ⓑ \$21 Ⓓ \$25

4. Keith drove 512 mi over two days. The first day, he drove for 6 hours. The second day, he drove 188 miles. What was Keith's average speed the first day?

Ⓐ 45 mi/h Ⓒ 54 mi/h

Ⓑ 52 mi/h Ⓓ 65 mi/h

5. Which expression is not equivalent to the other three?

Ⓐ $5(4y - 3) - 2y$

Ⓑ $-2y + 20y - 15$

Ⓒ $20y - 3 - 2y$

Ⓓ $18y - 15$

6. A large ice bucket contains 16 apple drinks and 24 orange drinks. Michael chooses a drink at random. What is the probability that he chooses an orange drink?

Ⓐ 0.24 Ⓒ 0.6

Ⓑ 0.4 Ⓓ 0.75

7. The line plot shows the ages of 10 of the 11 children in the waiting room at a clinic. The median age of all the children is 5. Which of these could be the age of the 11th child?

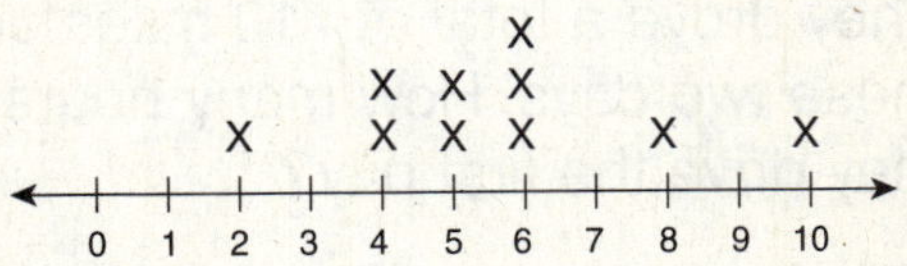

Ⓐ 5 Ⓒ 7

Ⓑ 6 Ⓓ 10

8. There is a one-time fee of \$29.50 to join Gym A and a fee of \$18 per month. At Gym B, the one-time fee is \$38.50. A 6-month membership at the two gyms costs the same. What is the monthly fee at Gym B?

Ⓐ \$14.25 Ⓒ \$20.75

Ⓑ \$16.50 Ⓓ \$29.33

9. Tuan has six state quarters in his pocket. Three are South Carolina quarters, two are Kentucky quarters, and one is an Ohio quarter. He chooses a quarter without looking. What is the probability that he does NOT choose a Kentucky quarter?

Ⓐ $\frac{1}{6}$ Ⓒ $\frac{2}{3}$

Ⓑ $\frac{1}{3}$ Ⓓ $\frac{5}{6}$

10. Which equation has a solution that is different from the solutions of the other three equations?

Ⓐ $3p - 2p - 1 = 3$

Ⓑ $-1 = 3 + 5p$

Ⓒ $\frac{p}{4} - \frac{p}{6} - \frac{1}{12} = \frac{1}{4}$

Ⓓ $3p = 4 + 2p$

11. On the first day of their road trip, the Ramirez family drove at an average speed of 45 mi/h. On the second day, they drove two hours more than the first day and they averaged 55 mi/h. They drove a total of 410 miles on these two days. How many hours did they drive the first day?

Ⓐ 3 hours Ⓒ 5 hours

Ⓑ 4.5 hours Ⓓ 5.2 hours

12. $\angle P$ is supplementary to $\angle Q$. $\angle P$ and $\angle R$ are vertical angles. Given that $m\angle Q = 128°$, what is $m\angle R$?

Ⓐ 38° Ⓒ 128°

Ⓑ 52° Ⓓ 142°

13. Which equation illustrates the Associative Property?

Ⓐ $8(k - 3) = 8k - 24$

Ⓑ $5a + 7b = 7b + 5a$

Ⓒ $-6x - 14y = -2(3x + 7y)$

Ⓓ $(2p + 9m) + 5 = 2p + (9m + 5)$

Open Response

14. Kate has a set of small photos that she is arranging on her bedroom wall. If she arranges them in 12 rows, she has 1 photo left over. If she arranges them in 10 rows of the same length, she has 11 photos left over.

a. Write an equation to determine the number of photos in a row.

b. How many photos are in each row?

c. What would happen if Kate arranged the photos in 9 rows of the same length? Explain.

15. The width of a rectangle is 3 inches less than its length. The perimeter of the rectangle is 54 inches.

a. Let the length of the rectangle be ℓ. Write an equation you can use to find the value of ℓ.

b. What are the length and width of the rectangle?

16. The table shows the daily high temperature and hot cider sales.

Hot Cider Sales			
High Temp (°F)	Number Sold	High Temp (°F)	Number Sold
10	40	40	30
20	30	45	20
30	22	55	15
30	27	60	25
30	35	70	15

a. Make a scatter plot of the data.

b. Describe the correlation, if any, in the scatter plot.

c. The owner of the cafe wants to predict the number of hot ciders that will be sold on a day when the high temperature is 85°. Explain how she can do this.

CHAPTER 12

Project
Graphing Lines: Life Lines

Career: Wildlife Ecologist

Whatever happened to the California parakeet and the passenger pigeon, two species of birds that once inhabited the United States? They are now as extinct as *Tyrannosaurus rex*. The primary focus of wildlife ecologists is to keep other animals from becoming extinct. They have been successful with the whooping crane, the largest wild bird in North America.

Before you start, work through the Whooping It Up worksheet on the next page.

In this chapter project . . .

You collect data on the population of a local animal.

What to do:

- Research local animals in your area. You may choose an endangered species or a more common animal.
- Keep track of population numbers of the animals over several months. You can do this through observation or by getting information from an official source, like the World Wildlife Fund or the U.S. Department of Fish and Wildlife.
- Graph the data you collected and try to draw a line to fit the data. If the data doesn't appear to be linear, break the time period into several segments.
- Make conclusions about whether the population is increasing, decreasing, or staying the same.

What to present:

- Prepare a poster including the graph of your data, your conclusions, facts about the animal. If you like, decorate the poster with pictures of the animal.

Where's the math?

- Graph population data in a line graph.
- Use a line of best fit to describe the data.

go.hrw.com
Chapter Project Online
KEYWORD: MT7 Ch12

Name ______________________ Date ____________ Class ____________

CHAPTER 12

Project Recording Sheet

Graphing Lines

Whooping it Up

Whooping Cranes reach almost five feet tall when standing and weigh about 14–16 pounds. The adult birds are snow white with black-tipped eight-foot wingspans. Each family group produces one or two offspring. The migration route from Wood Buffalo National Park in northern Alberta province to the Aransas National Wildlife Refuge in Texas is approximately 2400 miles long and only about 140 miles wide. The population of Whooping Cranes was likely never very large. The original population even before the European settlement of North America is estimated to be only 500 to 1400.

Year	1940	1945	1950	1955	1960	1965	1970	1975	1980	1985	1990	1995	2000
Cranes	15	17	33	28	36	44	57	49	79	82	142	129	202

1. Plot a graph of the wild population from 1940 to 2000.

2. Draw straight lines that best conform to the data. What are the slopes of the lines? Write an equation for each line.

____

____

3. What are your feelings about endangered species such as the Whooping Crane? Species of animals and plants have been disappearing from the Earth for millions of years. Should we try and protect all of them? Why or why not?

____

____

____

____

____

____

EXPLORATION

12A Graphing Linear Equations

Program of Studies and Core Content

MA-8-AT-S-PRF2, MA-8-AT-S-PRF3, MA-8-AT-S-PRF4, MA-8-AT-S-PRF5, MA-8-G-S-CG1, MA-08-3.3.1, MA-08-5.1.2, MA-08-5.1.5

The calculator screens show the graphs of $y = x + 1$ and $y = x + 3$.

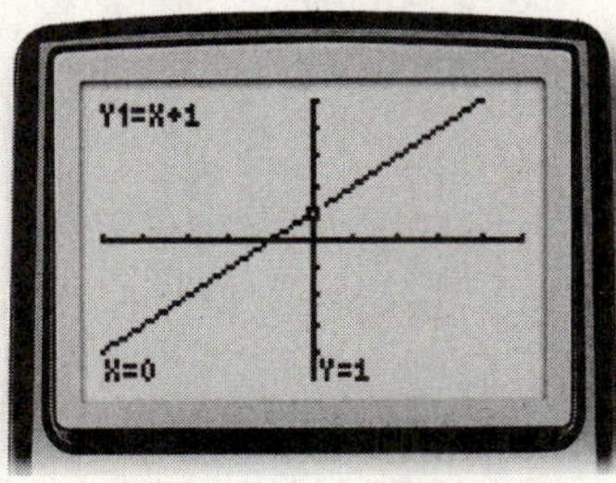

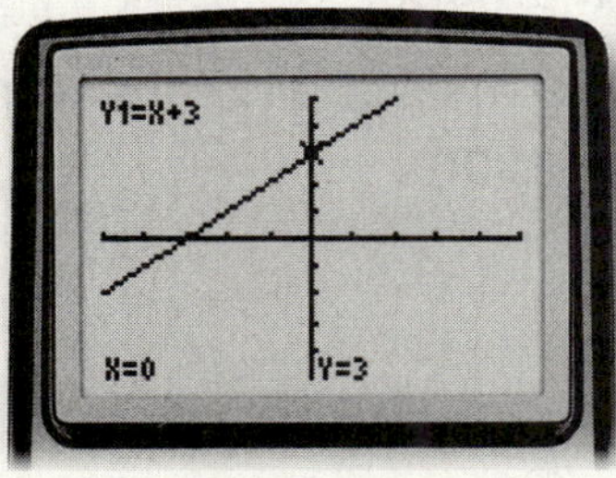

1. Why is the graph of $y = x + 3$ higher along the y-axis than the graph of $y = x + 1$?

The calculator screens show the graphs of $y = x + 2$ and $y = 2x + 2$.

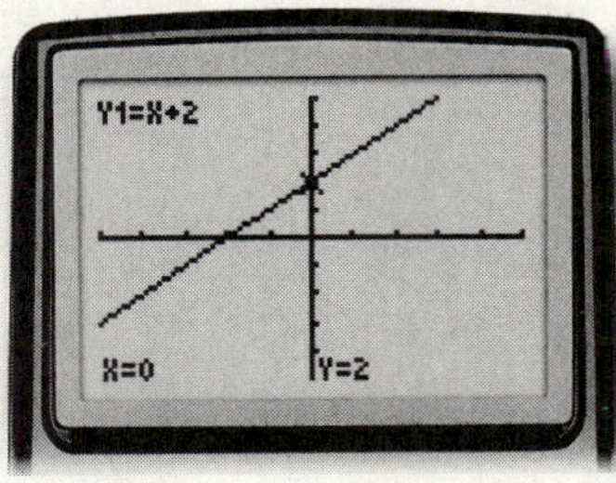

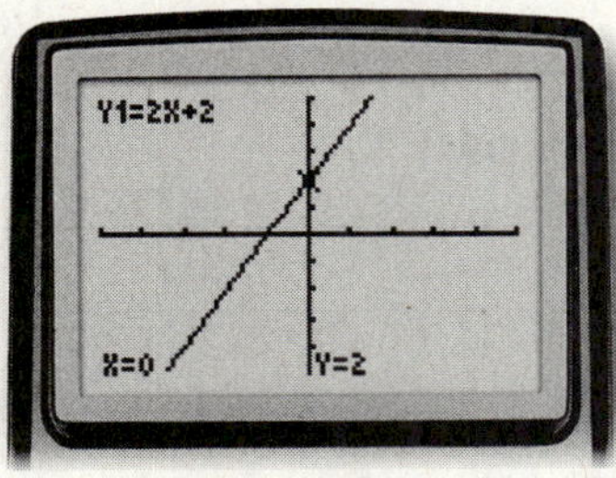

2. Why is the graph of $y = x + 2$ flatter than the graph of $y = 2x + 2$?

Think and Discuss

3. Discuss how the two graphs for Problem 1 are similar and different.

4. Discuss how the two graphs for Problem 2 are similar and different.

Name ______________________ Date __________ Class __________

LESSON 12A

Hands-On Lab

Graphing Linear Equations

If the graph of an equation is a straight line, the equation is known as a linear equation. A linear equation has no operations other than addition, subtraction, and multiplication of a variable by a constant. Also, a variable cannot be multiplied by another variable or appear in a denominator, and a linear equation cannot contain variables with exponents other than one.

The equation $y = 3x + 7$ meets all of these requirements. This equation is a linear equation.

Activity

1. Complete the following table.

x	1	2	3	4	5	6	8	10
$y = 3x + 7$	10							

2. Predict what the graph of the data you have generated would look like.

____

3. Graph the data from part 1.

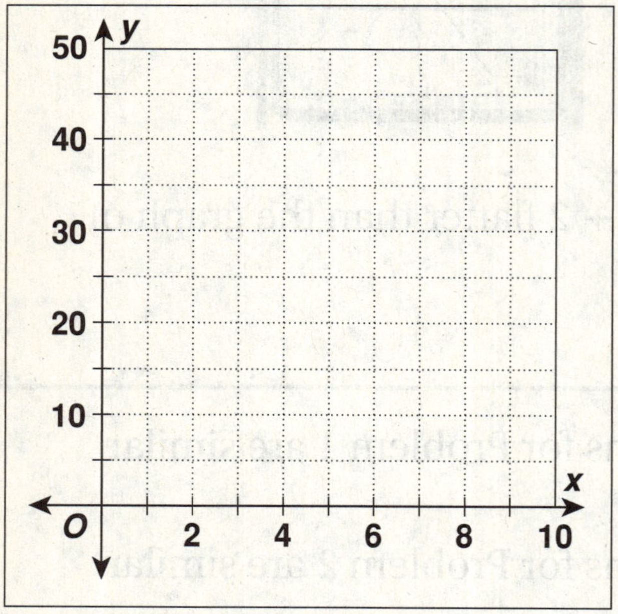

Name ______________________ Date ____________ Class ____________

LESSON 12A

Hands-On Lab

Graphing Linear Equations, continued

4. Complete the following table and draw a graph as in part 1.

x	1	2	3	4	5
$y = x^2$					

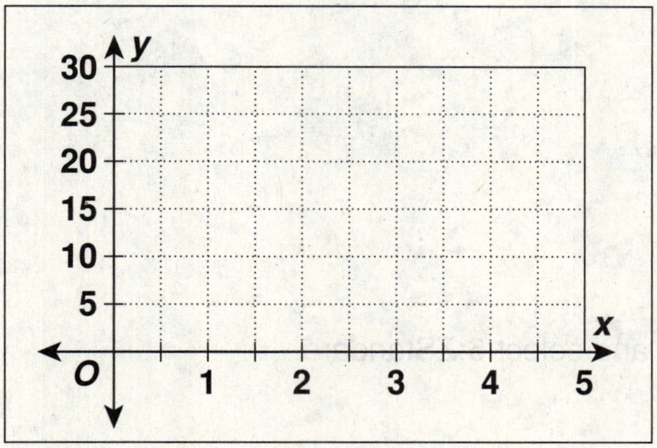

Explain why linear equations are called linear.

____

Think and Discuss

1. Which of the following are linear equations?

a. $y = 3x$ **b.** $x + 6 = y - 3$ **c.** $y = x^2$ **d.** $y = 3x^3 + 2x^2 + x$

________ ________ ________ ________

2. Is the equation $y = \sqrt{x + 2}$ a linear equation? Explain your answer.

____

____

Try This

1. Which of the following are linear equations?

a. $y = 2x - 7$ **b.** $x = y - 8$ **c.** $3y = x^2 + 3$ **d.** $y = \frac{3}{x}$

________ ________ ________ ________

Name ______________________________ Date ____________ Class ____________

LESSON **12A**
Technology Lab
Graphing Linear Equations

Activity

Graph the following equation on a graphics calculator:

$$y = x - 3$$

Step 1: Enter the equation by pressing Y= X,T,O,n – 3 ENTER.

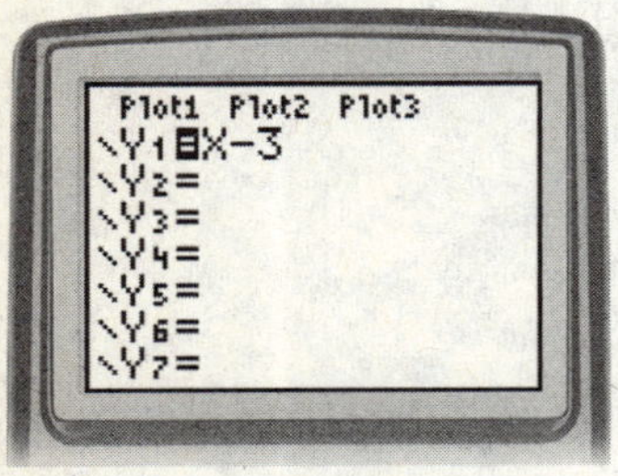

Step 2: Set viewing window. Press ZOOM and select 6:ZStandard.

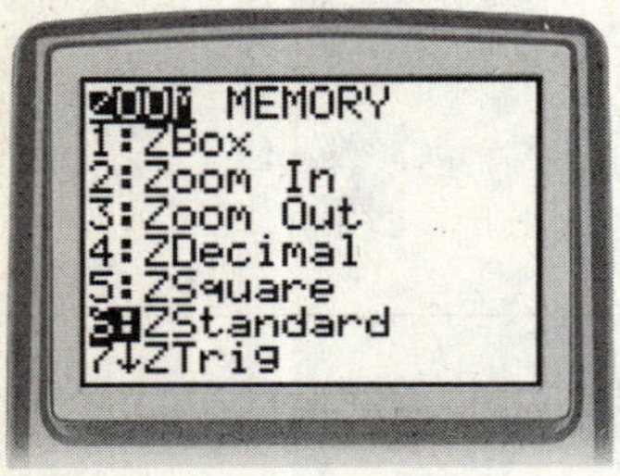

Step 3: Graph the equation. Press GRAPH if needed.

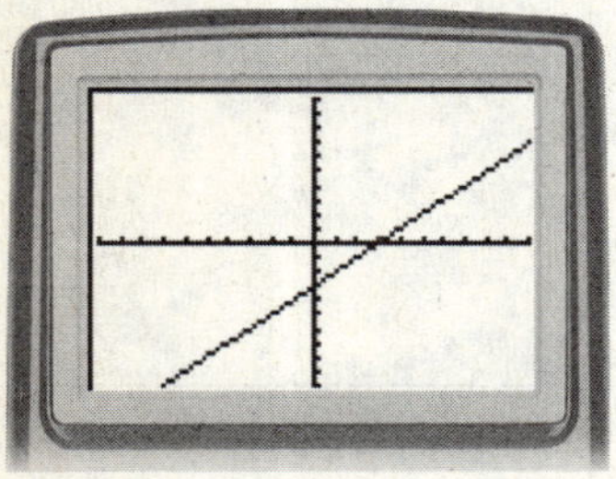

Notice that the graph of $y = x - 3$ is a straight line.

Name ______________________ Date __________ Class __________

LESSON 12A

Technology Lab

Graphing Linear Equations, continued

Think and Discuss

1. In the function table below, discuss how *x* is related to *y* and write an equation to show this relationship.

x	*y*
−6	−2
−3	1
1	5
7	11
9	13

____

____

Try This

1. Translate "*y* is 4 more than twice *x*" into an equation and use your graphing calculator to graph the equation.

____

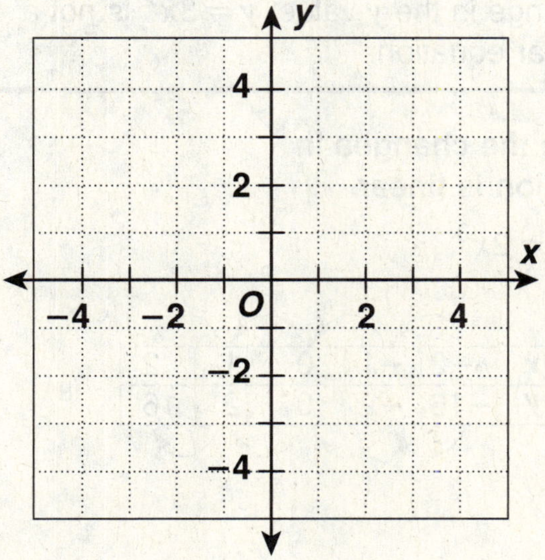

Name ______________________________ Date ______________ Class ______________

LESSON 12A

Review for Mastery

Graphing Linear Equations

The graph of a **linear equation** is a straight line.

The line shown is the graph of $y = \frac{3}{2}x + 1$.

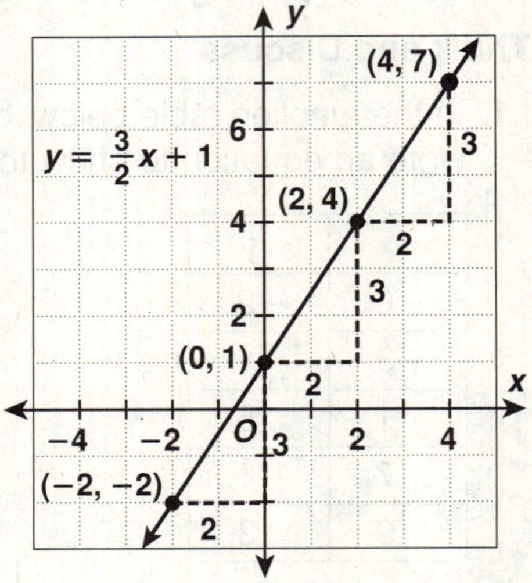

Each variable has an exponent of 1. There are no operations other than addition, subtraction, and multiplication of a variable by a constant. No variable is multiplied by another variable. No variable appears in a denominator.

All the points on the line are solutions of the equation.

Each time the *x*-value increases by 2, the *y*-value increases by 3. So, a constant change in the *x*-value corresponds to a constant change in the *y*-values.

$y = 3x - 4$

Changes in *x*: +1, +1, +1, +1

x	−2	−1	0	1	2
y	−10	−7	−4	−1	2

Changes in *y*: +3, +3, +3, +3

Since a constant change in the *x*-value corresponds to a constant change in the *y*-value, $y = 3x - 4$ is a linear equation.

$y = 3x^2$

Changes in *x*: +1, +1, +1, +1

x	−2	−1	0	1	2
y	12	3	0	3	12

Changes in *y*: −9, −3, +3, +9

Since a constant change in the *x*-value does not correspond to a constant change in the *y*-value, $y = 3x^2$ is not a linear equation.

Each equation has a table of solutions. Indicate the changes in *x*-values and in *y*-values. Tell whether the equation is linear.

1. $y = 2x - 5$

x	−2	−1	0	1	2
y	−9	−7	−5	−3	−1

____

2. $y = 2x^3$

x	−2	−1	0	1	2
y	−16	−2	0	2	16

____

Name ______________________ Date ____________ Class ____________

LESSON 12A

Review for Mastery

Graphing Linear Equations, continued

To graph a linear equation, make a table to find several solutions. Choose *x*-values that are easy to graph. Substitute each *x*-value into the equation to find the corresponding *y*-value. Plot your solutions and draw a line connecting them.

x	$\frac{1}{2}x - 3$	y	(x, y)
−4	$\frac{1}{2}(-4) - 3$	−5	(−4, −5)
−2	$\frac{1}{2}(-2) - 3$	−4	(−2, −4)
0	$\frac{1}{2}(0) - 3$	−3	(0, −3)
6	$\frac{1}{2}(6) - 3$	0	(6, 0)

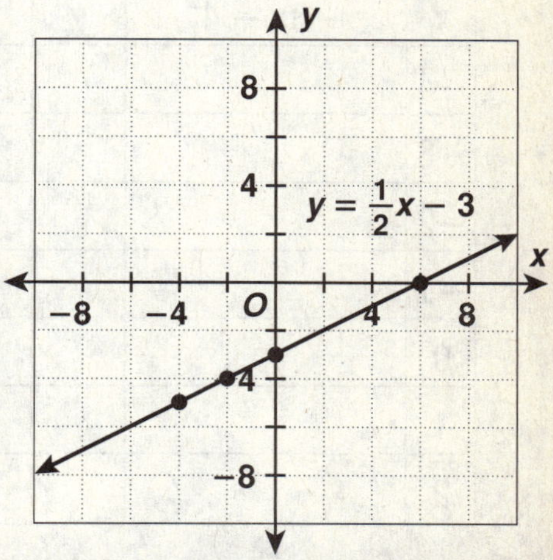

Complete the table for each equation and then graph the equation.

3. $y = 3x - 1$

x	$3x - 1$	y	(x, y)
−2	3(−2) − 1		(−2,)
0	3( ) − 1		(0,)
1	3( ) − 1		(1,)
3	3( ) − 1		(3,)

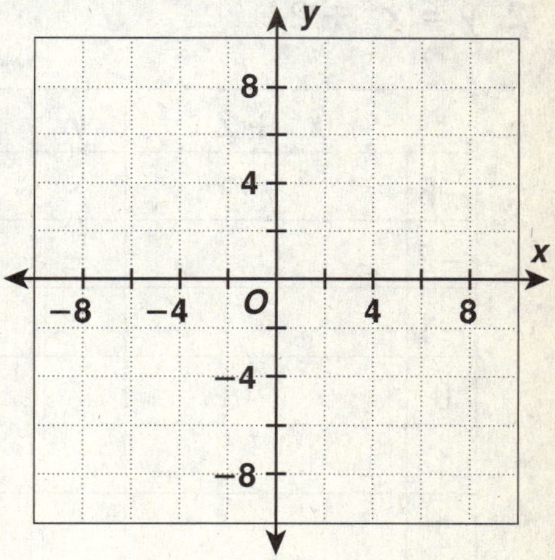

4. $y = -\frac{3}{2}x + 1$

x	$-\frac{3}{2}x + 1$	y	(x, y)
−2	$-\frac{3}{2}(\quad) + 1$		(−2, 4)
0	$-\frac{3}{2}(\quad) + 1$		
2	$-\frac{3}{2}(\quad) + 1$		
4	$-\frac{3}{2}(\quad) + 1$		

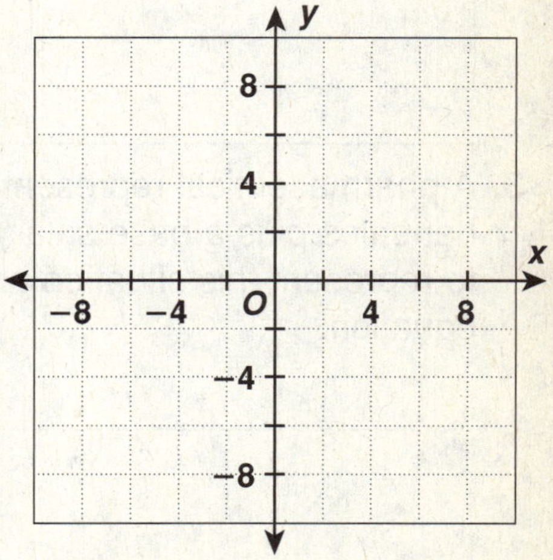

Name ______________________________ Date ____________ Class ____________

LESSON 12A

Homework and Practice

Graphing Linear Equations

Graph each equation and tell whether it is linear.

1. $y = -3x - 4$

x	$-3x - 4$	y	(x, y)
−4			
−3			
−2			
−1			
0			
1			
2			

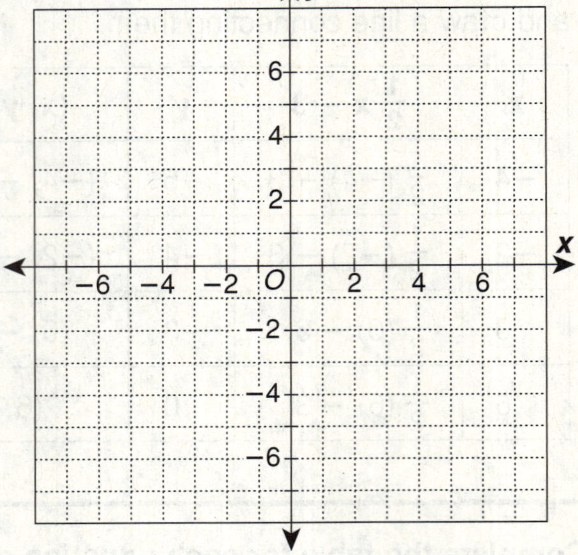

2. $y = x^2 + 2$

x	$x^2 + 2$	y	(x, y)
−3			
−2			
−1			
0			
1			
2			
3			

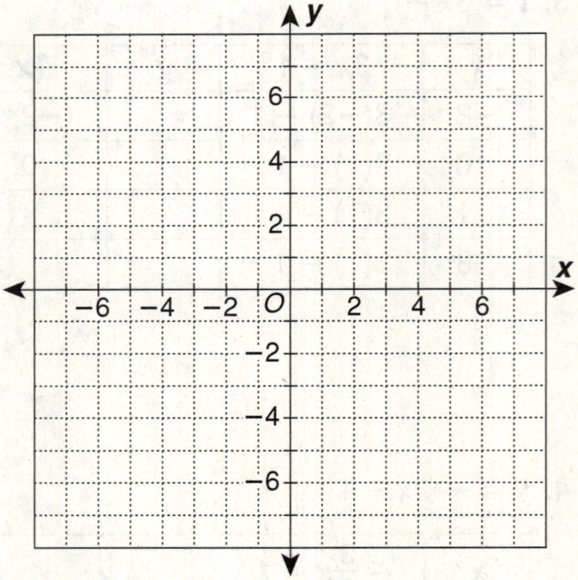

____

3. A pharmaceutical representative earns 5.5% commission on all her sales plus a base salary of $400 a month. Write an equation to represent this situation and then explain whether it is a linear equation.

____

EXPLORATION

Program of Studies and Core Content

MA-8-AT-S-PRF2, MA-8-AT-S-PRF6, MA-8-G-S-CG1, MA-8-G-S-CG2, MA-08-1.4.1, MA-08-3.3.1, MA-08-5.1.5

12B Slope of a Line

Joseph plots a graph of his 300-mile trip to the coast.

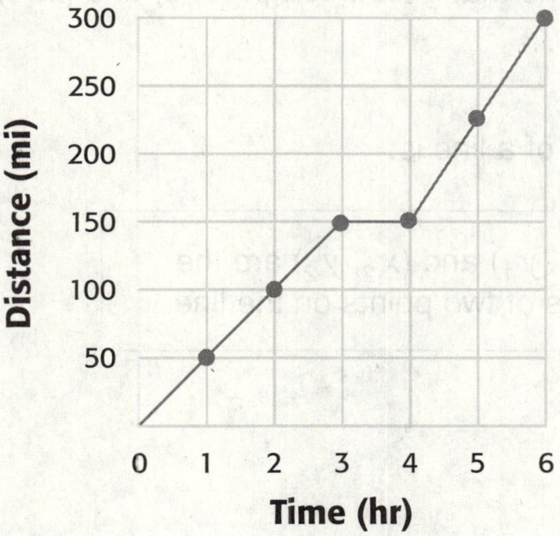

1. Calculate his average speed for the first 3 hours. (*Hint:* Divide the distance traveled during the first three hours by 3.)

2. What happens between hours 3 and 4?

3. Calculate his average speed for the last 2 hours. (*Hint:* Divide the distance traveled during the last two hours by 2.)

Think and Discuss

4. Discuss how you could find Joseph's average speed between hours 3 and 4.

5. Discuss how average speed is related to the shape of the graph.

Name ______________________ Date __________ Class __________

LESSON 12B

Technology Lab

Slope of a Line

Activity

Slope tells how much a line slants. When a line rises from left to right, the slope is positive; when a line falls from left to right, the slope is negative. Graph the line that goes through (0, 0) and (4, 4) on a graphing calculator.

Step 1: Find the slope.

The formula for the slope (m) of a line is:

$m = \dfrac{y_2 - y_1}{x_2 - x_1}$	Where (x_1, y_1) and (x_2, y_2) are the coordinates of two points on the line.

Substitute: $m = \dfrac{4 - 0}{4 - 0}$

$= \dfrac{4}{4} = 1$

Step 2: Graph the equation.

Since the line goes through (0, 0), its equation is in the form $y = mx$. You know the slope is 1 from Step 1, so the equation is $y = 1x$ or $y = x$.

Enter the equation in Y1 and press GRAPH.

Name ______________________ Date ____________ Class ____________

LESSON 12B

Technology Lab

Slope of a Line, continued

Think and Discuss

1. What happens to your graph if you make the value of *m* greater? What happens if the value of *m* approaches zero? Becomes negative?

___

___

___

___

Try This

1. Write and graph the equation of the line that goes through (0, 0) and (2, 6).

___

2. Write and graph the equation of the line that goes through (0, 0) and (−3, 9).

___

3. Find the slope of a line that passes through the points (4, 2) and (−6, −3).

___

Name ______________________ Date ____________ Class ____________

LESSON 12B

Review for Mastery

Slope of a Line

The **slope** of a line is a measure of its rate of change.

The slope of a straight line is a constant ratio, the "rise over run," or the **vertical change** over the **horizontal change.**

You can find the slope of a line by comparing any two of its points. The vertical change is the difference between the two *y*-values. And the horizontal change is the difference between the two *x*-values.

$$\text{slope} = \frac{y_2 - y_1}{x_2 - x_1}$$

point *A*: (3, 2) point *B*: (4, 4)

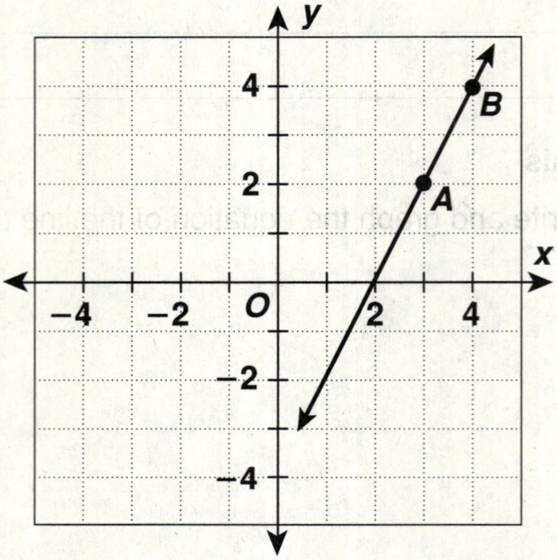

Make point *A* (x_1, y_1).
Make point *B* (x_2, y_2).

$$\text{slope} = \frac{4 - 2}{4 - 3}$$
$$= \frac{2}{1}, \text{ or } 2$$

So, the slope of the line is 2.

You can make point *A* (x_2, y_2) and point *B* (x_1, y_1).

$$\text{slope} = \frac{2 - 4}{3 - 4}$$
$$= \frac{-2}{-1}, \text{ or } 2$$

So, the slope remains 2.

Find the slope of the line that passes through each pair of points.

1. (1, 5) and (2, 6)

____

2. (0, 3) and (2, 7)

____

3. (2, 5) and (3, 4)

____

4. (6, 9) and (2, 7)

____

5. (6, 5) and (8, −1)

____

6. (7, −4) and (4, −2)

____

Name ______________________ Date ____________ Class ____________

LESSON 12B

Review for Mastery

Slope of a Line, continued

A straight line has a constant slope, so it shows a **constant rate of change**. The same change in *y* always results in the same change in *x*.

From point *C* to point *B*:

$$\frac{\text{change in } y}{\text{change in } x} = \frac{2}{3}$$

From point *B* to point *A*:

$$\frac{\text{change in } y}{\text{change in } x} = \frac{2}{3}$$

A curve doesn't have a constant slope, so it shows a **variable rate of change**.

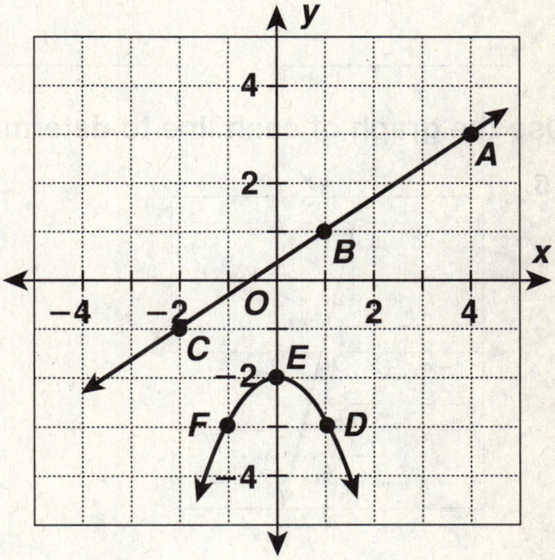

Between point *F* and point *E*, the curve has a positive slope.
Between point *E* and point *D*, the curve has a negative slope.

So, the curve has a variable rate of change.

Determine whether each graph shows a constant or a variable rate of change. Write *constant* or *variable*.

7.

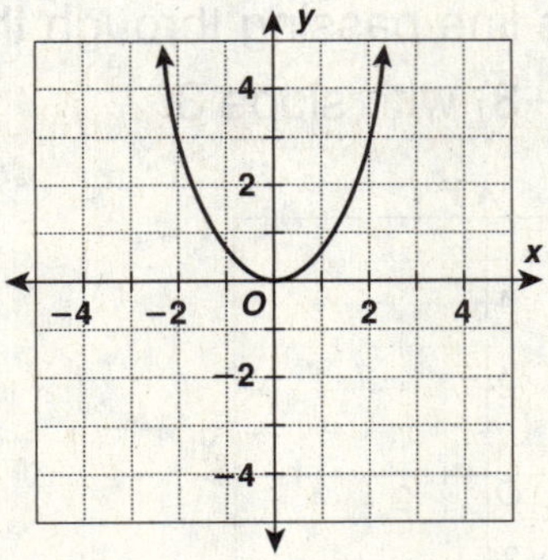

____

8.

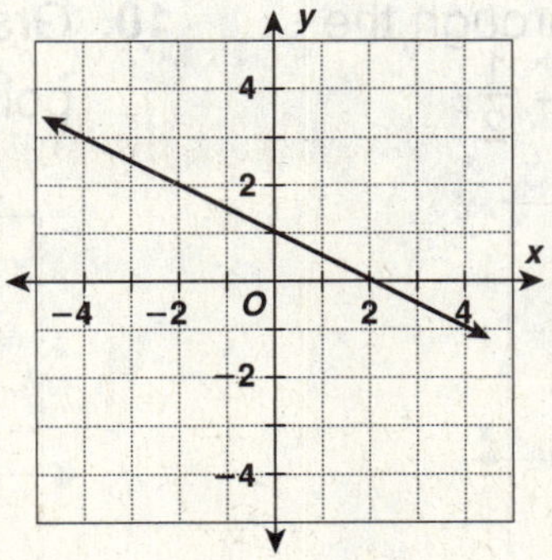

____

9.

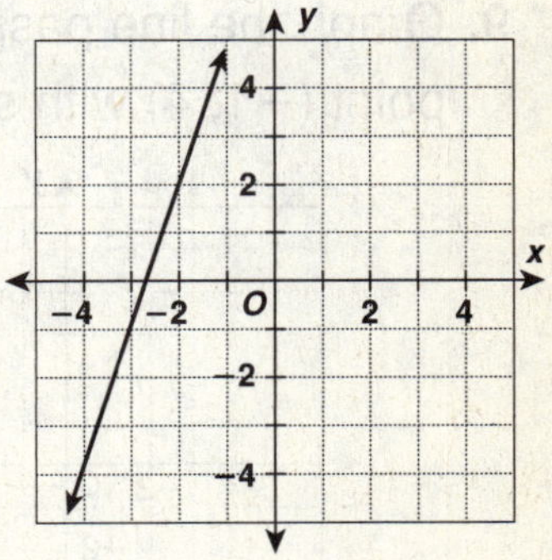

____

10. Explain how an increase in the *x*-value of a linear equation affects the *y*-value.

____

Name ______________________ Date ____________ Class ____________

LESSON **12B**

Homework and Practice

Slope of a Line

Find the slope of the line passing through each pair of points.

1. (5, 7), (6, 9) ____________

2. (2, 8), (8, 2) ____________

3. (−2, 6), (−4, −4) ____________

4. (−8, −7), (−9, −1) ____________

Use the graph of each line to determine its slope.

5.

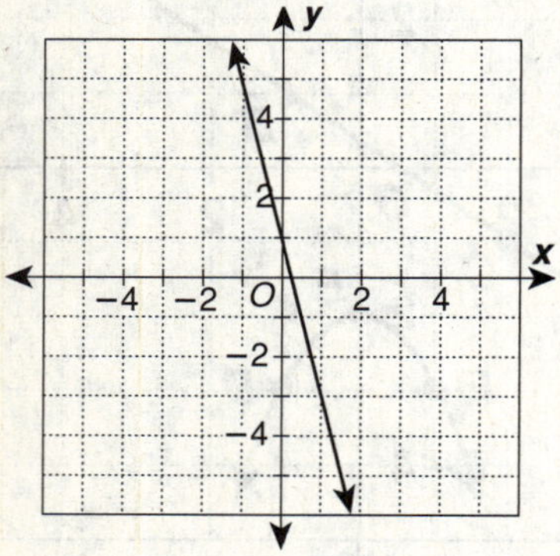

____

6.

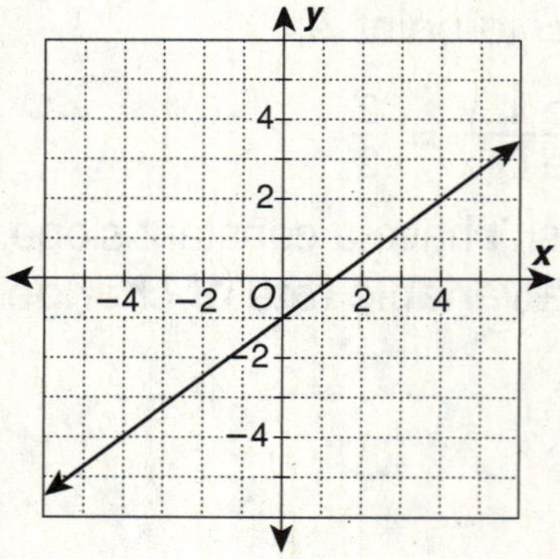

____

Tell whether the lines passing through the given points are parallel or whether they intersect.

7. line 1: (−1, −2), (−4, −8)
line 2: (−6, 7), (−7, 5)

____

8. line 1: (6, 8), (8, 10)
line 2: (−5, −9), (3, −9)

____

9. Graph the line passing through the point (−1, 4) with slope $-\frac{1}{2}$.

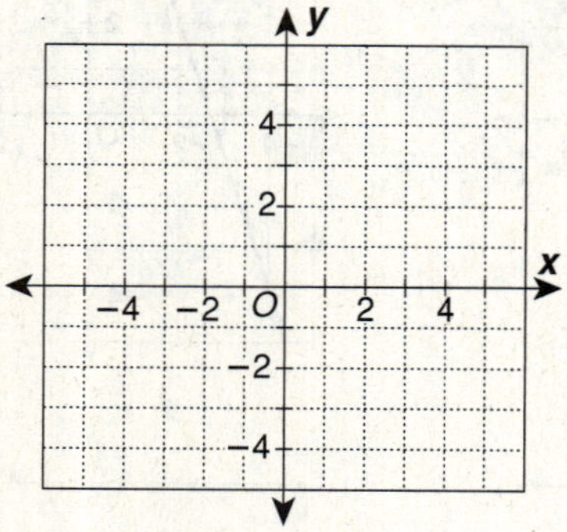

10. Graph the line passing through the point (1, −3) with slope 3.

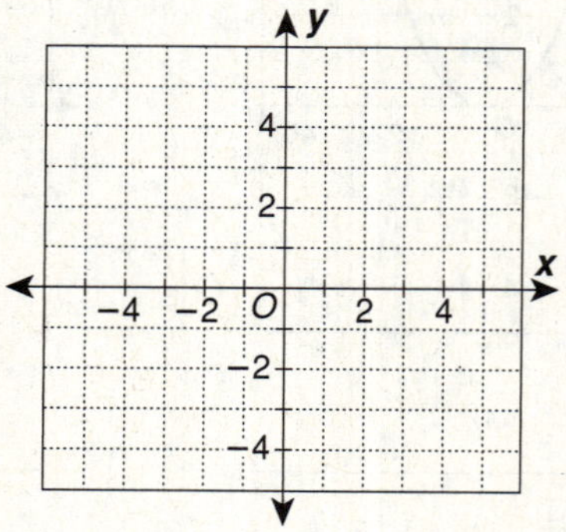

11. If an equation has a variable rate of change, can you assume that a constant increase in the *x*-value will result in a constant increase for the *y*-value? Explain.

____

EXPLORATION

Program of Studies and Core Content

MA-8-AT-S-EI1, MA-8-AT-S-PRF2, MA-8-AT-S-PRF5, MA-8-AT-S-VE03, MA-8-G-S-CG2, MA-08-3.3.1

12C Using Slopes and Intercepts

On the graph below, point *A* is called the ***y*-intercept** and point *C* is called the ***x*-intercept.**

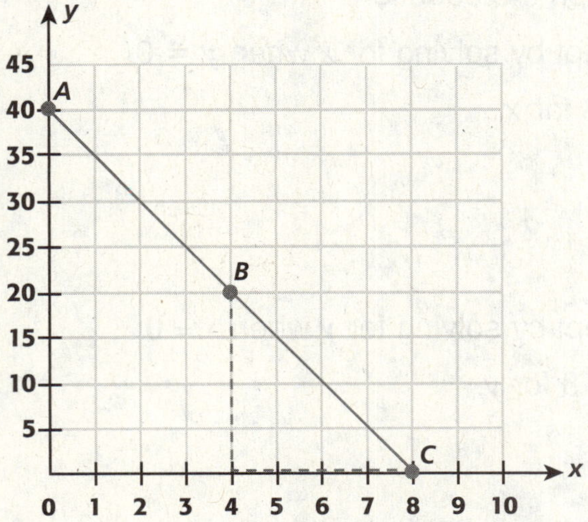

1. Make a table of values by finding the coordinates of points *A*, *B*, and *C* above.

Point	*x*	*y*
A		
B		
C		
D		

2. Plot one more point on the line and label it point *D*. List its coordinates in the table and then find the slope between *B* and *D*.

3. Next find the slope of the line from the *y*-intercept (point *A*) to the *x*-intercept (point *C*). What do you notice about the two slopes you found?

Think and Discuss

4. Explain how to calculate the slope of a line using the *x*- and *y*-intercepts.

5. Explain whether calculating the slope was easier using the *x*- and *y*-intercepts or using points *C* and *D*.

Name ________________________________ Date ___________ Class ___________

LESSON 12C

Hands-On Lab 1

Using Slopes and Intercepts

Activity 1

An equation in the form $Ax + By = C$, where A, B, and C are real numbers, is said to be in *standard form.*

Find the x-intercept and y-intercept of the linear equation $2x + y = 8$. Use the intercepts to graph the equation.

Step 1 Find the x-intercept by solving for x when $y = 0$.

Solve $2x + 0 = 8$ for x.

$x =$ _____

Step 2 Find the y-intercept by solving for y when $x = 0$.

Solve $2(0) + y = 8$ for y.

$y =$ _____

Step 3 Plot the x-intercept and y-intercept on the graph below. Draw the line of the equation through the two points.

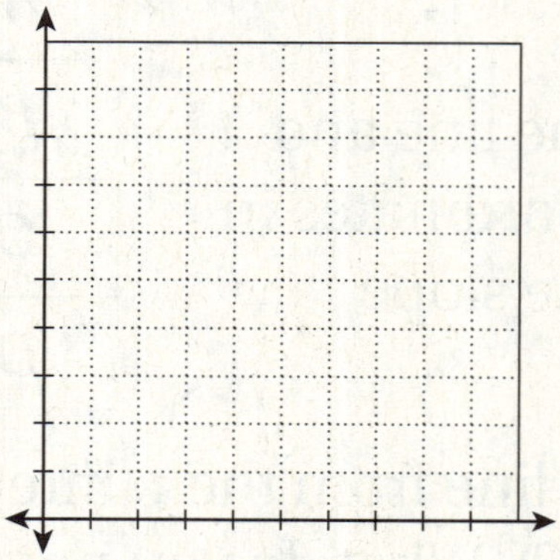

Step 4 Check your work by finding the coordinates of any other point on the line. (**Hint:** If possible, choose a point that has integer coordinates.) Substitute the x- and y-values of the point into the equation. If the coordinates of the point you chose are a solution to the equation, you can conclude that you graphed the line correctly. Show your work below.

Name ______________________ Date ____________ Class ____________

LESSON 12C

Hands-On Lab 1

Using Slopes and Intercepts, continued

Think and Discuss

1. How would you find the *x*-intercept and *y*-intercept of an equation that is not written in standard form, such as $2y = 10 - 5x$? Find the *x*-intercept and *y*-intercept of this equation.

____

____

____

____

2. Write the equation $4 - 3x = y$ in standard form.

____

Try This

1. Graph each of the following linear equations on the graph below by finding the x-intercept and y-intercept of each. How is the slope of the two lines different?

Equation 1: $x - 3y = 9$

Equation 2: $x + 3y = 9$

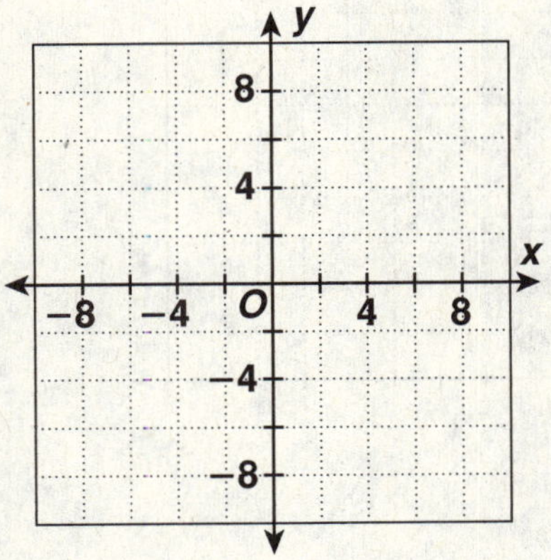

____

____

____

Name ______________________________ Date ___________ Class ___________

LESSON 12C

Hands-On Lab 2

Using Slopes and Intercepts

Activity 1

An equation in the form $y = mx + b$, where m is the slope and b is the y-intercept, is said to be in *slope-intercept form.*

Write the equation below in slope-intercept form and graph the line.

$4x + y = 12$

Step 1 Subtract $4x$ from both sides.

$-4x + 4x + y = 12 - 4x$

$y = 12 - 4x$

Step 2 Rewrite the equation to match slope-intercept form.

$y = -4x + 12$

What is the y-intercept of the equation?

____

What is the slope of the equation?

____

Step 3 To graph the equation, plot the y-intercept. Then use the slope of the line to plot another point on the line. The slope is negative so the line will slant downward from left to right.

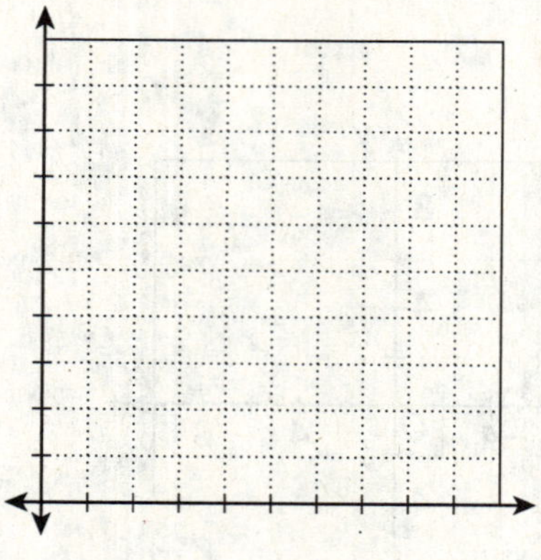

Name ______________________ Date __________ Class __________

LESSON 12C

Hands-On Lab 2
Using Slopes and Intercepts, continued

Think and Discuss

1. How can you find the equation of a line if you know two points on the line?

____

____

____

____

2. For every 1 hour that Emily babysits, she earns \$7.00. Write an equation in slope-intercept form to describe this situation.

____

Try This

1. Jaden's parents gave him \$50 to start a savings account. Each week he has to put \$5.00 of his allowance into his savings account. Write an equation in slope-intercept form for the line that describes how much money Jaden has in his savings account.

____

What is the slope of the line?

____

What is the *y*-intercept of the line?

____

Draw the line on the graph below.

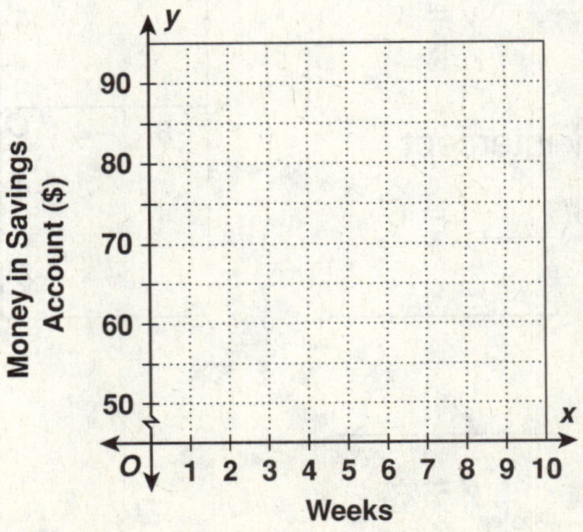

Name ______________________________ Date ____________ Class ____________

LESSON **12C**

Review for Mastery

Using Slopes and Intercepts

***x*-intercept**: the *x*-coordinate of the point at which a line crosses the *x*-axis

***y*-intercept**: the *y*-coordinate of the point at which a line crosses the *y*-axis

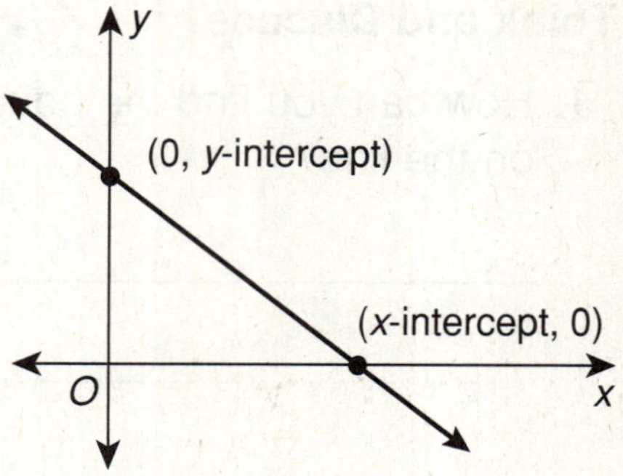

You can find the intercepts of a line from its equation. Then you can use the intercepts to graph the line.

Find the intercepts of the line $3x + 4y = 24$.

For the *x*-intercept, let $y = 0$.

$$3x + 4y = 24$$
$$3x + 4(0) = 24$$
$$3x + 0 = 24$$
$$3x = 24$$
$$\frac{3x}{3} = \frac{24}{3}$$
$$x = 8$$

The *x*-intercept is 8.

For the *y*-intercept, let $x = 0$.

$$3x + 4y = 24$$
$$3(0) + 4y = 24$$
$$0 + 4y = 24$$
$$4y = 24$$
$$\frac{4y}{4} = \frac{24}{4}$$
$$y = 6$$

The *y*-intercept is 6.

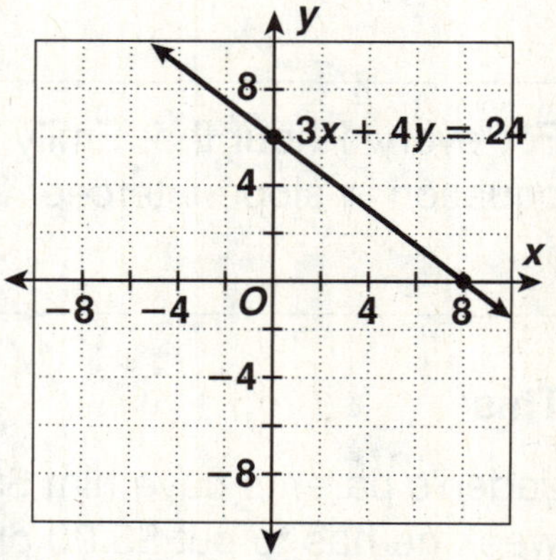

Find the intercepts of each line.
Draw both graphs on the same grid.

1. $2x + 3y = 12$

for *x*-intercept

$2x + 3(\_\_\_\_) = 12$

____

____

$x =$ ____

for *y*-intercept

$2(\_\_\_\_) + 3y = 12$

____

____

$y =$ ____

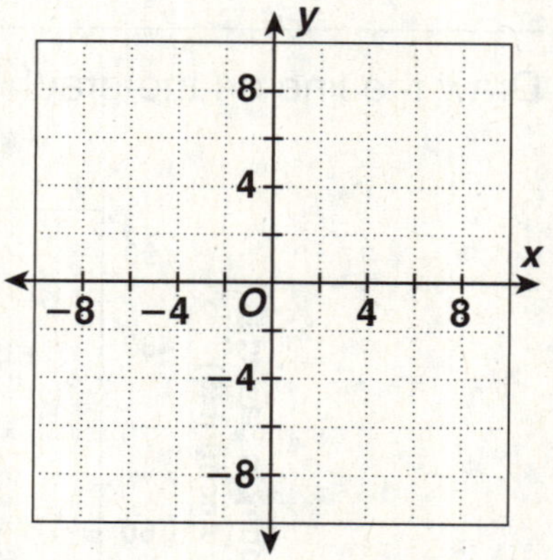

2. $6y - 3x = 6$

for *x*-intercept

$6(\_\_\_\_) - 3x = 6$

____

____

$x =$ ____

for *y*-intercept

$6y - 3(\_\_\_\_) = 6$

____

____

$y =$ ____

Name ______________________ Date __________ Class __________

LESSON 12C

Review for Mastery

Using Slopes and Intercepts, continued

slope-intercept form

$$y = mx + b$$

↑ slope ↖ *y*-intercept

In this form, the coefficient of *x* is the slope and the constant term is the *y*-intercept.

To rewrite an equation in slope-intercept form, *isolate y*.

$$2x + 3y = -12$$
$$-2x \quad\quad -2x \quad\quad \text{Subtract } 2x.$$
$$3y = -2x - 12$$
$$\frac{3y}{3} = \frac{-2}{3}x - \frac{12}{3} \quad\quad \text{Divide by 3.}$$
$$y = \frac{-2}{3}x - 4$$

So, $m = -\frac{2}{3}$ and $b = -4$.

Write each equation in slope-intercept form and then find the slope and *y*-intercept.

3. $3y = 4x + 15$

$\frac{3y}{3} = \frac{4}{3}x + \frac{15}{3}$ Divide by 3.

____

____

4. $3x - 2y = 6$

______ ______ Subtract 3*x*.

____

____

Given two points of a line, you can write its equation.

(2, 5) and (−1, −4)

$$\text{slope} = \frac{y_2 - y_1}{x_2 - x_1} = \frac{-4 - 5}{-1 - 2} = \frac{-9}{-3} = 3$$

To find *b*, substitute the slope and the values from one of the points into the slope-intercept equation.

$$y = mx + b \quad \rightarrow \quad 5 = 3(2) + b$$
$$5 = 6 + b$$
$$-1 = b$$

So, the equation for the line that passes through (2, 5) and (−1, −4) is $y = 3x - 1$.

Write the equation of the line that passes through each pair of points in slope-intercept form.

5. (2, 11) and (0, 3)

____

6. (−1, 3) and (4, −2)

____

7. (10, 1) and (6, −1)

____

Name ______________________ Date __________ Class __________

LESSON 12C
Homework and Practice
Using Slopes and Intercepts

Find the *x*-intercept and *y*-intercept of each line. Use the intercepts to graph the equation.

1. $x = y + 5$

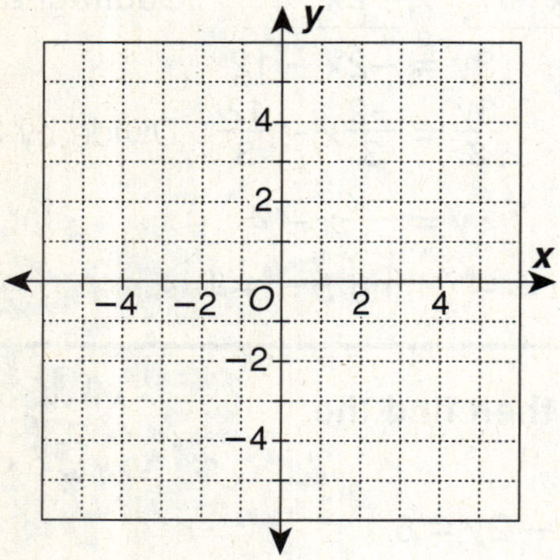

2. $3x - 2y = 6$

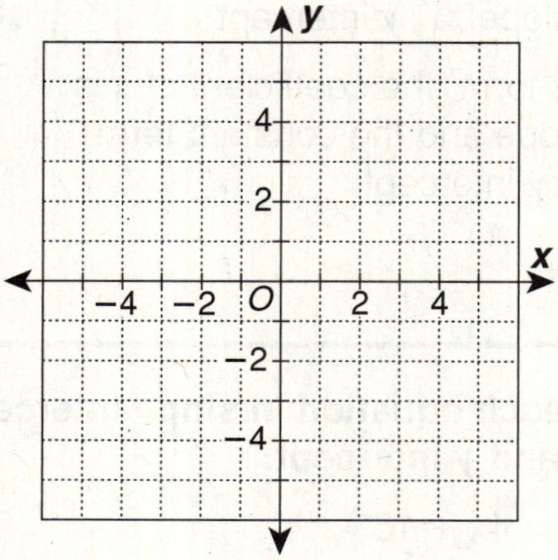

______________________ ______________________

______________________ ______________________

Write each equation in slope-intercept form and then find the slope and *y*-intercept.

3. $2x = y - 4$

4. $4x - 3y = -9$

5. $x - 4y = 8$

______________ ______________ ______________

______________ ______________ ______________

Write the equation of the line that passes through each pair of points in slope-intercept form.

6. (−1, 7), (4, −3)

7. (2, 5), (−8, 15)

8. (8, 4), (−10, −5)

______________ ______________ ______________

9. Thad's father gives him \$10 for a passing report card plus \$5 for every grade of A. Write an equation of a line in slope-intercept form to express *y*, the amount received with *x* grades of A. State the slope, *x*-intercept, and *y*-intercept of the equation.

____

Name ______________________ Date __________ Class __________

LESSON 12C

Homework and Practice

Using Slopes and Intercepts, continued

The cost of food for an honor roll dinner is $300 plus $10 per student. The cost of the food as a function of the number of students is shown in the graph. Write the correct answer.

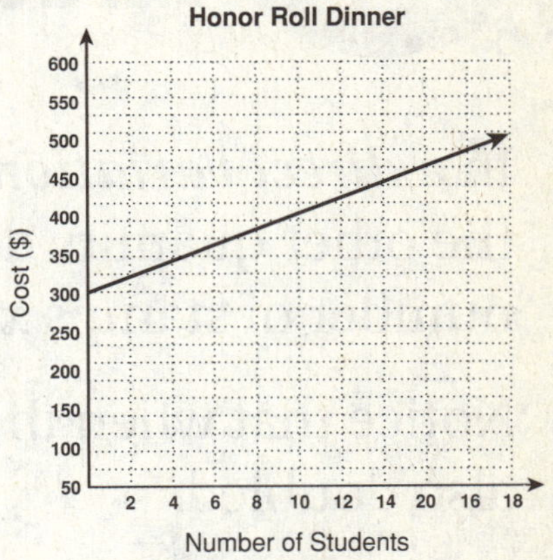

10. Write an equation that represents the cost as a function of the number of students.

____

11. Identify the slope and *y*-intercept and describe their meanings.

____

____

12. Find the cost of the food for 50 students. ______________________

Laura is on a two-day hike in the Smoky Mountains. She hiked 8 miles on the first day and is hiking at a rate of 3 mi/hr on the second day. Her total distance as a function of times is shown in the graph. Select the best answer.

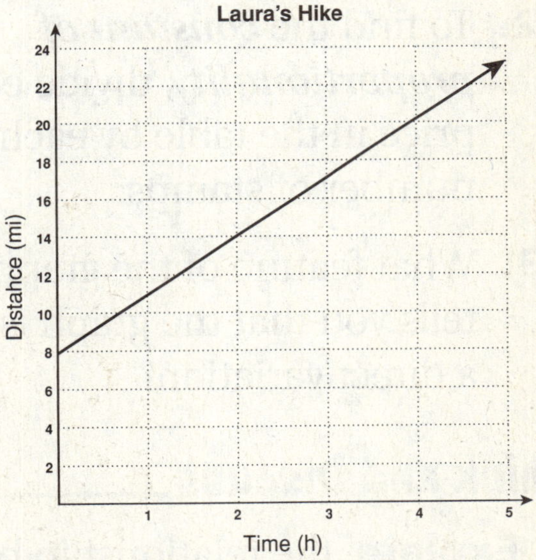

13. Which equation represents Laura's total distance as a function of time?

A $y = 3x$ **C** $y = 3x + 8$

B $y = 8x$ **D** $y = 8x + 3$

14. What does the slope represent?

F Laura's total distance after one day

G Laura's total distance after two days

H the number of miles Laura hiked per hour on the first day

J the number of miles Laura hiked per hour on the second day

15. What does the *y*-intercept represent?

F Laura's total distance after one day

G Laura's total distance after two days

H the number of miles Laura hiked per hour on the first day

J the number of miles Laura hiked per hour on the second day

16. What will be Laura's total distance if she hikes for 6 hours on the second day?

F 14 miles **H** 26 miles

G 18 miles **J** 28 miles

EXPLORATION

Program of Studies and Core Content
MA-8-G-S-CG2, MA-08-3.3.1, MA-08-5.1.5

12D Direct Variation

In a ***direct variation,*** when one quantity increases or decreases, the other quantity does the same. The table below shows the number of stamps x and the price y for each number.

Notice that when the number of stamps is doubled, the price is also doubled.

Stamps x	1	2	3	4	5	6
Price y	\$0.37	\$0.74	\$1.11	\$1.48	\$1.85	\$2.22

1. Use each pair of values (x, y) in the table to complete the graph.

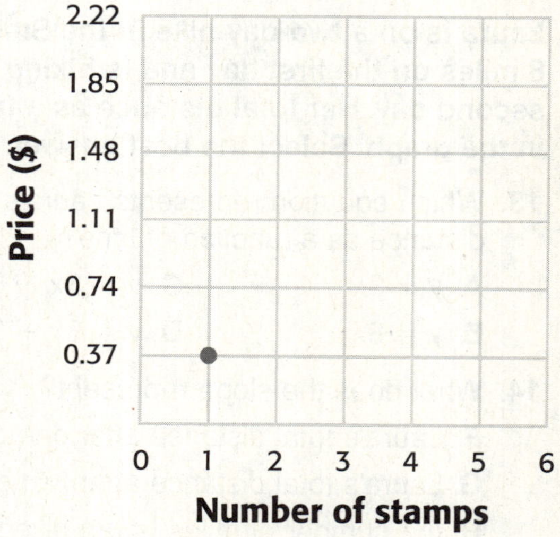

2. To find the ***constant of proportionality,*** divide each price in the table by each number of stamps.

3. What feature of the graph tells you that the graph is of a direct variation?

Think and Discuss

4. Explain the relationship between a constant of proportionality and the slope of the line.

5. Give a real-world example of a direct variation.

Name ______________________ Date __________ Class __________

LESSON **12D**

Review for Mastery

Direct Variation

Two data sets have **direct variation** if they are related by a constant ratio, the **constant of proportionality.** A graph of the data sets is linear and passes through (0, 0).

$y = kx$ **equation of direct variation,**
where k is the constant ratio

To determine whether two data sets have direct variation, you can compare ratios. You can also graph the data sets on a coordinate grid.

x	3	5	8
y	15	25	40

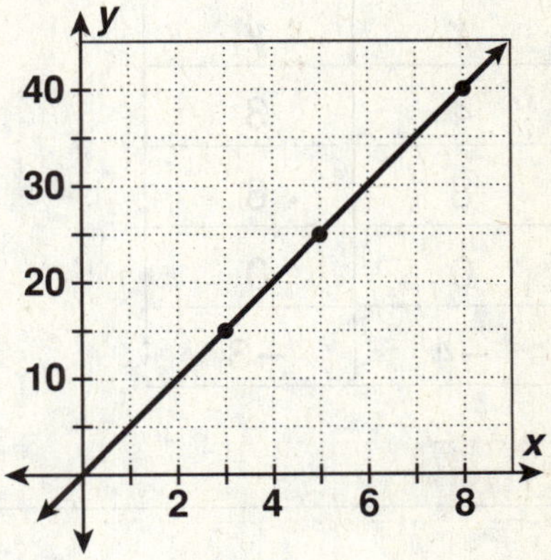

$\frac{y}{x} = \frac{15}{3} = \frac{25}{5} = \frac{40}{8} = \frac{5}{1}$ ⟵ constant ratio

$k = 5 \rightarrow y = 5x$

The graph of the data sets is linear and passes through (0, 0).

So, the data sets show direct variation.

Determine whether the data sets show direct variation. If there is a constant ratio, identify it and write the equation of direct variation. Plot the points and tell whether the graph is linear.

1.

x	1	2	4	8
y	8	4	2	1

constant ratio? ________________

If yes, equation. ________________

Is the graph linear? ________________

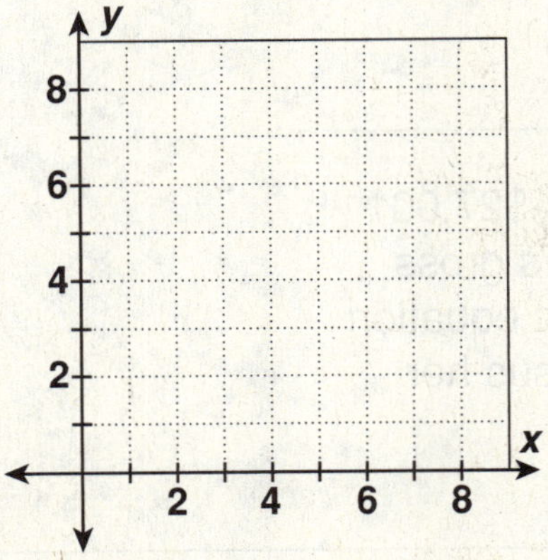

2.

x	0	2	3	5
y	0	20	30	50

constant ratio? ________________

If yes, equation. ________________

Is the graph linear? ________________

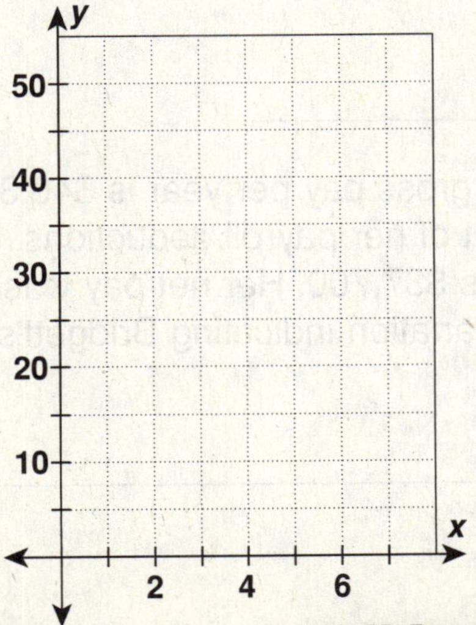

Name ______________________ Date ____________ Class ____________

LESSON 12D

Homework and Practice

Direct Variation

Make a graph to determine whether the data set shows direct variation.

1.

x	*y*
4	3
8	6
0	0
−4	−3

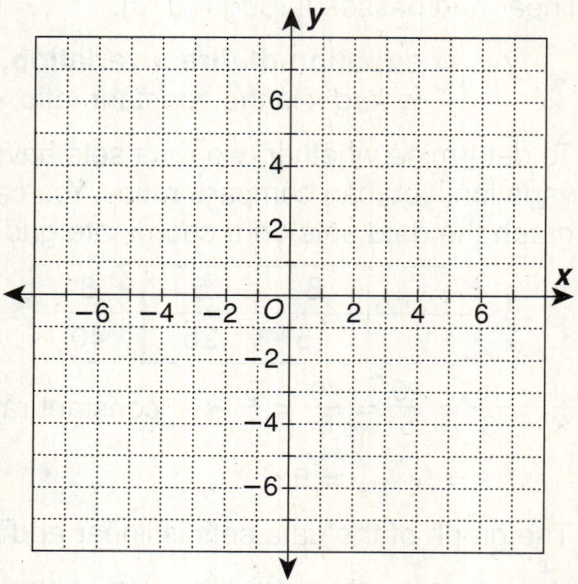

____

____

2. Write the equation of direct variation for exercise 1.

____

Find each equation of direct variation, given that *y* varies directly with *x*.

3. *y* is 27 when *x* is 3

____

4. *y* is 8 when *x* is −40

____

5. *y* is −54 when *x* is −12

____

6. *y* is 21 when *x* is 49

____

7. *y* is −31.5 when *x* is 14

____

8. *y* is 180 when *x* is −216

____

9. Bridgett's gross pay per year is $40,300. Her net pay is $27,001 as a result of her payroll deductions. Last year Bridgett's gross salary was $37,700. Her net pay was $25,259. Write an equation of direct variation indicating Bridgett's gross pay, *x*, versus her net pay, *y*.

____

EXPLORATION

Program of Studies and Core Content

MA-8-AT-S-EI1, MA-8-AT-S-PRF5, MA-8-AT-S-PRF3

12E Lines of Best Fit

The table shows student enrollment at a college by year. The enrollment numbers are graphed below.

Year	2003	2004	2005	2006	2007	2008	2009	2010
Enrollment	950	995	1011	1020	1035			

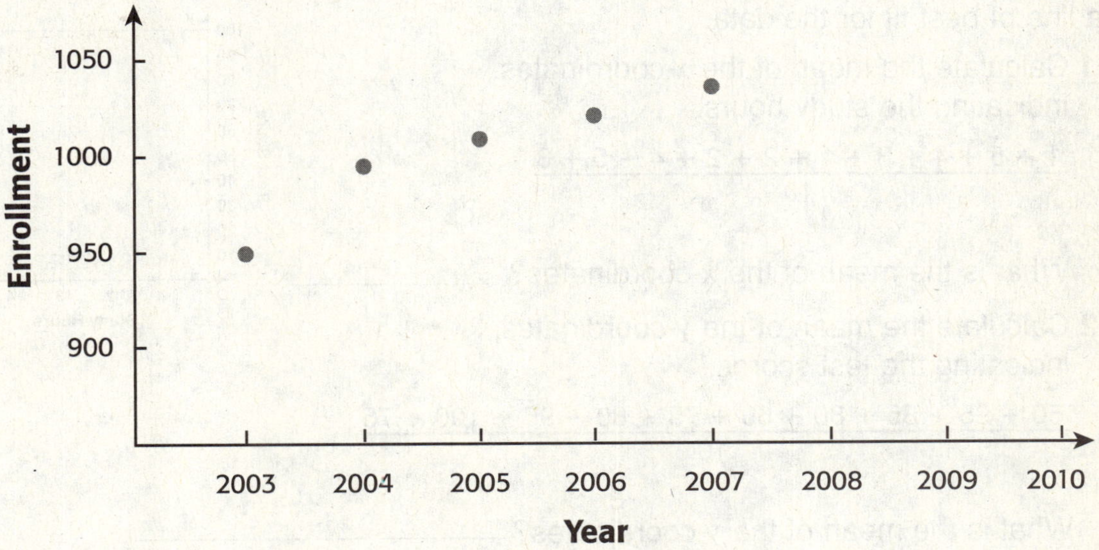

1. Find x_m, the mean of the x-values of the points on the graph.

2. Find y_m, the mean of the y-values of the points on the graph.

3. Plot (x_m, y_m). Lay the edge of a ruler on the graph through (x_m, y_m). Pivot the ruler around (x_m, y_m) and draw the line that you think is closest to the line of best fit.

Think and Discuss

4. Predict the enrollment for 2008, 2009, and 2010 by using the line of best fit.

Name ______________________ Date __________ Class __________

LESSON **12E**

Hands-On Lab

Lines of Best Fit

Activity

The table below shows the number of hours 10 students studied for a test and the scores they got on the test. Plot the data on the graph below.

Study Hours	1	5	4	3	1	2	2	4	5	3
Test Score	50	95	85	80	60	75	80	95	100	75

Find a line of best fit for the data.

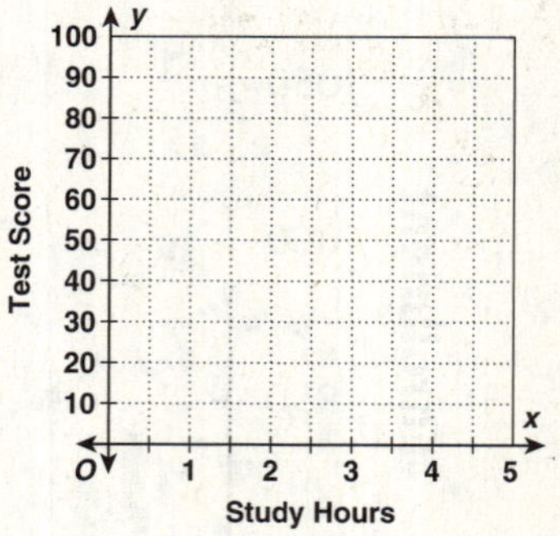

Step 1 Calculate the mean of the *x*-coordinates, indicating the study hours.

$$\frac{1 + 5 + 4 + 3 + 1 + 2 + 2 + 4 + 5 + 3}{10}$$

What is the mean of the *x*-coordinates? ____________

Step 2 Calculate the mean of the *y*-coordinates, indicating the test scores.

$$\frac{50 + 95 + 85 + 80 + 60 + 75 + 80 + 95 + 100 + 75}{10}$$

What is the mean of the *y*-coordinates? ____________

Step 3 Plot the point (x, y), where x is the mean of the *x*-coordinates and y is the mean of the *y*-coordinates.

Step 4 Estimate and plot the coordinates of another point on the line. Find the slope of the line as $\frac{y_2 - y_1}{x_2 - x_1}$. $m =$

____

Step 5 Use point-slope form, $y - y_1 = m(x - x_1)$. Substitute one of the points on the line for x_1 and y_1. Simplify to find the equation of a line of best fit.

____

Name ______________________ Date __________ Class __________

LESSON 12E

Hands-On Lab
Lines of Best Fit, continued

Think and Discuss

1. Can there be only one line of best fit for a graph? Explain why or why not.

____

____

____

2. According to a line of best fit for the scatter plot on the previous page, to get a 100 on the test, a student needs to study 5 hours. Explain why this is not true for every student.

____

____

Try This

1. A college compared high-school grade point averages of ten of its freshman students with their grade point averages their first year of college.The data is shown in the scatterplot.

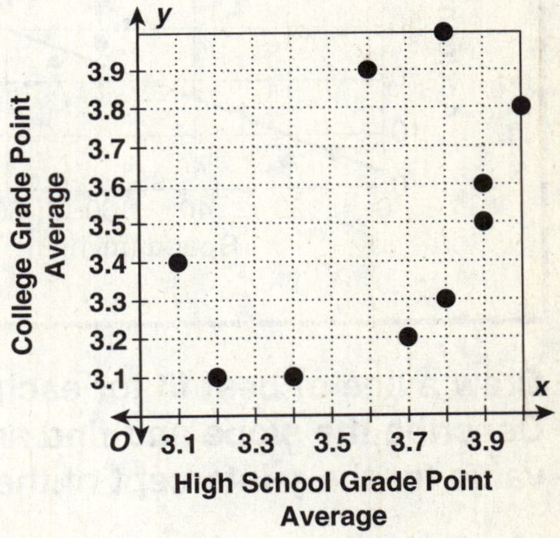

The data is

High School GPA	3.9	3.1	3.2	3.4	3.9	4.0	3.6	3.8	3.7	3.8
College GPA	3.5	3.4	3.1	3.1	3.6	3.8	3.9	4.0	3.2	3.3

Draw a line of best fit for the graph and find the equation of the line.

____

Equation of the line:

____

____

____

Name ______________________________ Date ____________ Class ____________

LESSON 12E

Review for Mastery

Lines of Best Fit

Line of best fit: a line drawn near the points on a scatter plot to show the trend between two sets of data

To draw a line of best fit:

- Draw the line through as many points as you can.
- Try to get an equal number of points above the line as below.
- Ignore any outliers.
- It may happen that none of the points lie on the line.

The line of best fit for the data in the scatter plot below slants up, indicating a positive correlation.

The slope of this line of best fit is positive and its *y*-intercept is about 3.

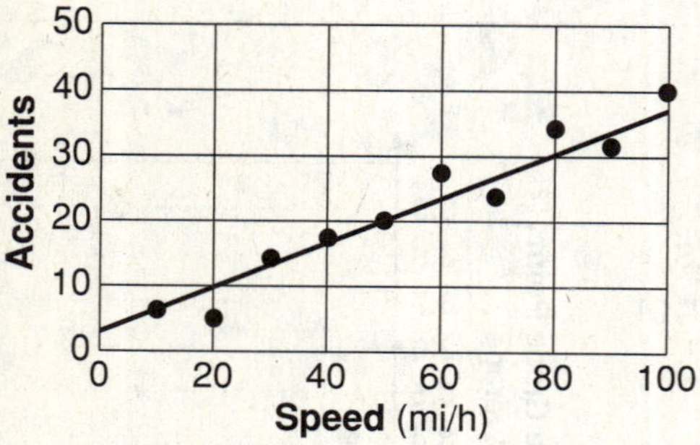

The line of best fit for the data in the scatter plot below slants down, indicating a negative correlation.

The slope of this line of best fit is positive and its *y*-intercept is about 28.

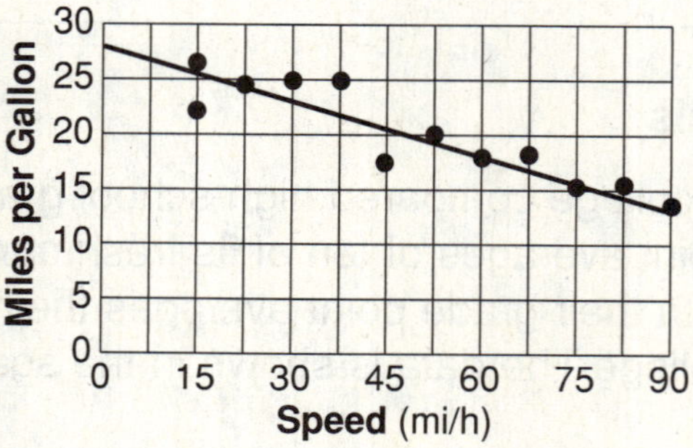

Draw a line of best fit for each graph.
Describe the slope and find an approximate value for the *y*-intercept of the line of best fit.

1.

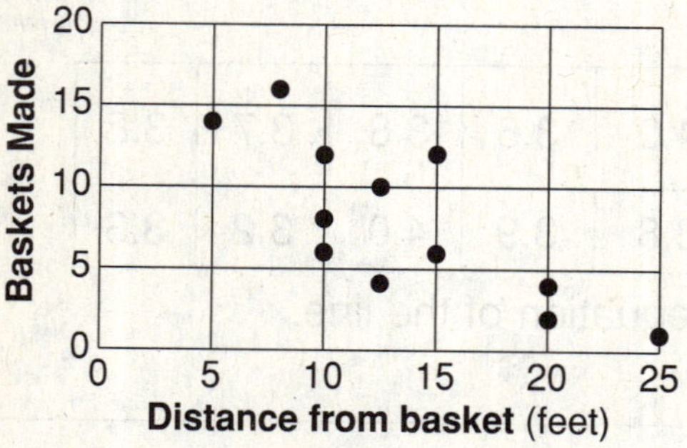

The line of best fit slants __________.

The slope of the line is __________.

The *y*-intercept is approximately ____.

2.

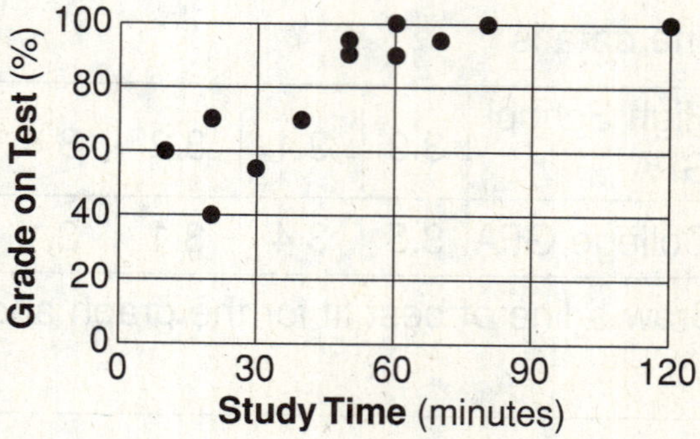

The line of best fit slants __________.

The slope of the line is __________.

The *y*-intercept is approximately ____.

Name ______________________ Date __________ Class __________

LESSON 12E

Review for Mastery

Lines of Best Fit, continued

To write an equation for a line of best fit, you can use the slope-intercept form, $y = mx + b$.

After you have drawn the line of best fit, estimate its slope from any two points on the line whose coordinates you can read.

$$\text{slope} = \frac{\text{vertical change}}{\text{horizontal change}}$$

slope $= -\frac{2}{1}$, y-intercept $= 10$

equation for the line of best fit:

$y = -2x + 10$

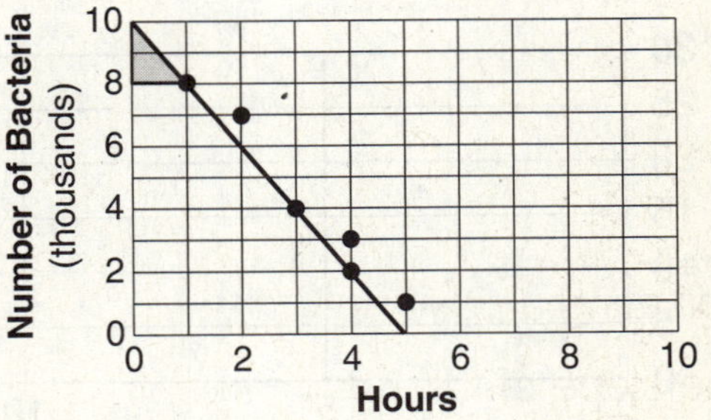

The data shows the number of bacteria present in a culture that has been treated with an anti-bacterial agent. The line of best fit can be used to predict the number of bacteria present after 2.5 hours.

$y = -2x + 10$

$y = -2(2.5) + 10$ Substitute $x = 2.5$.

$= -5 + 10$ Solve for y.

$= 5$

So, after 2.5 hours, the expected number of bacteria in the culture is approximately 5,000.

Find an equation for each line of best fit. Use the equation to answer the question.

3.

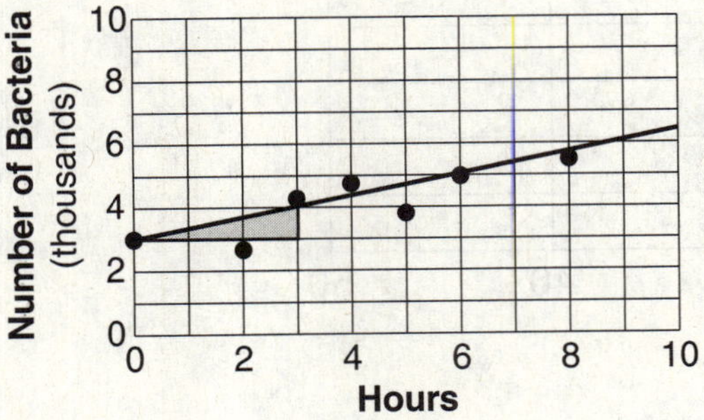

$m =$ ____, $b =$ ____,

equation: ____________

The expected number of bacteria after 10 hours is about ________.

4.

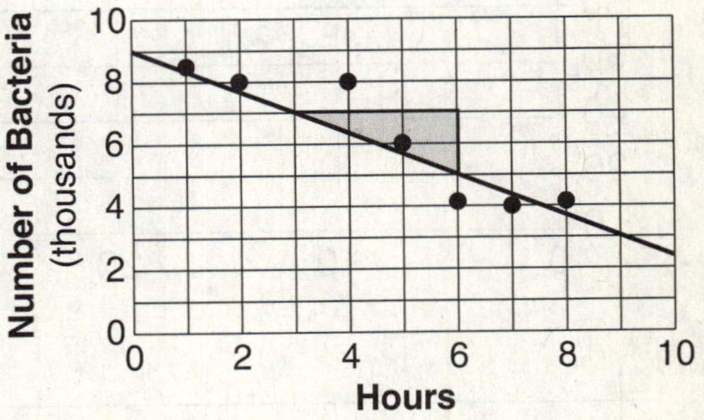

$m =$ ____, $b =$ ____,

equation: ____________

The expected number of bacteria after 10 hours is about ________.

Name ______________________ Date __________ Class __________

LESSON **12E**

Homework and Practice

Lines of Best Fit

Plot the data and find a line of best fit.

1.

x	2	4	6	8	10	12	14	16
y	12	14	18	16	20	21	25	26

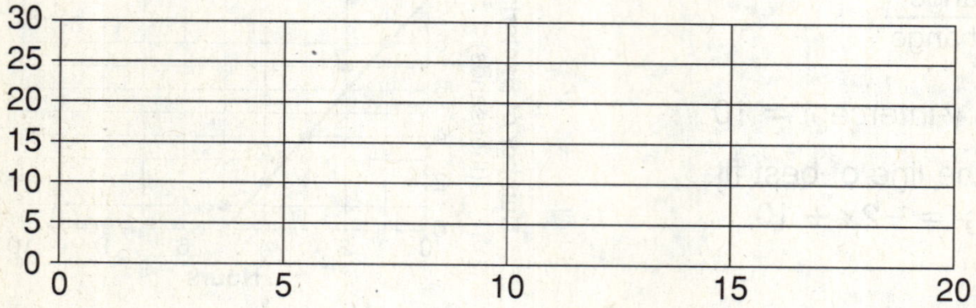

____

2.

x	5	10	18	20	25	30	35	40
y	135	120	105	90	94	56	50	30

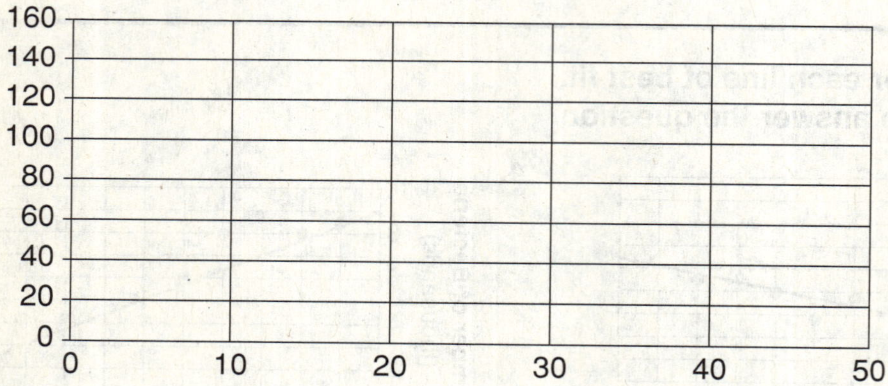

____

3. Nuxhall Company wants to predict their payroll costs for the next two years based on the salary cost over the past six years. Use the information in the table below and a line of best fit to estimate the payroll costs for the next two years.

x	Year 1	Year 2	Year 3	Year 4	Year 5	Year 6
y	$96,750	$104,490	$112,925	$125,150	$131,650	$148,825

____

Name ______________________ Date ____________ Class ____________

LESSON 12E

Homework and Practice

Lines of Best Fit, continued

Fawn is trying to improve her reading skills by taking a speed-reading class. She is measuring how many words per minute (wpm) she can read after each week of the class.

4. Graph a scatter plot using the given data.

Weeks	1	2	3	4	5
wpm	220	230	260	260	280

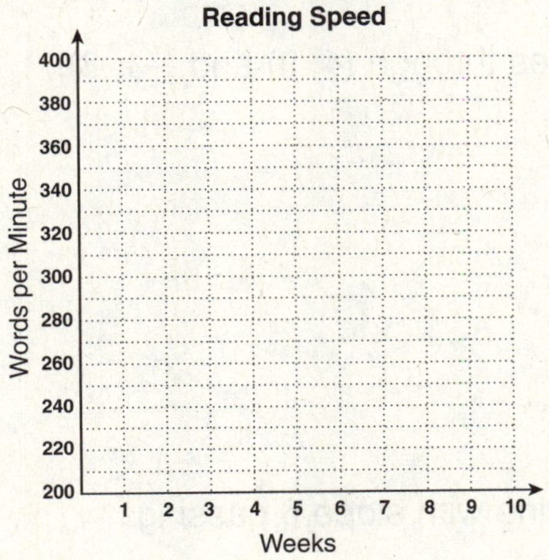

5. Describe the correlation illustrated by the scatter plot.

____

____

6. Draw a line of best fit and use it to predict the number of words per minute that Fawn will read after 8 weeks of this class.

____

7. Fawn is paying for this class each week out of her savings account. Identify the correlation between the number of classes and Fawn's account balance.

____

____

Choose the scatter plot that best represents the described relationship.

8. the distance a person runs and how physically tired that person is

A Graph 1 **C** Graph 3

B Graph 2 **D** Graph 4

9. the price of a new car and the number of hours in a day

F Graph 1 **H** Graph 3

G Graph 2 **J** Graph 4

10. a person's age and the amount of broccoli the person eats

A Graph 1 **C** Graph 3

B Graph 2 **D** Graph 4

11. the number of cats in a barn and the number of mice in that barn

F Graph 1 **H** Graph 3

G Graph 2 **J** Graph 4

Graph 1

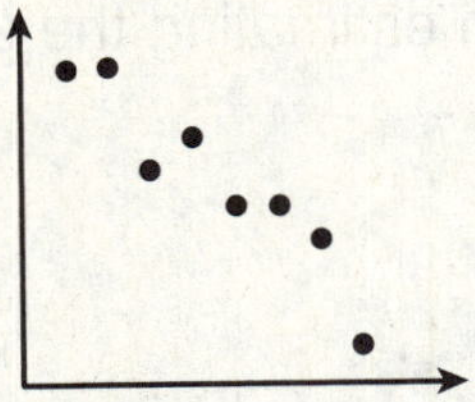

Graph 2

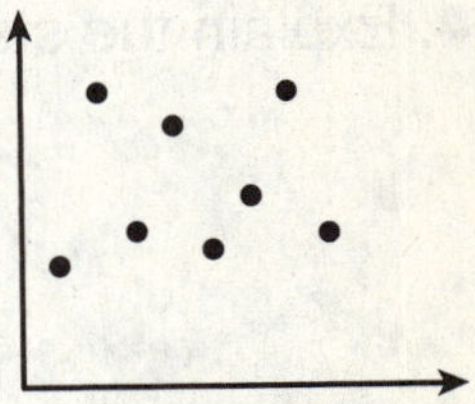

Graph 3

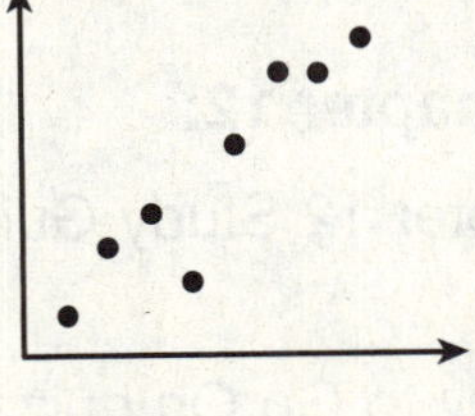

Graph 4

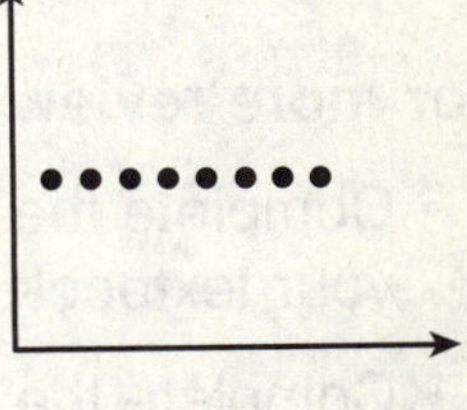

CHAPTER 12

Chapter Summary

Answer these questions to summarize the important concepts from Chapter 12 in your own words.

1. Tell whether $y = -5x + 4$ is a linear equation. Explain.

2. Explain how to find the slope of a line that passes through (4, 9) and (−1, 8).

3. Explain how to write the point-slope form of a line with slope 6 passing through (5, −2).

4. Explain the steps in estimating the equation of a line of best fit.

For more review of Chapter 12:

- Complete the Chapter 12 Study Guide and Review on pages 670–672 of your textbook.
- Complete the Ready to Go On quizzes on pages 648 and 664 of your textbook.

Name ______________________ Date __________ Class __________

CHAPTER 12

KCCT Prep

Multiple Choice

1. Alison graphs the following four equations. Which equation has a graph that is not a straight line?

(A) $y = -5$ (C) $y = -x^2$

(B) $y = 3x - 1$ (D) $y = \frac{5x}{4}$

2. The table shows the relationship between *x* and *y*. Which equation represents this relationship?

x	4	5	6	7
y	2	−1	−4	−7

(A) $y = x - 2$ (C) $y = \frac{1}{2}x$

(B) $y = 2x + 10$ (D) $y = -3x + 14$

3. Which expression is not equivalent to the other three?

(A) $3x + 8x + 4$ (C) $8x + 32 + 3x$

(B) $8(x + 4) + 3x$ (D) $11x + 32$

4. Which equation has the same solution as $\frac{4}{5}q + 1 = \frac{2}{5}$?

(A) $4q + 1 = 2$ (C) $4q + 5 = 2$

(B) $\frac{4}{5}q = -\frac{2}{5}$ (D) $\frac{4}{5}q = 1\frac{2}{5}$

5. Each morning, Katrina buys two bagels and a coffee drink. The coffee drink costs $2.50. At the end of a 5-day work week, she has spent a total of $27.50 on the bagels and coffee. What is the cost of the two bagels?

(A) $1.50 (C) $3.00

(B) $2.50 (D) $3.50

6. The table shows the amount of four different juices that are used to make a fruit punch. Which list shows the juices in order from the least amount to the greatest amount?

Fruit Punch Ingredients

Juice	Amount (cups)
Apple	$1\frac{3}{8}$
Cranberry	1.3
Orange	$1\frac{2}{5}$
Pineapple	1.25

(A) Pineapple, Orange, Cranberry, Apple

(B) Apple, Orange, Pineapple, Cranberry

(C) Pineapple, Cranberry, Apple, Orange

(D) Apple, Pineapple, Cranberry, Orange

7. What is the probability of rolling a number cube and rolling a number less than or equal to 5?

(A) $\frac{1}{2}$ (C) $\frac{3}{4}$

(B) $\frac{2}{3}$ (D) $\frac{5}{6}$

8. An ant crawls around the edge of the circular fountain shown below at a rate of 2 meters per minute. Which is the best estimate of the time it takes the ant to go once around the fountain?

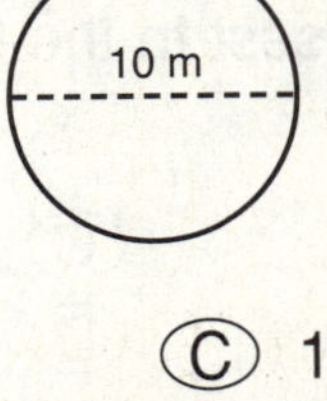

(A) 7.5 min (C) 15 min

(B) 10 min (D) 40 min

9. Which of the following is a rational number?

Ⓐ 3.1 Ⓒ $\sqrt{3}$

Ⓑ π Ⓓ $\sqrt{31}$

10. Which table shows a linear relationship?

Ⓐ

x	1	3	5	7
y	1	9	25	49

Ⓑ

x	4	6	8	10
y	8	12	16	20

Ⓒ

x	0	1	2	3
y	0	1	8	27

Ⓓ

x	4	5	10	20
y	5	4	2	1

11. Which statement about the number of edges, faces, and vertices is true for every prism?

Ⓐ Edges = Vertices

Ⓑ Vertices < Edges

Ⓒ Edges < Faces

Ⓓ Vertices = Faces

12. Suppose that $a > 0$ and $b < 0$. Which point lies in Quadrant III?

Ⓐ (a, b) Ⓒ $(a, -b)$

Ⓑ $(-a, b)$ Ⓓ $(-a, -b)$

13. A right triangle has legs that are 12 cm long and 7 cm long. Which of these is closest to the length of the hypotenuse?

Ⓐ 9.2 cm Ⓒ 14.1 cm

Ⓑ 13.9 cm Ⓓ 19.0 cm

14. Jonah earns $14.76 per hour. Which of these is the best estimate for the total amount he earns by working 8 hours per day for 5 days?

Ⓐ $120 Ⓒ $600

Ⓑ $550 Ⓓ $700

Open Response

15. A catering company uses the linear equation $C = 45 + 16p$ to find the total cost C of catering a party for p people.

a. What is the cost of catering a party for 19 people?

b. Joan places an order with the catering company. Then she finds out that there will be 3 additional people at the party. How will the total cost change? Explain.

c. Reynaldo has a party catered for some friends. The total cost is $237. Show how you can find the number of people at the party.

16. Mikiko drives her car on the highway at a constant speed. The table gives data about her trip.

Time (h)	2	4	7
Distance (mi)	116	232	406

a. What is Mikiko's speed in miles per hour?

b. Describe how an increase of 1 hour in the driving time affects the distance that Mikiko drives.

c. Mikiko claims she can drive 493 miles from Boston to Buffalo in less than 10 hours. Assume she drives at the speed you found in part **a** and assume she takes a 15-minute break after every 2 hours of driving. Assess Mikiko's claim and explain your response.

CHAPTER 13

Project

Sequences and Functions: Growing and Growing

Career: Bacteriologist

Bacteriologists study the growth and characteristics of microorganisms. They generally work in the fields of medicine and public health. Bacteria colonies grow very quickly. The rate at which bacteria multiply depends upon temperature, nutrient supply, and other factors.

Before you start, work through the Addition by Division worksheet on the next page.

In this chapter project . . .

You draw a fractal pattern.

What to do:

- Research common fractals, such as the Koch snowflake or the Sierpinski triangle.
- Choose a fractal and draw several stages. Label each stage and give the number of elements in that stage. For example, list the number of line segments for each stage of a Koch snowflake or the number of triangles in each stage of a Sierpinski triangle.

What to present:

- Your fractal drawing with labeled stages.
- The geometric sequence that describes the number of elements in each stage of your drawing. Give the starting value and the common ratio of the sequence.

Where's the math?

- Use the idea of self-similarity to draw a fractal.
- Use geometric sequences to describe the number of elements in each stage of a fractal.

go.hrw.com
Chapter Project Online
KEYWORD: MT7 Ch13

Name ______________________ Date __________ Class __________

CHAPTER 13

Project Recording Sheet

Sequences and Functions

Addition by Division

Bacteriologists study the growth and characteristics of microorganisms, the way they interact with their environment including human beings.

Bacteria are one-celled organisms which grow in number by the division of individual cells. Although some bacteria create illness and even death in people, bacteria have important roles in many natural cycles. How long it takes a bacteria population to grow depends upon factors such as temperature and nutrient supply. Doubling time is the time it takes a population to double at a constant rate of growth. The number of bacteria is 2^n where n is the number of times the population has doubled.

E. coli bacterium	
Conditions	**Doubling Time** (min)
Optimum temperature (37°C) and growth medium	20
Low Temperature (below 37°C)	40
Low Growth Medium	60
Low Temperature, Low Growth Medium	120

B. anthracis bacterium	
Conditions	**Doubling Time** (min)
Optimum temperature (37°C) and growth medium	45
Low Temperature (below 37°C)	75
Low Growth Medium	90
Low Temperature, Low Growth Medium	120

Create a graph that compares the growth of E. coli and B. anthracis for 12 hours under optimum conditions. Extend the curves to estimate the differences in the numbers of bacteria after one day.

____

Research: What other things can you discover that show geometric sequences of growth?

____

EXPLORATION

13A Terms of Arithmetic Sequences

Program of Studies and Core Content
MA-8-AT-S-PRF1, MA-08-5.1.1

In an *arithmetic sequence,* the difference between one term and the next is always the same.

The table shows the first four terms of an arithmetic sequence.

Term 1	Term 2	Term 3	Term 4	Term 5	Term 6	Term 7
10	16	22	28			

1. Find the difference between two consecutive terms.

2. Find the next three terms.

3. Find a short way to find the 20th term.

The table shows an arithmetic sequence with the first three terms missing.

Term 1	Term 2	Term 3	Term 4	Term 5	Term 6	Term 7
			88	84	80	76

4. Find the difference between two consecutive terms.

5. Find the first three terms.

Think and Discuss

6. Discuss the similarities and differences between the two sequences.

7. Explain your strategy for finding the 20th term in the first sequence.

Name ______________________ Date __________ Class __________

LESSON **13A**

Review for Mastery

Terms of Arithmetic Sequences

In an **arithmetic sequence,** the difference between terms is constant. The difference is called the **common difference.**

This is an arithmetic sequence with a common difference of 3.

2, 5, 8, 11, 14, ...

3 3 3 3

This is not an arithmetic sequence since there is no common difference.

2, 5, 9, 14, 20, ...

3 4 5 6

Complete to determine if each sequence is arithmetic.

1. 20, 16, 12, 8, 4, ...

−4 ____ ____ ____

arithmetic? ________

2. 1, 2, 4, 8, 16, ...

1 ____ ____ ____

arithmetic? ________

3. 0.1, 0.2, 0.3, 0.4, ...

____ ____ ____

arithmetic? ________

4. $\frac{1}{2}$, 1, $\frac{3}{2}$, 2, $\frac{5}{2}$, ...

____ ____ ____ ____

arithmetic? ________

5. 2, $\frac{3}{2}$, 1, $\frac{1}{2}$, 0, ...

____ ____ ____ ____

arithmetic? ________

6. 3, 1, 0, $-\frac{1}{2}$, $-\frac{1}{4}$, ...

____ ____ ____ ____

arithmetic? ________

You can use the common difference to find any term in an arithmetic sequence.

4, 6, 8, 10, 12, ...This arithmetic sequence has a common difference of 2.

2 2 2 2

This is the 1st term of the sequence.	4
For the 2nd term, add the common difference × 1	4 + 2 × 1 = 6
For the 3rd term, add the common difference × 2	4 + 2 × 2 = 8
For the 4th term, add the common difference × 3	4 + 2 × 3 = 10
For the 5th term, add the common difference × 4	4 + 2 × 4 = 12
For the *n*th term, add the common difference × (*n* − 1)	4 + 2 × (*n* − 1)

Complete to find the given term of the arithmetic sequence 4, 6, 8, 10, 12, ...

7. the 9th term

4 + 2 × _____ = _____

8. the 20th term

4 + 2 × _____ = _____

9. the 100th term

4 + 2 × _____ = _____

Name ______________________________ Date ____________ Class ____________

LESSON 13A

Review for Mastery

Terms of Arithmetic Sequences, continued

You can use a formula to find the nth term, a_n, of an arithmetic sequence with common difference d. $\mathbf{a_n = a_1 + (n - 1)d}$

Find the 20th term of this arithmetic sequence: 2, 5, 8, 11, 14, ...

$a_n = a_1 + (n - 1)d$

$a_{20} = 2 + (20 - 1)3$

$a_{20} = 2 + (19)3 = 2 + 57 = 59$

Complete to find the given term of each arithmetic sequence.

10. 28, 34, 40, 46, 52, ...
Find the 18th term.

$n =$ _____, $a_1 =$ _____, $d =$ _____

$a_{18} =$ _____ + (_____ − 1)_____

= _____ + _____ = _____

11. $\frac{1}{8}, \frac{1}{4}, \frac{3}{8}, \frac{1}{2}, \frac{5}{8}, \ldots$
Find the 25th term.

$n =$ _____, $a_1 =$ _____, $d =$ _____

$a_{25} =$ _____ + (_____ − 1)_____

= _____ + _____ = _____

You can use the same formula to find other missing information.

What term of the arithmetic sequence 0.25, 0.50, 0.75, ... is 6.5?

$a_1 = 0.25,\ d = 0.25,\ a_n = 6.5,\ n = ?$	Assign values to the variables.
$a_n = a_1 + (n - 1)d$	Substitute into the formula.
$6.5 = 0.25 + (n - 1)0.25$	Solve for n.
$6.5 = 0.25 + 0.25n - 0.25$	Multiply.
$6.5 = 0.25n$	Combine like terms.
$\frac{6.5}{0.25} = \frac{0.25n}{0.25}$	Divide.
$26 = n$	So, 6.5 is the 26th term.

Complete to answer the question.

12. What term of the arithmetic sequence 8, 16, 24, 32, ... is 112?

$a_1 =$ _____

$d =$ _____

$a_n =$ _____

So, 112 is the _____ term

$a_n = a_1 + (n - 1)d$

_____ = _____ + (n − 1)_____ Substitute.

______________________ Solve for n.

____

____

_____ = n

Name ______________________ Date ____________ Class ____________

LESSON 13A

Homework and Practice

Terms of Arithmetic Sequences

Determine if each sequence could be arithmetic. If so, give the common difference.

1. −19, −16, −13, −10, −7, …

____

2. 7, 14, 28, 56, 112, …

____

3. 8, 1, −6, − 13, − 20, …

____

4. 1, 2.2, 4.84, 10.648, 23.4256 …

____

5. $\frac{1}{10}, \frac{3}{5}, \frac{11}{10}, \frac{8}{5}, \frac{21}{10}, \ldots$

____

6. −10.4, −8.7, −7.1, −5.4, −3.7, …

____

7. $\frac{1}{8}, \frac{1}{4}, \frac{3}{8}, \frac{1}{2}, \frac{5}{8}, \ldots$

____

8. 91, 42, −7, −56, −105, …

____

9. −25.3, −18.5, −11.7, −4.9, 1.9, …

____

Find the given term in each arithmetic sequence.

10. 15th term: 35, 42, 49, 56, 63, …

____

11. 28th term: −41, −29, −17, −5, 7, …

____

12. 32nd term: 204, 158, 112, 66, 20, …

____

13. 97th term: −8, −22, −36, −50, −64, …

____

14. 45th term: 3, 5.9, 8.8, 11.7, 14.6, …

____

15. 61st term: 49, 31, 13, −5, −23, …

____

16. 74th term: 84, 62, 40, 18, −4, …

____

17. 37th term: $\frac{1}{9}, \frac{4}{9}, \frac{7}{9}, 1\frac{1}{9}, 1\frac{4}{9}, \ldots$

____

18. Write the first six terms in an arithmetic sequence that begins with −8.7 and has a common difference of −3.9.

____

19. Corrine starts a book collection by purchasing seven books. Each week she adds two new books. How many weeks before Corrine has 75 books in her collection?

____

Name ______________________ Date __________ Class __________

LESSON 13A

Homework and Practice

Terms of Arithmetic Sequences, continued

Find the indicated term of each arithmetic sequence.

20. Darnell has a job and is saving his paychecks each week.

Weeks	1	2	3	4
Savings	$130	$260	$390	$520

How much will Darnell have saved after 11 weeks?

____

21. A tube containing 3 ounces of toothpaste is being used at a rate of 0.15 ounces per day. How much toothpaste will be in the tube after one week?

____

22. A new car costs $13,000 and is depreciating by $900 each year. How much will the car be worth after 4 years?

____

23. Jessie is playing an arcade game that costs 50¢ for the first game and 25¢ to continue if she loses. How much will she spend on the game if she continues 9 times?

____

Use the graph below to answer questions 25–29. The graph shows the size of Ivor's ant colony over the first four weeks. Assume the ant population will continue to grow at the same rate. Select the best answer.

Ivor's Ant Farm

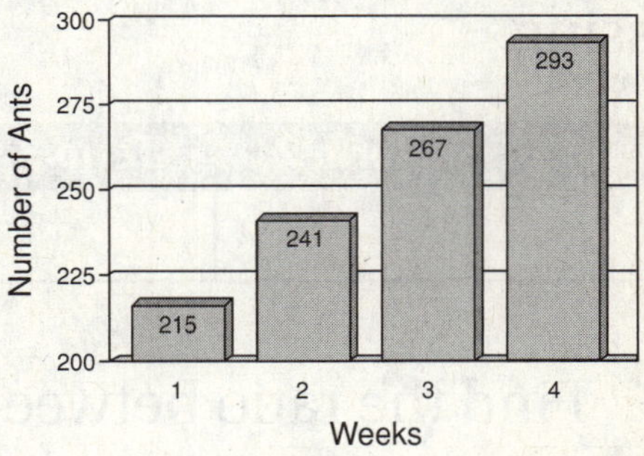

24. Which of the following shows how many ants Ivor will have in each of the next three weeks?

A 315, 341, 367

B 317, 343, 369

C 318, 334, 350

D 319, 345, 371

25. Which rule can be used to find how large the colony will be in *n* weeks?

F $a_n = 215 + 26n$

G $a_n = 215n + 26$

H $a_n = 215(n - 1) + 26$

J $a_n = 215 + 26(n - 1)$

26. How many ants will Ivor have in 27 weeks?

A 891 **C** 5616

B 917 **D** 5831

27. Ivor's ants weigh 1.5 grams each. How many grams do all of his ants weigh in 13 weeks?

F 660.5 **H** 722

G 683 **J** 790.5

28. When the colony reaches 1385 ants, Ivor's ant farm will not be big enough for all of them. In how many weeks will the ant population be too large?

A 45 **C** 47

B 46 **D** 48

EXPLORATION

13B Terms of Geometric Sequences

Program of Studies and Core Content
MA-8-AT-S-PRF1,
MA-08-5.1.1

In a *geometric sequence,* the ratio of one term to the next is always the same.

The table shows the first four terms of a geometric sequence.

Term 1	Term 2	Term 3	Term 4	Term 5	Term 6	Term 7
10	20	40	80			

1. Find the ratio between two consecutive terms.
2. Find the next three terms.
3. Find a short way to find the 20th term.

The table shows a geometric sequence with the first three terms missing.

Term 1	Term 2	Term 3	Term 4	Term 5	Term 6	Term 7
			150	75	37.5	18.75

4. Find the ratio between two consecutive terms.
5. Find the first three terms.

Think and Discuss

6. **Discuss** the similarities and differences between the two sequences.
7. **Explain** your strategy for finding the 20th term in the first sequence.

Name ______________________________ Date ____________ Class ____________

LESSON 13B

Hands-On Lab

Terms of Geometric Sequences

Activity 1

What is the common ratio of the geometric sequence in the table? Complete the table.

Term Number	a_1	a_2	a_3	a_4	a_5	a_6	a_7	a_8
Term	1	4	16	64				

Step 1 Divide the second term by the first term.

$4 \div 1 =$ _____

Step 2 Divide the third term by the second term.

$16 \div 4 =$ _____

Step 3 Divide the fourth term by the third term.

$64 \div 16 =$ _____

What is the common ratio? _____

Step 4 To complete the table, multiply a_4 by the common ratio to get a_5. Then multiply a_5 by the common ratio and so on.

Activity 2

Find the 12th term in the geometric sequence: 1, 3, 9, 27 …

Step 1 Find the common ratio, r, of the sequence.

$r = 3 \div 1 =$ _____

Step 2 Use the formula $a_n = a_1 \times r^{n-1}$ to find the 12th term.

$a_{12} = 1 \times 3^{11}$.

Step 3 Use a calculator to find 3^{11}.

$a_{12} =$ ____________

Name ______________________________ Date ____________ Class ____________

LESSON 13B

Hands-On Lab

Terms of Geometric Sequences, continued

Think and Discuss

1. Could the following sequence be geometric? Explain how you know.

512, 128, 32, 8, …

____

____

____

Try This

1. Could the following sequence be geometric? If so, what is the common ratio?

2, 5, 8, 11, 14, …

____

2. Find the next 4 terms in the sequence shown in the table.

Term Number	a_1	a_2	a_3	a_4	a_5	a_6	a_7	a_8
Term	2	$\frac{1}{2}$	$\frac{1}{8}$	$\frac{1}{32}$				

3. Find the 8th term of the geometric sequence. Show your work.

160, 80, 40, 20 …

Name ______________________________ Date ___________ Class ___________

LESSON **13B**

Technology Lab 1

Generate Arithmetic and Geometric Sequences, cont.

Graphing calculators can be used to explore arithmetic and geometric sequences.

Activity

1 The command **seq(** is used to generate a sequence.

a. Press [2nd] [STAT] **OPS 5:seq.**

The **seq(** command is followed by the rule for generating the sequence, the variable used in the rule, and the positions of the first and last terms in the sequence. To find the first 20 terms of the arithmetic sequence generated by the rule
5 + (*x* – 1) • 3, enter **seq(5 + (*x* – 1) • 3, *x*, 1, 20):**

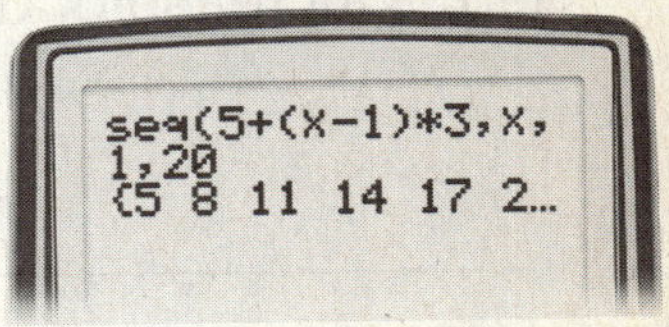

5 [+] [(] [X,T,θ,n] [–] 1 [)] [×]

3 [,] [X,T,θ,n] [,] 1 [,] 20 [ENTER]

b. You can see all 20 terms by pressing the right arrow key repeatedly. From the calculator display, the first term is 5, the second is 8, the third is 11, the fourth is 14, and so on.

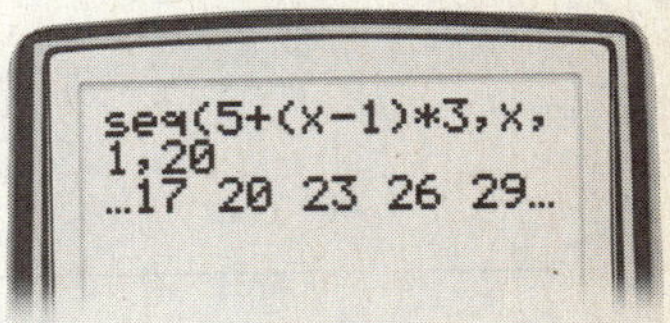

2 Consider the *geometric* sequence whose *n*th term is

$3\left(\frac{1}{4}\right)^{n-1}$. To use a graphing calculator to find the first 15 terms in fraction form, press [2nd] [STAT] (LIST) [>] **5:seq 3** [×] [(] 1 [÷] 4 [)] [^] [(] [X,T,θ,n] [–] 1 [)] [,] [X,T,θ,n] [,] 1 [,] 15 [MATH] **1:Frac** [ENTER] .

To see all 15 items, press the right arrow key repeatedly.

Name ______________________ Date __________ Class __________

LESSON 13B

Technology Lab 1

Generate Arithmetic and Geometric Sequences

Think and Discuss

1. Why is the seventh term of the sequence in **2** *not* displayed as a fraction?

____

____

____

Try This

Find the first 15 terms of each sequence. Tell if the consecutive terms increase or decrease.

1. $-4 + (n - 1) \cdot 7$

____

2. $2\left(\frac{1}{5}\right)^{n-1}$

____

____

3. 9, 14, 19, 24...

____

4. $2, \frac{2}{3}, \frac{2}{9}, \frac{2}{27}\cdots$

____

____

____

Name ______________________ Date __________ Class __________

LESSON 13B

Technology Lab 2

Terms of Geometric Sequences

Activity

Consider the following sequence of tile groupings.

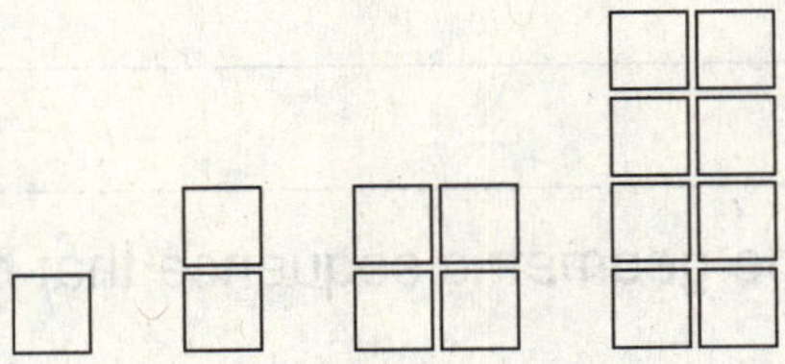

Write down the number of tiles in each grouping.

____

Starting with the single tile, as you look from one tile grouping to the next, what pattern do you see?

____

What type of sequence do these tile groupings represent? Why?

____

____

Draw the next two tile grouping in the sequence.

____

How many tiles will the 10th tile grouping have? Justify your answer.

____

____

Name ______________________________ Date ____________ Class ____________

LESSON 13B

Technology Lab 2

Terms of Geometric Sequences, continued

Think and Discuss

1. Do geometric sequences always increase? Justify your answer.

____

____

2. What is the 10th term in the geometric sequence that begins with $1, \frac{1}{2}, \frac{1}{4}, \ldots$?

____

Try This

1. Use circles, squares or other shapes to represent the first three terms of a geometric sequence. The sequence begins with the number $a_1 = 2$, and has a common ratio $r = 3$.

____

2. Find the 4th, 5th, and 10th terms in the sequence.

____

____

3. What is the *n*th term? Use the formula for the nth term: $a_n = a_1 r^{n-1}$.

____

____

4. What is the 20th term?

____

Name ______________________ Date __________ Class __________

LESSON 13B

Review for Mastery

Terms of Geometric Sequences

In a **geometric sequence,** the ratio of one term to the next is constant. The ratio is called **common ratio**.

This is a geometric sequence with a common ratio of $\frac{3}{1} = 3$.

2, 6, 18, 54, 162, …

$\frac{3}{1}$ $\frac{3}{1}$ $\frac{3}{1}$ $\frac{3}{1}$

This is not a geometric sequence since there is no common ratio.

2, 4, 2, 4, 2, …

$\frac{2}{1}$ $\frac{1}{2}$ $\frac{2}{1}$ $\frac{1}{2}$

Complete to determine if each sequence is geometric. Write *yes* or *no*.

1. $\frac{1}{125}, \frac{1}{25}, \frac{1}{5}, 1, 5, \ldots$

$\frac{5}{1}$ _____ _____ _____

geometric? _______

2. $\frac{1}{8}, \frac{1}{4}, \frac{3}{8}, \frac{1}{2}, \frac{5}{8}, \ldots$

$\frac{2}{1}$ _____ _____ _____

geometric? _______

3. $4, -2, 1, -\frac{1}{2}, \frac{1}{4}, \ldots$

_____ _____ _____ _____

geometric? _______

You can use the common ratio to find any term in a geometric sequence.

2, 8, 32, 128, … This geometric sequence has a common ratio of $\frac{4}{1}$ or 4.

4 4 4 4

This is the 1st term of the sequence.	2
Multiply it by the common ratio to get the 2nd term.	$2 \times 4 = 8$
Multiply it by the square of the common ratio to get the 3rd term.	$2 \times 4^2 = 32$
Multiply it by the cube of the common ratio to get the 4th term.	$2 \times 4^3 = 128$
To get the *n*th term, multiply the 1st term by the common ratio raised to the $(n - 1)$ power.	$2 \times 4^{n-1}$

Complete to find the given term of the geometric sequence 2, 8, 32, 128, …

4. 5th term

$2 \times 4^{-} =$ __________

5. 7th term

$2 \times 4^{-} =$ __________

6. 6th term

$2 \times 4^{-} =$ __________

7. 10th term

2 × ____ = __________

8. 8th term

2 × ____ = __________

9. 9th term

2 × ____ = __________

Name ______________________ Date ____________ Class ____________

LESSON 13B

Review for Mastery

Terms of Geometric Sequences, continued

You can use a formula to find the *n*th term, a_n, of a geometric sequence with common ratio *r*: $\mathbf{a_n = a_1 \cdot r^{n-1}}$

Find the 7th term of this geometric sequence: 3, 6, 12, 24, 48, ...

$a_n = a_1 \cdot r^{n-1}$ Find *r*. $r = \frac{6}{3} = 2$

$a_7 = 3 \cdot 2^{7-1}$ Substitute $n = 7$, $a_1 = 3$, $r = 2$.

$a_7 = 3 \cdot 2^{7-1} = 3 \cdot 2^6 = 3 \cdot 64 = 192$ The 7th term is 192.

Find the given term of each geometric sequence.

10. 1, 10, 100, 1000, . . .
Find the 9th term.

$n = 9$, $a_1 = 1$, $r = \frac{10}{1} =$ ______

$a_9 = 1 \cdot$ ______$^{9-1}$ $= 1 \cdot$ ______

$a_9 = 1 \cdot$ ______________

= ______________

11. 1.1, 1.21, 1.331, 1.4641, . . .
Find the 7th term.

$n = 7$, $a_1 = 1.1$, $r = \frac{1.21}{1.1} =$ ______

$a_7 = 1.1 \times$ ______ $= 1.1 \times$ ______

$a_7 = 1.1 \times$ ______________

= ______________

Which sequence has the greater 20th term? by how much?

1000, 1050, 1100, 1150, ...
arithmetic sequence, $d = 50$

$a_{20} = 1000 + (20 - 1)50$

$a_{20} = 1000 + (19)50 = 1000 + 950$

$a_{20} = 1950$

2, 4, 8, 16, ...
geometric sequence, $r = 2$

$a_{20} = 2 \cdot 2^{20-1}$

$a_{20} = 2 \cdot 2^{19} = 2 \cdot 524{,}288 = 1{,}048{,}576$

$a_{20} = 1{,}048{,}576$

The 20th term of this geometric sequence is greater by 1,046,626.

Determine whether each sequence is arithmetic or geometric and find its 15th term.

12. $2, 1, \frac{1}{2}, \frac{1}{4}, \ldots$

sequence is ______________

____

____

$a_{15} =$ ______________

13. $\frac{5}{2}, 3, \frac{7}{2}$

sequence is ______________

____

____

______________ $a_{15} =$ ______

Name ______________________ Date __________ Class __________

LESSON **13B**

Homework and Practice

Terms of Geometric Sequences

Determine if each sequence could be geometric. If so give the common ratio.

1. 3, 15, 75, 375, 1,875, ...

2. −36, −30, −24, −18, −12, ...

3. $3, \frac{3}{2}, \frac{3}{4}, \frac{3}{8}, \frac{3}{16}, \ldots$

4. 850, 170, 34, 6.8, 1.36, ...

5. 6, 33, 171, 861, 4,311, ...

6. 14, 98, 686, 4,802, 33,614, ...

7. 999, 333, 111, 99, 33, ...

8. $95, 38, 15\frac{1}{5}, 6\frac{2}{25}, 2\frac{54}{125}, \ldots$

9. 4, 22, 121, 665.5, 3,660.25, ...

Find the given term in each geometric sequence. If necessary, round to the nearest hundredth.

10. 8th term; 4, 12, 36, 108, 324, ...

11. 5th term; $a_1 = 155$, $r = 0.6$

12. 7th term; 4, 14, 49, 171.5, 600.25, ...

13. 12th term; $a_1 = 0.05$, $r = 9$

14. 16th term; $\frac{1}{3125}, \frac{1}{625}, \frac{1}{125}, \frac{1}{25}, \frac{1}{5}, \ldots$

15. 9th term; $\frac{7}{9}, 2\frac{1}{3}, 7, 21, 63, \ldots$

16. Nicholas has a picture that is 3 in. × 5 in. If he enlarges the photo by 25% five times, what are the dimensions of the new photo? Round to the nearest tenth of an inch?

Name ____________________ Date ____________ Class ____________

LESSON 13B

Homework and Practice

Terms of Geometric Sequences, continued

17. A ball is dropped from 400 feet. The table shows the height of each bounce.

Bounce	Height (ft)
1	280
2	196
3	137.2

Find the height of the ball on the 6th bounce. Round your answer to the nearest tenth of a foot.

____

18. A plant starts with 1 branch. Every year, each branch becomes 3 branches. A sketch of the plant for the first 3 years is shown. How many branches will the plant have in year 10?

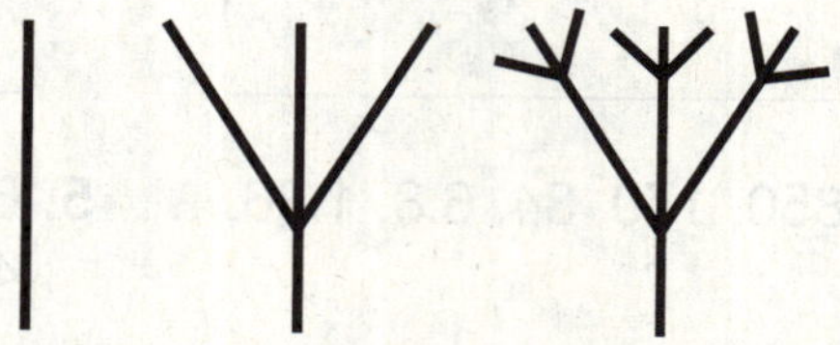

____

How many branches would the plant have in year 10 if the plant had 5 branches the first year? (Each branch still becomes 3 branches every year.)

____

19. Jeanette started selling bagels to offices in her area. Her sales for the first 3 months are shown in the table.

Month	Sales ($)
1	$200.00
2	$230.00
3	$264.50

If this trend continues, find the amount of Jeanette's sales in Month 8.

____

The table shows the number of houses in a new subdivision. Use the table to answer questions 20–23. Select the best answer.

Month	Houses
1	3
2	6
3	12
4	24

20. The number of houses forms a geometric sequence. What is r?

A 0.5 **C** 3
B 2 **D** 6

21. Assuming that the trend continues, how many houses would be in the subdivision in Month 6?

F 36 **H** 60
G 48 **J** 96

22. Management decides the subdivision is complete when the number of houses reaches 48. When will this happen?

A Month 5 **C** Month 7
B Month 6 **D** Month 8

23. Suppose the number of houses tripled every month. How many more houses would be in the subdivision in Month 4? (The number of houses in Month 1 is still 3.)

F 48 **H** 72
G 57 **J** 81

EXPLORATION

Program of Studies and Core Content

MA-8-AT-S-PRF1, MA-8-AT-S-PRF2, MA-8-AT-S-PRF3, MA-8-AT-S-PRF5, MA-8-AT-S-PRF6, MA-08-5.1.1

13C Linear Functions

Many everyday situations can be represented with *linear functions.*

For each situation, write a rule and complete the table.

1. A car beginning at time = 0 hours travels 60 miles per hour.

Input x (hr)	Rule $y =$ ______	Output y (mi)
0	$60 \cdot 0$	0
1	$60 \cdot 1$	60
2		120
3		
4		

2. A club that has \$2000.00 in its treasury plans to spend \$150.00 per week.

Input x (weeks)	Rule $y =$ ______	Output y (\$)
0		2000
1		1850
2		1700
3		
4		

Think and Discuss

3. Explain how you determined the rules in Problems 1 and 2.

Name ______________________ Date ____________ Class ____________

LESSON 13C

Review for Mastery

Linear Functions

The graph of a **linear function** is a straight line, so you can write a **rule** for a linear function in slope-intercept form. Use function notation to show that the output value, $f(x)$, corresponds to the input value, x.

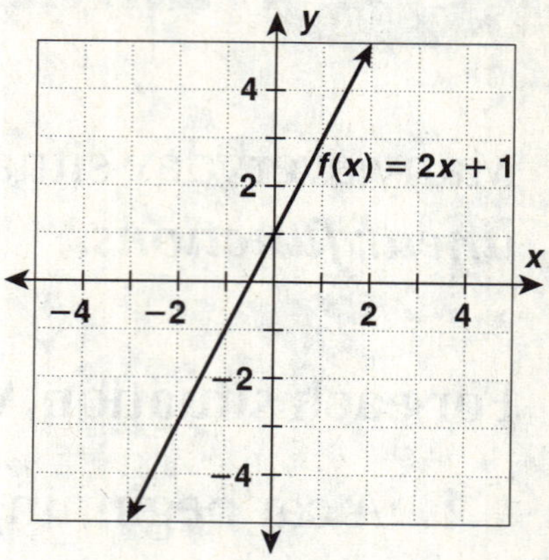

$f(x) = mx + b$

↑ slope ↖ y-intercept

You can find the slope and y-intercept of a linear function using a graph of the function or a table of its x-values and y-values.

x	y
−1	−1
0	1
1	3
2	5

The table shows the y-intercept, 1. $b = 1$

Substitute 1 for b into the slope-intercept form. $f(x) = mx + 1$

Substitute a pair of x- and y-values and solve. $5 = m \cdot 2 + 1$

$4 = 2m$

$2 = m$

So, the rule for the function is $f(x) = 2x + 1$.

Write the rule for the linear function.

1.

x	y
−2	−1
−1	1
0	3
1	5

$f(x) = mx + b$

$f(x) = mx +$ ____

$(x, y) = (1, 5)$ $5 = m($____$) +$ ____

____ $= m$

$f(x) =$ ____________

2.

x	y
−1	3
0	−1
1	−5
2	9

$f(x) = mx + b$

$f(x) = mx$ ______

$(x, y) = (-1, 3)$ ____ $= m($____$)$ ______

____ $= m$

$f(x) =$ ____________

Name ______________________ Date ____________ Class ____________

LESSON 13C

Review for Mastery

Linear Functions, continued

If a table does not contain the *y*-intercept, use two points to find the slope.

x	y
−2	−8
−1	−5
1	1
2	4

Use (1, 1) and (2, 4).

$m = \frac{4 - 1}{2 - 1} = \frac{3}{1} = 3$

Then substitute the value for *m* and the coordinates of any point, say (1, 1), into the equation $y = mx + b$ and solve for *b*.

$f(x) = mx + b$
$1 = 3(1) + b$
$1 = 3 + b$
$-3 \quad -3$
$-2 = b$

Use the values of *m* and *b* to write the function rule: $f(x) = 3x - 2$

Write the rule for the linear function.

3.

x	y
−3	−17
−1	−7
1	3
3	13

Using (1, 3) and (3, 13);

$m =$ ________ = ________ = ____

Using your value for *m* and (1, 3):

$f(x) = mx + b$

____ = ____ + b

____ = b

$f(x) =$ ______________

Use the graph to write the function rule.

4. From point *B*, read the *y*-intercept.

$b =$ ____

Write the coordinates of two points on the line.

A(____), *C*(____)

Use these points to find the slope.

$m =$ ____________ = ______ = ____

$f(x) =$ ______________

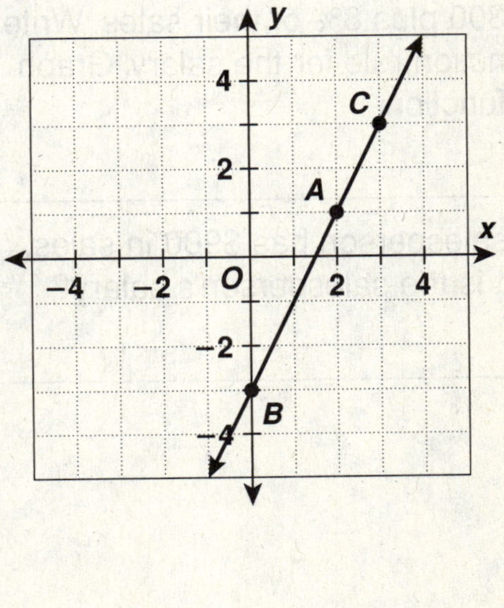

Name ______________________ Date __________ Class __________

LESSON 13C

Homework and Practice

Linear Functions

Write a rule for each linear function.

1.

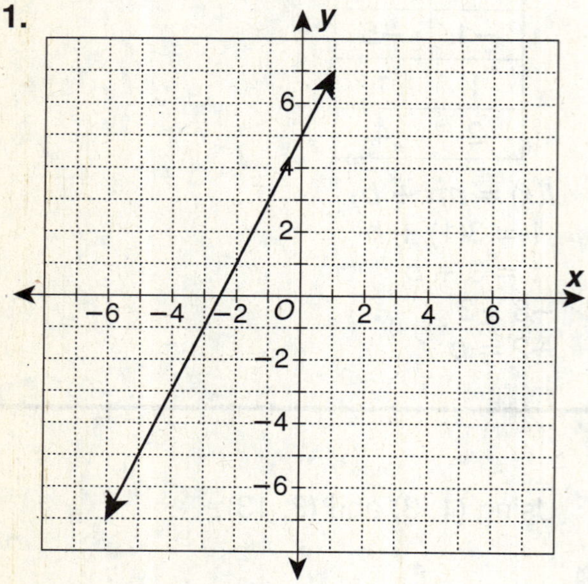

____

2.

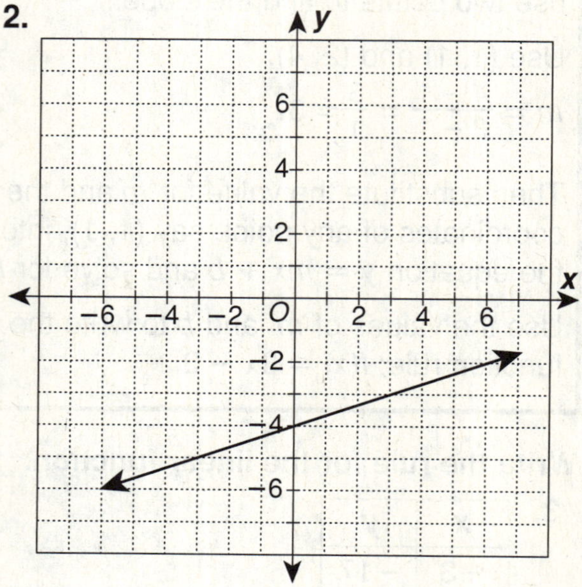

____

3.

x	−4	−2	3	5
$f(x)$	13	7	−8	−14

____

4.

x	−6	−4	2	8
$f(x)$	−10	−9	−6	−3

____

5. A salesperson is paid a base salary of $300 plus 8% of their sales. Write a function rule for the salary. Graph the function.

____

6. If a salesperson has $900 in sales, what is the salesperson's salary?

____

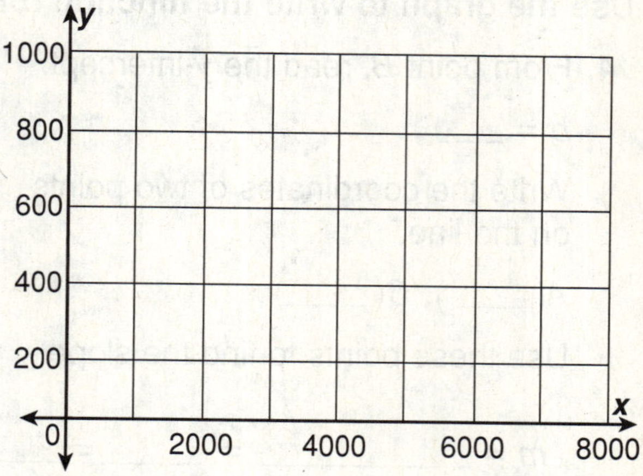

EXPLORATION

Program of Studies and Core Content

MA-8-AT-S-PRF2, MA-8-AT-S-EI2, MA-08-5.1.2

13D Quadratic Functions

The variable in a ***quadratic function*** is squared. The graph of $y = (x - 2)(x + 3)$ shows that the function has two x-intercepts, $x = -3$ and $x = 2$, and one y-intercept, $y = -6$.

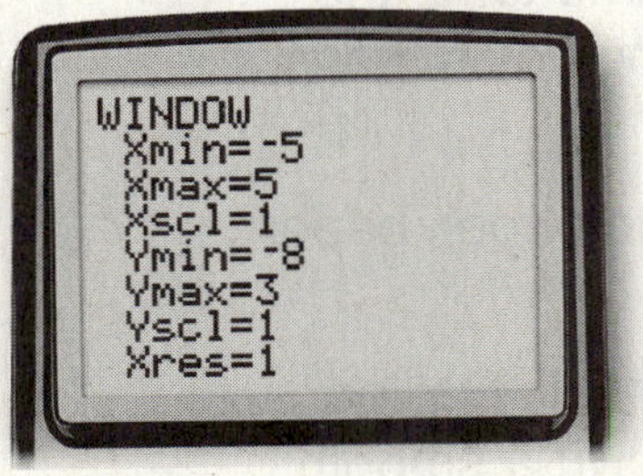

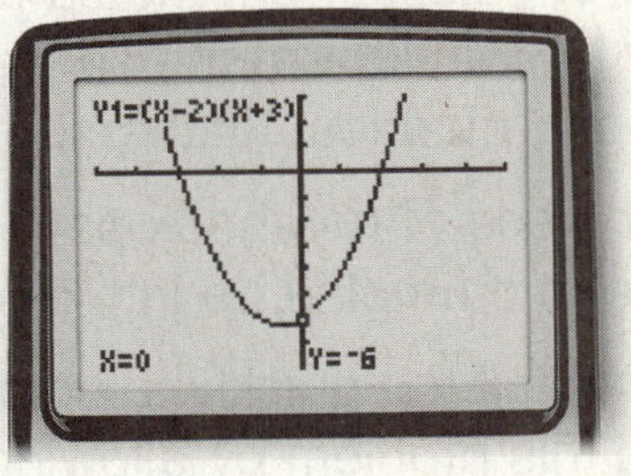

Graph each quadratic function on a graphing calculator. Use the given window settings to find the *x*- and *y*-intercepts.

1. $y = (x + 5)(x + 1)$

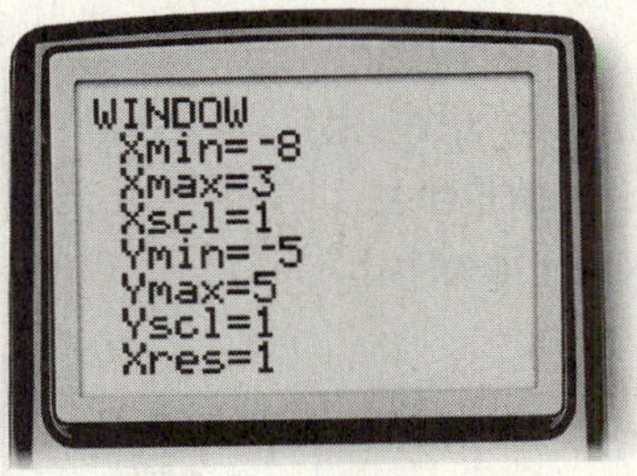

2. $y = (x - 4)(x - 2)$

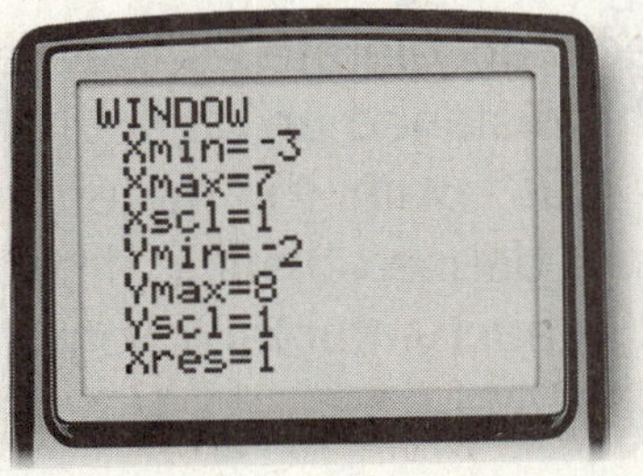

Think and Discuss

3. Discuss the relationship between the function rule and the intercepts, the points where the graph crosses the x- and y-axes.

Name ______________________________ Date ____________ Class ____________

LESSON 13D

Hands-On Lab 1
Quadratic Functions

Aristotle thought that heavier objects fall faster, but Galileo conducted a series of experiments showing that all objects accelerate at the same rate as they fall. The rate at which falling objects accelerate is known as the acceleration due to gravity.

Activity

1. Drop pairs of objects from the same height at the same time and see if they hit the floor at the same time. You can drop coins, pens, slips of paper, marbles, feathers, etc.
2. Drop a flat piece of notebook paper and a piece of notebook paper that has been crumpled up into a small ball.
3. Most objects fall too quickly to compare their rates of fall. Galileo got around this problem by allowing objects to roll down a smooth ramp. Construct a ramp by propping one end of a smooth board on a pencil or by lifting one side of your desk or table by a small amount. Allow various-sized marbles to roll down this ramp. To compare pairs of marbles, you will need to use a "starting gate" so that the marbles begin to roll at the same time. You can use a ruler or anything else that is straight. Place the marbles behind the gate and lift the gate.
4. Since objects accelerate as they fall, they fall farther during each second. The distance an object has fallen is related to the time of fall by $d = 4.9t^2$, with d in meters and t in seconds. Suppose a small, round, dense object is dropped from a tall tower. How far does it fall in each interval? Complete the following table.

t	1	2	3	4	5
t^2	1				
$d = 4.9t^2$	4.9				

5. Construct a graph of distance versus time for this object.

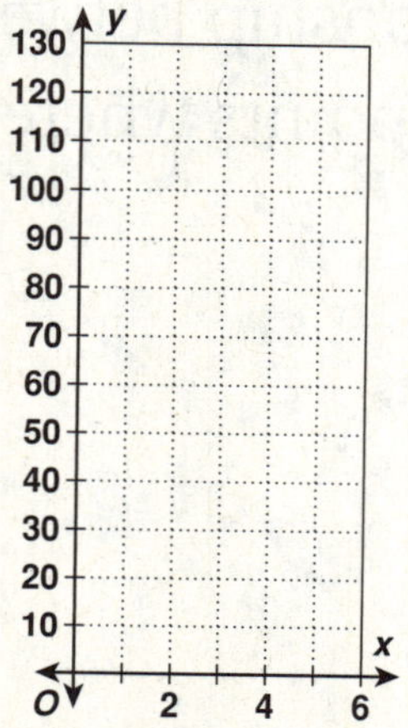

Name ______________________ Date ____________ Class ____________

LESSON 13D

Hands-On Lab 1

Quadratic Functions, continued

Think and Discuss

1. Do all objects fall at the same rate?

____

____

2. Did the flat sheet of paper and the crumpled sheet of paper fall at the same rate?

____

____

3. Do all small, dense objects fall at the same rate?

____

____

4. Did all of the marbles roll down the ramp at the same rate?

____

____

Try This

The kinetic energy of an object is given by the relation $KE = \frac{1}{2}mv^2$.

When the mass is given in kg and the speed is given in m/s, the kinetic energy will be in joules (J). Calculate the kinetic energy of the following objects.

1. mass of 1.0 kg and speed of 1.0 m/s ____________
2. mass of 3.5 kg and speed of 2.0 m/s ____________
3. mass of 3.5 kg and speed of 4.0 m/s ____________

Name ______________________ Date __________ Class __________

LESSON **13D**

Hands-On Lab 2

Quadratic Functions

Activity

All graphs of quadratic functions are related to the basic function $f(x) = x^2$, which is shown at the right. Create a table for each quadratic function below. Use the tables to graph the functions Compare the graphs.

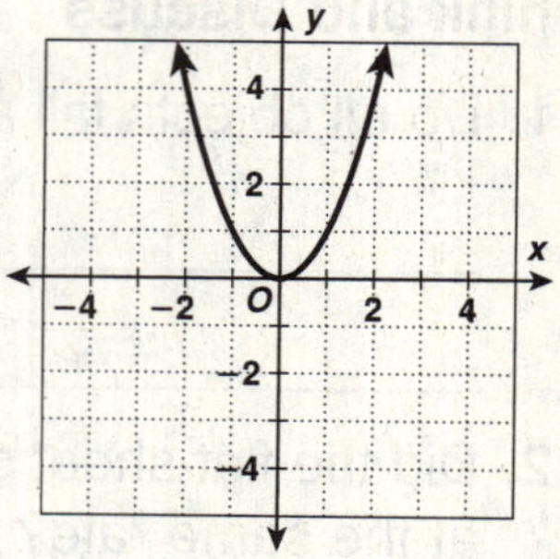

Step 1 Find $f(x)$ for each x value to complete the tables.

x	$f(x) = 2x^2$
−3	
−2	
−1	
0	
1	
2	
3	

x	$f(x) = x^2 + 2$
−3	
−2	
−1	
0	
1	
2	
3	

Step 2 Plot the points for each parabola on a separate sheet of graph paper or as a separate plot using a graphing calculator.

Step 3 Compare each graph to the graph of $f(x) = x^2$.

How does multiplying x^2 by 2 affect the parabola of the function $f(x) = x^2$?

____

____

How does adding 2 to the function $f(x) = x^2$ change the parabola?

____

____

Name ______________________ Date ____________ Class ____________

LESSON 13D

Hands-On Lab 2

Quadratic Functions, continued

Think and Discuss

1. Using what you learned on the previous page, predict what the graph of the function $f(x) = \frac{1}{2}x^2$ will look like compared to $f(x) = x^2$.

____

____

2. Predict what the graph of the function $f(x) = x^2 - 2$ will look like compared to $f(x) = x^2$.

____

____

Try This

The cross-section of the bottom of a boat is approximated by the graph of $f(x) = \frac{1}{2}x^2$. Complete the table for the function and use it to make a graph of the function.

x	$f(x) = \frac{1}{2}x^2$
−3	
−2	
−1	
0	
1	
2	
3	

If the boat is 4 m across, how deep is it?

____

Name ______________________ Date ____________ Class ____________

LESSON 13D

Technology Lab
Quadratic Functions

Activity

A square has sides with length *x*. How does the area of the square change when the length of the side is changed? Make a table of values.

Length *x* (cm)	area *y* (cm^2)
0	____
1	____
2	____
3	____
4	____
5	____

Write an equation expressing the relationship between the length of the side and the area of the square.

$y = x^2$

Use a graphing calculator or graphing software to graph the function, using only positive values for *x* (since the length of a square's side will always be positive).

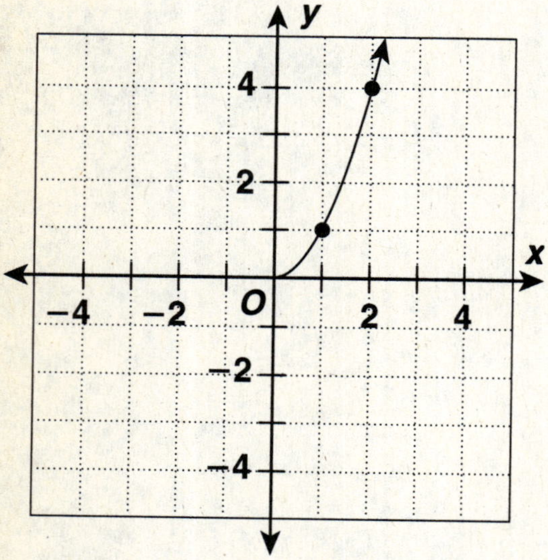

Notice that the graph of this line is curved, not straight. It is not a linear equation.

Name ______________________ Date __________ Class __________

LESSON 13D

Technology Lab

Quadratic Functions, continued

Think and Discuss

1. What if the equation in the activity above were not being used to describe the relationship between the length of a square's side and its area? How would the graph be different?

____

____

____

____

Try This

1. Graph the function $y = \frac{1}{2}x^2 + x$. Give the y–coordinate for point A (1, ?).

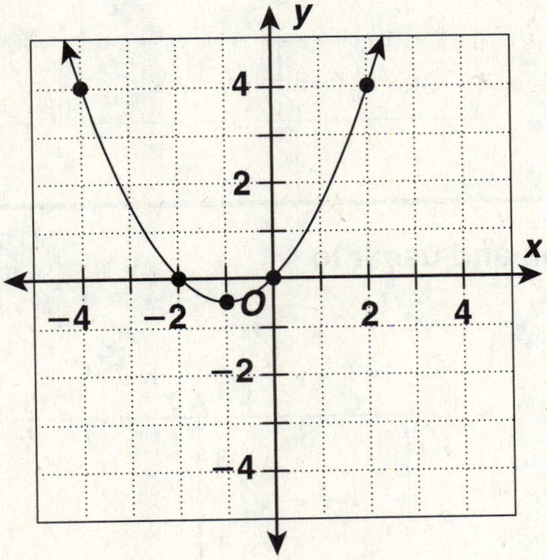

____

Name ______________________ Date ____________ Class ____________

LESSON **13D**

Review for Mastery

Quadratic Functions

A **quadratic function** has a variable that is squared.

general quadratic function $f(x) = ax^2 + bx + c$

square term (ax^2)   *y*-intercept (c)

$$f(x) = x^2 - 4x + 3$$

The graph of a quadratic function is a **parabola,** a curve that falls on one side of a turning point and rises on the other. You can make a table of a function's values and use them to graph the function.

x	$f(x) = x^2 - 4x + 3$
−1	$f(-1) = (-1)^2 - 4(-1) + 3 = 8$
0	$f(0) = 0^2 - 4(0) + 3 = 3$
1	$f(1) = 1^2 - 4(1) + 3 = 0$
2	$f(2) = 2^2 - 4(2) + 3 = -1$
3	$f(3) = 3^2 - 4(3) + 3 = 0$
4	$f(4) = 4^2 - 4(4) + 3 = 3$
5	$f(5) = 5^2 - 4(5) + 3 = 8$

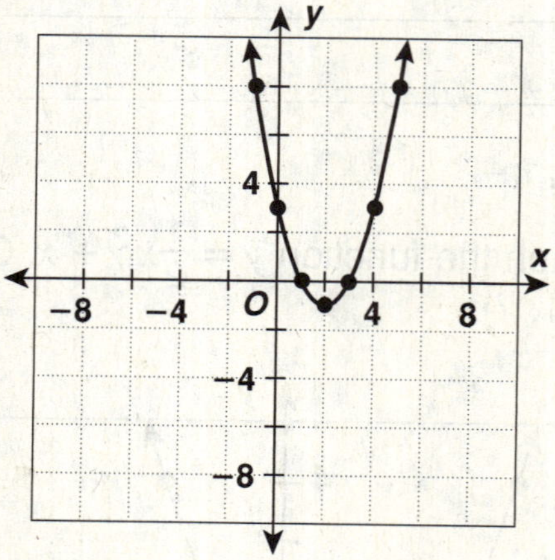

Complete the table for the quadratic function and use it to graph the function.

1. $f(x) = x^2 - 2x - 3$

x	$f(x) = x^2 - 2x - 3$
−2	$f(-2) = (\quad)^2 - 2(-2) - 3 =$
−1	$f(-1) = (\quad)^2 - 2(-1) - 3 =$
0	$f(0) = (\quad)^2 - 2(0) - 3 =$
1	$f(1) = (\quad)^2 - 2(1) - 3 =$
2	$f(2) = (\quad)^2 - 2(2) - 3 =$
3	$f(3) = (\quad)^2 - 2(3) - 3 =$
4	$f(4) = (\quad)^2 - 2(4) - 3 =$

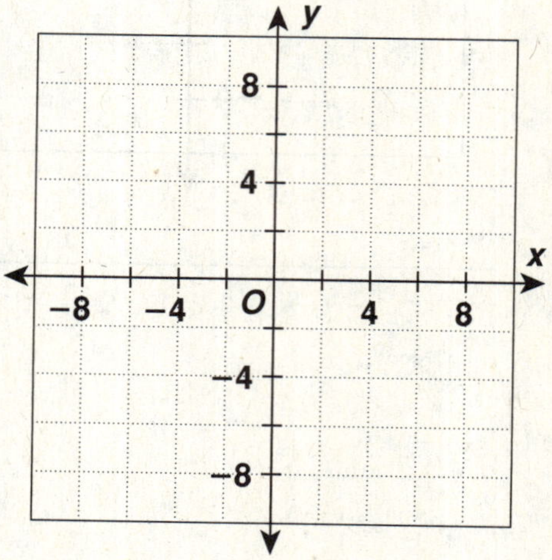

Name ______________________ Date __________ Class __________

LESSON **13D**

Review for Mastery

Quadratic Functions, continued

When a quadratic function is written as the product of two differences, you can read the two x-intercepts.

$$f(x) = (x - r)(x - s)$$

(r and s: x-intercepts)

For the quadratic function $f(x) = (x - 3)(x + 4)$, the x-intercepts are 3 and -4.

Identify the *x*-intercepts for each function.

2. $f(x) = (x - 4)(x - 7)$

4 and ______

3. $f(x) = (x + 1)(x - 5)$

-1 and ______

4. $f(x) = (x + 2)(x + 4)$

______ and ______

Complete the table for the quadratic function and use it to graph the function. Identify the *x*-intercepts and the *y*-intercept.

5. $f(x) = (x + 1)(x - 3)$

x	$f(x) = (x + 1)(x - 3)$
-2	$f(-2) = (-2 + 1)(-2 - 3) = 5$
-1	$f(-1) = (\quad + 1)(\quad - 3) =$
0	$f(0) = (\quad + 1)(\quad - 3) =$
1	$f(1) = (\quad + 1)(\quad - 3) =$
2	$f(2) = (\quad + 1)(\quad - 3) =$
3	$f(3) = (\quad + 1)(\quad - 3) =$
4	$f(4) = (\quad + 1)(\quad - 3) =$

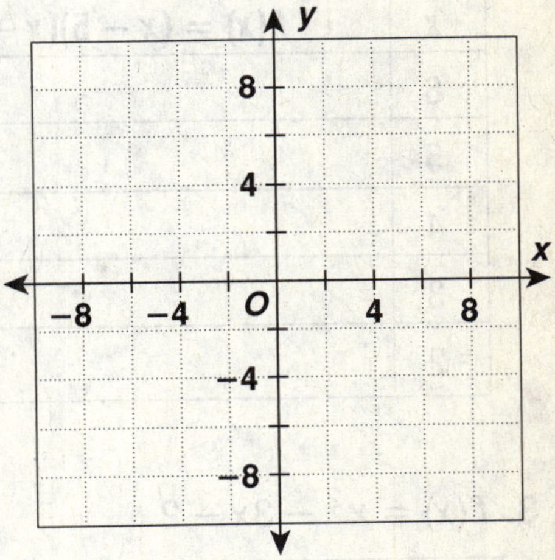

x-intercepts = ______ and ______

y-intercept = ______

6. During a softball game, Kay hit a fly ball. The function $f(x) = -16x^2 + 64x + 4$ describes the height of the softball in feet. Make a table of values for the function and then graph it.

x					
$f(x)$					

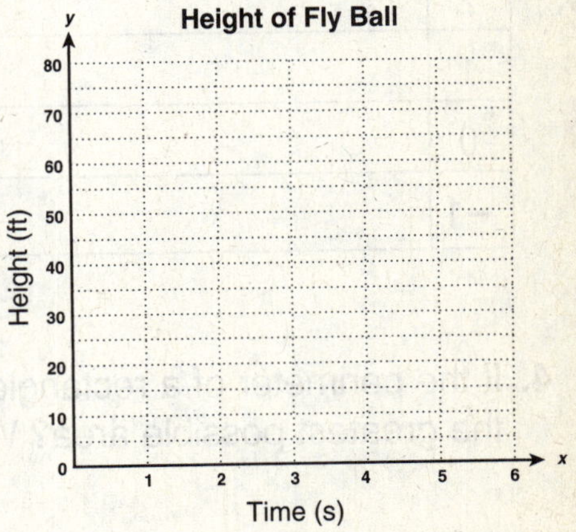

Name ______________________ Date ____________ Class ____________

LESSON **13D**

Homework and Practice

Quadratic Functions

Create a table for each quadratic function and use it to make a graph.

1. $f(x) = x^2 - 6$

x	$f(x) = x^2 - 6$
-3	$f(-3) = (-3)^2 - 6 = 3$
-1	
0	
2	
3	

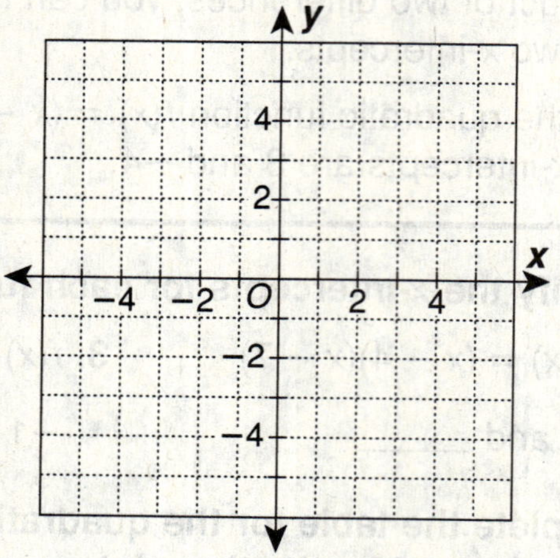

2. $f(x) = (x - 5)(x - 3)$

x	$f(x) = (x - 5)(x - 3)$
6	
5	
4	
3	
2	

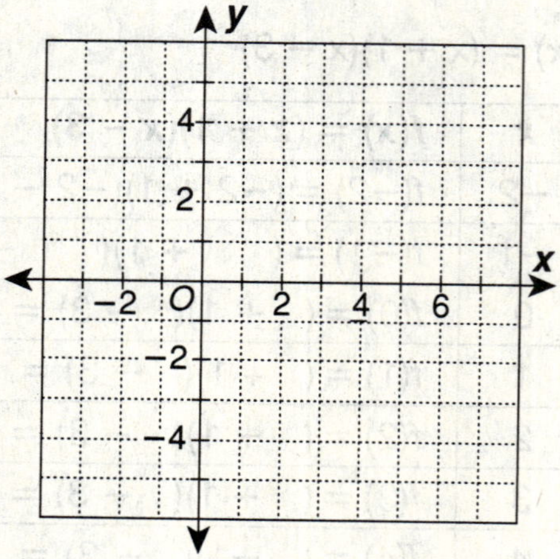

3. $f(x) = x^2 - 3x - 2$

x	$f(x) = x^2 - 3x - 2$
3	
2	
1	
0	
-1	

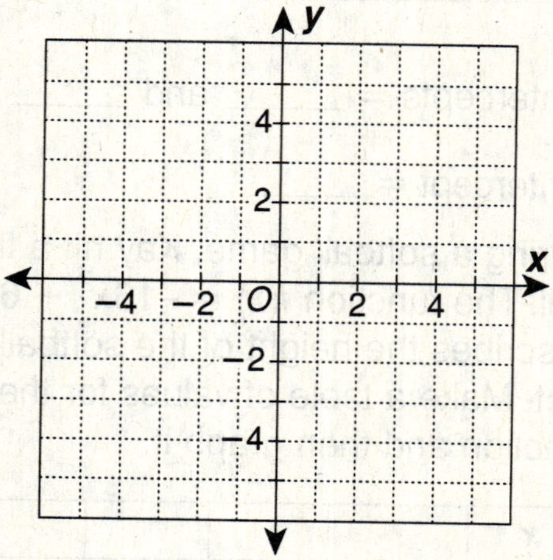

4. If the perimeter of a rectangle is 124 cm, find the dimensions of the rectangle with the greatest possible area? What is the greatest possible area?

____

CHAPTER 13

Chapter Summary

Answer these questions to summarize the important concepts from Chapter 13 in your own words.

1. Explain how to determine if the sequence is arithmetic.

5, 7, 9, 11, 13, 15, ...

2. Explain how to determine if the sequence is geometric.

120, 60, 30, 15, ...

3. Explain how to recognize the graph of a quadratic function.

4. Explain how to graph a linear function using intercepts.

For more review of Chapter 13:

- Complete the Chapter 13 Study Guide and Review on pages 722–724 of your textbook.
- Complete the Ready to Go On quizzes on pages 698 and 718 of your textbook.

Name ______________________ Date __________ Class __________

CHAPTER 13

KCCT Prep

Multiple Choice

1. Which of these is closest to 0.05?

Ⓐ 0.4%
Ⓑ 6%
Ⓒ $\frac{1}{6}$
Ⓓ 40%

2. What is the rule for the linear function shown in the graph?

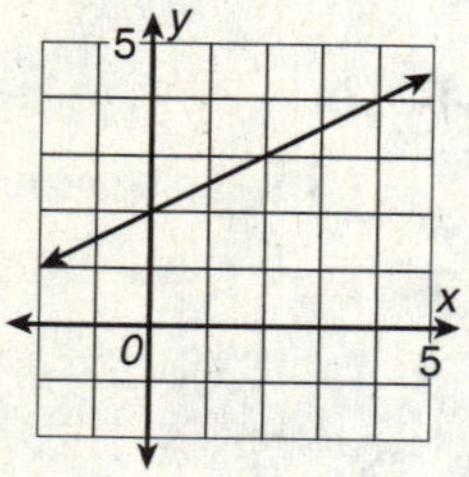

Ⓐ $f(x) = 2x + \frac{1}{2}$
Ⓑ $f(x) = \frac{1}{2}x - 2$
Ⓒ $f(x) = 2x + 2$
Ⓓ $f(x) = \frac{1}{2}x + 2$

3. Which of these is an example of an irrational number?

Ⓐ $\sqrt{3}$
Ⓑ $\sqrt{9}$
Ⓒ $\frac{3}{8}$
Ⓓ $\frac{1}{9}$

4. $\angle M$ is complementary to $\angle N$ and $m\angle M$ is 52°. $\angle P$ is supplementary to $\angle N$. What is $m\angle P$?

Ⓐ 38°
Ⓑ 52°
Ⓒ 128°
Ⓓ 142°

5. Nelda took four science exams. On three of the exams, she got scores of 75, 87, and 92. The mean of her four scores was 85.5. What score did she get on the fourth exam?

Ⓐ 82
Ⓑ 84
Ⓒ 85
Ⓓ 88

6. What is the rule for the linear relationship shown in the table?

x	−1	0	1	2
y	−2	1	4	7

Ⓐ $f(x) = 3x + 1$
Ⓑ $f(x) = x + 3$
Ⓒ $f(x) = -3x + 1$
Ⓓ $f(x) = -x + 3$

7. A ladder is leaning against a wall as shown. The bottom of the ladder is 2 ft from the wall. Which is closest to the length of the ladder?

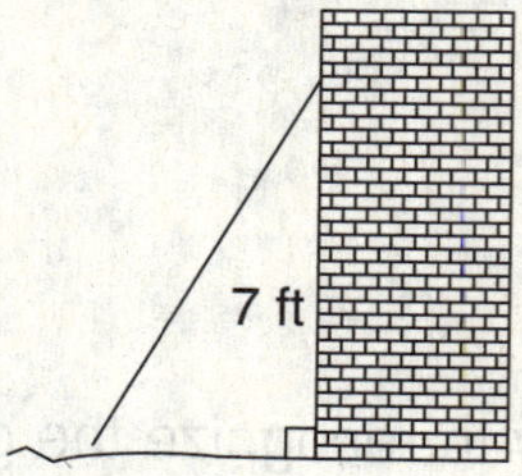

Ⓐ 7 ft
Ⓑ 7.3 ft
Ⓒ 7.9 ft
Ⓓ 9 ft

8. An aquarium in the shape of a rectangular prism has a volume of 2160 in^3. It has a length of 15 in. and a width of 12 in. What is the height of the aquarium?

Ⓐ 6 in.
Ⓑ 12 in.
Ⓒ 15 in.
Ⓓ 80 in.

9. Victoria is making a graph of the quadratic function $f(x) = 3x^2 + 4$. At which point should her graph intersect the *y*-axis?

Ⓐ (0, 3)
Ⓑ (3, 0)
Ⓒ (0, 4)
Ⓓ (4, 0)

10. For which quadratic function is $f(2)$ the greatest?

(A) $f(x) = 5x^2 - 3$
(B) $f(x) = x^2 + x + 1$
(C) $f(x) = 2x^2 + 8$
(D) $f(x) = 5x^2 - x$

11. A baseball uniform consists of a cap, a jersey, and pants. The caps come in 3 colors and the jerseys come in 2 colors. There are 24 different combinations of uniforms. How many colors of pants are there?

(A) 2 (C) 4
(B) 3 (D) 6

12. Lionel plots point M on a coordinate plane, then he moves 3 units right and 2 units up ending in Quadrant I. Which of these could be the coordinates of point M?

(A) (–1, –3) (C) (3, –4)
(B) (–2, –1) (D) (0, –3)

13. Which of these has the greatest number of edges?

(A) Rectangular prism
(B) Triangular prism
(C) Square pyramid
(D) Triangular pyramid

14. A pot that holds 4 quarts of liquid is filled with chicken broth. Judy removes six 1-cup servings from the pot. How many 1-cup servings remain in the pot?

(A) 6 (C) 16
(B) 10 (D) 26

Open Response

15. Selena wants to hire an electrician to do repairs in her house. She compares the two rates below.

ElectroFix Rates				
Hours	1	2	3	4
Cost	\$60	\$100	\$140	\$180
A-Plus Electric Rates \$50 plus \$25 per hour				

a. Write a linear function for the cost of hiring ElectroFix for x hours.
b. Write a linear function for the cost of hiring A-Plus Electric for x hours.
c. Show how you can solve an equation to find the number of hours for which both electricians have the same cost.
d. Selena thinks it will take at least $2\frac{1}{2}$ hours to do the required work at her house. Which electrician should she hire? Why?

16. Greg sets a pitching machine to throw a baseball straight up into the air. The height of the ball in feet after x seconds is given by the function $f(x) = -16x^2 + 80x$.

a. Complete a table of $f(x)$ values for $x = 0, 1, 2, 3, 4,$ and 5.
b. Graph the function.
c. Greg wants to know how high the baseball will go. Explain how he can use the graph to find out.
d. Suppose that the height of the ball after x seconds is given by $f(x) = -16x^2 + 48x$. Will the ball now stay in the air for a longer or shorter amount of time? Why?